ASTRONOMY IN FOCUS - XXIXA

AS PRESENTED AT THE IAU XXIX GENERAL ASSEMBLY, 2015

COVER ILLUSTRATION:

The two images of the city of Milan, Italy, taken from the International Space Station at just two years interval from each other, show the dramatic change in the streets' illumination due to the massive replacement of traditional sources to new technology devices (LED sources). The impact of this technological revolution on the night-sky light pollution as well as other relevant concerns for the protection of dark skies were discussed in the Focus Meeting 21 - "Mitigating Threats of Light Pollution & Radio Frequency Interference".

IAU SYMPOSIUM PROCEEDINGS SERIES

Chief Editor
THIERRY MONTMERLE, IAU General Secretary
*Institut d'Astrophysique de Paris,
98bis, Bd Arago, 75014 Paris, France
montmerle@iap.fr*

Editor
PIERO BENVENUTI, IAU Assistant General Secretary
*University of Padua, Dept of Physics and Astronomy,
Vicolo dell'Osservatorio, 3, 35122 Padova, Italy
piero.benvenuti@unipd.it*

INTERNATIONAL ASTRONOMICAL UNION

UNION ASTRONOMIQUE INTERNATIONALE

ASTRONOMY IN FOCUS XXIXA

AS PRESENTED AT THE IAU XXIX GENERAL ASSEMBLY, HONOLULU, HAWAII, UNITED STATES, 2015

Edited by

PIERO BENVENUTI

University of Padua, Dept of Physics and Astronomy

CAMBRIDGE UNIVERSITY PRESS
University Printing House, Cambridge CB2 8BS, United Kingdom
1 Liberty Plaza, Floor 20, New York, NY 10006, USA
10 Stamford Road, Oakleigh, Melbourne 3166, Australia

© International Astronomical Union 2016

This book is in copyright. Subject to statutory exception
and to the provisions of relevant collective licensing agreements,
no reproduction of any part may take place without
the written permission of the International Astronomical Union.

First published 2016

Printed in the UK by Bell & Bain, Glasgow, UK

Typeset in System LaTeX 2_ε

A catalogue record for this book is available from the British Library Library of Congress Cataloguing in Publication data

ISBN 9781107169814 hardback
ISSN 1743-9213

Table of Contents

Preface xiv

FM 1: Dynamical problems in Extrasolar planets science

Dynamical Problems in Extrasolar Planetary Science 3
    A. Morbidelli & N. Haghighipour

An Overview of Inside-Out Planet Formation 6
    J. C. Tan, S. Chatterjee, X. Hu, Z. Zhu & S. Mohanty

Planet migration and magnetic torques 14
    A. Strugarek, A. S. Brun, S. P. Matt & V. Reville

The Effect of Convective Overstability on Planet Disk Interactions 19
    H. Klahr & A. L. Gomes

Formation of the Planetary Candidates Observed by Kepler Mission 27
    S. Wang & J. Ji

Period Ratio Distribution of Near-Resonant Planets Indicates Planetesimal
    Scattering 30
    S. Chatterjee, S. O. Krantzler & E. B. Ford

Long-term orbital stability of exosolar planetary systems with highly eccentric
    orbits 38
    K. I. Antoniadou & G. Voyatzis

The Diversity of Low-mass Exoplanets Characterized via Transit Timing .. 40
    D. Jontof-Hutter, E. B. Ford, J. F. Rowe, J. J. Lissauer & D. C. Fabrycky

K2-19, The first K2 muti-planetary system showing TTVs 51
    S. C. C. Barros, J. M. Almenara, O. Demangeon, M. Tsantaki,
    A. Santerne, D. J. Armstrong, D. Barrado, D. Brown, M. Deleuil,
    J. Lillo-Box, H. Osborn, D. Pollacco, L. Abe, P. Andre, P. Bendjoya,
    I. Boisse, A. S. Bonomo, F. Bouchy, G. Bruno, J. Rey Cerda, B. Courcol,
    R. F. Díaz, G. Hébrard, J. Kirk, J. C. Lachurié, K. W. F. Lam,
    P. Martinez, J. McCormac, C. Moutou, A. Rajpurohit, J.-P. Rivet,
    J. Spake, O. Suarez, D. Toublanc & S. R. Walker

Tidal evolution of stars hosting massive close-in planets 57
    S. Ferraz-Mello, L. F. R. Moda, J. D. do Nascimento Jr. & E. S. Pereira

Search for Close-in Planets around Evolved Stars with Phase-curve variations and
    Radial Velocity Measurements 63
    T. Hirano, B. Sato, K. Masuda, O. M. Benomar, Y. Takeda, M. Omiya &
    H. Harakawa

Dynamical Effects of Stellar Companions 65
    S. Naoz

Spin-orbit alignment of exoplanet systems: how can Asteroseismology help us? . 71
    T. L. Campante

FM 2: Astronomical Heritage: Progressing the UNESCO-IAU Initiative

Focus Meeting 2, "Astronomical Heritage: Progressing the UNESCO–IAU Initiative" Introduction and overview.................................... 79
C. Ruggles & A. Sidorenko

The UNESCO Thematic Initiative "Astronomy and World Heritage" 83
A. Sidorenko

The IAU's involvement in the Astronomy and World Heritage Initiative: achievements and challenges 89
C. Ruggles

What makes astronomical heritage valuable? Identifying potential Outstanding Universal Value in cultural properties relating to astronomy 93
M. Cotte

Establishing the credibility of archaeoastronomical sites 97
C. Ruggles

Astronomy, Illumination and Heritage: the Arles-Fontvieille megalithic monuments and their implications for archaeoastronomy and world heritage 100
M. Saletta

Serial nominations for the AWH initiative: The paradigm of seven-stone antas and beyond... 102
J. Antonio Belmonte, C. G. García & M. Hoskin

As international as they would let us be 105
V. Trimble

Twentieth-century astronomical heritage: the case of the Brazilian National Observatory 106
C. H. Barboza

Canada's Dominion Astrophysical Observatory and the rise of 20th Century Astrophysics and Technology.. 109
J. E. Hesser, D. Bohlender & D. Crabtree

The Development of Mauna Kea as an Astronomical Site Panelists: John Jefferies, Ann Boesgaard, Alan Stockton, Eric Becklin, and Alan Tokunaga 112
T. M. Harmony & D. DeVorkin (moderators)

The AURA Observatory in Chile—part of the IAU/UNESCO Extended Case Study 114
M. G. Smith, R. C. Smith & P. Sanhueza

Site Protection Efforts at the AURA Observatory in Chile 115
R. C. Smith, P. Sanhueza & M. G. Smith

Protection of SAAO observing site against light and dust pollution 118
R. Sefako & P. Väisänen

How can UNESCO World Heritage Criteria be applied to the "Windows to the Universe" Sites?... 121
M. Cotte

"Route of astronomical observatories" project: Classical observatories from the Renaissance to the rise of astrophysics..... 124
G. Wolfschmidt

Are Historical Observations "Ancient" or "Modern"?..... 129
R. E. Griffin

BAO Plate Archive digitization, creation of electronic database and its scientific usage..... 130
A. M. Mickaelian, H. V. Abrahamyan, H. R. Andreasyan, N. M. Azatyan, S. V. Farmanyan, K. S. Gigoyan, M. V. Gyulzadyan, K. G. Khachatryan, A. V. Knyazyan, G. R. Kostandyan, G. A. Mikayelyan, E. H. Nikoghosyan, G. M. Paronyan & A. V. Vardanyan

Odyssey of Human Creative Genius: From Astronomical Heritage to Space Technology Heritage..... 134
O. Dluzhnevskaya & M. Marov

Nā Inoa Hōkū: Hawaiian and Polynesian star names..... 140
C. Ruggles, R. K. Johnson & J. K. Mahelona

Astronomical Heritage and Aboriginal People: Conflicts and Possibilities..... 142
A. Martín López

Gufa, a Unique Cultural Ritual—a Tale of a Forbidden Sun and a Girl..... 146
P. Shrestha

Astronomical Knowledge from Holy Books..... 148
S. V. Farmanyan, V. G. Devrikyan & A. M. Mickaelian

Astronomy in the City for Astronomy Education..... 150
R. M. Ros & B. García

FM 3: Scholarly Publication in Astronomy: Evolution or Revolution?

Evolution of Scholarly Publishing and Library Services in Astronomy Its Impact, Challenges, and Opportunities..... 157
H. Wesley & G. Sheshadri

Behind the Spam: A "Spectral Analysis" of Predatory Publishers..... 166
J. Beall

The data sharing advantage in astrophysics..... 172
B. F. Dorch, T. M. Drachen & O. Ellegaard

That over-used and much abused 4-letter word: DATA..... 176
E. Griffin

Evolution of Scholarly Publishing and Library Services in Astronomy Its Impact, Challenges, and Opportunities..... 179
H. Wesley & G. Sheshadri

FM 8: Statistics and Exoplanets

Introduction 187
    S. Aigrain & E. Feigelson

Stellar rotation period inference with Gaussian processes 191
    R. Angus, S. Aigrain & D. Foreman-Mackey

Dealing with activity in RV planet searches 193
    I. Boisse

An artificial Kepler dichotomy? Implications for the coplanarity of planetary
    systems 196
    T. Bovaird & C. H. Lineweaver

Kepler Reliability and Occurrence Rates 197
    S. Bryson & The Kepler Team

A probabilistic and automated tool for the vetting of transit candidates 198
    O. Demangeon & Pascal Bordé

Validation of transting planet candidates: a Bayesian view 200
    R. F. Díaz, J. M. Almenara & A. Santerne

Estimations of uncertainties of frequencies 201
    L. Eyer, J.-M. Nicoletti & S. Morgenthaler

Reliable inference of light curve parameters in the presence of systematics 202
    N. P. Gibson

Bayesian Planet Searches for the 10 cm/s Radial Velocity Era 205
    P. C. Gregory

Measuring the mass of Kepler-78b using nonparametric Gaussian process estimation 208
    S. K. Grunblatt, A. W. Howard & R. D. Haywood

Improving Accuracy of Quasars' Photometric Redshift Estimation by Integration
    of KNN and SVM 209
    B. Han, H. Ding, Y. Zhang & Y. Zhao

Advances in the *Kepler* Transit Search Engine 210
    J. M. Jenkins

Low-contrast pre-coronagraph for extra contrast of dark-hole 213
    J. Nishikawa, M. Oya, N. Murakami, M. Tamura, T. Kurokawa, Y. Tanaka
    & T. Kotani

Combining Transit and Radial Velocity Data 214
    L. A. Rogers

Astrometric exoplanet surveys in practice 217
    J. Sahlmann

Significance of periodogram peaks 219
    M. Süveges, L. Guy, S. Zucker & the Gaia CU7 team

Planet Frequency beyond the Snow Line from MOA-II Microlensing Survey.... 220
  D. Suzuki, D. P. Bennett & the MOA collaboration

An Analytic Model Approach to the Frequency of Exoplanets 221
  W. A. Traub

The Small Exoplanet Mass-Radius Relation: Quantifying the Astrophysical Scatter 223
  A. Wolfgang, L. A. Rogers & E. B. Ford

GATE (Gaia Transiting Exoplanets): Detecting Transiting Exoplanets with Gaia 224
  S. Zucker, L. Eyer, S. Hodgkin & G. Clementini

FM 9: Highlights in the Exploration of Small Worlds

Comet composition and Lab 227
  D. Bockelée-Morvan

Chemical diversity in the comet population................................. 228
  N. Biver & D. Bockelée-Morvan

Measuring the Distribution and Excitation of Cometary CH_3OH Using ALMA . 233
  *M. A. Cordiner, S. B. Charnley, M. J. Mumma, D. Bockelée-Morvan,
  N. Biver, G. Villanueva, L. Paganini, S. N. Milam, A. J. Remijan,
  D. C. Lis, J. Crovisier, J. Boissier, Y.-J. Kuan & I. M. Coulson*

Active Asteroids: Main-Belt Comets and Disrupted Asteroids................. 237
  H. H. Hsieh

Icy Dwarf Planets: Colored Popsicles in the Outer Solar System 241
  N. Pinilla-Alonso

The Chelyabinsk event.. 247
  J. Borovička

The asteroid-comet continuum from laboratory and space analyses of comet samples and micrometeorites 253
  *C. Engrand, J. Duprat, N. Bardin, E. Dartois, H. Leroux, E. Quirico,
  K. Benzerara, L. Remusat, E. Dobrică, L. Delauche, J. Bradley, H. Ishii,
  M. Hilchenbach & the COSIMA team*

Carbonaceous Material in Extra-terrestrial Matter.......................... 257
  Z. Martins

Water-Rock Differentiation of Icy Bodies by Darcy law, Stokes law, and Two-Phase Flow.. 261
  W. Neumann, D. Breuer & T. Spohn

Ice-gas interactions during planet formation 267
  K. I. Öberg

^{15}N fractionation in star-forming regions and Solar System objects 271
  E. S. Wirström, G. Adande, S. N. Milam, S. B. Charnley & M. A. Cordiner

FM 11: Global Coordination of Ground and Space Astrophysics and Heliophysics

Global Coordination: What are the Next Steps? 277
    D. Spergel & R. Williams

FM 12: Bridging Laboratory Astrophysics and Astronomy

FM12: A Focus Meeting on Bridging Laboratory Astrophysics and Astronomy . 283
    F. Salama, L. Mashonkina & S. Federman

Atomic Data for Stellar Nucleosynthesis 287
    C. Sneden, J. E. Lawler, E. A. Den Hartog & M. E. Wood

Atomic processes in optically thin plasmas 291
    J. S. Kaastra, L. Gu, J. Mao, M. Mehdipour, T. Raassen & I. Urdampilleta

K-shell transitions in L-shell ions with the EBIT calorimeter spectrometer 295
    N. Hell, G. V. Brown, J. Wilms, P. Beiersdorfer, R. L. Kelley,
    C. A. Kilbourne & F. Scott Porter

Quantitative spectroscopy of hot stars: accurate atomic data applied on a large
    scale as driver of recent breakthroughs 297
    N. Przybilla, V. Schaffenroth, M. F. Nieva & K. Butler

The molecular universe: from observations to laboratory and back 299
    E. F. van Dishoeck

Untangling the Formation Mechanisms of Biorelevant Molecules in the ISM with
    Photoionization Reflectron Time-of-Flight Mass Spectrometry 305
    M. Förstel & R. I. Kaiser

CO isotopologue ratios in the solar photosphere 307
    J. R. Lyons, E. Gharib-Nezhad & T. R. Ayres

Laboratory constraints on ice formation, restructuring and desorption 309
    K. I. Öberg

Interstellar dust modelling: Interfacing laboratory, theoretical and observational
    studies (The THEMIS model) 313
    A. P. Jones

Telescope Observations of Interstellar and Circumstellar Ices: Successes of and Need
    for Laboratory Simulations 317
    A. C. A. Boogert

AKARI NIR spectroscopy of interstellar ices 319
    T. Onaka, T. I. Mori, I. Sakon, F. Usui, R. Wu & T. Shimonishi

Comet composition and Lab 321
    D. Bockelée-Morvan & N. Biver

Laboratory and theoretical work applied to planetary atmospheres 325
    A. Coustenis

The THS experiment: Simulating Titan's atmospheric chemistry at low temperature (200 K) 327
    E. Sciamma-O'Brien, K. T. Upton, J. L. Beauchamp & F. Salama

Magnetic field generation, Weibel-mediated collisionless shocks, and magnetic reconnection in colliding laser-produced plasmas 329
    W. Fox, A. Bhattacharjee & G. Fiksel

Underground Nuclear Astrophysics in China............................. 333
    L. WeiPing for JUNA collaboration

Cherenkov Telescope Array: Unveiling the Gamma Ray Universe and its Cosmic Particle Accelerators 337
    E. M. de Gouveia Dal Pino (on behalf of the CTA Collaboration)

Recent Status of Multi-Dimensional Core-Collapse Supernova Models......... 340
    K. Kotake, K. Nakamura & T. Takiwaki

MACHETE: A transit Imaging Atmospheric Cherenkov Telescope to survey half of the Very High Energy γ-ray sky 345
    R. López-Coto, J. Cortina & A. Moralejo

Atomic and Molecular Databases, VAMDC................................ 347
    M. L. Dubernet, C. M. Zwölf, N. Moreau, Y. A. Ba & VAMDC Consortium

FM12p: Focus Meeting 12 Poster Session 349

FM 13: Brightness variations of the Sun and Sun-like stars

Properties of stellar activity cycles..................................... 355
    H. Korhonen

Sun-like Stars: magnetic fields, cycles and exoplanets..................... 360
    R. Fares

The Photometric Variability of Solar-Type Stars 365
    M. S. Giampapa

Solar influence on Earth's climate 372
    R. Thiéblemont & K. Matthes

FM 20: Astronomy for Development

Preface - Focus Meeting 20: Astronomy for Development 379
    K. Govender

The IAU Strategic Plan... 380
    G. Miley

Towards "Astronomy for Development"................................. 382
    K. Govender

An Evidence-Based Framework to Optimise Social Impact in Astronomy for Development 385
    E. Grant

Astronomy for a Better World: IAU Office of Astronomy for Development Activities to Grow and Advance Astronomy Education and Research at Universities in the Developing World........ 390
E. F. Guinan & K. Kolenberg

On formation and activity of IASS........ 392
A. S. Hojaev

The Role of Astronomy in Development: The Case of Uganda........ 393
E. Jurua

West African International Summer School for Young Astronomers........ 395
L. E. Strubbe & B. Okere

Web-based Teaching Radio Interferometer for Africa........ 397
C. Carignan & Y. Libert

Space Awareness: Inspiring A New Generation of Space Explorers........ 398
P. Russo

GalileoMobile, sharing astronomy with students and teachers around the world........ 399
S. Benitez-Herrera & P. F. Spinelli

Ten years of RELEA: Achievements and challenges for astronomy education development........ 401
P. S. Bretones, L. C. Jafelice & J. E. Horvath

Task Force 3 Discussions........ 403
S.-l. Cheung

Star Parties in Mexico extended to Colombia and China........ 405
S. Torres-Peimbert & J. Franco

Reflections on a Multi-stakeholder National Campaign in India around Comet ISON........ 406
P. Shastri (on behalf of the Eyes on ISON *campaign team)*

Dark Skies Africa........ 408
C. E. Walker, D. Tellez & S. M. Pompea

The East Asian Regional Office of Astronomy for Development........ 410
R. de Grijs, Z. Zhang & J. He

Strategies for Astronomy Development in the Southeast Asia........ 412
B. Soonthornthum

East African ROAD........ 414
K. Tekle

Southern African Office of Astronomy for Development: A New Hub for Astronomy for Development........ 416
M. S. Mutondo & P. Simpemba

The new Andean Regional Office of Astronomy for Development........ 418
F. Char & J. Forero-Romero

Portuguese Language Expertise Center for the OAD 420
    R. Doran, L. Canas, S. Anjos, T. Heenatigala, J. Retrê, J. Afonso &
    A. Alves

IAU South West Asian ROAD 422
    A. Mickaelian, N. Azatyan, S. Farmanyan & G. Mikayelyan

Divisions Panel Discussion: Astronomy for Development 424
    K. Govender, M. K. Hemenway, A. Wolter, N. Haghighipour, Y. Yan,
    E. F. van Dishoeck, D. Silva & E. Guinan

Unconference session at the IAU General Assembly 2015. 427
    T. S. Nava, R. Venugopal & S. Verdolini

List of Posters for Focus Meeting 20: Astronomy for Development 430

FM 21: Mitigating Threats of Light Pollution & Radio Frequency Interference

Session 21.1 – Observations, Advances in LED Technology, and Dark Sky
    Protection 435
    D. M. Duriscoe

Session 21.2 – Measurement of Light at Night. 444
    R. J. Wainscoat

Session 21.3 – Radio and Optical Site Protection 453
    R. Sefako

Session 21.4 – World Heritage and the Protection of Working Observatory Sites   463
    C. Ruggles

Session 21.5 – Light at Night and Protected Areas..................... 473
    Z. Yongheng

Session 21.6: Preserving Dark Skies and Protecting Against Light Pollution in a
    World Heritage Framework 480
    M. G. Smith

Session 21.7 – Education Programs Promoting Light Pollution Awareness and
    IYL2015. 490
    C. E. Walker

Session 21.8 – Challenges and Solutions to Light Pollution, RFI and Implementing
    IAU Resolution 2009 B5 500
    R. Green

Author index 506

Preface

A significative change has characterised the scientific programme of the XXIXth IAU General Assembly held in Honolulu, Hawai'i, from August 3rd to 14th, 2015. While the format of the usual 6 Scientific Symposia remained the same as in previous GAs, the meetings known as "Special Sessions (SpS)" and "Joint Discussions (JD)" were replaced by the new "Focus Meetings". Indeed the difference between SpS and JD had become rather vague, therefore creating uncertainties both for the proposers that had to choose the type of meeting, as well as for the members of the Evaluation Panel.

It was therefore decided to merge the two types of meeting in a single format: the Focus Meeting, aimed at discussing specific themes of great relevance in the current astronomical research scenario. The initial idea was to limit the duration of the Focus Meeting to no more than two days, in order to accept a larger number of proposals, covering a wide range of hot topics and giving the opportunity to a large number of prospective GA participants to contribute with their recent scientific results.

The new format was received with enthusiasm by the astronomical community and 43 Focus Meeting Proposals were submitted by the deadline of December 15th, 2013. The high quality of the proposals made the task of the Evaluation Panel particularly difficult and, because of that reason, 22 Focus Meetings were accepted: a larger number with respect to the initially proposed limit of 18.

The quality of the invited and contributed talks as well as the attendance at all the Focus Meetings testify the success of the overall GA Scientific Programme. The combination of the 6 large Symposia, that offered exhaustive reviews in their respective scientific areas, with the more specialised Focus Meetings, made the participation in the XXIXth General Assembly a refreshing and stimulating astronomical happening.

The novelty in the format of the GA scientific programme had consequences also in the publication of the Proceedings. Traditionally, two "Transaction" Volumes were produced as outcome of the IAU General Assembly: Transaction-A, that contained reports by the Divisions, Commissions, Working Groups and a summary of the SpS and JD meetings, and Transaction-B that contained more IAU "business" oriented reports, official speeches and documents as well as the updated list of IAU Members.

For the "transition" XXIXth GA, it was decided to maintain the publication of the Transaction-B Volume, being it a valuable formal document for the history of the Union, and to transform the Transaction-A Volume into the current "Astronomy in Focus" Volumes, that contains the most relevant contributions presented in the various Focus Meeting together with a summary of all the accepted papers and posters.

However, since the XXIXth GA marked also the completion of the revision of the IAU internal scientific structure, with the transformation of the previous 40 Commissions into the new 35 ones, it was decided to publish as well a "legacy" Transaction-A Volume that describes the overall restructuring of the scientific bodies of the Union and offers the opportunity to the Commissions to report about their achievements and future plans.

The large numbers of Focus Meetings and the desire to assign a reasonable number of pages to each of them, made it impractical to publish the Proceedings in a single Volume. In agreement with the Publisher, it was therefore decided to split them in two Volumes: Astronomy in Focus XXIXA and Astronomy in Focus XXIXB. The partition of the Focus Meetings between the two volumes is reported in Table 1 and 2.

In conclusion we can say that the change to the new format of the GA Scientific Programme has been successful. Naturally, as in all new experiments, there are lessons to be learned for the future. In particular it may be better to slightly decrease the

Table 1. Astronomy in Focus XXIXA

FM #	Chair	Title
FM 1	A. Morbidelli	Dynamical problems in Extrasolar planets science
FM 2	C. Ruggles	Astronomical Heritage: Progressing the UNESCO-IAU Initiative
FM 3	M. Bishop	Scholarly Publication in Astronomy: Evolution or Revolution?
FM 8	S. Aigrain	Statistics and Exoplanets
FM 9	D. Bockelee-Morvan	Highlights in the exploration of Small Worlds
FM 11	D. Spergel	Global Coordination of Ground and Space Astrophysics and Heliophysics
FM 12	F. Salama	Bridging Laboratory Astrophysics and Astronomy
FM 13	N. Krivova	Brightness variations of the Sun and Sun-like stars
FM 20	K. Govender	Astronomy for Development
FM 21	C. Walker	Mitigating Threats of Light Pollution & Radio Frequency Interference

Table 2. Astronomy in Focus XXIXB

FM #	Chair	Title
FM 5	J. Tauber	The Legacy of Planck
FM 6	A. Comastri	X-ray Surveys of the Hot and Energetic Cosmos
FM 7	C. Leitherer	Stellar Physics in Galaxies throughout the Universe
FM 10	Ch. Thoene	Stellar explosions in an ever-changing environment
FM 14	J. Lazio	The Gravitational Wave Symphony of Structure Formation
FM 15	S. Kwok	Search for water and life's building blocks in the universe
FM 16	B. Davies	Stellar Behemoths - Red Supergiants across the local Universe
FM 17	S. Jeffery	Advances in Stellar Physics from Asteroseismology
FM 18	E. Falgarone	Scale-free processes in the universe
FM 22	H. Ebeling	The Frontier Fields: Transforming our understanding of cluster and galaxy evolution

number of scheduled Focus Meetings and to impose a strict limit to their duration in order make the distinction between Symposia and FM more clear. In addition, after the selection of the FM, more attention should be paid in the planning and scheduling of the detailed scientific programme of each FM, avoiding possible overlapping and duplication of subjects.

All these indications, as well as those arising from the reading of the Astronomy in Focus Proceedings, will be taken into due account in the preparation of the Scientific Programme of the XXXth General Assembly.

Auf-wiedersehen in Vienna!

Piero Benvenuti
IAU General Secretary

FM1:
Dynamical problems in Extrasolar planets science

Dynamical Problems in Extrasolar Planetary Science

Alessandro Morbidelli[1] and Nader Haghighipour[2]

[1]Laboratory Lagrange, UCA, CNRS, OCA, Nice, France
email: morby@oca.eu;
[2]Institute for Astronomy, University of Hawaii-Manoa, HI
email: nader@ifa.hawaii.edu

The past few years have witnessed a large increase in the number of extrasolar planets. Thanks to successful surveys from the ground and from space, there are now over 1000 confirmed exoplanets and more then 3000 planetary candidates. More than 130 of these systems host multiple planets. Many of these systems demonstrate physical and orbital characteristics fundamentally different from those of our solar system. The challenges associated with the diversity of planetary systems have raised many interesting questions on planet formation and orbital dynamics.

The list of dynamical problems in extrasolar planetary science is long. Planets in mean motion resonances prompt the investigation of resonant dynamics in the framework of the general (i.e. non-restricted) three-body problem. The general problem is much more complicated than the restricted one, so that we have yet to achieve a global description of resonant dynamics. For instance, previous observations have identified several systems of giant planets in mean-motion resonances, whereas recent observations point to systems of small planets which seem to lie outside the resonance libration domains. While the physics of the processes governing planet-disk interactions suggest a perfect capture in a mean-motion resonance during planetary migration, the reason for these near-resonance configurations is not yet fully understood. Tidal interactions could have played a role in extracting planets from perfect resonance capture, and so might have planetesimal scattering or turbulence in the original proto-planetary disk.

The origin of "hot" exoplanets (planets with orbital periods of a few days) is also another mystery in extrasolar planetary science. It is still a matter of debate whether these planets reached their orbits by migrating through the proto-planetary disk, or they arrived at their current orbital configurations via scattering and tidal damping. Another unresolved problem is the origin of the surprisingly large eccentricities and/or inclinations (relative to the stellar equator) of many extrasolar planets. While planet instabilities, planet-disk interactions, and external perturbations from eccentric or inclined stars have been presented as viable options, the debate is still open on what is the dominant mechanism for the origin of these large eccentricities and inclinations.

In addition to addressing the above-mentioned problems, dynamical models also complement observations by allowing a better characterization of extrasolar planets. Dynamical maps have been very useful to constrain the orbits of multi-planet systems for which the uncertainties in the orbital parameters due to the observational errors are often much larger than the range of orbital configurations ensuring the long-term stability of the system. Another important phenomenon that is a characteristic of planet-planet interactions in multi-planet systems is Transit Time Variation (TTV). TTVs are now routinely detected in these systems and are used to determine mass of planetary bodies and confirm planet candidates. The power and success of this technique can be emphasized by recent discoveries and orbital determination of non-transiting planets through the analysis of the TTV signals of their transiting companions. The TTV method brings Celestial Mechanics back to the glorious time when Le Verrier and Adams predicted independently the existence and the position of Neptune from the analysis of the anomalies of the motion of Uranus. Last but not least, dynamical studies are essential to determine long-term orbital stability. Assessing planet stability is necessary to determine whether specific planets can maintain their orbits in the habitable zone for long durations of time, necessary for the development and evolution of life, to understand the long-term properties of their spin etc..

The 2015 IAU General Assembly provided a very timely occasion to organize a meeting on the advancements made in extrasolar planetary science with a focus on the formation, evolution, and characteristics of multiple planetary systems. This focus meeting was timely in the sense that it followed the end of the primary mission of the *Kepler* space telescope and the discovery of several thousands planets and planetary candidates. The aim of the focus meeting was to have a diverse scientific program, covering all topics related to planetary dynamics in multiple planetary systems and the challenges associated with them. The topic covered by the focus meeting included

- Planet formation;
- Planet migration;
- Secular dynamics and planet instabilities;
- Spin orbit alignment/misalignment;
- Mean-motion resonances and packed planetary systems;
- Transit time variations;
- Tidal evolution, and
- planets in binary systems.

The program consisted of a series of invited talks (each session had at least one invited talk) as well as a rich collection of contributed oral and poster presentations.

The collection of papers presented in this section is a representative of the diversity of the contributions that animated the meeting. The paper by Tan *et al.* proposes a new scenario of inside-out formation of planets, potentially capable of explaining the abundance of super-Earth planets on short-period orbits. Strugarek *et al.* and Klahr & Lobo Gomes present results of their studies of the effects of magnetic torques and convective over-stability on planet migration. Planet migration is also the subject of the contribution by Wang & Jianghui who investigate the distribution of orbital period ratios in the multi-planet systems observed by the *Kepler* telescope. The lack of a strong concentration of the orbital period ratios around resonant values is discussed by Chatterjee *et al.* and explained as a result of planetesimal scattering in the aftermath of gas removal. The paper by Antoniadou & Voyatzis investigates which configurations allow long-term stability in systems with planets on eccentric orbits, and papers by Jontof-Hutter *et al.* and Barros *et al.* describe the most recent results obtained from TTV analyses on *Kepler* multi-planet systems. Tidal evolution of planetary systems are discussed in several papers. Ferraz-Mello *et al.* discuss a model focused on the tidal evolution of stars with planets and Hirano *et al.* present a more observationally orientated analysis, reporting on a search for close-in planets around evolved stars which has the potential of testing tidal decay models. The paper by Naoz reviews the dynamics of planets in binary star systems and Campante *et al.* present a new method to assess the spin-orbit alignment of planets using asteroseismology.

As can be imagined, the organization of such a focused meeting and the preparation of its proceedings chapter is a large task beyond the capabilities of only two people. We had the pleasure of working with a great team of scientists who did not hesitate to help us at any stage of the work. We are indebted to the Scientific Organizing Committee: E. Ford,

S. Ida, J. Laskar, A. Levavalier des Etains, R. Mardling, T.A. Mictchenko, C. Terquem and J.L. Zhou, for helping us with the organization of this meeting, beginning from the preparation of the original proposal, to the selection of speakers, and to refereeing the papers for these proceedings. We are also thankful to many anonymous referees for their help and willingness to review all submitted articles. Finally, we would like to thank the authors for their participation in this focus meeting, their willingness to contribute to these proceedings, and for their cooperation and responsiveness during the editorial process.

We hope that this chapter, although only a representative of a fraction of what was presented at the meeting, will contribute to the intellectual stimulations and ongoing discussions on the fascinating problem of the origin of planetary systems.

Alessandro Morbidelli (*Former President of Commission 7: Celestial Mechanics and Dynamical Astronomy*)

Nader Haghighipour (*President of IAU Division F: Planetary Systems and Bioastronomy*)

An Overview of Inside-Out Planet Formation

Jonathan C. Tan[1], Sourav Chatterjee[2], Xiao Hu[3], Zhaohuan Zhu[4] and Subhanjoy Mohanty[5]

[1] Depts. of Astronomy & Physics, University of Florida, Gainesville, FL 32611, USA
email: jctan.astro@gmail.com
[2] CIERA, Physics and Astronomy, Northwestern University, Evanston, IL 60208, USA
[3] Dept. of Astronomy, University of Florida, Gainesville, FL 32611, USA
[4] Dept. of Astrophysical Sciences, Princeton University, Princeton, NJ 08544, USA
[5] Dept. of Physics, Imperial College, London, UK

Abstract. The *Kepler*-discovered Systems with Tightly-packed Inner Planets (STIPs), typically with several planets of Earth to super-Earth masses on well-aligned, sub-AU orbits may host the most common type of planets, including habitable planets, in the Galaxy. They pose a great challenge for planet formation theories, which fall into two broad classes: (1) formation further out followed by inward migration; (2) formation *in situ*, in the very inner regions of the protoplanetary disk. We review the pros and cons of these classes, before focusing on a new theory of sequential *in situ* formation from the inside-out via creation of successive gravitationally unstable rings fed from a continuous stream of small (∼cm-m size) "pebbles," drifting inward via gas drag. Pebbles first collect at the pressure trap associated with the transition from a magnetorotational instability (MRI)-inactive ("dead zone") region to an inner, MRI-active zone. A pebble ring builds up that begins to dominate the local mass surface density of the disk and spawns a planet. The planet continues to grow, most likely by pebble accretion, until it becomes massive enough to isolate itself from the accretion flow via gap opening. This reduces the local gas density near the planet, leading to enhanced ionization and a retreat of the dead zone inner boundary. The process repeats with a new pebble ring gathering at the new pressure maximum associated with this boundary. We discuss the theory's predictions for planetary masses, relative mass scalings with orbital radius, and minimum orbital separations, and their comparison with observed systems. Finally, we discuss open questions, including potential causes of diversity of planetary system architectures, i.e., STIPs versus Solar System analogs.

Keywords. formation — planets and satellites, protoplanetary disks

1. Introduction

Thousands of exoplanets have been discovered, especially by NASA's *Kepler* mission (e.g., Mullally *et al.* 2015), and most are in systems that are quite different from our own Solar System. In particular, a large percentage ($\gtrsim 30\%$) of low-mass stars are now thought to host Systems with Tightly-packed Inner Planets (STIPs). These usually have 3 or more detected planets of radii $\sim 1-10\,R_\oplus$ on orbital periods from ~ 1 to 100 days with a peak at ~ 10 to 20 days, i.e., orbital radii of ~ 0.1 AU (e.g., Fang & Margot 2012). Also, the systems are "tightly-packed," i.e., with period ratios near 1.5 to 3, equivalent to separations of ~ 10 to several tens of Hill radii, but are not on the verge of instability (as expected, since they are generally billions of years old). The period ratios are mostly non-resonant, with only $\sim 10\%$ piled-up just wide of first order resonances (mostly 2:1 and 3:2). They have a low dispersion in orbital inclination angles ($\lesssim 3°$). From the small subset of planets with dynamical mass measurements, we know that there is a wide range of mean densities of a factor of several, which indicates that some STIPs planets have accreted a H/He atmosphere that is a few % of the total mass. STIPs may host the most common kind of planet in the Universe and the most common type of habitable environments, which would be in STIPs around K and M main sequence stars.

The first theoretical scenario that has been proposed to explain STIPs involves formation of planets in the outer disk via the Core Accretion paradigm, followed by migration to the inner region (e.g., McNeil & Nelson 2010; Kley & Nelson 2012). Note that these models have generally assumed protoplanets are able to form from the outer disk, but have not explicitly model this step (c.f., Lambrechts & Johansen 2014; Levison *et al.* 2015; Bitsch *et al.* 2015), i.e., their initial conditions already involved massive protoplanets that are placed at quite arbitrary locations.

These models have faced several problems in reproducing the observed exoplanet systems. For example, McNeil & Nelson (2010) found it difficult to concentrate planets close to their host star to the degree observed in STIPs. Another major problem is that planets undergoing significant migration tend to become trapped in low-order mean motion resonances, which, as discussed above, are not a particular feature of the observed systems. This has then motivated other work to identify potential mechanisms of either reducing the efficiency of resonant trapping (Goldreich & Schlichting 2014) or to later move them out of resonance (e.g., Lithwick & Wu 2012; Rein 2012; Batygin & Morbidelli 2013; Chatterjee & Ford 2015).

As a very different alternative, *in situ* formation of the STIPs has been discussed by Chiang & Laughlin (2013) and modeled by Hansen & Murray (2012, 2013). However, this modeling again involves starting with a population of protoplanets (some as massive as 6 $M_\oplus$) that are initially distributed in a very concentrated region inside about 1 AU. After 10 Myr of collisional N-body evolution, Hansen & Murray found that oligarchic growth had led to planetary architectures similar to those of STIPs, including a relatively flat distribution of planetary masses with orbital radius. However, Ogihara *et al.* (2015)'s study, which is similar but also includes the effect of gas and resulting protoplanetary migration, leads to systems with planet masses that decline steeply with orbital radius, which are very different from the observed STIPs, and thus argues against this *in situ* oligarchic growth phase.

2. Inside-Out Planet Formation - Theoretical Summary

An overview of the Inside-Out Planet Formation (IOPF) model (Chatterjee & Tan 2014, hereafter CT14 or Paper I) is shown in Fig. 1. The first basic assumption is that there is efficient supply of "pebbles" drifting radially inwards to ∼0.1 AU from the outer disk. This radial drift is a well-known effect due to gas drag in regions where the gas disk derives some support from a radially decreasing pressure gradient causing its orbital speeds to be slightly sub-Keplerian (Weidenschilling 1977). Indeed this drift is so strong that it has long been recognized as a major problem for planetesimal formation, which is part of the so-called "meter-sized barrier." The radial drift of pebbles is assumed to be stopped at the local pressure maximum associated with the dead zone inner boundary (DZIB), i.e., where gas and pebbles both orbit at the Keplerian speed so that there is no headwind gas drag experienced by the pebbles. The location of this DZIB is assumed to be set by when gas temperatures reaches about 1,200 K, allowing thermal ionization of alkali metals Na and K (Umebayashi & Nakano 1988). These species should provide enough ionization to allow the magneto-rotational instability (MRI) (Balbus & Hawley 1991) to operate, which increases the disk's viscosity and so leads to reduced surface densities, volume densities and pressures compared to at the DZIB. A pebble ring then builds up at the DZIB, which can come to dominate the local mass surface density. A planet forms

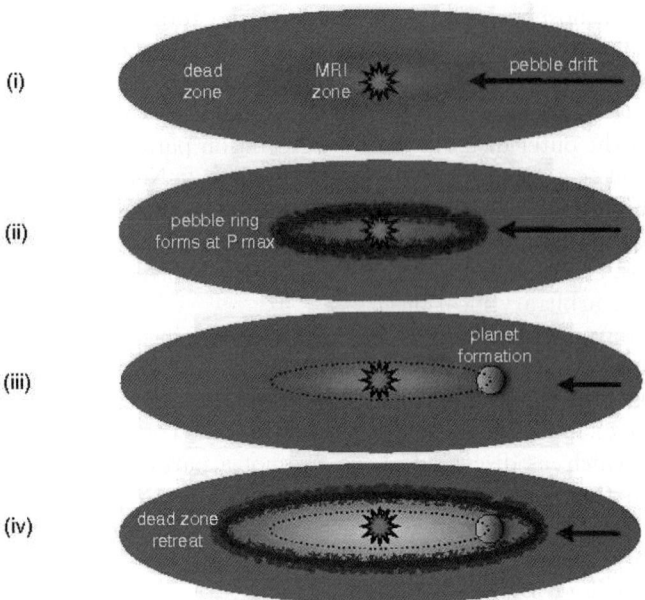

Figure 1. Schematic overview of Inside-Out Planet Formation (CT14). **(i) Pebble formation and drift to the inner disk.** Pebbles form via dust coagulation in the protoplanetary disk. Those with ∼cm to m sizes attain high radial drift velocities and quickly reach the dead zone inner boundary (DZIB), where they become trapped at the pressure maximum. **(ii) Pebble ring formation.** A ring of pebbles gradually builds up over a timescale set by the pebble formation and supply rate from the outer disk. **(iii) Planet formation and gap opening.** A planet forms from the pebble ring and continues to grow by pebble accretion until it becomes massive enough to open a gap. This shuts off pebble accretion and may also lead to reduction in gas supply to the inner disk, which would then dissipate by viscous clearing. **(iv) Dead zone retreat and subsequent pebble ring and planet formation.** Gap opening and potential viscous clearing of the inner disk lead to lower densities and greater penetration of X-ray photons from the protostar to the disk midplane, increasing its ionization fraction and thus activating the MRI. The inactive dead-zone retreats, along with the pressure maximum associated with its inner boundary. A new pebble ring starts to form at this location that forms a new planet. This cycle repeats, leading to sequential formation of a planetary system from the inside-out.

from this ring. The protoplanet grows without suffering significant migration. The next crucial stage is when the planet, which has been growing by pebble accretion, becomes massive enough to open a (potentially quite shallow) "gap" in the disk that is sufficient to move the local pressure maximum away from the planet, thus shutting off pebble accretion. At the same time, the reduction in gas density around the planet leads to increased ionization, perhaps also due to increased X-ray penetration from the protostar, activating the MRI and causing the DZIB to retreat outwards. This retreat can be self-propagating since increasing viscosity in the boundary region leads to further reductions in densities. However, this processes stabilizes relatively quickly and a new pebble ring begins to form at the pressure maximum at the retreated DZIB. This location will be at least several Hill radii from where the first planet formed, but could be significantly further away. The entire process repeats leading to the sequential formation of a compact, well-aligned planetary system from the inside-out.

In order to make quantitative estimates, CT14 adopted the Shakura & Sunyaev (1973) "α-disk" model framework for the structure of a steady, active accretion disk, i.e., in which the heating is dominated by accretion. Typical observed accretion rates of

T-Tauri stars (Alcala et al. 2014) and stars with transition disks (Manara et al. 2014) are $\sim 10^{-9}\ M_\odot\ \mathrm{yr}^{-1}$, with a dispersion of about a factor of 100. Transition disks in which there are gaps and holes in the very inner disk dust distribution inside ~ 1 AU may be particularly relevant for the IOPF model. Thus CT14 adopted $\dot{m} = 10^{-9}\ M_\odot\ \mathrm{yr}^{-1}$ as a fiducial value, i.e., $\dot{m}_{-9} = 1$, but consider potential variations of $\dot{m}_{-9} = 0.1$ to 10. For simplicity, CT14 also adopted a fixed opacity of 10 cm^2 g^{-1}, i.e., $\kappa_{10} = 1$, which is a typical value expected in inner protoplanetary disks (e.g., Zhu et al. 2009). The value of the α viscosity parameter in protoplanetary disks is quite uncertain. In DZIB regions, the simulations of Dzyurkevich et al. (2010) find effective viscosities equivalent to $\alpha \sim 10^{-4}$ to 10^{-3}, partly set by the propagation of turbulence outwards from the MRI-active region. CT14, with a focus on disk midplane conditions, adopted $\alpha = 10^{-3}$ as a fiducial value in the dead zone region (but we will see below that moderately smaller values may be preferred). In the MRI-active region, α is assumed to rise to much larger values $\sim 10^{-2}$ or more.

In the context of this accretion disk model, CT14 showed that the radial drift time of pebbles from the outer to inner disk was very short compared to expected disk lifetimes. Hu, Tan & Chatterjee (2014) presented more detailed calculations, including Stokes-limited pebble growth via sweep-up of small grains, finding that initially 1 mm-radius pebbles would reach the inner disk after only 2,000 or 40,000 yr if starting from 10 or 100 AU, respectively (this assumes the dead zone value of α extends to these scales). However, a quantitative estimate of the pebble production rate and thus the overall mass flux in pebbles to the inner disk has not yet been made for these models. Still, observations of disks are beginning to reveal both radial concentrations of dust with respect to gas (e.g, de Gregorio-Monsalvo et al. 2013) and increasing grain sizes in the inner regions (e.g., Pérez et al. 2012; Trotta et al. 2013), so a large mass flux of pebbles to inner disks remains a distinct and even likely possibility.

CT14 evaluated the location of the DZIB by the condition that disk midplane temperature reaches 1,200 K, finding a radius $r_{1200\mathrm{K}} = 0.1 \phi_{\mathrm{DZIB},0.1\mathrm{AU}} \kappa_{10}^{2/9} \alpha_{-3}^{-2/9} m_{*,1}^{1/3} \dot{m}_{-9}^{4/9}$ AU, with $\phi_{\mathrm{DZIB},0.1\mathrm{AU}} = 1.8$, i.e., a fiducial location of 0.18 AU (around a star of 1 $M_\odot$, i.e., $m_{*,1} = 1$). Note that the location of the DZIB increases for larger accretion rates. Hu et al. (2015) (Paper III) revisited the disk structure equations and adopted a slightly different choice for normalization of the vertical optical depth equation (or equivalently the definition of midplane conditions), which leads to an estimate of $r_{1200\mathrm{K}} = 0.13$ AU, i.e., $\phi_{\mathrm{DZIB},0.1\mathrm{AU}} = 1.3$. Mohanty & Tan (in prep.) considered the structure of a fully self-consistent MRI-active inner disk, finding α decreased rapidly due to Ohmic resistivity at a radius of ~ 0.1 to 0.2 AU.

CT14 discussed various potential mass scales of planet formation from the pebble ring, including the Toomre mass from a gravitationally unstable ring ($\sim 10^{-3}\ M_\oplus$ in the fiducial case) and the Toomre Ring mass (fiducial value of $\sim 1\ M_\oplus$). However, the most important mass scale is identified as being the gap-opening mass (Lin & Papaloizou 1993), $M_G = \phi_G 40 \nu m_*/(r^2 \Omega_K) \simeq 5.59 \phi_{G,0.5} \kappa_{10}^{1/5} \alpha_{-3}^{4/5} m_{*,1}^{3/10} \dot{m}_{-9}^{2/5} r_{0.1\mathrm{AU}}^{1/10}\ M_\oplus$, which is derived by considering the competition of the planet's gravity with the viscosity of the gas. Here the overall normalization, including choice of $\phi_G = 0.5$, is based on the numerical simulations of Paper III: at this mass scale the response of the disk to the presence of the planet leads to the pressure maximum being displaced outwards by about 5 R_H.

The mass scale for gap opening at the location of the DZIB set by midplane temperature of 1,200 K, which would be the mass of innermost, "Vulcan" planets in the IOPF model, has the following dependencies (Chatterjee & Tan 2015, hereafter CT15 or Paper II): $M_{p,1} = M_G(r_{1200\mathrm{K}}) = 5.59 \phi_{G,0.5} \phi_{\mathrm{DZIB},0.1\mathrm{AU}}^{-9/10} \alpha_{-3} r_{0.1\mathrm{AU}}\ M_\oplus$ (note the normalization here follows the Paper III disk model and is a factor of 0.745 smaller than in Paper II). This prediction is that inner planet mass scales linearly with orbital radius, does not depend on κ or m_*, but does depend on the value of α in the DZIB region.

The question of the potential migration of protoplanets as they are forming and opening gaps has been studied in Paper III, where we find that from 0.1 to 1 M_G, protoplanets are trapped very close to their formation location set by the initial pressure maximum (and associated gas surface density maximum) at the DZIB.

Subsequent retreat of the DZIB due to gap opening and increased X-ray penetration has been studied with simple, heuristic models in Paper III. Important parameters include the penetration depth of X-rays through the gas disk that allows MRI activation and the width the transition zone from the MRI-active to inactive regions. Paper III presented simple example models of this process that could lead to DZIB (and thus pressure maximum) retreat by several tens of Hill radii of the already-formed planet.

3. Inside-Out Planet Formation - Observational Summary

CT14 noted that if the dead zone inner boundary is set due to thermal ionization of alkali metals at $\sim 1,200$ K, then its expected location in disks with accretion rates of $\sim 10^{-9}$ $M_\odot$ yr^{-1} (i.e., similar to those of observed stars with transition disks) is estimated to be ~ 0.1 AU. This is very similar to the sizes of the observed orbits of the STIPs planets. The expected mass scale for gas gap opening is several Earth masses, assuming $\alpha \sim 10^{-3}$, which is again similar to the STIPs planet masses. However, it should be noted that most of these mass estimates are quite uncertain, since they are based on an assumed mass (or density) versus size relation (e.g., Lissauer et al. 2011) and it is clear from the planets with dynamical mass measurements that at a given mass there is actually a wide range of densities of a factor of about 5 (CT15). Fig. 2 plots the masses and orbital radii of the most recent census of STIPs planets, where mass has been estimated from the piecewise power law fit to the mass-size relation of STIPs with dynamical mass measurements (PL3 of CT15). The analytic values for gap opening masses assume $\alpha_{-3} = 0.205$, which is based on a comparison of only the innermost, Vulcan planets (CT15), discussed below. We see in Fig. 2 that the mass scales and orbital locations of STIPs planets are consistent with gap opening masses near DZIBs in disks with typical observed accretion rates.

CT14 also examined the dependence of planet mass with orbital radius. Around a given star and for a constant accretion rate, the gap opening mass is expected to scale as $M_G \propto r^{k_M}$ with $k_M = 0.1$, i.e., a relatively flat scaling. Examining the 4, 5 and 6-planet systems known at the time, CT14 found that masses scaled as power laws with orbital radius with indices of $k_M = 0.92 \pm 0.63$, 0.78 ± 0.64 and 0.50, respectively, with the quoted uncertainty being the dispersion in the samples. For the handful of systems that had dynamical mass measurements of their planets, CT14 found $k_M = 1.0 \pm 2.1$ (average of 6 systems) and $k_M = 0.47 \pm 2.7$ (average of all adjacent planet pairs). In summary, the data are consistent with the scaling predicted by the gap opening mass, although there is a hint that observed planet masses increase more steeply with orbital radius. However, for the trends derived from the STIPs that are lacking dynamical mass measurements, the results are strongly influenced by the choice of planet density with orbital radius: a systematic trend of denser inner planets, due either to pebble composition during formation or effects of subsequent evaporation, would tend to lower the values of k_M.

Papers I & III examined the orbital spacings between adjacent planet pairs, normalized by the Hill radius of the innermost planet, $R_{H,i}$, i.e., $\phi_{\Delta r,i} \equiv \Delta r_i / R_{H,i}$, where $\Delta r_i = r_{i+1} - r_i$. In the IOPF model we expect the first gap opening event and potential clearing

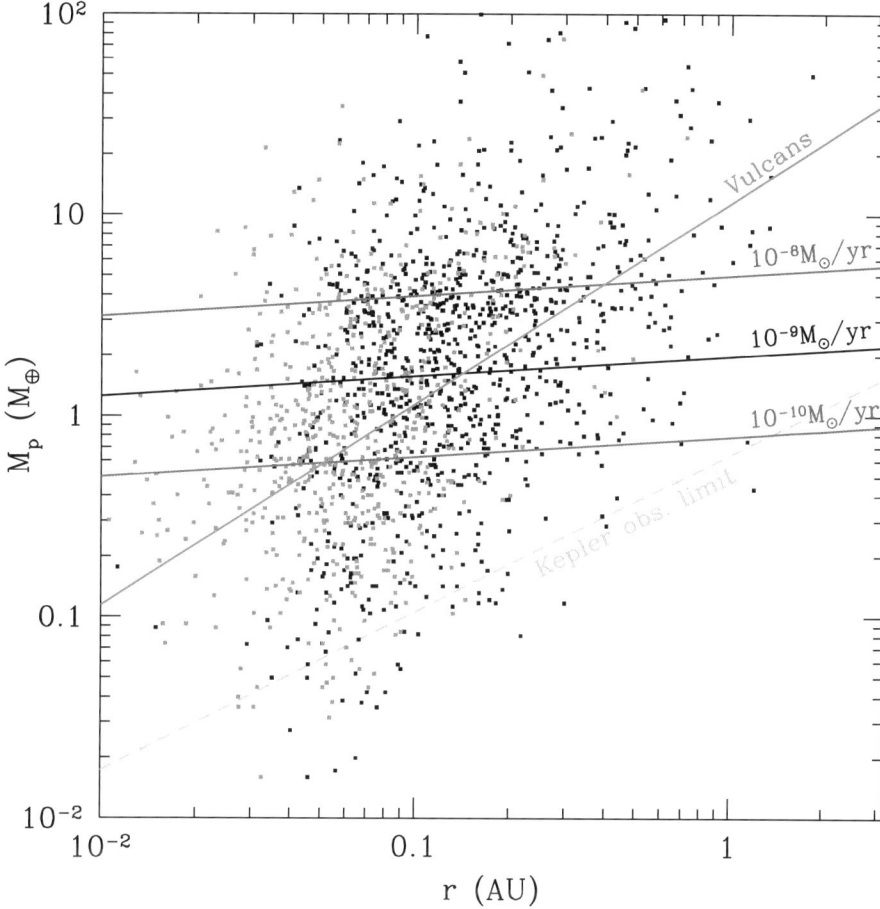

Figure 2. Planet mass versus orbital radius for 1,656 STIPs planets, with 664 innermost, "Vulcan" planets shown in magenta. The gap opening mass, M_G, for disks with $\alpha = 2.05 \times 10^{-4}$ and $\dot{m}_* = 10^{-10}, 10^{-9}, 10^{-8}\, M_\odot\,\text{yr}^{-1}$ are shown with the solid red, black and blue lines, respectively. The solid magenta line shows $M_G(r = r_{1200\text{K}})$ for these disks (with $\phi_{\text{DZIB},0.1\text{AU}} = 1$), which is the Inside-Out Planet Formation model prediction for Vulcan planet masses. The dashed green line shows an approximate estimate for *Kepler's* detection limit (S/N=7) for the median $K_p = 14.5$ host star and the CT15 PL3 mass-size relation.

of the inner disk to lead to relatively larger DZIB retreat, perhaps due to increased X-ray penetration from the protostar. However, subsequent planet formation events would have a more incremental effect on the disk structure and may be expected to lead to more modest retreats and thus smaller normalized orbital separations. In Papers I & III, the latter using improved STIPs planet mass estimates from Paper II, we do find statistically significant differences in the distributions of $\phi_{\Delta r,1}$ compared to those of the other planet pairs (which themselves have indistinguishable distributions). The distribution of $\phi_{\Delta r,1}$ peaks at larger values ~ 20 to 40, while $\phi_{\Delta r,2}$, $\phi_{\Delta r,3}$ & $\phi_{\Delta r,4}$ peak at $\lesssim 20$. For example, in systems with $\geqslant 3$ planets, the probability that the observed distributions of $\phi_{\Delta r,1}$ & $\phi_{\Delta r,2}$ are drawn from the same underlying distribution is only 9×10^{-4}, and restricting to systems with $\geqslant 4$ planets the probabilities are only $\sim 10^{-6}$ that $\phi_{\Delta r,1}$ has the same distribution as $\phi_{\Delta r,2}$ or $\phi_{\Delta r,3}$ (Paper III). These differences are interesting observational results that, in the context of IOPF, impose constraints on models of DZIB retreat.

CT15 tested the predicted mass versus orbital radius scaling of innermost, Vulcan planets, which is shown by the magenta line in Fig. 2. For each host star of a detected Vulcan, a planet with the appropriate gap opening mass, $M_G(r_{1200\rm K})$, with $r_{1200\rm K}$ set equal to the current observed orbital radius, was modeled. These planets were given densities (and thus sizes) randomly sampled from the distribution functions fitted to observed STIPs planets with dynamical mass measurements. It was then checked whether they would have been detected by *Kepler* and, if not, a new density was sampled. The "observational masses" of the simulated planets were then evaluated from the empirical piecewise power law mass-size relation. These observational masses thus have a random scatter compared to the input masses. Power laws of the form $M_{p,1}/M_\oplus = p_0 r_{\rm AU}^{p_1}$ were then fit to the real and simulated Vulcans. The real Vulcans have $p_0 = 7.8 \pm 1.5$ and $p_1 = 0.72 \pm 0.17$. The simulated Vulcans can achieve a similar value of p_0 if $\alpha_{-3} = 0.205$ (including normalization to the Paper III disk model). They have $p_1 \simeq 0.94 \pm 0.17$, i.e., slightly shallower than the actual input value of $p_1 = 1$. Overall, this comparison shows that the observed Vulcans have a mass versus orbital radius relation that is consistent with the scaling predicted by gap opening at the DZIB. The normalization of the viscosity parameter is also consistent with theoretical expectations (Dzyurkevich *et al.* 2010).

4. Conclusions and Open Questions

Systems with Tightly-packed Inner (Earth to Super-Earth) Planets (STIPs) are very common. They may have formed further out in the disk and then migrated inwards, but such models face a number of challenges in reproducing observed architectures. Alternatively, the planets may have formed *in situ*. Inside-Out Planet Formation (IOPF) is a new *in situ* formation model that embraces the large mass flux of pebbles predicted by Weidenschilling (1977) as a meter-sized barrier for planetesimal formation. IOPF assumes these pebbles are trapped at the pressure maximum associated with the dead zone inner boundary (DZIB) set by thermal ionization of Na and K at $\sim 1,200$ K. A pebble ring forms and grows to dominate the local mass surface density and form a planet. Gap opening by this planet is the key process that shuts off pebble accretion and leads to DZIB retreat. A new pebble ring forms once retreat is stabilized and the process repeats.

Features of this model include that the meter-sized barrier for planetesimal formation is not a particular problem as it is for standard Core Accretion planet formation models. IOPF predicts planets with masses of $\sim$few $M_\oplus$ are created on tightly-packed, aligned orbits at distances of ~ 0.1 AU, consistent with observed systems. It predicts a flat scaling of planet mass with orbital radius, again consistent with observed systems. Orbital spacings should be at least ~ 5 Hill radii of the inner planet, but will typically be larger due to DZIB retreat. The Hill-normalized spacing from first to second planet is expected to be larger than subsequent spacings, again as observed. Innermost, "Vulcan" planets have a particularly simple, linear mass versus orbital radius relation. This is also consistent, in both its normalization and scaling, with the observed Vulcans.

Many open questions remain to be addressed, including estimates of the pebble supply rate to the inner disk, which sets the rate of IOPF. Also the question of the onset of IOPF and the prior history of the disk at earlier stages, when the accretion rates were likely larger: in particular, why does IOPF require its onset to coincide with inner disk accretion rates of $\sim 10^{-9}\,M_\odot\,{\rm yr}^{-1}$? While we expect limited migration of the first planet when it is forming (Paper III), how much migration occurs in later stages, including due to planet-planet interactions? Can small amounts of H/He gas be accreted to planets forming via IOPF, i.e., under very warm conditions? Does atmospheric evaporation play any significant role in altering the properties of planets that may have formed by IOPF?

If IOPF is a valid model of planet formation, why has it not occurred in some systems, such as our Solar System? Possible reasons may include processes that sometimes truncate pebble supply to the inner disk, such as metallicity-dependent efficient planetesimal formation via the streaming instability (Youdin & Goodman 2005) leading to early giant planet formation (e.g., Lambrechts & Johansen 2014). Alternatively, there could be variation due to processes that maintain ionization and MRI activity to much larger scales than expected by thermal ionization (e.g., enhanced levels of cosmic rays or radionuclides) or due to processes that may completely suppress the MRI and thus remove the DZIB pressure trap in some circumstances (such as the Hall Effect and its dependence on global disk B-field orientation).

Acknowledgements

JCT and SC acknowledge support from NASA ATP grant NNX15AK20G. JCT and SM acknowledge support from a Royal Society International Exchange grant IE131607.

References

Alcalá, J. M., Natta, A., Manara, C. F. et al. 2014, *A&A*, 561, 2
Balbus, S. A. & Hawley, J. F. 1991, *ApJ*, 376, 214
Batygin, K. & Morbidelli, A. 2013, *AJ*, 145, 1
Bitsch, B., Lambrechts, M., & Johansen, A. 2015, *A&A*, in press (arXiv:1507.05209)
Chatterjee, S. & Ford, E. B. 2015, *ApJ*, 803, 33
Chatterjee, S. & Tan, J. C. 2014, *ApJ*, 780, 53 (CT14, Paper I)
Chatterjee, S. & Tan, J. C. 2015, *ApJ*, 798, L32 (CT15, Paper II)
Chiang, E. & Laughlin, G. 2013, *MNRAS*, 431, 3444
de Gregorio-Monsalvo, I., Ménard, F., Dent, W. et al. 2013, *A&A*, 557, 133
Dzyurkevich, N., Flock, M., Turner, N. J., Klahr, H., & Henning, T. 2010, *A&A*, 515, 70
Fang, J. & Margot, J. L. 2012, *ApJ*, 761, 92
Goldreich, P. & Schlichting, H. E. 2014, *AJ*, 147, 32
Hansen, B. & Murray, N. 2012, *ApJ*, 751, 158
Hansen, B. & Murray, N. 2013, *ApJ*, 775, 53
Hu, X., Tan, J. C., & Chatterjee, S. 2014, *IAUS*, 310, 66 (arXiv:1410.5819)
Hu, X., Zhu, Z., Tan, J. C., & Chatterjee, S. 2015, *ApJ*, submitted (arXiv:1508.02791)
Kley, W. & Nelson, R. P. 2012, *ARA&A*, 50, 211
Lambrechts, M. & Johansen, A. 2014, *A&A*, 572, 107
Levison, H. F., Kretke, K. A., & Duncan, M. J. 2015, *Nature*, 524, 322
Lin, D. N. C. & Papaloizou, J. C. B. 1993, in *Protostars and Planets III*, ed. E. H. Levy & J. I. Lunine (Tucson, AZ: Univ. Arizona Press), 749
Lissauer, J. J., Ragozzine, D., Fabrycky, D. C., et al. 2011, *ApJS*, 197, 8
Lithwick, Y. & Wu, Y. 2012, *ApJL*, 756, L11
Manara, C. F., Testi, L., Natta, A. et al. 2014, *A&A*, 568, 18
McNeil, D. S. & Nelson, R. P. 2010, *MNRAS*, 401, 1691
Mullally, F., Coughlin, J. L., Thompson, S. E. et al. 2015, *ApJS*, 217, 31
Ogihara, M., Morbidelli, A., & Guillot, T. 2015, *A&A*, 578, 36
Pérez, L. M., Carpenter, J. M., Chandler, C. J. et al. 2012, *ApJ*, 760, 17
Rein, H. 2012, *MNRAS*, 427, L21
Shakura, N. I. & Sunyaev, R. A. 1973, *A&A*, 24, 337
Trotta, F., Testi, L., Natta, A., Isella, A., & Ricci, L. 2013, *A&A*, 558, 64
Umebayashi, T. & Nakano, T. 1988, *PThPS*, 96, 151
Weidenschilling, S. J. 1977, *MNRAS*, 180, 57
Youdin, A. N. & Goodman, J. 2005, *ApJ*, 629, 459
Zhu, Z., Hartmann, L., Gammie, C., & McKinney J. C. 2009, *ApJ*, 701, 620

Planet migration and magnetic torques

A. Strugarek[1,2], A. S. Brun[2], S. P. Matt[3] and V. Reville[2]

[1] Université de Montréal, C.P. 6128 Succ. Centre-Ville, Montréal, QC H3C-3J7, Canada
email: strugarek@astro.umontreal.ca
[2] CEA-Saclay, IRFU/SAp, Gif-sur-Yvette, France.
[3] Department of Physics & Astronomy, University of Exter, EX2 4QL, UK

Abstract. The possibility that magnetic torques may participate in close-in planet migration has recently been postulated. We develop three dimensional global models of magnetic star-planet interaction under the ideal magnetohydrodynamic (MHD) approximation to explore the impact of magnetic topology on the development of magnetic torques. We conduct twin numerical experiments in which only the magnetic topology of the interaction is altered. We find that magnetic torques can vary by roughly an order of magnitude when varying the magnetic topology from an aligned case to an anti-aligned case. Provided that the stellar magnetic field is strong enough, we find that magnetic migration time scales can be as fast as ~ 100 Myr. Hence, our model supports the idea that magnetic torques may participate in planet migration for some close-in star-planet systems.

1. Introduction

The diversity of masses, sizes and orbits of known exoplanets has prompted recent efforts in the scientific community to explore the broad range of interactions that can exist between planets and their host stars (see Cuntz *et al.* 2000). In addition to tidal interactions (e.g. Mathis *et al.* 2013), planets orbiting inside the stellar wind Alfvén radius can magnetically interact with their host (Ip *et al.* 2004; Cohen *et al.* 2010; Strugarek *et al.* 2014). These interactions could lead to an angular momentum transfer between the planet and the star (Lovelace *et al.* 2008; Laine & Lin 2011; Strugarek *et al.* 2014), resulting in a substantial planetary migration and participating in the dynamical (in)stability of the system. Among the star-planet interaction (SPI) models that have been developed, MHD simulations combine state of the art numerical models of cool stars magnetospheres and winds (Matt *et al.* 2012; Réville *et al.* 2015) with simplified models of planets (e.g., Cohen *et al.* 2014; Strugarek *et al.* 2014, and references therein). These global, dynamical models enable us to assess the effects of SPI in a self-consistent manner, by taking into account the full interaction channel from the planetary magnetosphere down to the lower stellar corona.

In a recent paper (Strugarek *et al.* 2015), we have developed MHD simulations of magnetic star-planet interactions in three dimensions. We address here the important question of the physical origin of the magnetic torque, and in particular the key role magnetic topology plays in determining the strength of the torque. We first briefly describe our methodology in Section 2, and detail how the magnetic torques are sustained in Section 3. We conclude in Section 4 by showing that for sufficiently strong stellar wind magnetic fields, the migration time-scale associated with magnetic torque is comparable with typical tidal migration time-scales.

2. Three-dimensional models of star-planet magnetic interaction

We model magnetic star-planet interactions with global numerical simulation using the ideal MHD approach. We use the modular code PLUTO (Mignone *et al.* 2007) to solve the MHD equations with a standard HLL Riemann solver coupled to a second-order Runge-Kutta method for the time integration. The soleinoidality of the magnetic field is enforced with a constrained method transport (see Strugarek *et al.* 2015). We discretize space with a cartesian grid in which a star and a planet are added as internal boundary conditions. At the stellar surface a magnetized stellar wind is imposed. We enforce a given magnetic field at the planetary boundary. We solve the MHD equations in the orbital rotating frame, such that the planet is fixed in the simulation grid. The star is positioned at the center of the simulation grid, and is hence as well fixed in the grid (the stellar rotation rate is corrected to account for the orbital rotating frame). The simulations are carried on a $490 \times 355 \times 355$ grid, with an resolution of $0.03\, R_\star$ at the stellar boundary and a resolution of $0.06\, R_P$ in the vicinity of the planet.

The modelled star is considered to be a typical cool star with a coronal temperature of 10^6 K, with a relatively strong magnetic field such that the Alfvén speed v_A at the base of the corona is equal to the escape velocity $v_{\rm esc}$ (for details see Strugarek *et al.* 2015). The star is considered to slowly rotate ($v_{\rm rot} = 3.03\, 10^{-3}\, v_{\rm esc}$) and its wind is characterized by a large average Alfvén radius $\langle R_A \rangle = \sqrt{\dot{J}/\Omega_\star \dot{M}} \sim 18$ (see, e.g., Matt *et al.* 2012).

We consider a close-in planet located at $R_{\rm orb} = 5\, R_\star$. The orbit is supposed to be circular since such close-in planet is likely to be tidally locked. We further assume that the planet possess an intrinsic dipolar magnetic field sufficiently large to retain a magnetosphere. The planetary magnetic field is chosen to be 10 times larger than the wind magnetic field at the planetary surface. We neglect any kind of planetary outflows (see Matsakos *et al.* 2015, for a complete discussion about such outflows) to focus on the effect of magnetic topology on the star-planet interaction itself. We consider the two cases of aligned and anti-aligned dipolar planetary fields.

We illustrate those two configurations in Figure 1 where a three-dimensional representation of the interaction is shown. In the aligned case (left panel), the magnetic topology allows the polar magnetic field lines of the planet (grey lines) to reconnect with the wind magnetic field lines (coloured lines). In the anti-aligned case (right panel), the planetary magnetosphere remains closed due to the incompatible topology of the two magnetic fields. We recall that only the orientation of the planetary field has been changed between the two cases, leaving all the other parameters untouched.

3. Magnetic torques and planet migration

The torques that develop in close-in star-planet systems can be separated into different physical contributions through (see Strugarek *et al.* 2015)

$$\mathcal{T} = \mathcal{T}_{\rm ram} + \mathcal{T}_{\rm Coriolis} + \mathcal{T}_{\rm Magnetic\, tension} + \mathcal{T}_P + \mathcal{T}_{\rm Magnetic\, pressure}. \qquad (3.1)$$

In Figure 2 we display these various contributions on the spherical cut symbolized by the white transparent sphere in Figure 1. The torques can be positive (red) or negative (blue), which respectively correspond to a loss or a gain of angular momentum for the orbiting planet, *i.e.* an infall or an expulsion of the planet. The torques are shown in Mollweide projection, with the central black circle delimiting the day-night meridian. As a result, the rightmost part of the projection corresponds to the upstream side of the interaction, and the leftmost side to the downstream side. In the aligned case the total

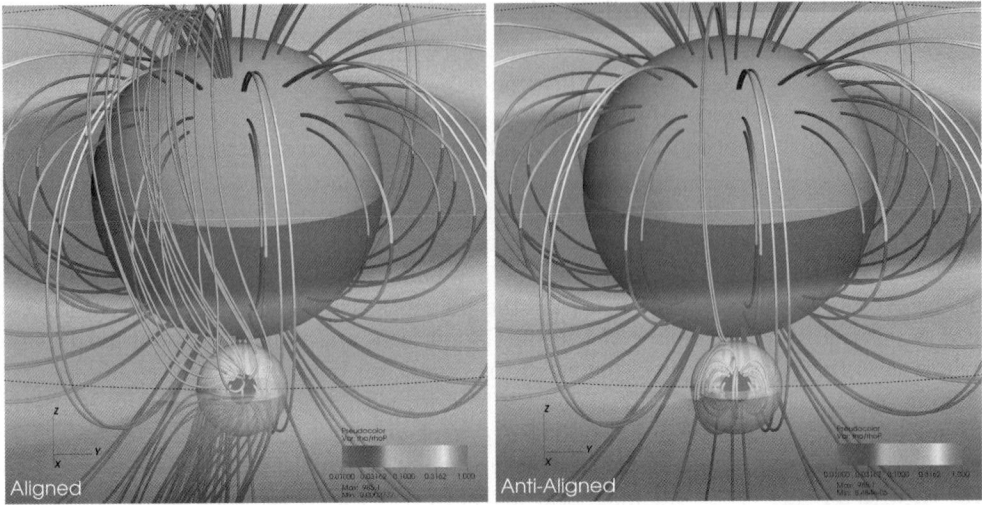

Figure 1. 3D renderings of the two star-planet interaction simulations. The stellar wind magnetic field lines are color-coded with the magnetic field strength, and the magnetic field lines connected to the planet are shown in grey. The stellar surface is represented by the orange sphere, and the planetary surface by the blue sphere. On the ecliptic the orbit is labeled by the black dashed line, and the logarithm of the density is shown by the transparent colormap. The transparent white sphere around the planet labels the integration sphere on which the torque components are calculated in Figure 2.

torque (bottom right panel) is completely dominated by magnetic tension (upper right panel) and magnetic pressure (bottom middle panel). Conversely, magnetic tension play almost no role in the anti-aligned case and the total torque is completely dominated by the magnetic pressure of the ambient wind magnetic field.

The magnetic pressure contribution is composed of positive and negative torques up and downstream, as expected from a stream-obstacle type of interaction. In the anti-aligned case, the obstacle can hence be simply approximated by the almost-spherical magnetosphere of the planet. In the aligned case, the magnetic tension is dominated by a positive torque downstream. The tension originates from the polar magnetic field lines that can be identified in Figure 1 (grey lines). In addition to the magnetosphere of the planet itself, these field lines act as a second obstacle and thus participate to the angular momentum exchange between the planet and its environment. The overall torque is obtained by integrating the various contributions over the integration sphere. As a result, only a small torque remains in the anti-aligned case, while magnetic tension provides a much stronger torque in the aligned case. The key role of topology clearly appears: in the aligned case the effective obstacle is completely different from the obstacle in the anti-aligned case, which results in a much weaker torque in the later case.

The migration time-scale associated with those torques is given by

$$t_P = \frac{2J_P}{\mathcal{T}}, \qquad (3.2)$$

where the planetary orbital angular momentum is $J_P = M_P \left(GM_\star R_{\rm orb}\right)^{1/2}$. The migration time-scale depends directly on the density normalization chosen for the stellar wind (see Strugarek *et al.* 2015, for a complete discussion), which controls the mass-loss rate and the magnetic field strength of the wind in physical units. For a surface stellar magnetic field $B_\star \sim 10\,G$, the migration time-scale is of the order of $1.4\,10^7$ Myr in the

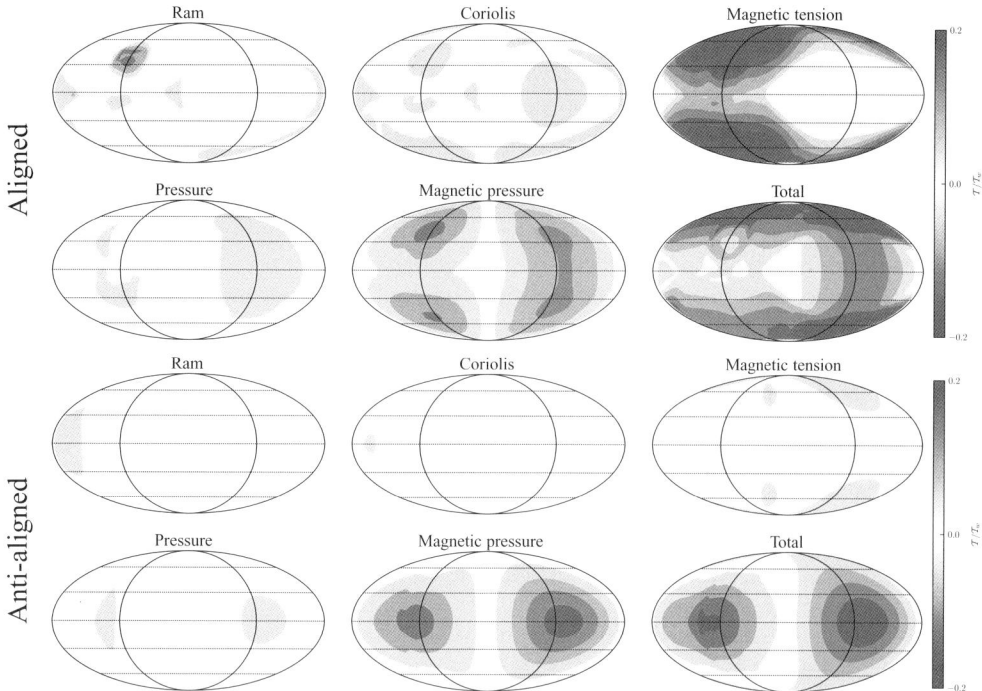

Figure 2. Contributions to the torque applied to the close-in planet. The terms are shown on the integration sphere of radius $0.3\,R_P$ enclosing the planet with a Mollweide projection. The black circle in the center of each projection corresponds to the day-night delimitation. The upstream side is located on the right of each plot, and the downstream side on the left. The contributions are normalized by the stellar wind torque $\mathcal{T}_w$. Red denotes angular momentum extraction from the planet, while blue denotes deposition of angular momentum.

aligned case and $8.4\,10^7$ Myr in the anti-aligned case. As a result, for such moderate large-scale stellar magnetic fields, the magnetic torque does not induce any significant migration of the planet. On the contrary, for a surface stellar magnetic field $B_\star \sim 1\,kG$, the migration time-scale drops by 4 orders of magnitude (hence a time-scale of $1.4\,10^3$ Myr in the aligned case). In this case magnetic torques are able to compete with tidal torques for planet migration.

In conclusion, close-in planets can migrate due to magnetic torques provided the stellar magnetic field is sufficiently strong and the planet is sufficiently close. In addition, the magnetic topology has a key impact on the strength of the torque: by simply reversing the orientation of the planetary magnetic field, the torque is decreased by a factor of 6.

4. Conclusions

We have demonstrated how magnetic topology changes close-in star-planet interactions using three-dimensional global numerical simulations. We developed twin simulations of close-in star-planet magnetic interactions in which we only changed the orientation of the planetary field. In the so-called aligned case, we showed that the effective planetary obstacle is much greater than the planetary magnetosphere alone. As a result, the magnetic torque is found to be 6 larger than in the conventional anti-aligned case. The magnetic torques are also shown to be sufficiently strong to compete with tidal torques for planet migration, provided the stellar magnetic field is sufficiently strong. These results confirm the first-order estimates that were derived in Strugarek et al. (2014) with simpler 2.5 numerical simulations. We are currently running a more extensive set of 3D models to empirically refine the torque scaling laws with orbital radius, magnetic field strengths and magnetic topology that were proposed in Strugarek et al. (2014).

Real stars posses much more complex magnetic fields than the simple dipolar configuration we considered here. In reality close-in planets are likely to interact with different local magnetic configurations along their orbit. Assessing what average magnetic torque results from a magnetic topology varying along the orbit will require dedicated 3D simulations tackling the dynamical aspects of a planet orbiting in a non-homogenous corona. Indeed, the time-scale on which the equilibrated configurations modelled in this work establish depends on the resistivity of the magnetospheric plasma of the planet, and on its reconnection efficiency with the stellar wind magnetic field. The numerical model presented in this work provides a solid basis for further, more realistic studies of star-planet magnetic interactions in which these dynamical aspects could be explored.

References

Cohen, O., Drake, J. J., Glocer, A., et al. 2014, *ApJ*, 790, 57
Cohen, O., Drake, J. J., Kashyap, V. L., Sokolov, I. V., & Gombosi, T. I. 2010, *ApJ*, 723, L64
Cuntz, M., Saar, S. H., & Musielak, Z. E. 2000, *ApJ*, 533, L151
Ip, W.-H., Kopp, A., & Hu, J.-H. 2004, *ApJ*, 602, L53
Laine, R. O. & Lin, D. N. C. 2011, *ApJ*, 745, 2
Lovelace, R. V. E., Romanova, M. M., & Barnard, A. W. 2008, *MNRAS*, 389, 1233
Mathis, S., Alvan, L., & Remus, F. 2013, EAS Publications Series, 62, 323
Matsakos, T., Uribe, A., & Königl, A. 2015, *A&A*, 578, A6
Matt, S. P., MacGregor, K. B., Pinsonneault, M. H., & Greene, T. P. 2012, *ApJL*, 754, L26
Mignone, A., Bodo, G., Massaglia, S., et al. 2007, *ApJS*, 170, 228
Réville, V., Brun, A. S., Matt, S. P., Strugarek, A., & Pinto, R. F. 2015, *ApJ*, 798, 116
Strugarek, A., Brun, A. S., Matt, S. P., & Réville, V. 2014, *ApJ*, 795, 86
—. 2015, Submitted to ApJ

The Effect of Convective Overstability on Planet Disk Interactions

Hubert Klahr[1] and Aiara Lobo Gomes[2]

Max-Planck-Institut für Astronomie,
Königstuhl 17, 69117 Heidelberg, Germany
[1]email: klahr@mpia.de
[2]email: gomes@mpia.de

Abstract. We run global two dimensional hydrodynamical simulations, using the PLUTO code and the planet-disk model of Uribe *et al.* 2011, to investigate the effect of the convective overstability (CO) on planet-disk interactions. First, we study the long-term evolution of planet-induced vortices. We found that the CO leads to smoother planetary gap edges, thus weaker planet-induced vortices. The main result was the observation of two generation of vortices, which can pose an explanation for the location of the vortex in the Oph IRS48 system. The lifetime of the primary vortices, as well as the birth time of the secondary vortices are shown to be highly dependent on the thermal relaxation timescale. Second, we study the long-term evolution of the migration of low mass planets and assess whether the CO can prevent the saturation of the horseshoe drag. We found that the disk parameters that favour slow inward or outward migration oppose the amplification of vortices, meaning that the CO does not seem to be a good mechanism to prevent the saturation of the horseshoe drag. On the other hand, we observed a planetary trap, caused by vortices formed in the horseshoe region. This trap may be an alternative mechanism to prevent the fast type I migration rates.

Keywords. Accretion disks, hydrodynamics, instabilities - Methods: numerical - Planetary systems: protoplanetary disks

1. Introduction

The magnetorotational instability (MRI, Balbus & Hawley 1991) is the most invoked mechanism to explain turbulence in protoplanetary disks (PPDs). Nevertheless, hydrodynamical (HD) instabilities are also shown to be able to produce turbulence, especially in regions of the disk with too low ionisation levels to couple efficiently to the magnetic fields, so called Dead Zones. For instance, vortices generated by the Rossby wave instability (RWI, Lovelace *et al.* 1999) are good candidates to create vortices at the dead zone edge (Lyra & Mac Low 2012). The baroclinic instability (BI, Klahr & Bodenheimer 2003) leads also to vortex formation and thermal relaxation or diffusion are important ingredients for vortex amplification (Petersen *et al.* 2007a,b). The non-linear phase of baroclinic vortex amplification is coined as subcritical baroclinic instability (SBI, Lesur & Papaloizou 2010), whereas the linear phase is the convective overstability (CO, Klahr & Hubbard 2014). Hence we claim that BI, SBI and CO are all the same phenomenon, i.e. radial convection in a rapidly rotating medium with the proper thermal relaxation, and in call the phenomenon CO. CO does occur for disks with superadiabatic radial stratification, that is pressure gradient ($\beta_P = \frac{d \log P}{d \log R}$) and entropy gradients ($\beta_S = \frac{d \log S}{d \log R}$) pointing in the same direction. In the two-dimensional vertically integrated case they can be derived from the radial temperature ($\beta_T = \frac{d \log T}{d \log R}$) and surface density profile ($\beta_T = \frac{d \log T}{d \log R}$) to $\beta_P = \beta_T + \beta_\Sigma$ and using the adiabatic index of $\gamma = 1.345$ to $\beta_S = \beta_T - (\gamma - 1)\beta_\Sigma$. The maximum growth-rates for optimal thermal relaxation are then proportional to the local pressure scale height $h = H/R$ and proportional to the radial buoyancy frequency $N^2 = -\beta_S \beta_P h^2$. Thus as long as surfacedensity and temperature fall with radius and the surfacedensity slope is shallower than $\beta_\Sigma = \frac{\beta_T}{(\gamma-1)}$ then the disk is radially buoyant and provided that the thermal relaxation occurs on timescales on the order of the local orbital period the instabilities will develop and create vortices (see Raettig et al. 2013 for a detailed exploration). For a temperature profile of $\beta_T = -0.5$ the radial surface density profile has to be shallower than $\beta_\Sigma > -1.44$, values that can easily be found in disks around young stars (Andrews et al. 2010).

Whereas the detailed properties of global 3D turbulence driven by the CO is still poorly understood for the lack of vertically stratified simulations, it is obvious the flow is dominated by large scale vortices and large scale density waves in between, with the latter providing Reynolds stresses on the order of $\alpha \approx 10^{-3}$ (Lyra & Klahr 2011; Raettig et al. 2013).

Here, we discuss the effect of the CO/SBI on planet-disk interactions for high and low mass planets, i.e. planets that can open a gap or can not. The formation and evolution of vortices induced by high mass planets have recently been thoroughly investigated (e.g., Zhu & Stone 2014; Fu et al. 2014; Les & Lin 2015). Vortices are important structures in several contexts. They are good candidates to trap dust particles, preventing them to quickly drift towards the central star and allowing them to grow to bigger sizes (Barge & Sommeria 1995; Klahr & Bodenheimer 2006). Therefore they are a good channel to overcome the radial drift barrier (Whipple 1972) for planet formation. They may also explain angular momentum transport through the disk dead zone, which is essential to allow for inward accretion (e.g., Meheut et al. 2012). The exchange of angular momentum between waves in the inner and outer side of the vortex position leads to a negative net flux of angular momentum and therefore a positive flux of mass. Finally, vortices may also trap planet cores, slowing down the core migration towards the central star, easing the timescale problem of the rapid inward drift of small objects (e.g., Koller et al. 2003; Ou et al. 2007; Li et al. 2009; Yu et al. 2010; Regály et al. 2013; Ataiee et al. 2014).

Type I migration is another aspect of planet-disk interactions worth investigating in this context. Low mass planets typically experience type I migration (Ward 1997; Tanaka et al. 2002), which is governed by the Lindblad torque (Ward 1986, 1988) and the horseshoe drag (e.g., Paardekooper & Papaloizou 2009; Casoli & Masset 2009). The former being the gravitational torque due to spiral density waves launched by the planet. The latter being the torque exerted by fluid elements executing a U-turn around the planet. A problem related to type I migration is the timescales predicted by isothermal models, which are much shorter than the typical disk lifetimes (Ward 1986). It would not be possible to explain the observed mass-distance distribution of exoplanets with such fast migration rates (Alibert et al. 2004; Ida & Lin 2008; Mordasini et al. 2009). Recent works have shown that non-isothermal effects can increase the horseshoe drag, which leads to slower migration rates or even the reversal of the migration direction (e.g., Morohoshi & Tanaka 2003; Paardekooper & Mellema 2006; Baruteau & Masset 2008). Nevertheless, there are still questions related to this subject. How can the saturation of the horseshoe drag be prevented, such that slow type I migration is sustained? This is an important question to address. In the next sections we summarise the main results and conclusions for simulations considering high and low mass planets.

2. High Mass Planets

High mass planets are here defined as planets massive enough to open a gap in the disk and thus will be subject of TypeII migration. The critical mass lies around the mass

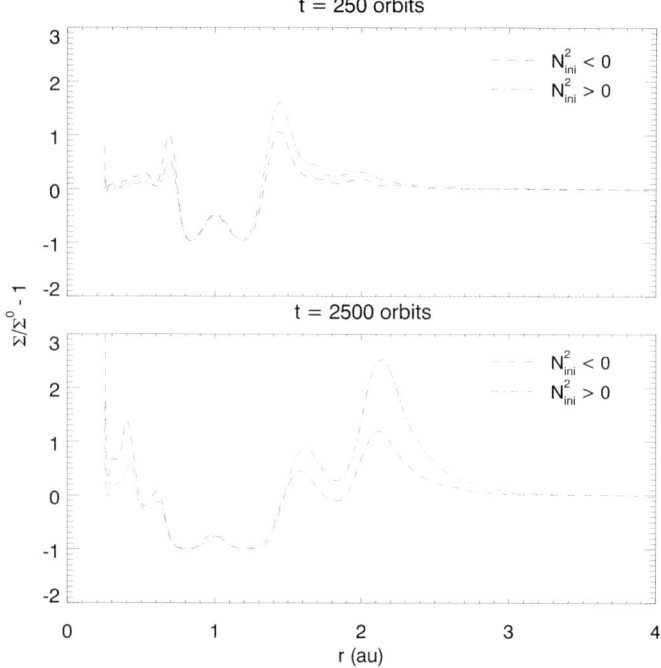

Figure 1. Surface density profiles averaged in azimuth for a disk with a thermal relaxation time of $\Omega\tau = 0.1$ in the presence a Jupiter mass planet. The red dashed line shows the results for a initially convective overstable disk profile, whereas the blue dotted-dashed line for a initially convective stable disk profile. In the upper plot one finds one vortex (after 250 orbits) and after 2500 orbits in lower plot there are two vortices, with the density perturbation stronger for the stable stratified disk.

of Saturn (in detail depending on the disk temperature), i.e. 100 $M_\oplus$. For the present study we use a mass of 300 $M_\oplus$ i.e. a Jupiter mass planet.

Dust asymmetries have recently been observed in several transition disks, e.g., in the Oph IRS 48 system (van der Marel et al. 2013). These features may be explained by the presence of large-scale vortices born at the outer edge of a planetary gap. Motivated by these observations, several recent studies have focused on the formation and evolution of planet-induced vortices (e.g., Zhu & Stone 2014; Fu et al. 2014; Les & Lin 2015). Questions regarding the lifetimes of these features and whether they are able to trap dust particles are fundamental to check if they can explain these recent observations. We performed two-dimensional (2D) global inviscid HD simulations to investigate planet-induced vortices (Lobo Gomes et al. 2015). A Newtonian cooling approach (thermal relaxation) was used to evolve the disk temperature structure. The main aim was to assess the importance of the CO for the development of planet-induced vortices. We found that the CO leads to smoother planetary gap edges, therefore weaker vortices are generated, see Figure 1. An interesting outcome of this study was the observation of a second generation of vortices. The first generation of vortices is formed in the outer wall of the planetary gap due to the triggering of the RWI.

The RWI is here responsible for the initial creation of a vortex whereas the CO will govern its evolution, i.e. amplify or damp the vortex depending on the local radial stratification and the thermal relaxation time.

The merged primary vortex induces accretion, which depletes the mass on its orbit. This process creates a surface density enhancement beyond the primary vortex position.

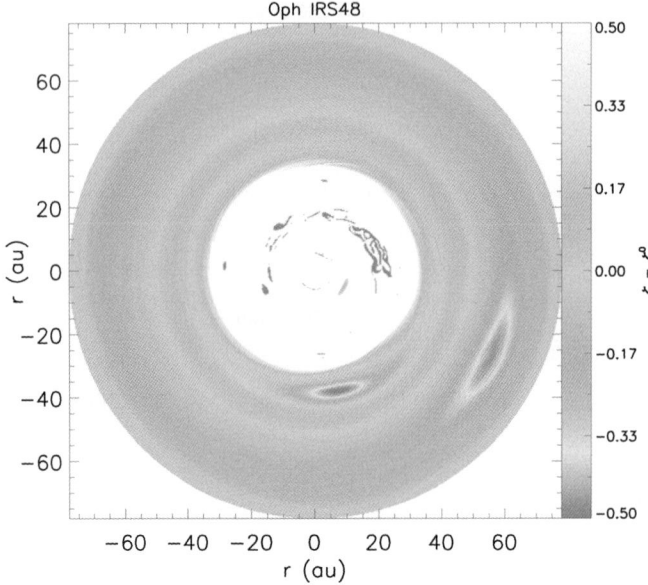

Figure 2. Potential vorticity with the Keplerian profile subtracted after 700 planetary orbits. The color bar was truncated from −0.5 to 0.5 in order to obtain a higher contrast.

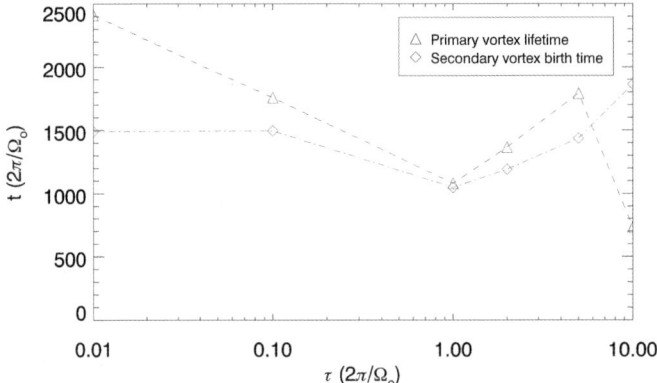

Figure 3. The lifetime of the primary vortex (red dashed line) and the birth time of the secondary vortex (blue dotted-dashed line) as a function of the thermal relaxation timescale.

The second generation of vortices arises in this surface density enhancement, indicating that the bump in this region is sufficient to trigger the RWI once more. We modelled the Oph IRS48 system with the objective of checking whether the second generation of vortices would also be formed. The hypothesis that a planet-induced vortex is the reason for the asymmetric feature in this system has a major problem: the location of the asymmetry cannot be explained by a vortex at the planetary gap edge, since the vortex location is very far from the possible planet location. We observed a second generation of vortices in the model of this system, therefore posing a promising explanation for the location of the vortex, see Figure 2.

We also studied the dependence of the vortices lifetimes and birth times as a function of the thermal relaxation timescale. The primary vortices' lifetimes, as well as the secondary vortices' birth times, were found to be dependent on the disk thermal relaxation

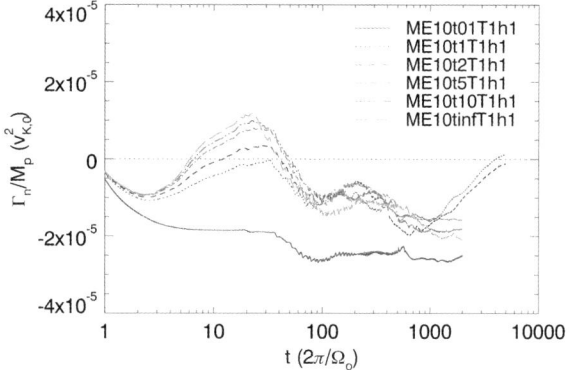

Figure 4. Cumulative averaged torques as a function of time. The different lines show the total torques for a $10 M_\oplus$ planet, $\beta_T = 1.0$, $h = 0.1$, and the thermal relaxation timescales from bottom to top (solid brown to dashed pink): $\Omega\tau = [0.1, 1.0, 2, 5, 10, \infty]$. See Table 1 in Lobo Gomes et al. (submitted) for the details of the runs.

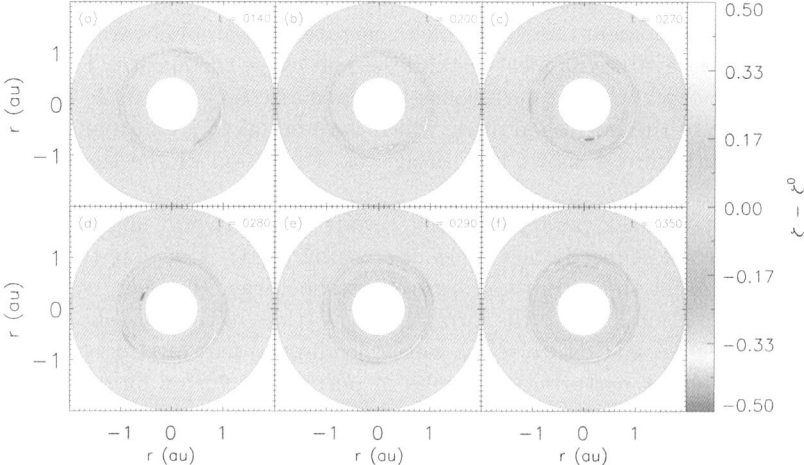

Figure 5. Temporal evolution of the potential vorticity with Keplerian profile subtracted for simulation ME10t1T1h1 with $\Omega\tau = 1$ (see Table 1 in Lobo Gomes et al. (submitted) for the details of the runs).

timescale, which is in agreement with previous studies (Fu et al. 2014; Les & Lin 2015). We used thermal relaxation times ranging from $\Omega\tau = 0.01$ to $\Omega\tau = 10$ (see Fig. 3) and found the strongest secondary vortex for $\Omega\tau = 0.1$. Les & Lin (2015) explained that the nonmonotonic behavior is due to the fact that the vortex lifetime depends on (i) the decay timescale of the RWI, which decreases for increasing $\Omega\tau$, and (ii) the vortex growth time, which increases for values of up to $\Omega\tau = 5.0$ and then decreases for larger values. Therefore the cooling properties of the disk play a crucial role for the generation and maintenance of these structures.

3. Low Mass Planets

We used global 2D-HD simulations, with the same cooling approach as in the previous section, to study the unsaturation of the horseshoe drag for type I migration (Lobo

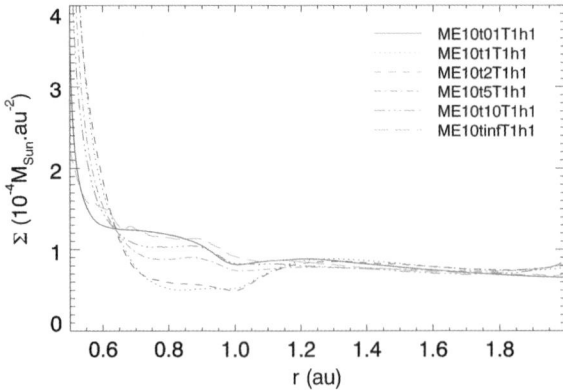

Figure 6. Surface density profiles averaged in azimuth at 2000 orbits and for a 10 $M_\oplus$ planet, $\beta_T = 1.0$, $h = 0.1$, and the thermal relaxation timescales from solid brown to dashed pink (same as Fig.4): $\Omega\tau = [0.1, 1.0, 2, 5, 10, \infty]$.

Gomes *et al.*, submitted). Radial stratification in combination with thermal relaxation can lead to the development of a CO, which generates and amplifies vortices (e.g., Klahr & Hubbard 2014). Hydrodynamical instabilities can cause turbulence. The main goal of this study was to check whether turbulence-triggered viscosity due to the CO can sustain the unsaturation of the horseshoe drag, which is essential to balance or counteract the Lindblad torque. Figure 4 shows the torques as a function of the thermal relaxation timescale.

We found that the disk parameters that favour slow/outward migration oppose the amplification of vortices (see Lobo Gomes *et al.*, submitted). Therefore this is not a good mechanism to prevent the saturation of the horseshoe drag. However, we observed that the planet perturbation leads to the formation of vortices in the horseshoe region. These vortices get amplified and excite spiral waves, leading to fast inward vortex migration. The continuous planet perturbation results in permanent vortex formation, see Figure 5. The process of constant creation and amplification of vortices, that later migrate to the inner disk, depletes the matter in the inner side of the planet orbit, see Figure 6. The planet remains trapped in the outer edge of this under-density. This result is a new promising mechanism to prevent the fast type I migration rates in regions of the disk that are convective overstable.

4. Conclusions

The CO was found to play a role for both high and low mass planet cases. For the former, the presence of the CO leads to the formation of weaker planet-induced vortices. The further evolution of the vortices, however, is not strongly dependent on the CO. The formation of a second generation of vortices was observed, nonetheless it is not linked to the presence of the CO. This second generation of vortices may explain the vortex in the Oph IRS48 system. We found also a strong dependence of the vortices lifetimes and birth times with the thermal relaxation timescale, in agreement with previous results. For the latter, the CO does not prevent the saturation of the horseshoe drag, in contrast, the disk parameters that favour CO seem to oppose the horseshoe drag unsaturation. Nevertheless, the formation of a planet trap was observed for the cases in which the CO was in action. The effect of continuous vortex formation and shedding inside of the planet appears to be a new interesting mechanism to trap planets. Detailed studies on how the efficiency depends of the susceptibility of the disk to CO (detailed slopes of T and Sigma) will have to follow including the possibilities that vortices do form spontaneously outside the planet orbit or even being triggered by additional planets.

Acknowledgments

H. Klahr and A. Lobo Gomes would like to thank financial support from the Deutsche Forschungsgemeinschaft (DFG), grant n. KL 1469/9-1. A. Lobo Gomes is also grateful for the support by the International Max Planck Research School at Heidelberg (IMPRS- HD). The simulations were performed on the THEO cluster at the Rechenzentrum Garching (RZG) of the Max Planck Society.

References

Andrews, S. M., Wilner, D. J., Hughes, A. M., Qi, C., & Dullemond, C. P. 2010, *ApJ*, 723(2), 1241-1254
Asphaug & Benz (1996)]Asphaug Alibert, Y., Mordasini, C., & Benz, W. 2004, *A&A*, 417, L25
Ataiee, S., Dullemond, C. P., Kley, W., Regály, Z., & Meheut, H. 2014, *A&A*, 572, A61
Balbus, S. A. & Hawley, J. F. 1991, *ApJ*, 376, 214
Baruteau, C. & Masset, F. 2008, *ApJ*, 672, 1054
Barge, P. & Sommeria, J. 1995, *A&A*, 295, L1
Casoli, J. & Masset, F. S. 2009, *ApJ*, 703, 845
Fu, W., Li, H., Lubow, S., & Li, S. 2014, *ApJL*, 788, L41
Ida, S. & Lin, D. N. C. 2008, *ApJ*, 673, 487
Klahr, H. H. & Bodenheimer, P. 2003, *ApJ*, 582, 869
Klahr, H. & Bodenheimer, P. 2006, *ApJ*, 639, 432
Klahr, H. & Hubbard, A. 2014, *ApJ*, 788, 21
Koller, J., Li, H., & Lin, D. N. C. 2003, *ApJL*, 596, L91
Les, R. & Lin, M.-K. 2015, *MNRAS*, 450, 1503
Lesur, G. & Papaloizou, J. C. B. 2010, *A&A*, 513, A60
Li, H., Lubow, S. H., Li, S., & Lin, D. N. C. 2009, *ApJL*, 690, L52
Lobo Gomes, A., Klahr, H., Uribe, A. L., Pinilla, P., & Surville, C. 2015, *ApJ*, 810, 94
Lobo Gomes, A., Klahr, H., Surville, C. & Uribe, A. L. submitted, *ApJ*
Lovelace, R. V. E., Li, H., Colgate, S. A., & Nelson, A. F. 1999, *ApJ*, 513, 805
Lyra, W. & Mac Low, M.-M. 2012, *ApJ*, 756, 62
Lyra, W. & Klahr, H. 2011, *A&A*, 527, A138
Meheut, H., Yu, C., & Lai, D. 2012, *MNRAS*, 422, 2399
Mordasini, C., Alibert, Y., & Benz, W. 2009, *A&A*, 501, 1139
Morohoshi, K. & Tanaka, H. 2003, *MNRAS*, 346, 915
Ou, S., Ji, J., Liu, L., & Peng, X. 2007, *ApJ*, 667, 1220
Paardekooper, S.-J. & Mellema, G. 2006, *A&A*, 459, L17
Paardekooper, S.-J. & Papaloizou, J. C. B. 2009, *MNRAS*, 394, 2283
Petersen, M. R., Julien, K., & Stewart, G. R. 2007a, *ApJ*, 658, 1236
Petersen, M. R., Stewart, G. R., & Julien, K. 2007b, *ApJ*, 658, 1252
Raettig, N., Lyra, W., & Klahr, H. 2013, *ApJ*, 765(2), 115
Regály, Z., Sándor, Z., Csomós, P., & Ataiee, S. 2013, *MNRAS*, 433, 2626
Tanaka, H., Takeuchi, T., & Ward, W. R. 2002, *ApJ*, 565, 1257
Uribe, A. L., Klahr, H., Flock, M., & Henning, T. 2011, *ApJ*, 736, 85
van der Marel, N., van Dishoeck, E. F., Bruderer, S., et al. . 2013, *Science*, 340, 1199
Ward, W. R. 1986, *Icarus*, 67, 164
-. 1988, *Icarus*, 73, 330

-. 1997, *Icarus*, 126, 261

Whipple, F. L. 1972, in *From Plasma to Planet*, ed. A. Elvius (New York: Wiley Interscience Division), 211

Yu, C., Li, H., Li, S., Lubow, S. H., & Lin, D. N. C. 2010, *ApJ*, 712, 198

Zhu, Z. & Stone, J. M. 2014, *ApJ*, 795, 53

Formation of the Planetary Candidates Observed by Kepler Mission

Su Wang and Jianghui Ji

Key Laboratory of Planetary Sciences, Purple Mountain Observatory,
Chinese Academy of Sciences, Nanjing 210008, China
email: wangsu@pmo.ac.cn, jijh@pmo.ac.cn

Abstract. The Kepler Mission which launched in 2009 March focuses on detecting potentially habitable terrestrial-sized planets. To date the Kepler mission has released more than 4000 planetary candidates. There are plenty of planet pairs trapped near the first order mean motion resonance (MMR). From the statistical results of numerical simulations based on the formation scenario we proposed for the planetary configurations near 3:2 and 2:1 MMRs, we find that the proportions of period ratios close to 1.5 and 2.0 can arrive at 14.5% and 26.0%, respectively. This scenario may explain the formation of Kepler candidates pairs in near 3:2 and 2:1 MMRs.

Keywords. (stars:) planetary systems: formation, methods: numerical

1. Introduction

Up to 2015 September, there are ~ 4696 planetary candidates with ~ 1889 confirmed planets released from the Kepler mission data. Among them, 473 multiple planetary systems exist. The analysis results on the planets from the first 16 months released show that plenty planet pairs are near MMRs in 222 three-planet systems (Fabrycky *et al.* 2014) especially the first orders, $\sim 7.0\%$ near 3:2 MMR (the period ratio in the range of [1.45, 1.54]) and $\sim 18.0\%$ near 2:1 MMR (the period ratio in the range of [1.83, 2.18]). Near 4:2:1 and 3:2:1 MMRs chains may appear in multiple planetary systems.

Wang *et al.* (2012) has explained the formation of planetary system KOI-152 which contains three planets in near 4:2:1 MMRs. The formation scenario of such system is supposed as: firstly, planets form far away from the central star; then they undergo orbital migration due to the effect of the gas disk until they go to the inner region of the system, in the process they are trapped into MMRs; at last, with the tidal effect raised by the central star they are out of MMRs to be a configuration similar to KOI-152 with three planets in near 4:2:1 MMRs (Terquem & Papaloizou 2007, Lithwick & Wu 2012, Wang *et al.* 2012, Batygin & Morbidelli 2013, Delisle & Laskar 2014, Wang & Ji 2014). The formation scenario may explain the formation of planet pairs near 3:2 and 2:1 MMRs. The properties of star such as the accretion rate ($\dot{M}$) and the magnetic field (B_*), the disk model especially the gas density profile, the speed of orbital migration (with a reduce factor f_1 of type I migration), the masses (m) and numbers of planets in the system are the main factors that influence the formation of final planetary system configuration.

2. Numerical simulations and statistical results

In the numerical simulation model, we consider a solar-like star with three planets (P1 represents the innermost one, P2 is the middle planet, and P3 displays the outmost one)

around in the system. All three planets embed in the gas disk initially. There are four groups of simulations with different masses of planets. According to the isolation mass (Ida & Lin 2004), the mass of planet is proportional to $a^{3/4}$, where a is the semi-major axis of planet. Thus, in the first group, we choose the masses of three planets increased in turn, they are $m_{p1} = 5\ M_\oplus$, $m_{p2} = 10\ M_\oplus$, and $m_{p3} = 15\ M_\oplus$, respectively. In the second group, we set planets with two kinds of masses: three planets with equal masses 5 $M_\oplus$, and $m_{p1} = m_{p2} = 5\ M_\oplus$ with an outer larger planet $m_{p3} = 10\ M_\oplus$. We also consider the masses of planets in other two groups based on the statistical results on planetary candidates from the Kepler data. The masses ratios are 1:1:1.5, 2:1:7.5, 12:1:0.2, and 2:0.2:4, respectively. In the simulations, the star accretion rate is in the range of $[0.1, 2.5]$ $\times 10^{-8}\ M_\odot \mathrm{yr}^{-1}$, the magnetic field of star is from 0.5 to 2.5 KG, the reduction factor of type I migration f_1 is chosen to be 0.01, 0.03, 0.1, 0.3, and 1. Three planets are at the orbital periods of 100, 250, and 600 days for $f_1 = 0.01$ and 0.03, while they are at 140, 500, and 1450 days for $f_1 \geqslant 0.1$ at the beginning of the simulations. Initially all three planets are in coplanar and near-circular orbits, the mean anomaly, the argument of the pericenter, and the longitude of ascending node are generated randomly in the range of $[0°, 360°]$. The gas disk profile is $\Sigma_g \propto r^{-1}$ (Pringle 1981) and will deplete in a timescale of 10^6 yr. In the disk, there are two locations with gas density changed sharply and the locations are related to the magnetic field and the star accretion rate. Migrating planets may be stopped at these two locations (Masset et al. 2006, Kretke et al. 2009, Wang et al. 2012). In our simulations, the masses of all planets are less than 100 $M_\oplus$, besides the gravitational effects among them, they suffer type I migration and the gas damping force before the gas dissipated.

We carried out 1020 runs of simulations with different parameters ($\dot{M}$, B_*, f_1, m) and potentially planets survived in the systems. The main results are shown in Figure 1. Panel (a) shows the overall results from the simulations compared with the observation. In both data, two peaks are obtained and when $\dot{M} = 1 \times 10^{-9}\ M_\odot\ \mathrm{yr}^{-1}$ they have similar distribution roughly. Panel (b1) and (b2) show the fraction of planet pairs near 2:1 or 3:2 MMRs changing with the star accretion rate and migration speed, respectively. The early stage of star with high accretion rate, $\dot{M} = 2 \times 10^{-8}\ M_\odot\ \mathrm{yr}^{-1}$, is good to the formation of near 2:1 MMRs, while the late stage may be responsible for the formation of near 3:2 MMRs, $\dot{M} = 0.1 \times 10^{-8}\ M_\odot\ \mathrm{yr}^{-1}$. $f_1 \geq 0.1$ and $f_1 \geq 0.3$ favor the formation of near 2:1 and 3:2 MMRs, respectively. In our simulations, we also test the influence of additional planets which potentially survive in the system. Panel (c1) and (c2) show the main results. In panel (c1), if there are three planets P1, P2, and P3 in the system, they will be in near 4:2:1MMRs at the end of the simulation, here shows the result that there are additional planet E1 locating inside the planet P1 and E2 residing outside the planet P3 with other conditions the same with the previous simulation, 2:1 MMRs are not affected, the Laplacian resonance is not disrupted easily. But in panel (c2) for the system which are involved in two pairs of 3:2 MMRs for only three planets exist, the results are very different, 3:2 MMRs will disintegrate at ~ 0.2 Myr.

3. Discussion

Using the formation scenario we supposed, the peak near 2:1 and 3:2 MMRs can be obtained. The near 4:2:1 MMRs are relatively stable, such configuration may be common in the exoplanetary systems. The near 3:2:1 MMRs may be appeared in the system with more than three planets. In the future work, we will consider the effect of mass growing process on the formation scenario and the configuration formation with giant planets in the system.

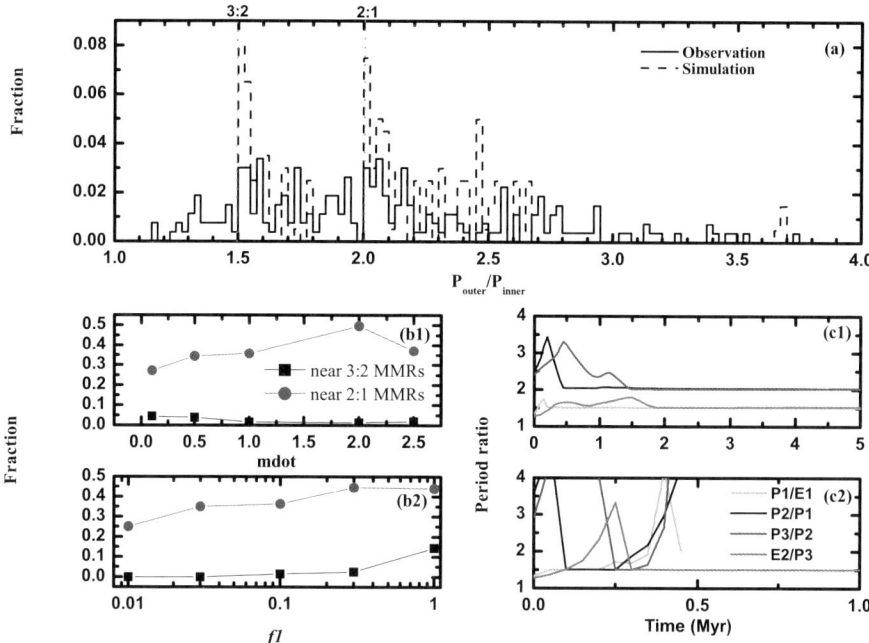

Figure 1. (a) The statistical results of the simulations compared with the observation data. The solid line displays the observation results and the dashed line represents the simulation results. (b1) and (b2) show the fraction that two planets pair in 2:1 MMRs or 3:2 MMRs differs with the star accretion rate and the orbital migration speed, respectively. (c1) The effect of additional planets on the system with planet pairs which would in 2:1 MMRs if there are only three planets in the system. (c2) The effect of additional planets on the system with planet pairs which would in 3:2 MMRs if there are only three planets in the system.

Acknowledgements

This work is financially supported by National Natural Science Foundation of China (Grants No. 11203087, 11573073, 11273068), the Natural Science Foundation of Jiangsu Province (Grant No. BK20151607), the Strategic Priority Research Program on Space Science (Grant No. XDA04060901), and the Foundation of Minor Planets of Purple Mountain Observatory.

References

Batygin, K. & Morbidelli, A. 2013, *AJ*, 145, 1
Delisle, J. B. & Laskar, J. 2014, *A&A*, 570, L7
Fabrycky, D. C., Lissauer, J. J., Ragozzine, D., et al. 2014, *ApJ*, 790, 146
Ida, S. & Lin, D. N. C. 2004, *ApJ*, 604, 388
Kretke, K. A. et al. 2009, *ApJ*, 690, 407
Lithwick, Y. & Wu, Y. 2012, *ApJL*, 756, 11
Masset F. S. et al. 2006, *ApJ*, 642, 478
Pringle, J. E. 1981, *ARA & A*, 19, 137
Terquem, C. & Paploizou, J. C. B. 2007, *ApJ*, 654, 1110
Wang, S., Ji, J. H., & Zhou, J. L. 2012, *ApJ*, 753, 170
Wang, S. & Ji J. H. 2014, *ApJ*, 795, 85

Period Ratio Distribution of Near-Resonant Planets Indicates Planetesimal Scattering

Sourav Chatterjee[1], Seth O. Krantzler[1] and Eric B. Ford[2,3]

[1] Center for Interdisciplinary Exploration and Research in Astrophysics
Northwestern University
2145 Sheridan Road, Evanston, IL 60208, USA
email: sourav.chatterjee@northwestern.edu

[2] Department of Astronomy & Astrophysics
The Pennsylvania State University
525 Davey Laboratory, University Park, PA 16802, USA

[3] Center for Exoplanets and Habitable Worlds
The Pennsylvania State University
525 Davey Laboratory, University Park, PA, 16802, USA

Abstract. An intriguing trend among *Kepler*'s multi-planet systems is an overabundance of planet pairs with period ratios just wide of mean motion resonances (MMR) and a dearth of systems just narrow of them. In a recently published paper Chatterjee & Ford (2015; henceforth CF15) has proposed that gas-disk migration traps planets in a MMR. After gas dispersal, orbits of these trapped planets are altered through interaction with a residual planetesimal disk. They found that for massive enough disks planet-planetesimal disk interactions can break resonances and naturally create moderate to large positive offsets from the initial period ratio for large ranges of planetesimal disk and planet properties. Divergence from resonance only happens if the mass of planetesimals that interact with the planets is at least a few percent of the total planet mass. This threshold, above which resonances are broken and the offset from resonances can grow, naturally explains why the asymmetric large offsets were not seen in more massive planet pairs found via past radial velocity surveys. In this article we will highlight some of the key findings of CF15. In addition, we report preliminary results from an extension of this study, that investigates the effects of planet-planetesimal disk interactions on initially non-resonant planet pairs. We find that planetesimal scattering typically increases period ratios of non-resonant planets. If the initial period ratios are below and in proximity of a resonance, under certain conditions, this increment in period ratios can create a deficit of systems with period ratios just below the exact integer corresponding to the MMR and an excess just above. From an initially uniform distribution of period ratios just below a 2:1 MMR, planetesimal interactions can create an asymmetric distribution across this MMR similar to what is observed for the *Kepler* planet pairs.

Keywords. scattering, methods: n-body simulations, methods: numerical, planets and satellites: general, planetary systems, planetary systems: protoplanetary disks

1. Introduction

NASA's *Kepler* mission has revolutionized our understanding of planetary systems, their occurrence rate, multiplicity and physical properties. One trend apparent among this new class of small planets was a-priori quite unexpected from traditional theories; there is a statistically significant excess of planet pairs with period ratios slightly wide of first order mean motion resonances such as 2:1 and 3:2, and a dearth of them just narrow of these resonances (Lissauer *et al.* 2011, Fabrycky *et al.* 2014, Steffen & Hwang 2015; Figure 1). Interestingly, this trend is absent in planets that were previously discovered via radial velocity (RV) surveys (e.g., Butler *et al.* 2006). Smooth gas-disk migration can trap planets in MMRs. However, such resonant planets are expected to have period ratios with very small offsets from the integer ratio corresponding to the MMR, $\epsilon \equiv P_2/P_1 - (j+1)/j \sim \pm 10^{-3}$. Indeed, adjacent planet pairs discovered via past RV surveys show period ratio distribution with a distinct excess at the expected period ratio for the 2:1 MMR with very small ϵ, consistent with the expectations from smooth gas-disk driven migration (Lee & Peale 2002, Butler *et al.* 2006, Armitage 2013). In contrast, the near-resonant *Kepler* planet pairs are likely not in actual resonance (Veras & Ford 2012). Nevertheless, the overall close to uniform distribution away from resonance, and the mysterious asymmetric abundance across MMRs, such as 2:1 and 3:2, indicate that the *Kepler* planet pairs somehow knew about these resonances. However, some other process has driven them wide of the resonance and created this asymmetry. Planet-planet scattering can break resonances, however, they bring dramatic changes in the planetary orbits, often making them highly eccentric, which is inconsistent with the multi-transiting architecture of the *Kepler* systems (e.g., Ford & Rasio 1996, Chatterjee *et al.* 2008). The large ($\epsilon \sim 10\%$) positive offsets in the near-resonant planet pairs observed by *Kepler* thus has generated a lot of interest.

It is generally believed that these planets were initially trapped in a MMR. Subsequently, some dissipative process drove them wide of their initial period ratios. The most likely dissipative mechanism responsible for the observed trend is still a matter of debate. The proposed dissipative mechanisms include dissipation from tide (Lithwick & Wu 2012, Batygin & Morbidelli 2013, Delisle & Laskar 2014), turbulence in protoplanetary disk (Rein 2012), and scattering with a planetesimal disk (Moore *et al.* 2013; CF15). Although, the most well studied, the tidal dissipation mechanism is also the most debated. Lee *et al.* (2013), Silburt & Rein (2015) argue that even under generous assumptions, the large observed positive ϵ for most *Kepler* planet pairs near a MMR cannot be explained by tides alone. It was also suggested that in-place mass growth of a planet via planetesimal accretion can lead to formation of an over density of particles just wide of a MMR (Petrovich *et al.* 2013). However, planetesimal accretion typically lead to migration of the planet making the in situ growth assumption questionable. It was also suggested that the observed period ratio distribution may be explained due to overstable libration of the *Kepler* planets due to gas-disk driven migration coupled with eccentricity damping (Goldreich & Schlichting 2014). However, Hands *et al.* (2014), Deck & Batygin (2015) present an opposing view.

In this article, we will focus on planet-planetesimal disk interactions as the mechanism for the observed asymmetric period ratio distribution across 2:1. Planet-planetesimal disk interactions have been well studied in other contexts, especially for the outer Solar system, and is generally believed to be a natural consequence of the core-accretion paradigm of planet formation (e.g., Fernandez & Ip 1984, Ida *et al.* 2000, Gomes *et al.* 1004, Kirsh *et al.* 2009). In §2 we will briefly describe our numerical setup. In §3 we will highlight our key results. First, we will highlight the findings of CF15 with some additional details. We will also present results from new simulations involving interactions of initially non-resonant planets with a planetesimal disk. Finally, in §4 we will summarize our results and discuss the implications.

2. Numerical Setup

The simulations presented in this article are from two distinct sets. One investigates the effects of planetesimal scattering on the orbits of two planets initially trapped in a 2:1 MMR (CF15). We call this `Set 1`. The other investigates the effects of planetesimal

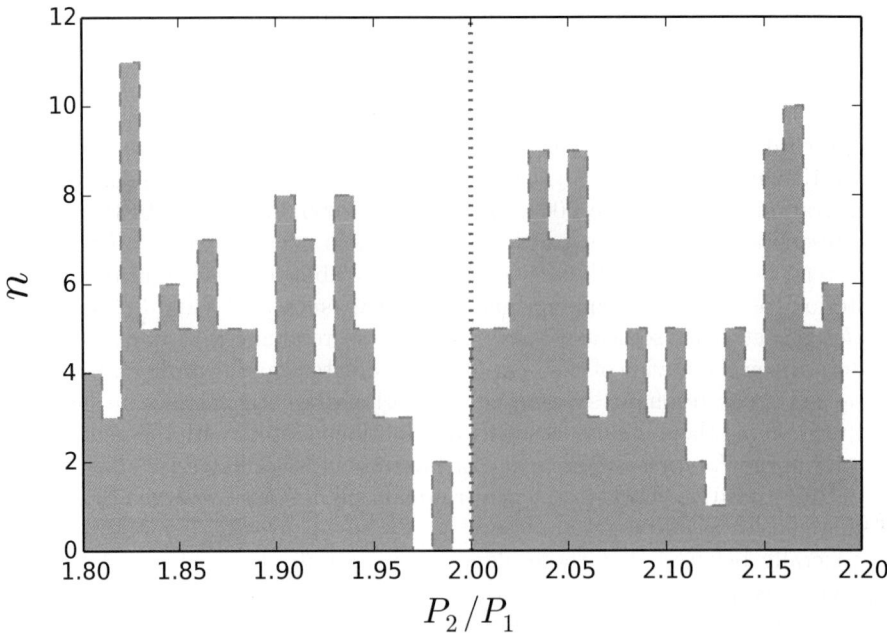

Figure 1. Period ratio distribution of adjacent planet pairs discovered by *Kepler* close to the period ratio expected for a 2:1 MMR. There is a dearth of systems with period ratios just narrow of the resonance and an excess of systems just wide of the resonance. The vertical (red-dotted) line shows the exact position of a period ratio of 2. *Kepler* data was extracted from NASA's exoplanet archive.

scattering on planet orbits that are not initially trapped in 2:1 MMR but, are just narrow of the MMR. We call this Set 2.

The details of the numerical setup for Set 1 are described in CF15. However, for completeness, we will briefly describe the key aspects. In general, the physical picture we have in mind is that while a gas disk is present, gas-disk interactions may trap two planets into 2:1 MMR. Once the gas disk is depleted, the resonant planets can freely interact with a residual planetesimal disk. We are interested in the effects of the latter interactions. Thus, we are interested in a system that initially was dissipative and transitions into a N-body. Ideally, planets, planetesimals, and a gas disk should be modeled together with all physics included, however, this full problem is computationally impractical. Hence, we generate plausible initial conditions for the stage of planet-planetesimal disk interactions in two steps. First, we use an analytic $\dot{a}$ and $\dot{e}$ prescription to trap two planets in 2:1 MMR (Lee & Peale 2002). Second, we create planetesimal orbits consistent with the presence of the planets in the following way. The structure of the residual planetesimal disk after gas-disk depletion is uncertain. Nevertheless, we use planetesimal disk profiles described by $d\Sigma/da \propto a^\alpha$, where Σ and a are the surface density and distance from the star for the planetesimals, respectively. At the epoch of gas disk depletion, the planetesimal disk profile would not remain a simple power-law. Instead, the planets would alter the planetesimal disk densities near them by dynamically scattering or accreting some of the nearby planetesimals that are on orbits unstable even with the stabilization provided by dissipation from a gas disk (e.g., Matsumura *et al.* 2010). To imitate this effect we embed the resonant planets in a planetesimal disk with a power-law profile given by $d\Sigma/da \propto a^\alpha$. We treat all planetesimals as test particles. We let the planets alter the disk for at least $\sim 10^2$ orbits of the outer planet. We collect the properties of the planetesimals that

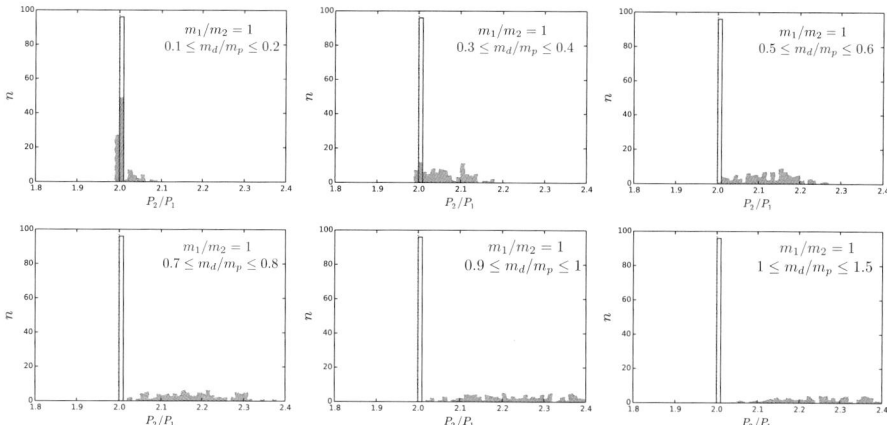

Figure 2. The initial (black) and final (blue filled) period ratio distributions for models with equal mass planets and planetesimal disk profile given by $\alpha = -3/2$ (a subset of models presented in CF15). Each panel shows results from models with a different m_d/m_p value, listed in each panel. In all cases, the planet pairs are initially trapped in a 2:1 MMR. As a result, the initial period ratios are always close to 2. Depending on m_d/m_p the final period ratio distribution changes. For $m_d/m_p < 0.3$ resonance is not broken for most planet pairs. As m_d/m_p increases, so fraction of systems for which resonance is broken and the highest ϵ attained, both increase.

survive this phase and create a database of allowed planetesimal orbits for each planet pair. We call this the clean-up stage. We randomly choose 2×10^3 orbits from this database, assign masses to the planetesimals according to the assumed planetesimal disk mass to planet mass ratio. We evolve the resonant planets with the planetesimals until the period ratio of the planets' orbits stop changing ($\sim 10^5$ years). CF15 has varied the planetesimal disk profile by changing the power-law index α between -2.5 to 3, planet-planet mass ratios m_1/m_2 between 0.1 to 10, and the planetesimal disk to planet mass ratio ($m_d/(m_1 + m_2) \equiv m_d/m_p$) between 0.1 to 1.5.

We have started investigating the effects of a residual planetesimal disk on initially non-resonant planets. In this set, Set 2, we closely follow the prescriptions of CF15 summarized above. However, we choose planet pairs with initial orbits such that the initial period ratio is between 1.8 and 2, just narrow of the 2:1 MMR. The inner planet's semimajor axis is kept at $a_1 = 0.1$ AU, a typical value for the *Kepler* planets. We generate 50 systems such that the period ratio P_2/P_1 is between 1.8 and 2. The eccentricities of the planets' orbits are drawn from a Rayleigh distribution with scale 0.005 (e.g. Hadden & Lithwick 2014). The orbital inclinations (I) are drawn uniformly in $\cos I$ with I between -0.1 and $0.1°$. Planets have equal mass and are equal to the mass of Neptune (M_N) and have Neptune-like densities. For each pair of planetary orbits created this way, we generate 4 random realizations varying the phase angles in their full range. The initial planetesimal disk profile is given by $d\Sigma/da \propto a^{-3/2}$. The planetesimal disk edges are set at orbits with period $P_1/3$ and $3P_2$. For each planet pair, we perform the initial cleaning up of planetesimals exactly the same way as prescribed by CF15. For each case, a database of orbits is generated. We randomly select 2×10^3 orbits and give each planetesimal a mass $m_{pl} = 1 \times 10^{-3}$ M_N, such that $m_d/m_p = 1$. We integrate the planet-pair and planetesimals using the Bulirsch-Stoer integrator included in Mercury (Chambers 1999). We stop our integrations at 2×10^4 year, equivalent to $\approx 6 \times 10^5$ of the inner planet's initial orbital period. We confirm that the majority of the planetesimal interactions happen much earlier than our chosen integration stopping time.

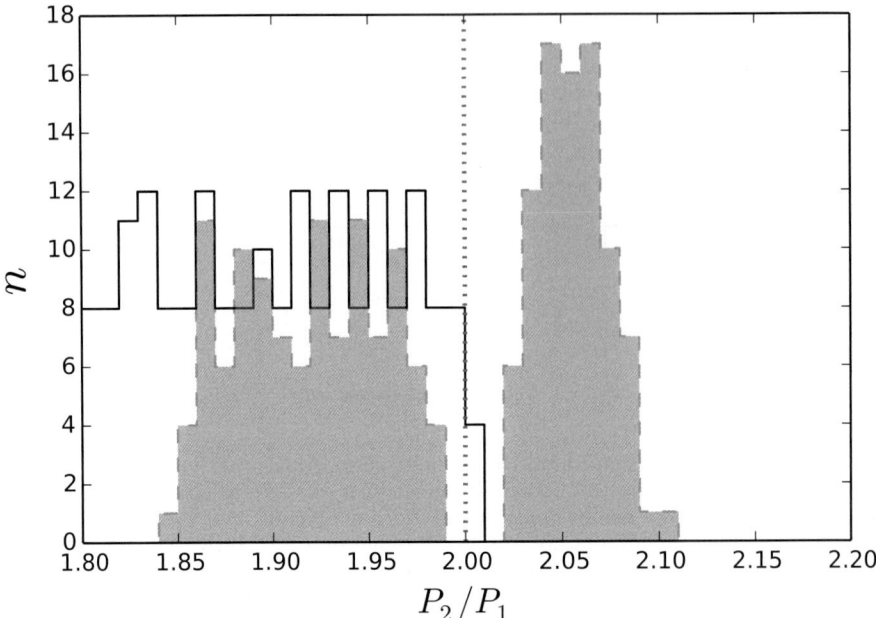

Figure 3. Initial (black) and final (blue filled) period ratio distributions for the set of simulations including initially non-resonant planets and a planetesimal disk (Section 2). Initially $m_1/m_2 = 1$, $\alpha = -3/2$, and $m_d/m_p = 1$. As a result of planetesimal interactions the period ratios generally increase. As the pairs approach 2:1 MMR from the inside, the resonance is skipped and the period ratio increases to a value > 2. Thus, a dearth of planet pairs with period ratios just smaller than 2 and an excess of pairs with period ratios just higher than 2 are created.

3. Results

CF15 has investigated the effects of planetesimal interactions on the orbits of planet pairs initially trapped in a 2:1 MMR. The key results of CF15 are as follows. If the total mass of planetesimals that had strong interactions with the planets is high enough to break resonance, then planet-planetesimal interactions naturally increase the period ratio. The final offset from the MMR depends strongly on the ratio of the total mass of planetesimals that interacted with the planets and the planet mass. When resonance is broken, offset ϵ can have large positive values. If the resonance is not broken due to insufficient mass in nearby planetesimals, ϵ remains small ($\sim 10^{-3}$) and can have both positive or negative values. As a result, it is easier to break resonance and create large positive ϵ for low-mass planet pairs typical of those discovered by *Kepler*, compared to the much higher mass planet pairs discovered via past RV surveys. Figure 2 shows the initial and final period ratio distribution from a subset of simulations presented in CF15. As the ratio m_d/m_p increases, so does the fraction of systems where the resonance is broken and also the value of the highest ϵ the planet pairs can attain.

CF15 results suggest that planetesimal interactions with resonant planet pairs can naturally redistribute these planet pairs wide of the initial resonance. However, CF15 do not directly address the dearth of planet pairs with period ratios just narrow of 2:1. We have started a systematic study of the effects of planetesimal scattering on planet pairs that are initially not in resonance, rather has period ratios slightly smaller than 2 (Set 2). This is equivalent to a scenario where planet pairs do not go through significant migration in a gas disk. Thus, period ratios below and up to 2 are populated uniformly before planetesimal scattering can take place. Figure 3 shows the initial and final period ratio

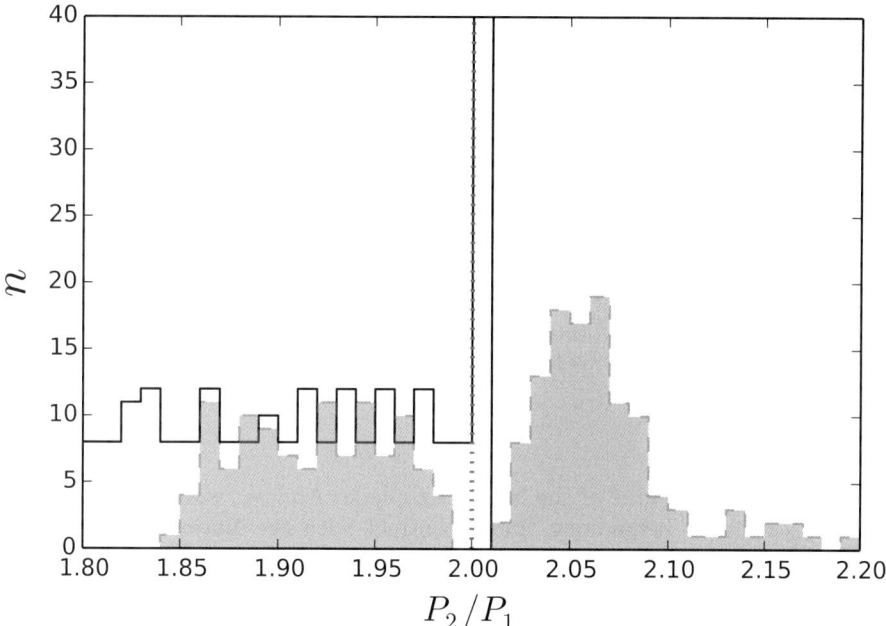

Figure 4. The same as Figure 3 but combining models from CF15 with $m_1/m_2 = 1$, $\alpha = -3/2$, and $m_d/m_p \geqslant 0.3$ and those from Set 2 (§2).

distributions for a large (2×10^2) set of simulations, each involving two planets and 2×10^3 planetesimals (§2). Initially, the planet pairs have period ratios uniformly spaced between 1.8 and 2. Interactions with planetesimals from the disk generally increase the period ratios. Interestingly, as some of the planet pairs approach the 2:1 MMR via planetesimal driven divergent migration, they tend to skip the resonance and get deposited wide of the resonance. As a result, a dearth of systems is created with period ratios slightly below 2 and an excess of systems is created with period ratios slightly above 2.

In Figure 4 we combine the models from CF15 with $m_1/m_2 = 1$, $\alpha = -3/2$ and $m_d/m_p \geqslant 0.3$ with those from Set 2. Although the relative initial abundances between initially resonant and non-resonant planets is somewhat ad-hoc, the combined period ratio distribution is qualitatively very similar to what is observed across the 2:1 MMR for *Kepler*'s adjacent planet pairs. Hence, immediately after gas disk dispersal, if planet pairs have period ratios distributed uniformly below 2:1 with some excess of pairs at 2:1, and there is sufficient mass in nearby planetesimals in a residual planetesimal disk, then after planetesimal interactions the final period ratio distribution will, at least qualitatively, be very similar to what is observed of the *Kepler* planet pairs near 2:1 (Figure 1).

4. Discussion

In this article we summarize the key results of CF15. In addition, we present preliminary results from a new set of simulations with initially non-resonant planets and a planetesimal disk. Our results suggest that planetesimal scattering may be responsible for both the excess of planet pairs just wide of the 2:1 MMR and the dearth of systems just narrow of it.

We made several simplifying assumptions in this article. For example, in the new set of simulations with non-resonant planet pairs, we assume that the initial period ratios are distributed uniformly below 2 and all the way up to 2. This of course is not necessarily true in reality. Sufficiently far away from a MMR the period ratios of *Kepler*'s planet pairs appear to be random. However, they may not have been so immediately after gas disk dispersal. In reality, there may have been a dearth of systems just narrow of a MMR simply because some systems, while migrating within a gas disk, got trapped into the MMR. In that sense, our assumption of uniform period ratio distribution below the 2:1 MMR is the most conservative one. Even if after gas disk dispersal the period ratios are uniformly distributed narrow of the 2:1 MMR, planetesimal scattering from a sufficiently massive disk can create a dearth of systems with period ratios just below 2. These systems, in turn, pile up with period ratios slightly above 2. Encouraged by our preliminary results, we are exploring this problem more thoroughly by covering a larger parameter space and obtaining a more detailed understanding of the evolution of the planet pairs as they cross the 2:1 MMR from inside out via planetesimal driven migration.

Acknowledgements

This research has made use of the NASA Exoplanet Archive, which is operated by the California Institute of Technology, under contract with the National Aeronautics and Space Administration under the Exoplanet Exploration Program. We thank Frederic A. Rasio for helpful comments and discussions. The CF15 simulations were done using the High Performance computing (HPC) center at University of Florida. The simulations involving initially non-resonant planets were done using the HPC Quest at CIERA, Northwestern University. SC was partially supported by NASA grant NNX12AI86G. SOK thankfully acknowledges NASA's summer research grant. EBF was supported in part by NASA Kepler Participating Scientist Program award NNX14AN76G. The Center for Exoplanets and Habitable Worlds is supported by the Pennsylvania State University, the Eberly College of Science, and the Pennsylvania Space Grant Consortium.

References

Armitage, P. J. 2013 *Astrophysics of Planet Formation, Cambridge University Press*
Butler, R. P., Wright, J. T., Marcy, G. W. et al. 2006 *ApJ*, 646, 505
Batygin, K. & Morbidelli, A. 2013 *AJ*, 145, 1
Chambers, J. E. 1999 *MNRAS*, 304, 793
Chatterjee, S., Ford, E. B., Matsumura, S., & Rasio, F. A. 2008 *ApJ*, 686, 580
Chatterjee, S. & Eric B. Ford 2015, *ApJ*, 803, 33
Deck, K. M. & Batygin, K. 2015 *ApJ*, 810, 119
Delisle, J. B. & Laskar, J. 2014 *A&A*, 570, 7
Fabrycky, D. C., Lissauer, J. J., Ragozzine, D. et al. 2014, *ApJ*, 790, 146
Fernandez, J. A. & Ip, W. H. 1984 *Icarus*, 58, 109
Ford, E. B. & Rasio, F. A. 1996 *Science*, 274, 954
Goldreich, P. & Schlihting, H. E. 2014 *ApJ*, 147, 32
Gomes, R. S., Morbidelli, A., & Levison, H. F. 2004 *Icarus*, 170, 492
Hadden, S. & Lithwick, Y. 2014 *ApJ*, 787, 80
Hands, T. O., Alexander, R. D., & Dehnen, W. 2014 *MNRAS*, 445, 749
Ida, S., Bryden, G., Lin, D. N. C., & Tanaka, H. 2000 *ApJ*, 534, 428
Kirsh, D. R., Duncan, M., Brasser, R., & Levison, H. F. 2009 *Icarus*, 199, 197
Lee, M. H. & Peale, S. J. 2002 *ApJ*, 567, 596
Lee, M. H., Fabrycky, D., & Lin, D. N. C. 2013 *ApJ*, 774, 52
Lissauer, J. J., Ragozzine, D., Fabrycky, D. C. et al. 2011, *ApJS*, 197, 8
Lithwick, Y. & Wu, Y. 2012 *ApJL*, 756, 11
Matsumura, S., Thommes, E. W., Chatterjee, S., & Rasio, F. A. 2010 *ApJ*, 714, 194

Moore, A., Hasan, I., & Quillen, A. C. 2013 *MNRAS*, 432, 1196
Petrovich, C., Malhotra, R., & Tremaine, S. 2013 *ApJ*, 770, 24
Rein, H. 2012 *MNRAS*, 427, 21
Silburt, A., & Rein, H. 2015 *MNRAS*, 453, 4089
Steffen, J. H., & Hwang, J. A. 2015 *MNRAS*, 448, 1956
Veras, D. & Ford, E. B. 2012 *MNRAS*, 420, 23

Questions and Comments:

QUESTION: In your simulations you have ignored planetary migration. If the planets migrate in a gas disk it will clear a larger gap in the planetesimal disk. As a result, after gas dispersal there won't be many planetesimals near the planets and interactions will be rare. Can you comment on that?

CHATTERJEE: Indeed, planets would migrate in a gas disk. So would the planetesimals. In fact, the planetesimals are expected to migrate faster than the planets. As a result, following gas dispersal, the surface density of planetesimals may be enhanced exterior to the planets. That is why we considered a wide variety of planetesimal surface density profiles, including those for which the surface density increases with the distance from the star.

Of course, the ideal way to study this problem is to simulate gas disks, planets, and planetesimals all together including all relevant physical effects. Such simulations are unfortunately numerically impractical. Hence we are forced to adopt a scheme that mimics the expected configuration of a system that was dissipative initially and then became pure N-body. In particular, we constructed our initial conditions in two steps, first trapping planets in resonance, and then removing planetesimals that would be unstable on short timescales. We believe that this could mimic the configuration of planets and planetesimal disks at the epoch of gas dispersal, at least qualitatively. A larger gap in planetesimals would slow down subsequent interactions with the planets, but as long as there are enough planetesimals to interact with the planets, the planetary orbits will diverge leading to a ratio of orbital periods greater than that of the exact resonance. Given that the general outcome of planet-planetesimal interactions is unchanged for the wide range in explored planetesimal disk properties, we believe that the details of planetesimal disk structures are unlikely to change our results qualitatively.

Long-term orbital stability of exosolar planetary systems with highly eccentric orbits

Kyriaki I. Antoniadou[1] and George Voyatzis[2]

Sect. of Astrophysics, Astronomy and Mechanics, Dept. of Physics,
Aristotle University of Thessaloniki, 54124, Greece
[1] email:kyant@auth.gr,
[2] email:voyatzis@auth.gr

Abstract. Nowadays, many extrasolar planetary systems possessing at least one planet on a highly eccentric orbit have been discovered. In this work, we study the possible long-term stability of such systems. We consider the general three body problem as our model. Highly eccentric orbits are out of the Hill stability regions. However, mean motion resonances can provide phase protection and orbits with long-term stability exist. We construct maps of dynamical stability based on the computation of chaotic indicators and we figure out regions in phase space, where the long-term stability is guaranteed. We focus on regions where at least one planet is highly eccentric and attempt to associate them with the existence of stable periodic orbits. The values of the orbital elements, which are derived from observational data, are often given with very large deviations. Generally, phase space regions of high eccentricities are narrow and thus, our dynamical analysis may restrict considerably the valid domain of the system's location.

Keywords. dynamical stability, periodic orbits, planetary systems.

With a vast amount of multiple extrasolar systems being discovered over the last two decades, we intend to address the following question: How can highly eccentric resonant exoplanets survive? A dynamical mechanism is offered by mean motion resonances (MMR), which can provide phase protection. Our study is based on this mechanism and aims to show its efficiency. Our methodology breaks down in two steps:

(*a*) We compute families of periodic orbits in the planar general three body problem, which indicate the exact position of a MMR (e.g. Hadjidemetriou(2006), Antoniadou & Voyatzis(2013)). We classify them in different configurations (e.g. Antoniadou & Voyatzis(2014)) and compute their linear horizontal and vertical stability. The vertical critical orbits (*vco*) represent the bifurcation points that generate spatial periodic orbits (see Fig. 1a). We justify the MMR and symmetric or asymmetric configuration via librations of resonant angles and apsidal difference about particular angles (Fig. 1b) and depict the planetary evolution (Fig. 1c).

(*b*) We construct maps of dynamical stability by creating grids defined by two variables (keep the rest orbital elements fixed) and compute the *de-trended* FLI (Voyatzis (2008)) for $t_{max} = 250$Ky and finally, we visualize its value. The de-trended FLI takes either small values (visualized by dark colors) suggesting regular motion, or very large values (visualized by yellow color) indicating chaotic motion. The distinction between these cases is very sharp when the integration time of orbits is sufficiently large. Certainly, regular orbits guarantee long-term stability of planetary orbits.

In the following figures, we present results for the 2/1 resonant dynamics of the two-planet system HD 82943b,c (see also Tan *et al.*(2013), Baluev & Beaugé(2014)). Maps are computed for the symmetric configuration, which is defined by the resonant angles $\theta_i = 2\lambda_2 - \lambda_1 - \varpi_i = 0$, $i = 1, 2$, where λ_i are the mean longitudes and ϖ_i the longitudes

Table 1. Data for the planets HD 82943b,c from Tan et al.(2013).

Planets	Mass(M_{Jup})	a (AU)	e	i (deg)	ϖ (deg)	M (deg)
HD 82943 c	4.78	1.000	0.425	19.4	133	256
HD 82943 b	4.8	1.5951	0.203	19.4	107	333

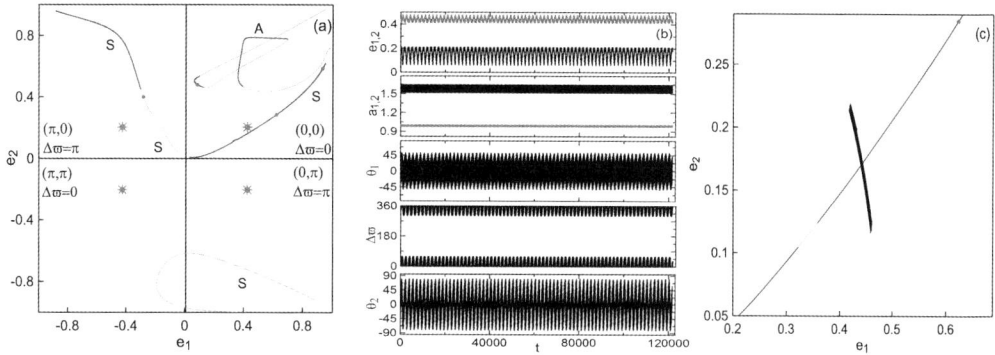

Figure 1. (a) Families of symmetric (S) and asymmetric (A) periodic orbits in 2/1 MMR for the planetary mass ratio of HD 82943. Blue (red) segments indicate stable (unstable) orbits. Red stars depict $e_{b,c}$, while magenta dots the vco. (b) Evolution of orbital elements and resonant angles. (c) Projection of the trajectory on the eccentricities plane and the family of periodic orbits.

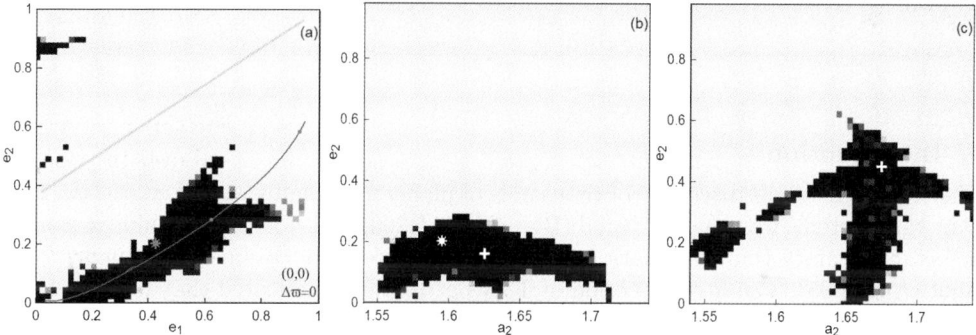

Figure 2. Maps of dynamical stability on: (a) grid plane (e_1, e_2) for a_2=1.595, where the family of periodic orbits and the collision line (bold gray curve) are also presented, (b) (a_2, e_2) grid plane for $e_1 = 0.425$, where the resonant periodic orbit is indicated by the cross symbol and the location of HD 82943b,c by a star and (c) as previously, but for $e_1 = 0.8$. In all cases we use the normalized value $a_1 = 1$.

of pericenter. Our analysis shows that the system is located in a region of dynamical stability, which is expanded around a stable periodic orbit. This periodic orbit is vertically stable, thus we may conclude that introducing small mutual inclination the planetary orbits will remain stable.

We have applied the above methodology to the extrasolar systems HD 3651, HD 7449, HD 89744 and HD 102272 and a complete paper is in preparation.

References

Antoniadou, K. I. & Voyatzis, G. 2014, *Astrophys. Space Sci.*, 349, 657
Antoniadou, K. I. & Voyatzis, G. 2013, *Celest. Mech. Dyn. Astr.*, 115, 161
Baluev, R. V. & Beaugé, C. 2014, *MNRAS*, 439, 673
Hadjidemetriou, J. D. 2006, *Celest. Mech. Dyn. Astr.*, 95, 225
Tan, X., Payne, M. J., Lee, M. H., et al. 2013, *ApJ*, 777, 101
Voyatzis, G. 2008, *ApJ*, 675, 802

The Diversity of Low-mass Exoplanets Characterized via Transit Timing

Daniel Jontof-Hutter[1], Eric B. Ford[1], Jason F. Rowe[2],
Jack. J. Lissauer[3] and Daniel C. Fabrycky[4]

[1] Dept. of Astronomy, Pennsylvania State University,
University Park, PA-16802, United States
email: dxj14@psu.edu
[2] Université de Montréal
[3] NASA Ames Research Center
[4] University of Chicago

Abstract. Transit timing variations (TTV) in multi-transiting systems enables precise characterizations of low-mass planets and their orbits. The range of orbital periods and incident fluxes with detailed TTV constraints complements the radial velocity sample for low-mass planets, pushing exoplanet characterization to the regime sub-Earth size planets and out to Mercury-like distances. This has revealed an astonishing diversity in the density of super-Earth mass planets. We summarize these and other contributions to exoplanet science from TTVs.

Keywords. Dynamics, Exoplanets, Multi-planet systems

1. Introduction

The *Kepler* mission's discovery of thousands of planet candidates within multi-transiting planetary systems was unanticipated (Borucki *et al.* 2008), although their value in causing detectable transit timing variations (TTVs) was predicted (Holman & Murray 2005; Agol *et al.* 2005). The first detection of TTVs at Kepler-9 indicated that radial velocity confirmations (RV) would be complemented by TTVs, using just the Kepler time series photometry (Holman *et al.* 2010).

The confirmation of many planet candidates with TTVs soon followed (Ford *et al.* 2011, 2012a; Steffen *et al.* 2012a). This was possible because the detection of anti-correlated TTV signals is difficult to attribute to anything other than two interacting planets orbiting the same star. The rapid confirmation of the multi-transiting planet systems with TTVs has greatly enhanced the return of the *Kepler* mission. In some cases, even the absence of a TTV signal has enabled planet candidates to be validated where low mass stellar companions or brown dwarfs would induce detectable TTVs (Lissauer *et al.* 2014; Rowe *et al.* 2014).

Using just the light curves, these confirmations were possible far quicker than could be expected of ground-based radial velocity follow-up (Fabrycky *et al.* 2012; Ford *et al.* 2012b; Steffen *et al.* 2013). Furthermore, detailed studies of individual systems including Kepler-11 (Lissauer *et al.* 2011a) and Kepler-30 (Sanchis-Ojeda *et al.* 2012) permitted the characterization of low-mass planets around faint targets. Characterization via transit timing has benefited not only from the abundance of compact multi-planet systems, but also the abundance near mean motion resonances (Lissauer *et al.* 2011b; Lithwick & Wu 2012; Fabrycky *et al.* 2014), which causes coherent perturbations and strong TTV signals.

In parallel to these data-driven developments, the progress in analytical models to explain the TTVs has continued. Lithwick *et al.* (2012) developed a solution for TTVs for planet pairs near first-order mean motion resonances. This solution enabled initial mass estimates and for dozens of planet pairs near first-order resonances (Wu & Lithwick 2013; Xie 2013; Hadden & Lithwick 2014).

In the absence of detectable frequencies in addition to the near resonance terms (Nesvorný & Morbidelli 2008), the solution of Lithwick *et al.* (2012) highlights the degeneracy between dynamical masses and the relative free eccentricities between interacting planets, but also enabled an initial characterization of the distribution of orbital eccentricities among *Kepler*'s multi-planet systems. Both the sample of TTV detections and the distribution of transit duration ratios indicate a low eccentricity dispersion among *Kepler*'s multis (Hadden & Lithwick 2014; Fabrycky *et al.* 2014).

The transit durations also imply a low inclination dispersion (Fabrycky *et al.* 2014). While strong TTV signals can detect moderate inclinations (Nesvorný *et al.* 2013), among low-mass planets TTVs are insensitive to mutual inclinations to first order (Nesvorný & Vokrouhlický 2014). While this has made it justifiable to reduce the number of free parameters in characterizing TTV systems, the price is that the light curve gives information on significant mutual inclinations only in rare cases, where signals are very strong or where planet-planet eclipses are detected during transit (Ragozzine & Holman 2010; Masuda *et al.* 2013).

In addition to the coherent TTV signal caused by a near resonant orbital period ratio, the effect of each conjunction at the synodic frequency causes so-called "chopping" in the TTV signal, breaking the mass-eccentricity in cases of high signal TTVs. In such cases, there are analytical solutions correct to first-order in eccentricity not only near mean motion resonances (Deck & Agol 2015; Agol & Deck 2015).

2. Low-mass transiting planets characterized via RV and TTV

Figure 1 highlights how TTVs complement the RV sample. Neither RV nor TTV sample the entire range of *Kepler*'s period-radius distribution, because of the detection biases in both techniques. For super-Earth mass planets, most RV detections have orbital periods of a few days or less. TTV signals, on the other hand, scale with orbital period, and detections begin at a few days. In the *Kepler* sample, the maximum orbital period for planets with detected TTVs is limited only by the 4 year photometric baseline. Kepler-87 c with an orbital period of 191 days, has the longest orbital period with a measured mass from TTVs (Ofir *et al.* 2014). This is one of several TTV systems pushing low-mass exoplanet characterization beyond Mercury-like distances.

TTVs have also characterized the lowest-mass transiting exoplanets, including the first sub-Earth sized planet with a measured density (Jontof-Hutter *et al.* 2015). Kepler-138 b has a size and mass consistent with Mars, and is the innermost of three transiting exoplanets orbiting an M-dwarf. With orbital periods at $\sim$ 10 days, $\sim$14 days, Kepler-138 b (0.5 $R_\oplus$ in size) and Kepler-138 c (1.2 $R_\oplus$) orbit near the 4:3 (first-order) mean motion resonance, and at $\sim$23 days, Kepler-138 d (1.2 $R_\oplus$) is near the 5:3 (second-order) mean motion resonance with Kepler-138 c. All three planets show TTVs consistent with the periods expected of their near resonant orbital periods. The TTVs and a sample of dynamical fits to the TTVs are shown in Figure 2.

Kipping *et al.* (2014) analyzed the TTVs of the outer pair and found that the outer two planets have mass ratio of $\sim$3. Using the complete 17 quarters of *Kepler* data and a three-planet model to the TTVs, Jontof-Hutter *et al.* (2015) found a similar mass ratio between the outer pair. The middle planet has a likely rocky composition, while the

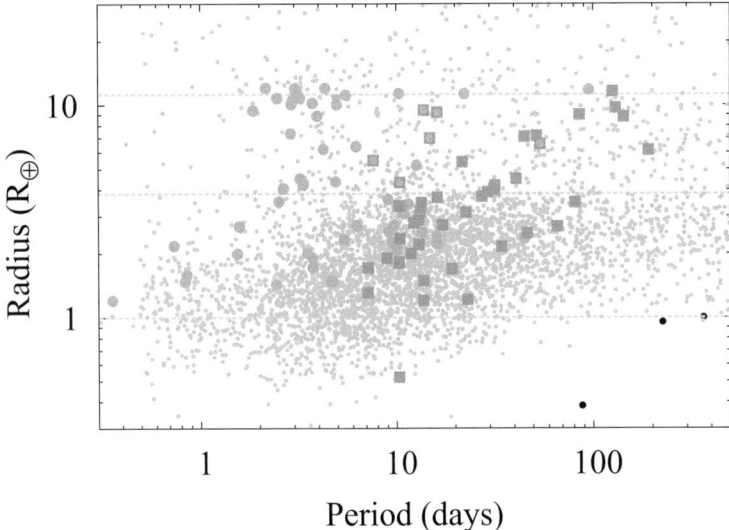

Figure 1. *Kepler*'s period-radius distribution. The grey points are the complete sample of planet candidates without uncertainties. The green circles are the subsample with published radial velocity detections, and orange squares denote the TTV sample with detailed dynamical fits. Some transiting planets have both RV and TTV data.

outer planet's composition is consistent with a mixture of rock and hydrogen or rock and water.

Kepler-138 c and Kepler-138 d are typical of the wide range in density seen in super-Earths, as shown in Figure 3. Kepler-78 b, characterized by RV, has a similar size and mass to Kepler-138 c (Howard *et al.* 2013; Pepe *et al.* 2013). These characterizations probe the boundary between rocky planets and super-Earth mass planets that retain deep atmospheres.

Dressing *et al.* (2015) show evidence of a pile-up at rock-like bulk densities up to $1.6R_\oplus$ among the strongest mass detections including Kepler-93 b, Kepler-36 b (Carter *et al.* 2012), CoRoT-7 b (Haywood *et al.* 2014), 55 Cnc e (Gillon *et al.* 2012), Kepler-10b (Batalha *et al.* 2011), Kepler-99 b and Kepler-406 b (Marcy *et al.* 2014). However, strong mass upper limits on Kepler-11 b, Kepler-11 f (Lissauer *et al.* 2013), Kepler-20 b (Gautier *et al.* 2012) and Kepler-231 c (Kipping *et al.* 2014) imply that low densities, and therefore significant volumes of volatiles is possible among planets larger than $1.6R_\oplus$. This is consistent with the findings of Rogers (2015), that the majority of planets larger than $1.6R_\oplus$ are not primarily rocky by volume.

3. RV and TTVs: a complementary sample of mass and composition

Most of the known low-density super-Earth mass planets have been characterized from TTV analysis, including all six planets known to orbit Kepler-11 (Lissauer *et al.* 2011a, 2013). The volumes of these planets can be explained with small mass fractions in a deep atmosphere of H/He. However, the measured masses from TTVs of Kepler-79 d (Jontof-Hutter *et al.* 2014), Kepler-51 b-d (Masuda 2014) and Kepler-87 c (Ofir *et al.* 2014) (all less than $10M_\oplus$, despite sizes larger than $6R_\oplus$) require significant fractions of their mass in their atmospheres. These planets reveal extraordinary variety in the bulk

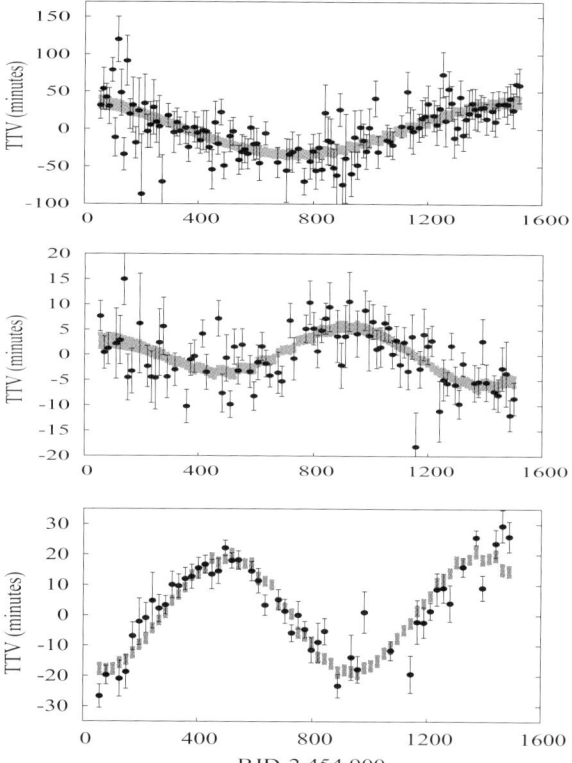

Figure 2. Model fits to the transit times of all transiting planets at Kepler-138. The black points are the measured differences between each transit time and a calculated linear fit to the transit times. In color are the 1σ range of simulated transit times from posterior samples of our model parameters. (Adapted from Jontof-Hutter *et al.* 2015)

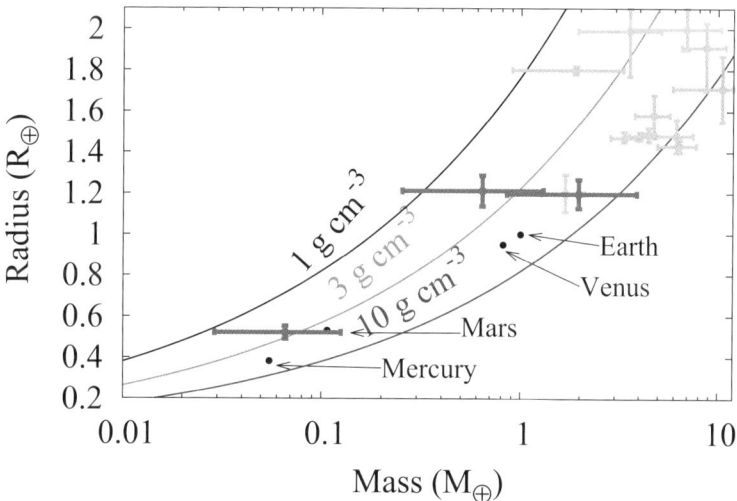

Figure 3. Mass-radius diagram of planets smaller than $2.1 R_\oplus$. In red are the three transiting planets of Kepler-138, compared to Mercury, Venus, Earth and Mars (black points). In grey are other well-characterized exoplanets. The colored curves mark contours of constant bulk density, at 1 g cm^{-3}, 3 g cm^{-3} and 10 g cm^{-3}.

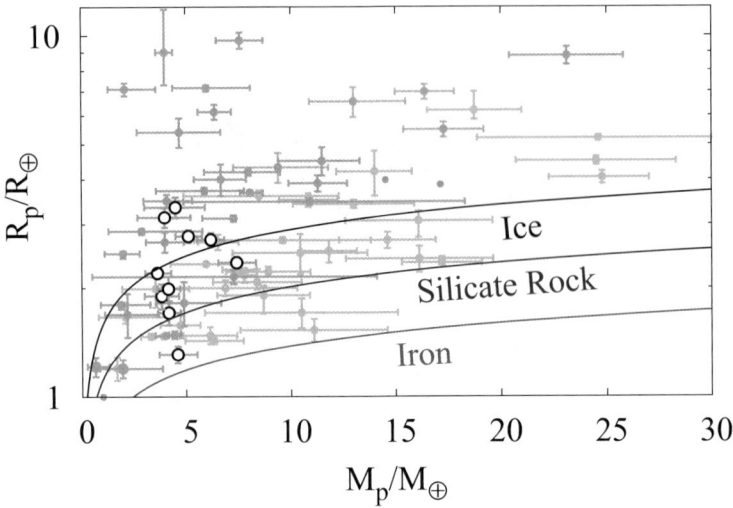

Figure 4. Mass-radius diagram of planets less than $25M_\oplus$. The grey points are planets of the Solar System. In green are planets characterized via RV, and in orange are the TTV sample. Planets shown in orange and green are constrained by RV and TTV data. The colored curves mark solutions for planets of pure iron, silicate rock, or water (Fortney *et al.* 2007). The open circles mark ten new additions to the mass radius diagram with TTVs that confirm the diverse range in density in the 3-8 $M_\oplus$ mass range.

density of super-Earth mass planets. At $6M_\oplus$ and $7R_\oplus$, Kepler-79 d has a bulk density below 0.1 g cm^{-3}, and is likely significantly more than 10% H/He by mass. All three planets transiting Kepler-51 are even less dense.

Figure 4 shows the mass-radius diagram for low-mass planets, highlighting the division between the RV and TTV sample, including six unpublished additions (Jontof-Hutter *et al.* 2015). While there is some overlap where there is both RV and TTV data– including Kepler-18 (Cochran *et al.* 2011) and Kepler-89 (Weiss *et al.* 2013; Masuda *et al.* 2013), the TTV planets appear systematically larger and therefore less dense in the sub-Neptune mass range than the RV planets.

There are biases in both techniques that contribute to this separation. In RV, for a given planetary size, there is a detection bias towards higher masses. TTVs, however, are strongly biased towards larger planets with deeper transit signals in a given mass range. Larger planets have more precisely measurable transit times and are the best candidates for TTV analysis. The bias is indirect, since the TTVs detected in a large planet are used to infer the masses of its neighbors, not itself. Nevertheless, in the complex TTV signals of systems with multi-planet interactions, like Kepler-79, the model as a whole benefits from the precisely measured transit times of the largest planet.

Another bias in the density distributions observed by each technique stems from the separation of the two samples shown in Figure 1. The low mass planets detected by RV all have high incident fluxes, and many of the low mass planets that are detectable are too hot to retain deep atmospheres (Lopez & Fortney 2013a; Owen & Jackson 2012). Hence there is a bias towards high densities in the RV sample. The TTV sample on the other hand, extends to distances where low-mass planets can retain deep atmospheres. However, it remains unknown whether or not the extreme low-density super-Earth mass planets are common, or whether they are limited to planets near the threshold of mass-loss over a small range of distances.

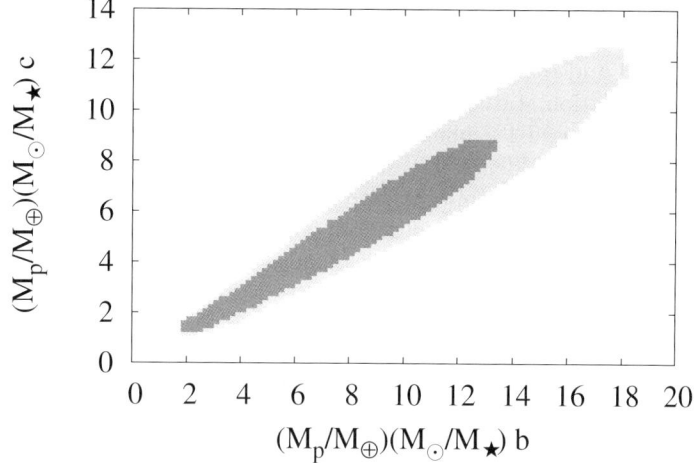

Figure 5. Joint posteriors for planetary masses relative to the host star for Kepler-29 b and c. 95.4% credible intervals are in light grey, 68.3% credible intervals are in dark grey. The dynamical masses are expressed relative to the host star, in approximate Earth-masses.

There may be additional detection biases in both RV and TTV based on planet formation and stellar properties. Most of the RV targets are cool stars, and RV surveys of the Kepler multi-planet systems have focused on size-limited surveys. TTV studies have focused on the most compact high multiplicity systems. Whether these planets truly represent all low-mass exoplanets within 1 AU remains to be seen.

Despite these biases, the RV and TTV samples overlap. This increases the scope of what we can study in multi-planet systems. In many cases, the mass posteriors from TTV analysis of planet pairs are strongly correlated, such that an additional constraint on either planetary mass from RV would enable both to be precisely characterized. An example is shown Figure 5, where the masses of Kepler-49 b and c are both well-constrained by TTVs and likely below the mass of Neptune. However, the ratio of the two masses is even better constrained, and thus additional constraints on the mass of the inner planet from RV would enable both planets to be further characterized.

4. Precise Eccentricities

In compact multi-planet systems, TTVs are particular sensitive to small changes in orbital eccentricity. Hence, where degeneracies between mass and eccentricity can be broken, precise eccentricities can be inferred (Deck & Agol 2015), including in the regime where $e < 0.1$. Lissauer *et al.* (2013) found upper limits on the eccentricities of Kepler-11 d and Kepler-11 e below ~ 0.02. Other exoplanets with precise limits on orbital eccentricities below ~ 0.05 include Kepler-30 b-d (Sanchis-Ojeda *et al.* 2012), Kepler-89 c-e Masuda *et al.* (2013), Kepler-79 b-e (Jontof-Hutter *et al.* 2014), Kepler-51 b-d (Masuda 2014), Kepler-88 b and c (Nesvorný *et al.* 2013), and Kepler-46 b (Nesvorný *et al.* 2012). However, there are few other examples of such precise constraints on orbital eccentricity from published TTV analyses, and it is unclear whether a large sample of precise eccentricities will come from the *Kepler* dataset. In some cases, precise individual eccentricities cannot be inferred from the transit times, even though relative eccentricities are well constrained. Jontof-Hutter *et al.* (2015) found that within the bounds of having well-constrained relative eccentricities, correlated posteriors in eccentricity vector components permit many solutions of high eccentricity. In such cases, the requirement of long-term stability cannot rule out the high eccentricity solutions, since the solutions are apsidally aligned and remain stable due to strong secular interactions.

Nevertheless, population statistics from the TTV sample can be used to infer eccentricity distributions. Using TTV phases and amplitudes for planet pairs near first-order mean motion resonance, Hadden & Lithwick (2014) inferred an eccentricity distribution consistent with a Rayleigh distribution with a scale parameter $\sigma_e = 0.018$. They found moderate evidence of higher eccentricities among smaller planets, although this could be due to larger timing errors on smaller planets and a TTV detection bias to high eccentricity for low mass planets.

5. Stellar Parameters

Planetary masses and radii cannot be characterized without precise constraints on the properties of the host, particularly the stellar radius. For the nearest planetary systems, interferometry directly constrains the stellar radius to the great precision (e.g., 55 Cancri, Gillon *et al.* 2012). Among targets in the *Kepler* field, asteroseismology detections on Kepler-10 (Batalha *et al.* 2011), Kepler-36 (Carter *et al.* 2012), Kepler-68 (Gilliland *et al.* 2013), and Kepler-93 (Dressing *et al.* 2015), have permitted uncertainties on stellar radius down to just 1–2%. This remains the gold standard for targets in the *Kepler* field, although detections are limited to evolved stars and the brightest dwarf stars.

Transit light curves provide an independent constraint on the bulk density of the star, and hence its mass and radius (Seager & Mallén-Ornelas 2003; Kipping 2010). The constraint depends on orbital period, transit depth, the duration of ingress or egress and the total transit duration (all of which come directly from the light curve), as well as orbital eccentricity. In most cases if $e \lesssim 0.1$, one has no choice but to assume the orbits are circular, although this adds an error of $\sim e \sin \omega$ to the inferred stellar radius.

In select cases, precise constraints on eccentricity vectors from TTVs enables more precise stellar radii measurements. For both Kepler-11 and Kepler-79 (Lissauer *et al.* 2013; Jontof-Hutter *et al.* 2014), stellar and hence planetary radii were constrained to a precision of $\sim 2\%$, enabling meaningful constraints on the planetary bulk densities.

6. A mass-radius-flux trend for low-mass exoplanets

To account for the effect of flux on the observed density of planets, we compare planet radii to worlds made of pure silicate rock of the same mass. Planets larger than a purely rocky world must retain some significant volume of volatiles in the form of ices and gases. However, planets that are larger and less dense than a world made of pure water ice must retain deep atmospheres of H/He.

Figure 6 highlights the effect of incident flux on the range of densities observed in low mass exoplanets. The wide range in bulk densities among super-Earth mass planets makes a simple mass-radius fit to the data sensitive to a small number of planets with wide scatter in radius. The transition from a wide range of densities at low incident flux to the sample of predominantly rocky hot super-Earths is consistent with atmospheric mass loss (Owen & Wu 2013).

The majority of systems with transiting planets and anti-correlated TTVs have yet to be analyzed for detailed characterizations (Mazeh *et al.* 2013), and many of the hosts require additional data and modeling to further constrain their properties. Hence, the addition of more precise TTV mass characterizations on the mass-radius diagram is

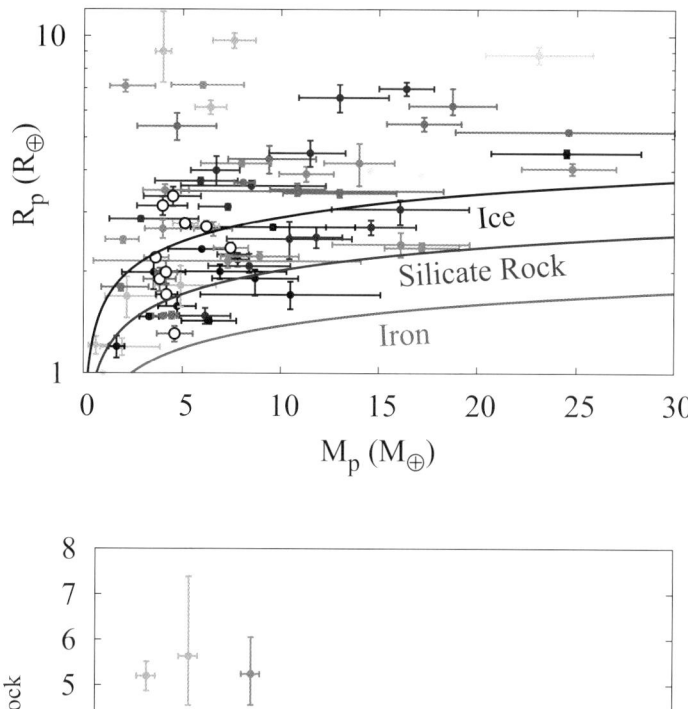

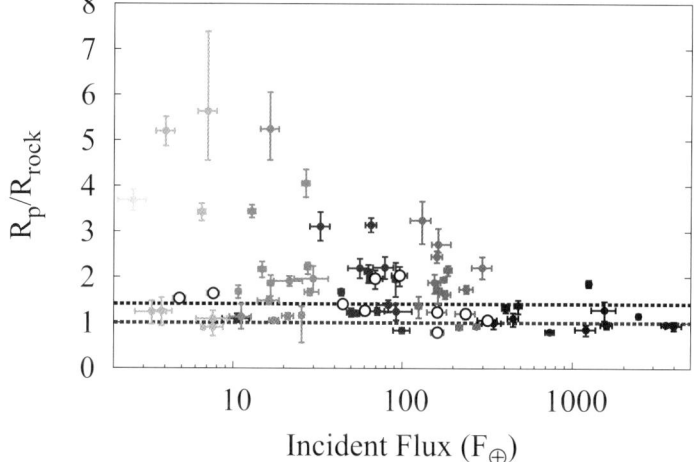

Figure 6. Mass-radius-flux diagram of planets less than $25 M_\oplus$. The colors in the top panel are coded by incident flux, which are illustrated in the lower panel. The lower panel compares planet radii to a planet made of pure silicate rock. Above the brown line, planets are less dense than rock and above the blue line, less dense than water. The range in planetary densities in this mass range is widest at low incident flux, while at high incident flux, most of the characterized planets are likely rocky. (Adapted from Jontof-Hutter et al. 2015, submitted)

certain. These will place additional constraints on mass-loss and formation scenarios for a wide range of stellar types and system architectures.

7. Non-transiting planets

Measuring planetary densities is only possible with TTVs if interacting planets are both transiting. TTVs caused by non-transiting perturbers enables the detection of additional planets, often orbiting further out than the known transiting planet. There are several single transiting systems in the *Kepler* dataset with significant TTVs likely caused by non-transiting planets (Mazeh et al. 2013).

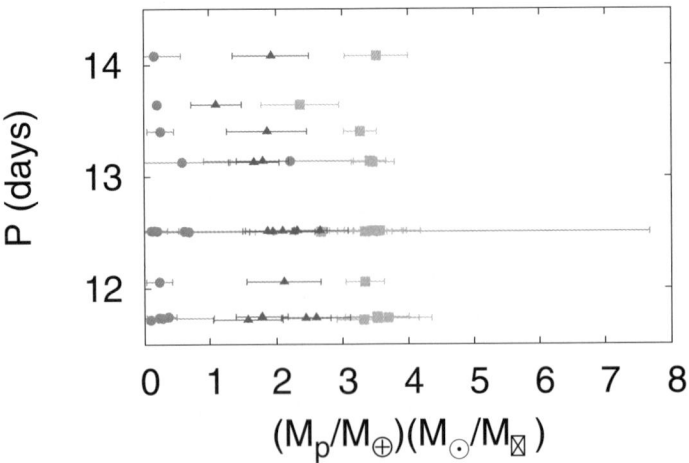

Figure 7. Solutions from a three-planet model for the TTVs of Kepler-50. The two known planets, Kepler-50 b (red triangles) and Kepler-50 c (green squares) have their masses compared to a non-transiting perturber (blue circles) for many potential orbital periods of the perturber. Although the parameters of the non-transiting planet are poorly constrained, the masses of the two transiting super-Earths are very well-constrained.

The strongest TTV detections permit a unique solution for the orbital period of a non-transiting perturber. Nesvorný et al. (2012) found a unique solution for a perturber of Kepler-46 b, attributable to a Saturn-mass perturber orbiting farther out. In another example, Nesvorný et al. (2013) analyzed the TTVs of Kepler-88 b (KOI-142.01) to predict the mass and orbital period of a jovian-mass planet that was soon confirmed with radial velocity observations (Barros et al. 2014).

For lower signal-to-noise TTVs (and lower mass planets), a detected periodicity can be attributed to a perturber at multiple locations near different resonances, with weak constraints on the perturbing mass. Kepler-50 has two transiting super-Earth sized planets with strongly detected TTVs (Steffen et al. 2013). While a two-planet model fails to fit the TTVs, a three planet model permits a close fit. In this case, there is no unique solution for the orbit period, phase or mass of the third planet. However, there are still useful constraints on the transiting planets. Figure 7 shows the potential orbital period of a third planet and the inferred planetary masses for all three objects for a range of solutions that fit the transit timing data and are stable for at least 100 kyr. Despite the poor constraints on the non-transiting perturber, these preliminary results imply that robust masses for the two transiting planets may be inferred.

8. Conclusion

Transit timing studies have yielded precise masses, precise orbital eccentricities and have even improved stellar parameters for transiting exoplanet systems, complementing the sample of transiting exoplanets confirmed via RV. In probing low-mass planets beyond the range of RV, TTVs have pushed mass characterization towards the regime of the terrestrial planets of the Solar System. The same studies however, have shown that terrestrial planets like our own are not typical of low mass planets within 1 AU. Rather super-Earth mass planets show a diversity in composition from rocky to Saturn-size gaseous worlds on the threshold of atmospheric mass loss. This fundamental result is one of many contributions to *Kepler*'s legacy. Exoplanet science will continue to benefit from the contribution of TTVs into era of *TESS* and *PLATO*, to detect, confirm and characterize transiting planets and the architectures of multi-planet systems.

References

Agol, E. & Deck, K. 2015, ArXiv e-prints
Agol, E., Steffen, J., Sari, R., & Clarkson, W. 2005, *MNRAS*, 359, 567
Barros, S. C. C., et al. 2014, *A&A*, 561, L1
Batalha, N. M., et al. 2011, *ApJ*, 729, 27
Borucki, W., et al. 2008, in IAU Symposium, Vol. 249, IAU Symposium, ed. Y.-S. Sun, S. Ferraz-Mello, & J.-L. Zhou, 17–24
Carter, J. A., et al. 2012, *Science*, 337, 556
Cochran, W. D., et al. 2011, *ApJS*, 197, 7
Deck, K. M. & Agol, E. 2015, *ApJ*, 802, 116
Dressing, C. D., et al. 2015, *ApJ*, 800, 135
Dumusque, X., et al. 2014, *ApJ*, 789, 154
Fabrycky, D. C., et al. 2012, *ApJ*, 750, 114
—. 2014, *ApJ*, 790, 146
Ford, E. B., et al. 2011, *ApJS*, 197, 2
—. 2012a, *ApJ*, 750, 113
—. 2012b, *ApJ*, 756, 185
Fortney, J. J., Marley, M. S., & Barnes, J. W. 2007, *ApJ*, 659, 1661
Gautier, III, T. N., et al. 2012, *ApJ*, 749, 15
Gilliland, R. L., et al. 2013, *ApJ*, 766, 40
Gillon, M., et al. 2012, *A&A*, 539, A28
Hadden, S. & Lithwick, Y. 2014, *ApJ*, 787, 80
Haywood, R. D., et al. 2014, *MNRAS*, 443, 2517
Holman, M. J. & Murray, N. W. 2005, *Science*, 307, 1288
Holman, M. J., et al. 2010, *Science*, 330, 51
Howard, A. W., et al. 2013, *Nature*, 503, 381
Jontof-Hutter, D., Lissauer, J. J., Rowe, J. F., & Fabrycky, D. C. 2014, *ApJ*, 785, 15
—. 2015, *Nature*, 785, 15
—. 2015, *ApJ* submitted
Kipping, D. M. 2010, *MNRAS*, 407, 301
Kipping, D. M., Nesvorný, D., Buchhave, L. A., Hartman, J., Bakos, G. Á., & Schmitt, A. R. 2014, *ApJ*, 784, 28
Lissauer, J. J., et al. 2011a, *Nature*, 470, 53
—. 2011b, *ApJS*, 197, 8
—. 2013, *ApJ*, 770, 131
—. 2014, *ApJ*, 784, 44
Lithwick, Y. & Wu, Y. 2012, *ApJL*, 756, L11
Lithwick, Y., Xie, J., & Wu, Y. 2012, *ApJ*, 761, 122
Lopez, E. D., & Fortney, J. J. 2013a, *ApJ*, 776, 2
Marcy, G. W., et al. 2014, *ApJS*, 210, 20
Masuda, K. 2014, *ApJ*, 783, 53
Masuda, K., Hirano, T., Taruya, A., Nagasawa, M., & Suto, Y. 2013, *ApJ*, 778, 185
Mazeh, T., et al. 2013, *ApJS*, 208, 16
Nesvorný, D. & Morbidelli, A. 2008, *ApJ*, 688, 636
Nesvorný, D., Kipping, D. M., Buchhave, L. A., Bakos, G. Á., Hartman, J., & Schmitt, A. R. 2012, *Science*, 336, 1133
Nesvorný, D., Kipping, D., Terrell, D., Hartman, J., Bakos, G. Á., & Buchhave, L. A. 2013, *ApJ*, 777, 3
Nesvorný, D. & Vokrouhlický, D. 2014, *ApJ*, 790, 58
Ofir, A., Dreizler, S., Zechmeister, M., & Husser, T.-O. 2014, *A&A*, 561, A103

Owen, J. E. & Jackson, A. P. 2012, *MNRAS*, 425, 2931
Owen, J. E. & Wu, Y. 2013, *ApJ*, 775, 105
Pepe, F., *et al.* 2013, *Nature*, 503, 377
Ragozzine, D. & Holman, M. J. 2010, ArXiv e-prints
Rogers, L. A. 2015, *ApJ*, 801, 41
Rowe, J. F., *et al.* 2014, *ApJ*, 784, 45
Sanchis-Ojeda, R., *et al.* 2012, *Nature*, 487, 449
Seager, S. & Mallén-Ornelas, G. 2003, *ApJ*, 585, 1038
Steffen, J. H., *et al.* 2012a, *MNRAS*, 421, 2342
—. 2013, *MNRAS*, 428, 1077
Weiss, L. M., *et al.* 2013, *ApJ*, 768, 14
Wu, Y. & Lithwick, Y. 2013, *ApJ*, 772, 74
Xie, J.-W. 2013, *ApJS*, 208, 22

K2-19, The first K2 muti-planetary system showing TTVs

S. C. C. Barros[1,2], J. M. Almenara[3,4], O. Demangeon[1], M. Tsantaki[2],
A. Santerne[2], D. J. Armstrong[5], D. Barrado[6], D. Brown[5],
M. Deleuil[1], J. Lillo-Box[6], H. Osborn[5], D. Pollacco[5], L. Abe[7],
P. Andre[8], P. Bendjoya[7], I. Boisse[1], A. S. Bonomo[9], F. Bouchy[1],
G. Bruno[1], J. Rey Cerda[10], B. Courcol[1], R. F. Díaz[10],
G. Hébrard[11,12], J. Kirk[5], J.C. Lachurié[8], K. W. F. Lam[5],
P. Martinez[8], J. McCormac[5], C. Moutou[13,1], A. Rajpurohit[1],
J.-P. Rivet[7], J. Spake[5], O. Suarez[7], D. Toublanc[8,14] and S. R. Walker[5]

[1] Aix Marseille Université, CNRS, LAM (Laboratoire d'Astrophysique de Marseille) UMR 7326, 13388, Marseille, France email: susana.barros@astro.up.pt

[2] Instituto de Astrofísica e Ciências do Espaço, Universidade do Porto, CAUP, Rua das Estrelas, PT4150-762 Porto, Portugal

[3] Univ. Grenoble Alpes, IPAG, F-38000 Grenoble, France

[4] CNRS, IPAG, F-38000 Grenoble, France

[5] Department of Physics, University of Warwick, Gibbet Hill Road, Coventry, CV4 7AL, UK

[6] Departamento de Astrofísica, Centro de Astrobiología (CSIC-INTA), ESAC campus 28691 Villanueva de la Cañada (Madrid), Spain

[7] Laboratoire Lagrange, UMR7239, Université de Nice Sophia-Antipolis, CNRS, Observatoire de la Cote d'Azur, F-06300 Nice, France

[8] Observatoire de Belesta en Lauragais - Assoc. Astronomie Adagio 30 Route de Revel 31450 Varennes, France

[9] INAF - Osservatorio Astrofisico di Torino, via Osservatorio 20, 10025, Pino Torinese, Italy

[10] Observatoire Astronomique de l'Universite de Genève, 51 chemin des Maillettes, 1290 Versoix, Switzerland

[11] Institut d'Astrophysique de Paris, UMR7095 CNRS, Université Pierre & Marie Curie, 98bis boulevard Arago, 75014 Paris, France

[12] Observatoire de Haute-Provence, Université d'Aix-Marseille & CNRS, 04870 Saint Michel l'Observatoire, France

[13] CNRS, Canada-France-Hawaii Telescope Corporation, 65-1238 Mamalahoa Hwy., Kamuela, HI 96743, USA

[14] Universite de Toulouse, UPS-CNRS, IRAP, 9 Av. colonel Roche, 31028 Toulouse, France

Abstract. In traditional transit timing variations (TTVs) analysis of multi-planetary systems, the individual TTVs are first derived from transit fitting and later modelled using n-body dynamic simulations to constrain planetary masses. We show that fitting simultaneously the transit light curves with the system dynamics (photo-dynamical model) increases the precision of the TTV measurements and helps constrain the system architecture. We exemplify the advantages of applying this photo-dynamical model to a multi-planetary system found in K2 data very close to 3:2 mean motion resonance, K2-19. In this case the period of the larger TTV variations (libration period) is much longer (>1.5 years) than the duration of the K2 observations (80 days). However, our method allows to detect the short period TTVs produced by the orbital conjunctions between the planets that in turn permits to uniquely characterise the system. Therefore, our method can be used to constrain the masses of near-resonant systems even when the full libration curve is not observed.

1. Introduction

Transit timing variations (TTVs) are caused by the mutual gravitational interaction of planets which perturb each others' orbit. These are larger when the planets are close to mean-motion resonances (MMR; Escude et al. 2002, Holman et al. 2005, Agol et al. 2005). In near-resonant systems the resonant angles which measure the displacement of the longitude of the conjunction from the periapsis of each planet, circulate (or librate) over a period much longer than the orbital period of the outer planet, called the libration period or super period (Lithwick et al. 2012). Analyses of the long term TTVs libration curve in *Kepler* transiting multi-planetary systems have allowed to derive dynamic planetary masses and helped confirm the planetary nature of many candidates e.g. Holman et al. (2010), Lissauer et al. (2011), Steffen et al. (2012).

After the failure of two out of four of the reaction wheels of the *Kepler* satellite the pointing accuracy was severely degraded. Cleaver engineering allowed the continuation of the mission in a new configuration named K2 (Howell et al. 2014). K2 observes 4 fields a year close to the Ecliptic. The short duration of the observations of each field (~ 80 days) does not favour the detection of TTVs amongst planetary candidates discovered in these observations.

K2-19 (EPIC201505350) is a multi-planetary system detected in the K2 Campaign 1 (C1) data by Armstrong et al. (2015). The K2 observations show 2 transiting planets, one with an orbital period $P_b \sim 7.92$ days and radius $R_b = 7.23 \pm 0.60\,R_\oplus$ and a companion close to the 3:2 MMR with an orbital period $P_c \sim 11.91$ days and radius $R_b = 4.21 \pm 0.31\,R_\oplus$. The closeness to resonance implied that K2-19 was a good candidate for TTV and the brightness of the host star allowed follow-up transit observations from the ground. Approximately 200 days after the end of the K2 C1, a ground based transit was obtained showing TTVs of the inner planet with an amplitude of 1 hour, allowing the authors to validate the system. Here we present a photo-dynamical analysis Carter et al. (2012) of the K2-19 system and allow us to detect TTVs in the K2 data alone.

2. Method

2.1. Observations

K2-19 was observed during Campaign 1 of the K2 mission between 2014 June 3 and 2014 August 20 spanning ~ 80 days. We downloaded the pixel data from the Mikulski Archive for Space Telescopes (MAST)† and used a modified version of the CoRoT imagette pipeline (Barros et al. 2014) to extract the light curve. We corrected the flux dependence with position due to the loss of pointing stability following Barros et al. in prep.

Furthermore, we obtained 10 spectroscopic observations of K2-19 from 2015 February 21 to 2015 April 25 with the SOPHIE spectrograph mounted on the 1.93m telescope at the Observatoire de Haute-Provence (Perruchot et al. 2011, Bouchy, et al. 2013). From these we derive radial velocities using a similar method to Santerne et al. (2012). Unfortunately, the radial velocity measurements are not precise enough to detect the stellar reflex velocity due to any of the planets. The spectroscopic observations were also used to derive the host stellar parameters following the methodology described in Tsantaki et al. (2013). We obtained $T_{\rm eff} = 5390 \pm 180$ K, $\log g = 4.42 \pm 0.34$ dex, $\xi_t =$

† $http://archive.stsci.edu/kepler/data\_search/search.php$

$1.02 \pm 0.24\,\mathrm{km\,s^{-1}}$, and $[Fe/H] = 0.19 \pm 0.12\,\mathrm{dex}$, hence it is a K-dwarf. These were used to derive the stellar mass and radius by interpolating the stellar evolution models using the MCMC described in Díaz *et al.* (2014) . We obtained $M_* = 0.918^{+0.086}_{-0.070}$ M$_\odot$ and $R_* = 0.926^{+0.19}_{-0.069}$ R$_\odot$.

2.2. Photo-dynamical model

All the transits and radial velocities were modelled simultaneously with an n-body dynamical integrator that accounts for the gravitational interactions between all components of the system. We use the MERCURY n-body integrator Chambers (1999) to compute the 3 dimensional position and velocity of all system components as a function of time. We assume that only the host star and two planets are present. The stellar velocity projected onto the line-of-sight is used to model the observed radial velocities. To model the transits, we use the Mandel & Agol (2002) transit model parametrised by the planet-to-star radius ratio, quadratic limb darkening coefficients for each filter and using the sky projection of the planet-star separation computed from the output of MERCURY. This photo-dynamical model is coupled to a Monte Carlo Markov Chain (MCMC) routine, described in detail in Díaz *et al.* (2014) , in order to derive the posterior distribution of the parameters.

To minimise correlations between the model parameters which prevent adequate exploration of the parameter space, we used the Huber *et al.* (2013) parametrisation. We run 46 independent MCMC chains and combined the results as described by Díaz *et al.* (2014) . Initially more than 100 chains were started at random points to explore the parameter space. It was found that the chains converged to the same region of the parameter space. To explore this region of the parameter space we launched the 46 independent MCMC chains . The chains were thinned by 970 which was the maximum correlation length of the parameters and merged as described by Díaz *et al.* (2014) resulting in a final merged chain with 3500 independent points. Further details about the photo-dynamic method can be found in Almenara *et al.* (2015).

3. Results

We find that K2-19b has a mass of 41.6 ± 17 $M_\oplus$ and radius of $7.34 \pm 0.69\,R_\oplus$ and K2-19c has a mass of $22.0^{+10}_{-4.3}$ $M_\oplus$ and a radius of $4.52 \pm 0.46\,R_\oplus$. In particular the dynamic parameters that do not depend on stellar models are well constrained: $q_+ = \frac{M_{p,b}+M_{p,c}}{M_*} = 0.000193^{+0.00011}_{-0.000031}$ and $q_p = \frac{M_{p,c}}{M_{p,b}} = 0.56 \pm 0.19$.

3.1. Transit timing variations

To derive the transit times, we calculate the mid point between the first and fourth transit contact using the MERCURY dynamic model output. Therefore, our transit time measurements include information on the system architecture and dynamics and as such are better constrained than direct measurement of the transit times in the light curve.

We find significant TTVs for the K2 observations. These are shown in Figure 1 where the chopping is clearly visible. So this method will be useful for short duration observations like K2, TESS and CHEOPS.

Armstrong *et al.* (2015) predicted that the resonant timescale of the system is $\sim$ 5 years and hence it is not detectable with the current observation baseline. Therefore, the phase curve cannot be used to constrain the planetary masses e.g. (Lithwick *et al.* 2012). However, we detect the short period chopping signal at the much shorter synodic timescale. In our case the detection of the chopping signal at the short synodic timescale

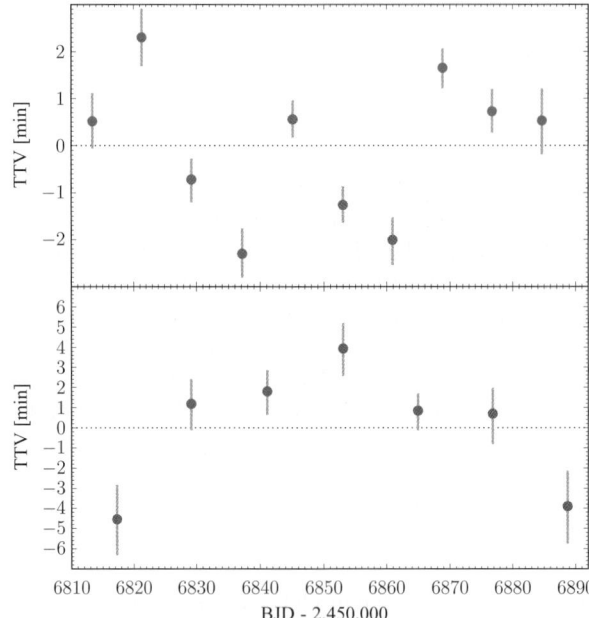

Figure 1. Transit times derived from K2 observations with the photo-dynamic model after removing a linear ephemeris.

allows us to directly determine planetary masses. This can be intuitively understood using the equations derived by Nesvorný & Vokrouhlický (2014), Deck & Agol (2015) although our system might not obey their model assumptions. We also find that, as expected, without the detection of the libration period the orbital eccentricities are poorly constrained.

In (Barros et al. 2015) a photodynamic analysis of the K2 data and 3 follow-up transits of planet b is presented. Our model predicts the times of the follow-up ground based transits to be $2457082.65858^{+0.076}_{-0.094}$, $2457090.57608^{+0.082}_{-0.10}$, $2457098.49117^{+0.09}_{-0.11}$ respectively for epochs 34, 35 and 36 which are within 1 sigma of the measured values presented in (Barros et al. 2015). Therefore, we conclude that our system solution is robust and it is not significantly affected by spurious TTV due to systematics or spots.

4. Comparison of photo-dynamic model with traditional TTV computation

For comparison, we computed the transit times directly from the K2 light curves using a procedure similar to what is described in Barros et al. (2013). For each planet the transits were fitted simultaneously ensuring the same transit shape. In Figure 2, we compare the derived transit times using the photo-dynamic model and the transit times derived with a standard procedure. To compute the ephemeris we use only the values of the observed transit times derived with the photo-dynamic model. For planet b we derived the ephemeris: T_b (BJD) = 7.921101(69)× Epoch + 2456813.3767(21) and for planet c T_c (BJD) = 11.90727(58)×Epoch + 2456817.2755(22). For each planet the respective and same ephemeris was subtracted from the transit times derived with both methods so that we could directly compare them.

We find that the difference of the transits times for both methods is less than 3σ hence the transit times from both methods agree. The higher discrepancy is found for

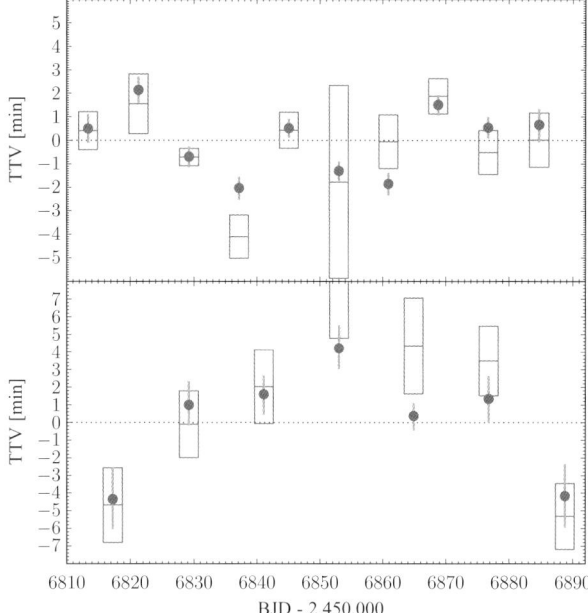

Figure 2. Comparison of the TTVs derived by the photo-dynamic model (as circles) with TTVs derived using a standard transit fitting (as boxes with the size of the 1 sigma error) for planet b (top panel) and planet c (bottom panel). For each planet we use the respective ephemeris derived using the photo-dynamic estimated values of only the observed transits which are marked in red for planet b and in blue for planet c.

epoch 3 of planet b (the 4th data point in Figure 2). The transit at epoch 3 shows signs of systematic noise. It has been shown by Barros *et al.* (2013) that, in this case, the errors of the transit times are underestimated, therefore, a difference of 3σ is not surprising. Using our photo-dynamic method, we obtain the double of the precision of the transit times as compared to the traditional method that does not include the dynamical constrains. For the K2-19 system the difference increases the significance of the TTVs for planet b and planet c, even in the short duration of the K2 observations, allowing us to better constrain the system architecture. The application of the photo-dynamic method to all available transits of the K2-19 system is presented in Barros *et al.* 2015 submitted.

5. Take home messages

• Applying a photodynamical model leads to a better constrain on the system parameters compared with traditional TTV methods.

• Detecting short period TTVs (chopping) in K2-19 allowed to constrain the system without long time coverage. This will be very important for the analysis of the short duration K2 data and future observations with CHEOPS and TESS.

References

Agol E., *et al.* 2005, *MNRAS*, 359, 567
Almenara J. M. *et al.* , 2015, *MNRAS*, in press
Armstrong D. J., *et al.*, 2015a, *MNRAS*, in press
Barros S. C. C. *et al.* 2013, *MNRAS* 430, 3032
Barros S. C. C. *et al.*, 2014, *A&A* 569, A74

Barros S. C. C. et al., 2015, *MNRAS* 454, 426
Borucki W. J., et al., 2010, *Science* 327, 977
Bouchy F., et al., 2009, *A&A* 505, 853
Bouchy F. et al. 2013, *A&A* 549, A49
Carter et al. 2012, *Science* 337, 556
Chambers J. E., 1999, *MNRAS* 304, 793
Deck K. M. & Agol E., 2015, *ApJ* 802, 116
Díaz R. F., et al. 2014, *MNRAS* 441, 983
Holman M. J., et al., 2010, *Science* 1195778
Howell S. B., et al., 2014, *PASP* 126, 398
Huber D., et al., 2013, *Science* 342, 331
Lissauer, J. J., Fabrycky, D. C., Ford, E. B., et al. 2011 *Nature*, 470, 53
Lithwick Y. & Wu Y., 2012, *ApJL* 756, L11
Lithwick Y., Xie J., & Wu Y., 2012, *ApJ* 761, 122
Mandel K., Agol E., 2002, 580, L171
Nesvorný D. & Beaugé C., 2010, *ApJL*, 709, L44
Nesvorný D. & Vokrouhlický D., 2014, *ApJ* 790, 58
Nesvorný D. et al. 2012, *Science* 336, 1133
Perruchot S., et al., 2011, *in Society of Photo-Optical Instrumentation Engineers (SPIE) Conference Series.* p. 15
Santerne A., et al., 2012, *A&A* 545, A76
Steffen et al. 2012, *MNRAS* 421, 2342
Tsantaki M. et al. 2013, *A&A* 555, A150

Tidal evolution of stars hosting massive close-in planets

S. Ferraz-Mello[1], L. F. R. Moda[1], J. D. do Nascimento Jr.[2] and E. S.Pereira[1]

[1] Instituto de Astronomia, Geofísica e Ciências Atmosféricas, Universidade de São Paulo, Brasil
email: sylvio @ usp.br

[2] Dep. Fisica Teórica e Experimental, Universidade Federal do Rio Grande do Norte, Natal 59072-970, Brasil

Abstract. Close-in massive planets transfer angular momentum to their host stars and influence their rotation through the torques associated with the tides raised on the star by the planet. For a star hosting a hot Jupiter, the limit of distance below which tidal torques cannot be neglected grows from $a \sim 0.04$ to $a \sim 0.07$ AU as the mass of the planet grows from 0.5 to $4 M_{\rm Jup}$.

Close-in massive planets transfer angular momentum to their host stars and influence their rotation, which will evolve differently from what it would be in absence of the close-in companion. If the star is not active and is single, the rotation speed remains high, and close to its primordial state. But if a non-active star has a close-in massive companion, its rotation evolves towards period values close to the orbital period of the companion.

When the star is active, the situation is more complex. In that case, a single star loses rotation speed and its rotation period grows proportionally to the square root of the time (Skumanich law; see Skumanich, 1972). However, if it has a massive close-in companion, the star's rotation evolves under the control of two factors: (1) The stellar magnetic wind, which causes a loss of the stellar rotational angular momentum (see Bouvier, 2013); and (2) the tides, which transfer angular momentum from the orbital motion of the companion to the stellar rotation. The scenario is then the following: In the initial stage, the star has a very fast rotation resulting from the contraction of the primordial cloud from which it formed, but looses speed very fast. The rate of variation of the angular rotation velocity is proportional to the cube of its value. As the rotation decreases, the variation becomes slower up to reach a situation in which the loss of angular momentum due to the magnetic wind braking is compensated by the angular momentum tidally transferred to the star from the orbit of the companion. This tidal interaction also affects the orbital parameters of the companion, and this equilibrium situation is dynamical: the companion is slowly falling toward the star and the proximity of the two bodies enhances all tidal effects, including the acceleration of the stellar rotation. The companion may eventually fall on the star and quickly transfer its whole angular momentum to the star's rotation. From this moment on, the stellar rotation resumes its evolution as in a single star. This resetting is important because, if the rules of gyrochronology are later applied to this star seeking to determine its age, they will result into values much smaller than the actual age of the star. In general, the rules of gyrochronology used to determine the age of single stars (see Barnes 2007, Brown 2014) are not valid for one star having (or having had) a close-in massive companion.

In this communication, we present the results of some calculations of the evolution of the rotation of stars hosting massive close-in companions taking into account the joint effects of tides and magnetic braking and we discuss the limits beyond which the star's rotation evolution is independent of the existence of a planet around it. The experiments reported here were designed to enhance the influence of the tidal star-planet interaction on the rotation of the host stars and were limited to massive planets and to the so-called coplanar model in which the stellar rotation axis is perpendicular to the planetary orbital plane. However, the same model can be used in more extended studies dealing, for instance, with the distribution of the stellar rotational periods vs. orbital periods, which are currently available for a great number of KEPLER objects (McQuillan et al. 2013; Teitler and Königl, 2014).

In the experiments, we have used Ferraz-Mello's creep tide theory (Ferraz-Mello, 2013, 2015) to compute the tidal evolution of the system and the transfer of angular momentum from the orbit of the companion to the star's rotation (or vice-versa in the initial stages, when the star is spinning much faster than the companion's orbital motion), and the Bouvier et al.s formulas (Bouvier et al. 1997) for the angular momentum leakage due to the stellar wind, which reproduces fairly well the Skumanich law and the observed rotations of the stars in open clusters of several ages.

We have considered one Jupiter-like planet, initially in orbit around an active star of mass 1.1 $M_\odot$, the upper limit of the range of validity of Bouvier et al.s formulas for the leakage of angular momentum. In each experiment, we considered 4 initial (current) semi-major axes: 0.03, 0.04, 0.05, 0.06 AU (orbital periods equal to 1.81, 2.79, 3.89 and 5.12 days, resp.) and 3 values of the tidal relaxation factor: 20, 40 and 80 s^{-1}. These values of the relaxation factor are the same used in the experiments reported in Ferraz-Mello et al. (2015) and are in the range of values determined there by comparing the expected evolution of the stellar rotation under the joint effect of tides and braking to its observed rotational period, for several host G stars of known isochronal age. These relaxation factors correspond to the quality factor Q in the interval $1 - 6 \times 10^6$ (if $k_2 = 0.2$), which is inside the range of dissipation values determined by Hansen's (2012) from the analysis of the survival of short-period planets and of the limits for circularization of planets with highly eccentric orbits. They correspond to a less dissipative scenario than the one adopted by Jackson et al. (2008) in earlier studies.

We choose to show here the results of the experiments for one host star whose initial (current) rotation period is 10 days. These results show the same main features as the results from experiments with stars endowed of faster or slower rotation. First, it is shown that the transfer of angular momentum from the orbit always happen when $a \leqslant 0.04$ AU. When $a = 0.03$ AU, the companion falls on the star in less than 2 Gyr. The transfer of angular momentum of the companion's orbit to the star accelerates its rotation and the period eventually decreases to much less than 10 days (see fig. 1). The falls are not always clearly visible in fig. 1, but one may find then at the times corresponding to the end of the decrease of the stellar rotation period. At the end of the life of the companion, the stellar rotation is strongly accelerated. When $a = 0.04$ AU, the behavior is similar, but the remaining lifetime of the companion is much longer. When $a \geqslant 0.05$ AU, the transfer of angular momentum from the orbit to the star is negligible. The corresponding curves are the envelope of the evolution tracks shown in fig. 1: the uppermost curve for $t > 0$, and lowest for $t < 0$. For more massive planets these limits are larger. For instance, for a 4 $M_{\rm Jup}$-planet, the transfer is significant up to $a = 0.06$ AU.

The evolution tracks shown in fig. 1 for $t < 0$ allow us to have an estimate of the age of the star. For instance, a star with the same mass (1.1 $M_\odot$) and current orbital period (10 d) and with a very close companion in circular orbit, may be expected to have an age a little less than 1 Gyr. If the companion is in an eccentric orbit, the estimated ages are spread over a larger interval. For instance, when the initial (current) eccentricity is $e = 0.2$, the age of a star with the above fixed characteristics may be $\sim$1.5 Gyr when

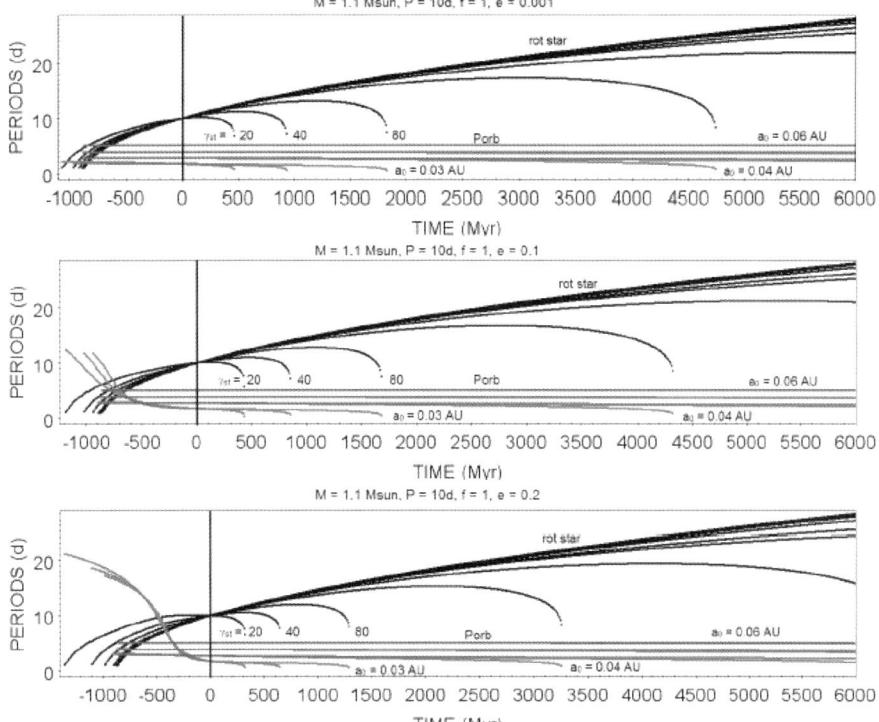

Figure 1. Evolution of the orbital period of one hot Jupiter (red) and the corresponding evolution of the rotational period of its active host star (black). Masses: 1 M_{Jup} and 1.1 $M_\odot$ resp. Initial (current) semi-major axes: 0.03 to 0.06 AU. Initial (current) eccentricities 0 (top), 0.1 (mid) and 0.2 (bottom). Initial (current) rotational period of the star: 10 days. Relaxation factors $\gamma = 20, 40$ and 80 s^{-1}.

$a = 0.03$ if the star is very dissipative ($\gamma = 20$ s^{-1}). If the planet is located farther away, say $a = 0.04 - 0.05$ AU, the estimate is again close to 1 Gyr.

The other curves shown in fig. 1 correspond to the orbital period of the planet which, in the conditions of the experiments done, is always decreasing. This means that, in the past, the companion is expected to have been moving in a more distant orbit. The figures shown for $t < 0$ may be considered as possible past evolutionary tracks of the system. However, these tracks are critically dependent on the adopted initial (current) parameters, mainly the orbit eccentricity and the stellar rotational period. The actual evolution for a given star may have been very different.

Similar calculations were done for several values of the planetary mass showing the variation of the limit value from $a \sim 0.04$ for a hot Jupiter with mass 0.5 M_{Jup} to $a \sim 0.07$ AU for super Jupiters with mass $4M_{Jup}$.

When the considered stars are not active, the behavior is much simpler because the stellar rotation drifts towards values close to the orbital period of the companion. Since these stars are generally more massive, the simulations in the absence of magnetic braking considered a host star of 1.3 $M_\odot$. Again, the initial (current) semi-major axes were fixed at 0.03, 0.04, 0.05, 0.06 AU (orbital periods equal to 1.66, 2.56, 3.58 and 4.71 days, resp.). Various values for the initial stellar rotational period were adopted. As before, the effects were visible only for initial semi-major axes $a \leqslant 0.04$ AU. For $a \geqslant 0.05$ AU, the stellar rotational period remains almost constant. When $a \leqslant 0.04$, the planet transfers

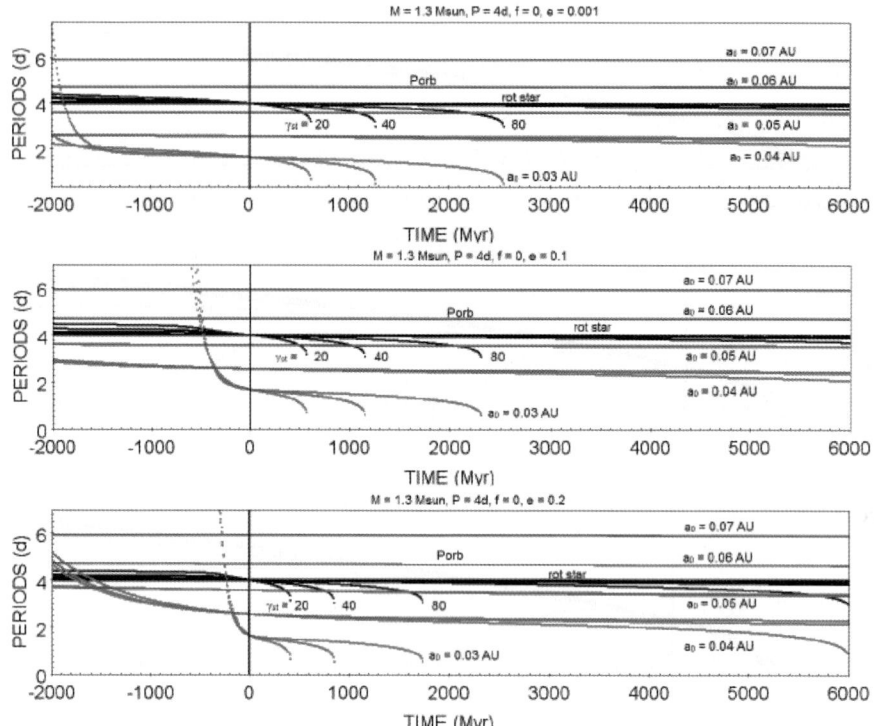

Figure 2. Same as fig. 1 but in the case of a non-active host star with mass 1.3 M$_\odot$ and initial (current) rotational period 4 days.

angular momentum to the rotation of the star, accelerating it. The star's rotation period decreases monotonically. The companion eventually falls on the star.

Fig. 2 shows the evolution of the orbital period of the hot Jupiter and the rotational period of a non active star whose current period is 4 days. Larger periods are not shown. In fact, the rotations measured by the space telescope KEPLER of the stars in the clusters NGC 6811 and NGC 6815 (Meibom et al., 2011, 2015) show that stars in this mass range have their rotational periods limited to a few days, even for ages as large as 2.5 Gyr.

We have used here the same tidal relaxation values used for solar-like stars (20, 40 and 80 s^{-1}) but it is worth recalling that dissipation in F stars is not as well constrained as in G stars and larger values of γ may be possible (Barker and Ogilvie, 2009). The technique used to determine the tidal relaxation factor in active stars with $m \leqslant 1.1$ M$_\odot$ cannot be used for the non-active stars because all well observed F stars with massive close-in companions have the rotation period almost equal to the orbital period, within the standard errors of their determinations, making impossible an estimation of the past evolution of the stellar rotation for different relaxation factors. The only exception is CoRoT-3 (Ferraz-Mello, in prep.). In this case the faster stellar rotation and the strong interaction between the star and the companion, a brown dwarf, allowed us to constrain the stellar tidal relaxation factor to the interval $40 - 70$ s^{-1}, which corresponds, for this star, to $Q \sim 6 - 9 \times 10^6$ (for $k_2 = 0.2$). This result agrees with Barker and Ogilvie (2009) statement that in lower mass F stars, the dissipation is similar but slightly weaker than that of solar-type stars.

Finally, for the sake of completeness, fig. 3 shows the variation of the eccentricity in the two non-circular cases of fig. 1 (In the cases of fig.2, the eccentricity variation is similar).

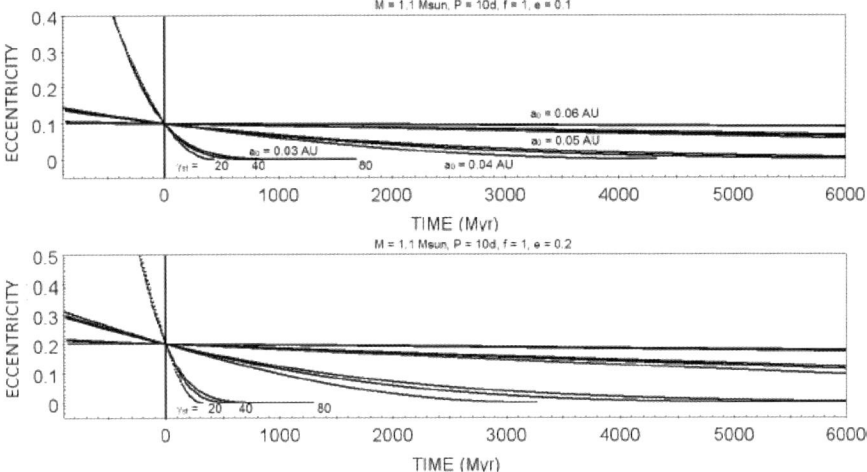

Figure 3. Evolution of the orbital eccentricity of the hot Jupiter in the solutions shown in fig. 1 for the initial (current) eccentricities 0.1 (top) and 0.2 (bottom).

This figure shows the tendency to circularization of all cases, which is as more important as the planet is closer to the star. The projection to the past when $a = 0.03$ leads to very high eccentricities indicating that we should not expect to find eccentric systems in the case of too close planets. When the current system is already circularized, it is not possible to make an estimation of the past evolution of the eccentricity, because it is not possible to know when the planet reached the circular orbit.

Conclusion

We reported in this communication some results of an extended series of calculations of the tidal evolution of planetary systems, in which the companion is a massive close-in planet. The results extend similar ones published in Ferraz-Mello et al. (2015) and allow to evaluate the limits beyond which the consideration of tidal effects cannot be neglected since they influence both the evolution of the planet orbit and the host star rotation. For systems with Jupiter-like planets, the rules of gyrochronology used for single stars can only be applied when the semi-major axis of the Jupiter-like planet is larger than a given limit. The reported simulations indicate that this limit grows from $a \sim 0.04$ to $a \sim 0.07$ AU as the mass of the planet grows from 0.5 to 4 $M_{\rm Jup}$. Comparing these values with the data of the known exoplanets, we see that about 200 of the currently known transiting hot Jupiters and brown dwarfs may affect significantly the rotation of their host stars.

Acknowledgement

This investigation was supported by grants CNPq 306146/2010-0, FAPESP 2014/13407-4 and 2015/15154-9 and by INCT Inespaço procs. FAPESP 2008/57866-1 and CNPq 574004/2008-4.

References

Barker, A. J. & Ogilvie, G. I. 2009. *MNRAS* **395**, 2268
Barnes, S. A. 2007, *ApJ*, **669** id. 1167
Bouvier, J., Forestini, M., & Allain, S. 1997 *A&A*, **326**, 1023
Bouvier, J. 2013, *EAS Publ. Series* Vol. **62**, 143
Brown, D. J. A. 2014, *MNRAS* **442**, 1844

Ferraz-Mello, S. 2013, *Celest. Mech. Dyn. Astr.* **116**, 109-140 (arXiv: 1204.3957)
Ferraz-Mello, S. 2015, *Celest. Mech. Dyn. Astr.* **122**, 359-389 (arXiv: 1505.05384)
Ferraz-Mello, S., Santos, M., Folonier, H., Csizmadia, S., Nascimento Jr, J. D. D., & Pätzold, M. 2015 *ApJ* **807**, id. 78 (arXiv: 1503.04369).
Hansen, B. M. S. 2012. *ApJ* **757**: 6
Jackson, B., Greenberg, R., & Barnes, R. 2008. *ApJ* **678**, 1396
McQuillan, A., Mazeh, T., & Aigrain, S. 2013. *ApJ Letters*, **775**, L11.
Meibom, S., Barnes, S. A., Latham, D. W. *et al.* 2011, *ApJ Letters*, **733**, L9
Meibom, S., Barnes, S. A., Platais, I. *et al.* 2015 *Nature* **517**, 589
Skumanich, A. 1972, *ApJ*, **171**, 565
Teitler, S. & Königl, A., 2014. *ApJ* **786**: 139

Search for Close-in Planets around Evolved Stars with Phase-curve variations and Radial Velocity Measurements

Teruyuki Hirano[1], Bun'ei Sato[1], Kento Masuda[2], Othman Michel Benomar[2], Yoichi Takeda[3], Masashi Omiya[3] and Hiroki Harakawa[3]

[1] Tokyo Institute of Technology,
2-12-1 Ookayama, Meguro-ku, Tokyo 152-8551, Japan
email: hirano@geo.titech.ac.jp

[2] The University of Tokyo

[3] National Astronomical Observatory

Abstract. Tidal interactions are a key process to understand the evolution history of close-in exoplanets. But tidals still have a large uncertainty in their prediction for the damping timescales of stellar obliquity and semi-major axis. We have worked on a search for transiting giant planets around evolved stars, for which few close-in planets were discovered. It has been reported that evolved stars lack close-in planets, which is often attributed to the tidal evolution and/or engulfment of close-in planets by the hosts. Meanwhile, *Kepler* has detected a certain fraction of transiting planet candidates around evolved stars. Confirming the planetary nature for these candidates is especially important since the comparison between the occurrence rates of close-in planets around main sequence stars and evolved stars provides a unique opportunity to discuss the final stage of close-in planets. With the aim of confirming KOI planet candidates around evolved stars, we measured precision radial velocities (RVs) for evolved stars with transiting planet candidates using Subaru/HDS. We also developed a new code which simultaneously models and fits the observed RVs and phase-curve variations in the Kepler data (e.g., transits, stellar ellipsoidal variations, and planet emission/reflected light). As a result of applying the global fit to KOI giants/subgiants, we confirmed two giant planets around evolved stars (Kepler-91 and KOI-1894), as well as revealed that KOI-977 is more likely a false positive.

Keywords. planets and satellites: individual (KOI-977, KOI-1894, KOI-2133)

1. Introduction

It has been reported that evolved stars lack close-in planets, which is often attributed to the tidal evolution and/or engulfment of close-in planets by the hosts. In the past year, we have worked on a search for giant planets around evolved stars, for which few close-in planets were discovered. We resort to both radial velocity (RV) measurements and analysis of "phase-curve variations" for planetary candidates detected by *Kepler*. Phase-curve variations are comprised of Doppler boosting, ellipsoidal variations, and reflected/emission light from the companion. From a precise phase-folded light-curve, we can extract the system parameters such as companion's mass and scaled semi-major axis, etc. Among the three effects above, ellipsoidal variation is particularly important for close-in planets around evolved stars, since its amplitude is approximately proportional to the cube of stellar radius. Combining the analysis of phase-curves (including transits) with an RV measurement would enable a more precise determination of system parameters. We apply this technique to three KOI (Kepler Object of Interest) systems below.

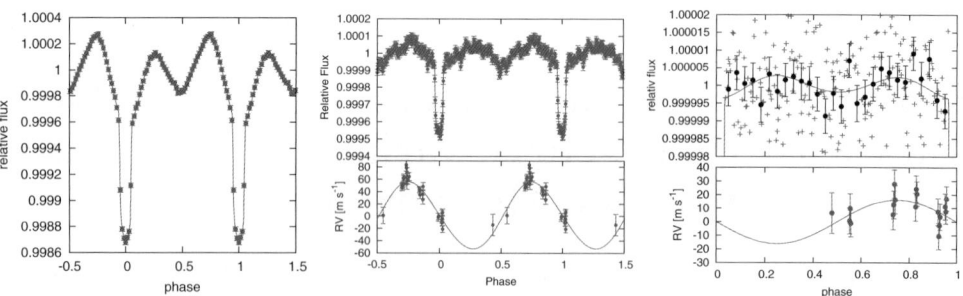

Figure 1. Phase-curve for KOI-977

Figure 2. Phase-curve for Kepler-91

Figure 3. Phase-curve for KOI-1894

2. Analyses and Results

KOI-977: False Positive. We first focused on KOI-977, which is reported to be a red giant in the Kepler Input Catalog (KIC), with a close-in ($P = 1.35$ days) giant planet candidate. In order to confirm the planetary nature of KOI-977.01, we performed 1) RV measurements, 3) asteroseismic analysis, and 4) phase-curve analysis of Kepler data. Consequently, though the folded light-curve exhibits a clear pattern of phase-curve variations, our stellar analyses by spectroscopy and asteroseismology suggest that the estimated radius is too large for a star with a close-in planet; the estimated semi-major axis of the planet candidate KOI-977.01 is ~ 0.027 AU, which is smaller than the stellar radius ($R_\star \sim 0.093$ AU). Combining this fact with the small RV variation revealed by the Subaru/HDS observation, KOI-977.01 is very likely a false positive, meaning that the transit-like signal is caused by an eclipsing binary, which is not identical to the red giant for which we measured stellar radius and RVs. On the assumption that KOI-977 is comprised of a red giant and an eclipsing binary, we searched for a solution to the observed phase-curve. Figure 1 shows the observed phase-curve (red) and its best-fit model (blue) by our MCMC fit. The resulting mass and radius ratios of the eclipsing binary range between $0.11 < M_1/M_2 < 0.28$ and $0.13 < R_1/R_2 < 0.17$, respectively. These values correspond to those of an F star and an early M star (Hirano et al. 2015).

Kepler-91 and KOI-1894: Real Planets. Two other systems were targeted by our search for giant planets around evolved stars: Kepler-91 (KOI-2133) and KOI-1894 (Kepler-91b was reported to be a real planet by Lillo-box et al. 2013 before our publication). These two stars have similar properties (e.g., $R_\star = 6.3 R_\odot$ for Kepler-91 and $R_\star = 8.6 R_\odot$ for KOI-1894) and similar planetary candidates (jovian planets with $P = 6.2$ days and $P = 5.3$ days, respectively). In order to gain the system parameters as accurately as possible, we combined the RV data by Subaru/HDS and Kepler public light-curve and performed a global analysis. The results of the fits confirmed the planetary nature for both candidates; The best-fit models as plotted in Figures 2 and 3 suggest that Kepler-91b is a sub-Jupiter mass planet ($M_p = 0.64 \pm 0.05 M_J$) while KOI-1894b is more likely a super-Neptune planet ($M_p = 0.14 \pm 0.08 M_J$) (Sato et al. 2015).

References

Hirano, T., Masuda, K., Sato, B., Benomar, O., Takeda, Y., Omiya, M., Harakawa, H., & Kobayashi, A. 2015, *ApJ*, 799, 9

Lillo-Box, J., Barrado, D., Moya, A., Montesinos, B., Montalban, J., Bayo, A., Barbieri, M., Regulo, C., Mancini, L., Bouy, H., & Henning, T. 2014, *A&A*, 562, A109

Sato, B., Hirano, T., Omiya, M., Harakawa, H., Kobayashi, A., Hasegawa, R., Takarada, T., Kawauchi, K., & Masuda, K. 2015, *ApJ*, 802, 57

Dynamical Effects of Stellar Companions

Smadar Naoz

Department of Physics and Astronomy, University of California, Los Angeles, CA 90095, USA
email: snaoz@astro.ucla.edu

Abstract. The fraction of stellar binaries in the field is extremely high (about $40\% - 70\%$ for $M > 1$ M$_\odot$ stars), and thus, given this frequency, a high fraction of all exoplanetary systems may reside in binaries. While close-in giant planets tend to be found preferentially in binary stellar systems it seems that the frequency of giant planets in close binaries ($< 100 - 1000$ AU) is significantly lower than in the overall population. Stellar companions gravitational perturbations may significantly alter the planetary orbits around their partner on secular timescales. They can drive planets to large eccentric orbits which can either result in plunging these planets into the star or shrinking their orbits and forming short period planets. These planets typically are misaligned with the parent star.

Keywords. (stars:) planetary systems, (stars:) binaries

1. Introduction

Most, if not all, stars are born in binaries or higher multiples and many of them stay in a binary configuration throughout their stellar lifetime ($\sim 40 - 70\%$ for $\gtrsim 1$ M$_\odot$, e.g., Raghavan *et al.* 2010). Thus, it seems likely that a stellar companion will significantly influence the planet formation and evolution in such a system. Furthermore, observations suggest that short period giant planets are more likely to have a far away companion (Knutson *et al.* 2014;Ngo *et al.* 2015;Wang *et al.* 2015). While on the other hand, some studies suggest that the frequency of giant planets in close binaries ($< 100 - 1000$ AU) is significantly lower than in the overall population (Wang *et al.* 2015), possibly due to a truncation or influence from the stellar companion.

The general dynamical behavior of multi-planetary systems in stellar binaries can be exceedingly complex and the many body interactions may exhibit highly irregular behavior. Even when in the regime dynamical interactions can be treated secularly, (i.e., orbital average, long term interaction) the system's dynamical evolution can be very complicated. In particular, apsidal precession arising from planet–planet gravitational interactions is in competition with binary companion forced evolution to either suppress or facilitate eccentricity excitations (e.g., Innanen *et al.* 1997;Takeda *et al.* 2008;Farr & Naoz in Prep.). In light of this phenomenological richness, here we begin by examining the simplest of the unsolved problems - namely, a setup wherein a single planet is perturbed by a binary companion. As will be discussed below, even within the context of this seemingly trivial configuration, complex dynamical evolution associated with pumping of extremely high eccentricities can arise, potentially leading to the production of close-in planetary systems that exhibit spin-orbit misalignments.

In the three-body approximation, dynamical stability requires a hierarchical configuration, in which the *inner binary* - in our case the primary star and a planet, is orbited by a third stellar companion on a much wider orbit (although we focus on a companion star, it can also be a planet or a brown dwarf). In the latter case, proximity to low-order mean-motion resonances becomes unlikely, the secular approximation (i.e., phase averaged, long term evolution) can be applied. The gravitational potential is then expanded

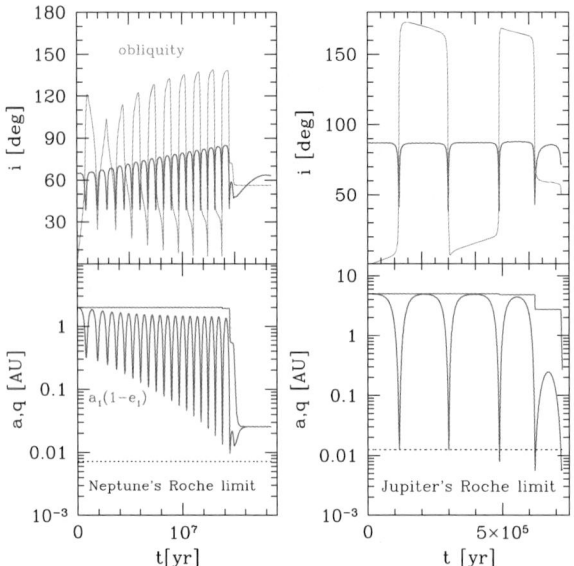

Figure 1. Effects of a companion: Tidal disruption of a planet (right panel) and circularization and shrinking the orbits due to tides (left panel). The systems' mutual inclination (red line), and obliquity (magenta lines) are depicted in the top panels, while the bottom panels shows the semi-major axes (red lines) and pericenter distances (blue lines) in AU. Also shown in dashed lines are the pericenter at which tidal disruption takes place. *Left panels* show a Neptune-size planet around an M dwarf star (0.32 $M_\odot$) where a Neptune mass planet is initialized at $a_1 = 2$ AU, and zero eccentricity. The companion here is a brown dwarf (10 M_j) at 50 AU with eccentricity of 0.52. The mutual inclination is set initially to be $65°$, and the argument of periapsis of both orbits is set to zero initially. *Right panels* show a Jupiter mass planet around a sun like star, set initially on a circular orbit at a 5 AU with zero eccentricity. The companion here is a sun like star at at 200 AU with 0.75 eccentricity. The mutual inclination is set initially to be $87°$, and the argument of periapsis of both orbits is set to zero initially.

in semi-major axis ratio, a_1/a_2 (Kozai 1962;Lidov 1962), where a_1 is the semi-major axis of the inner orbit and a_2 is the semi-major axis of the outer orbit. This ratio is a small parameter due to the hierarchical configuration. The lowest order of approximation, which is proportional to $(a_1/a_2)^2$ is called the quadrupole–level of approximation.

In early studies of high-inclination secular perturbations, the outer orbit was assumed to be circular and in addition, it was assumed that one of the inner binary members is a massless test particle (Kozai 1962;Lidov 1962). In this situation, the quadrupole–level of approximation, is valid (Naoz et al. 2011,2013a). The resulting Hamiltonian admits a constant of motion that corresponds to the component of the inner orbit's angular momentum along the z-axis (which is parallel to the total angular momentum). Naoz et al. (2011,2013a) showed that relaxing either one of these assumptions, i.e., allowing for an eccentric outer perturber, or going beyond the test particle approximation, leads to qualitative different behavior. This case requires the next level of approximation, called the octupole–level of approximation (Harrington 1968,1969;Ford et al. 2000;Blaes et al. 2002), which is proportional to $(a_1/a_2)^3$. Accordingly, recent developments in this field have shown that these systems have rich and exciting dynamics (e.g., Naoz et al. 2011,2013a;Lithwick& Naoz 2011;Katz et al. 2011;Li et al. 2014a).

In the octupole– level of approximation, initially moderate variations in orbital eccentricity can be driven to extremely high values (Ford et al. 2000;Naoz et al. 2013a;Teyssandier et al. 2013;Li et al. 2014a). In addition, the planet's inclination with respect to the

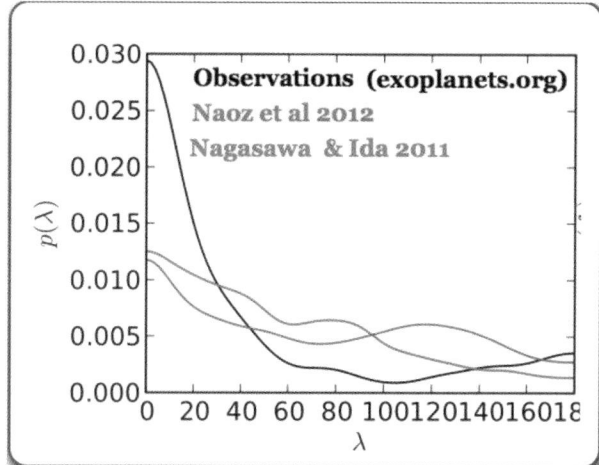

Figure 2. Projected Hot Jupiters obliquity We consider the observed distribution (updated as for 2012, black line) *exoplanets.org*, and the projected obliquity from Naoz *et al.* (2012) Monte-Carlo EKL simulations (green line) and planet–planet scattering adopting Nagasawa and Ida 2007, simulations. Bayesian analysis suggests that $\sim 30\%$ of the population is consistent with EKL formation mechanism. Furthermore, focusing on the misaligned population, the EKL mechanism seems to contributes about 60% to 80%. Figure adopted from Naoz *et al.* (2012).

total angular momentum can flip from prograde ($< 90°$) to retrograde ($> 90°$) (Naoz *et al.*2011,2013a). This process was coined as the *Eccentric Kozai–Lidov* (EKL) mechanism, and it was shown that it can be used as a great tool to understand different astrophysical settings.

2. Gravitational Perturbations from a Companion on the Orbital Configuration of Giant Planets

As already mentioned above, during the system evolution, the EKL mechanism can induce large eccentricity excitations on the planet's orbit. herefore, on one hand, the small pericenter can lead the planet to tidally disrupt. On the other hand, the tidal forces can shrink and circularize the planets orbit (see Figure 1 left and right panels, respectively). Significant fraction of Jupiters that are initially set at 5 AU stay in their birth place, and undergo moderate eccentricity and inclination oscillations. A small fraction of the systems (which depends on the tidally disrupt distance, e.g., Petrovich 2015) survive as Hot Jupiters.

Hot Jupiters formed via gravitational perturbations from a companion star have an interesting consequence: they are likely to be misaligned with their parent star. This was first predicted by Fabrycky and Tremaine (2007) and Wu *et al.* (2007), for a quadrupole–level calculation, including GR and tides. The consequence of using the quadrupole–level is that only a narrow part of the initial configurations is accessible for hot Jupiter generation. In particular, they found that Hot Jupiters are more likely to form in systems orbited by a nearly perpendicular stellar companion (e.g., Fabrycky and Tremaine 2007). Also, for the quadrupole–level to be applicable the companions orbits must be circular (Naoz *et al.* 2013a).

The EKL mechanism can tap into larger parts of the parameter space which may result in an exciting behavior such as large eccentricity excitations and flipping of the Jupiter orbit with respect to the total angular momentum (e.g., Naoz *et al.* 2011,2013a;Teyssandier

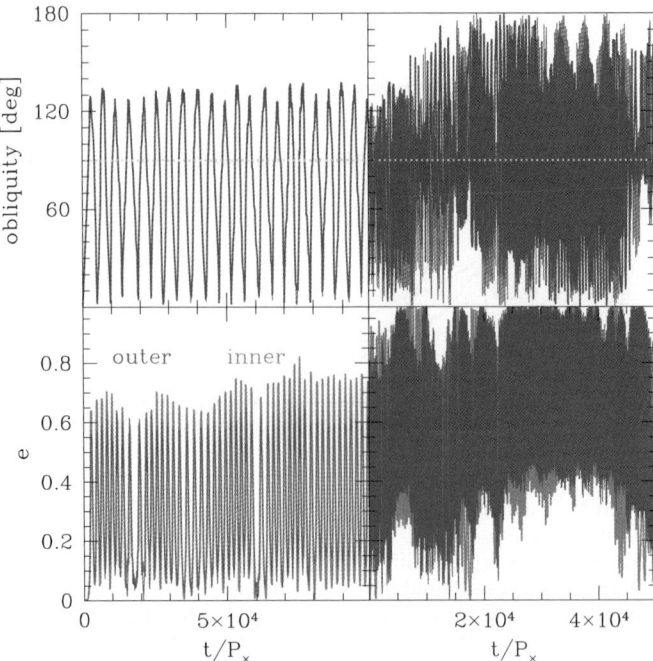

Figure 3. Eccentricity and obliquity evolution for two planet system in the presence of a companion. The left hand panels shows a system for which the outer planet separation is $0.021 \times a_c$, where a_c is the separation of the companion. The companion eccentricity here is 0.3. The system exhibit stable eccentricity oscillations. The right panels describes a system for which the outer planet separation is $0.06 \times a_c$. The companion eccentricity here is 0.5. In both cases the inner planet is located at half of the distance of the outer plant, and the mass ratio is $5 \times 10^{-5}, 3 \times 10^{-5}, 1$ (inner and outer planets to companion, respectively).

et al. 2013;Li et al. 2014a). A Monte-Carlo calculation extended to the EKL mechanism (as well as GR† and tides) showed that indeed the final (and initial) orbital inclinations and obliquity cover large range of the parameter space (Naoz et al. (2012) and Petrovich (2015), and see Figure 2 for the projected obliquity distribution). Furthermore, Naoz et al. (2012) conducted a bayesian analysis assuming that the obliquity distribution is a contribution of three processes, the first is an unknown process which results in an aligned component to the distribution, the second is an EKL mechanism and lastly, a planet–planet scattering process (adopting Nagasawa and Ida 2007). This analysis suggested that a large fraction of the observed high obliquity Hot Jupiter is consistent in forming via the EKL mechanism‡. In particular is seems that about 60% to 80% of systems with large obliquities are consisted with EKL mechanism formation scenario. Out of the entire observed population (which has a large aligned component), EKL is consistent with $\sim 30\%$ of the observed systems while planet–planet scattering contributes about $\sim 10\% - 20\%$. These results are independent on the formation rates and are based only on the obliquity distribution. The values are consistent with observations that suggest that Hot Jupiters are likely to have a far away companion (e.g., Knutson et al. 2014;Ngo et al. 2015;Wang et al. 2015). However, we note that Ngo et al. 2015 did not find a

† General relativity plays an important role, as it can in some cases suppress eccentricity excitations and in others excite them (e.g., Naoz et al. 2013b).
‡ Note that Storch and Lai (2014) and Storch et al. (2015) showed, in the frame work of hierarchical triple systems, that the behavior of the obliquity angle is chaotic and is very sensitive to the stellar spin, assumed here to be 25 d.

correlation with the misaligned systems. Thus, further investigation requires to explain this discrepancy.

3. Going Beyond the Three Body System

As briefly alluded to in the introduction, further phenomenological richness arises when planet- planet interactions are introduced into the picture. We investigate the secular interactions of two planet system in the presence of a stellar companion, using a multi-body secular gravitational interaction code. We use the Gauss's averaging method which is a phase-averaged calculation for which the gravitational interactions between non-resonant orbits are treated as being equivalent to angular momentum exchange among massive wires, whose line density in inversely proportional to the orbital velocity (see Touma *et al.* 2009). The strength of this approach (and thus the code) is that it is not limited to low inclinations and eccentricities or hierarchical configurations. We note however, that the secular approximation breaks if the planets are within few Hill radii of each other, or if they are in mean motion resonances (e.g., Takeda *et al.*2008). In general this method can be used to explore different many body secular effects, for example the evolution of a particle disk in the presence of a perturber (Batygin 2012).

Figure 3 shows two examples of two planet system evolution in the presence of a stellar companion, considering point mass Newtonian dynamics. These examples have different initial conditions as specified in the caption. The example depicted on the left panels shows a regular time evolution, that exhibits mutual eccentricity oscillations between the orbits. The gravitational forces between the two planets cause the planets angular momenta to precess together around the total angular momentum. Thus, the obliquity angle of each planet (defined here as a tilt angle compare to the initial two planets angular momenta) simply precesses around the inner two angular momenta (known as precession of the nodes), the latter precess around the total angular momentum. Similar behavior was noted in Innanen *et al.* (1997), and it can eventually produce high obliquity planets (e.g., Li *et al.* 2014b). The right panels on Figure 3 shows a chaotic evolution for both the eccentricity and the tilt angle of both planets. The two planets react independently to the outer companion gravitational perturbations. Here the system probably becomes unstable. These two examples demonstrate how relatively mild configurational changes in the initial setup of the system can lead to qualitatively different dynamical evolutions.

Acknowledgements

We thank Konstantin Batygin for thoroughly reading and refereeing the text. S.N. acknowledges partial support from a Sloan Foundation Fellowship.

References

Batygin, K., 2012, A primordial origin for misalignments between stellar spin axes and planetary orbits *Nature*, **491**, 418–420.
Blaes, O., M. H. Lee, & A. Socrates, 2002, The Kozai Mechanism and the Evolution of Binary Supermassive Black Holes. *ApJ*, **578**, 775–786.
Ford, E. B., B. Kozinsky, & F. A. Rasio, 2000, Secular Evolution of Hierarchical Triple Star Systems. *ApJ*, **535**, 385–401.
Harrington, R. S., 1968, Dynamical evolution of triple stars. *AJ*, **73**, 190–194.
—, 1969, The Stellar Three-Body Problem. *Celestial Mechanics*, **1**, 200–209.
Innanen, K. A., J. Q. Zheng, S. Mikkola, & M. J. Valtonen, 1997, The Kozai Mechanism and the Stability of Planetary Orbits in Binary Star Systems. *AJ*, **113**, 1915.

Knutson, H. A., B. J. Fulton, B. T. Montet, M. Kao, H. Ngo, A. W. Howard, J. R. Crepp, S. Hinkley, G. Á. Bakos, K. Batygin, J. A. Johnson, T. D. Morton, & P. S. Muirhead, 2014, Friends of Hot Jupiters. I. A Radial Velocity Search for Massive, Long-period Companions to Close-in Gas Giant Planets. *ApJ*, **785**, 126.

Katz, B., S. Dong, & R. Malhotra, 2011, Long-Term Cycling of Kozai-Lidov Cycles: Extreme Eccentricities and Inclinations Excited by a Distant Eccentric Perturber. *PRL* 107, 18.

Kozai, Y., 1962, Secular perturbations of asteroids with high inclination and eccentricity. *AJ*, **67**, 591.

Lidov, M. L., 1962, The evolution of orbits of artificial satellites of planets under the action of gravitational perturbations of external bodies. *planss*, **9**, 719–759.

Nagasawa, M. and S. Ida, 2011, Orbital Distributions of Close-in Planets and Distant Planets Formed by Scattering and Dynamical Tides. *ApJ*, **742**, 72.

Li, G., S. Naoz, B. Kocsis, & A. Loeb, 2014a, Eccentricity Growth and Orbit Flip in Near-coplanar Hierarchical Three-body Systems. *ApJ*, **785**, 116.

Li, G., S. Naoz, F. Valsecchi, J. A. Johnson, & F. A. Rasio, 2014b, The Dynamics of the Multi-planet System Orbiting Kepler-56. *ApJ* **794**, 131.

Lithwick, Y. & S. Naoz, 2011, The Eccentric Kozai Mechanism for a Test Particle. *ApJ*, **742**, 94.

Naoz, S., 2016, The Eccentric Kozai-Lidov Effect and Its Applications. *ARAA*, submitted.

Naoz, S. & D. C. Fabrycky, 2014, Mergers and Obliquities in Stellar Triples. *ApJ*, **793**, 137.

Naoz, S., W. M. Farr, Y. Lithwick, F. A. Rasio, & J. Teyssandier, 2011, Hot Jupiters from secular planet-planet interactions. *Nature*, **473**, 187–189.

—, 2013a, Secular dynamics in hierarchical three-body systems. *MNRAS*, **431**, 2155–2171. .

Naoz, S., B. Kocsis, A. Loeb, & N. Yunes, 2013b, Resonant Post-Newtonian Eccentricity Excitation in Hierarchical Three-body Systems. *ApJ*, **773**, 187.

Naoz, S., W. M. Farr, & F. A. Rasio, 2012, On the Formation of Hot Jupiters in Stellar Binaries. *ApJ*, **754**, L36.

Ngo, H., H. A. Knutson, S. Hinkley, J. R. Crepp, E. B. Bechter, K. Batygin, A. W. Howard, J. A. Johnson, T. D. Morton, & P. S. Muirhead, 2015, Friends of Hot Jupiters. II. No Correspondence between Hot-jupiter Spin-Orbit Misalignment and the Incidence of Directly Imaged Stellar Companions. *ApJ*, **800**, 138.

Petrovich, C. 2015, Steady-state Planet Migration by the Kozai-Lidov Mechanism in Stellar Binaries. *ApJ*, **799**, 27.

Raghavan, D., H. A. McAlister, T. J. Henry, D. W. Latham, G. W. Marcy, B. D. Mason, D. R. Gies, R. J. White, & T. A. ten Brummelaar, 2010, A Survey of Stellar Families: Multiplicity of Solar-type Stars. *ApJS*, **190**, 1–42.

Storch, N. I., K. R. Anderson, & D. Lai, 2014, Chaotic dynamics of stellar spin in binaries and the production of misaligned hot Jupiters. *Science*, **345**, 1317–1321.

Storch, N. I. & D. Lai, 2015, Chaotic dynamics of stellar spin driven by planets undergoing Lidov-Kozai oscillations: resonances and origin of chaos. *MNRAS*, **448**, 1821–1834.

Takeda, G., R. Kita, & F. A. Rasio, 2008, Planetary Systems in Binaries. I. Dynamical Classification. *ApJ*, **683**, 1063–1075.

Teyssandier, J., S. Naoz, I. Lizarraga, & F. A. Rasio, 2013, Extreme Orbital Evolution from Hierarchical Secular Coupling of Two Giant Planets. *ApJ*, **779**, 166.

Wang, J., D. A. Fischer, E. P. Horch, & J.-W. Xie, 2015, Influence of Stellar Multiplicity On Planet Formation. III. Adaptive Optics Imaging of Kepler Stars With Gas Giant Planets. *ApJ*, **806**, 248.

Spin-orbit alignment of exoplanet systems: how can Asteroseismology help us?

Tiago L. Campante[1,2]

[1] School of Physics and Astronomy, University of Birmingham,
Edgbaston, Birmingham, B15 2TT, UK
email: campante@bison.ph.bham.ac.uk

[2] Stellar Astrophysics Centre (SAC), Department of Physics and Astronomy,
Aarhus University, Ny Munkegade 120, DK-8000 Aarhus C, Denmark

Abstract. Measuring the obliquities of exoplanet-host stars provides invaluable diagnostic information for theories of planetary formation and migration. Most of these results have so far been obtained by measuring the Rossiter–McLaughlin effect, clearly favoring systems that harbor hot Jupiters. While it would be extremely helpful to extend these measurements to long-period and multiple-planet systems, it is also true that the latter systems tend to involve smaller planets, making it ever so difficult to apply such techniques. Asteroseismology provides a powerful method of determining the inclination of the stellar spin axis from an analysis of the rotationally-induced splittings of the oscillation modes. This provides an estimate of the obliquity independently of other methods. The applicability of the asteroseismic method is determined by the stellar properties and not by the signal-to-noise ratio of the transit data. Here we present a recap of the spin-orbit geometry, explain how the asteroseismic method works, and review previous applications of the method to exoplanet-host stars.

Keywords. asteroseismology, methods: statistical, planetary systems, planets and satellites: general, stars: solar-type, techniques: photometric

---

1. Introduction

The spin-orbit angle ψ has been recognized as an important diagnostic of theories of planet formation, migration, and tidal evolution. For practical reasons, most obliquity measurements to date have been for systems harboring hot Jupiters (e.g., Albrecht et al. 2012). In order to study the dynamical histories of planetary systems across a wider range of architectures, it is, however, imperative to extend these measurements to systems with smaller planets, longer-period planets, and multiple planets. An alternative technique for measuring the obliquities of planetary systems, one that does not depend on the signal-to-noise ratio (S/N) of the transit data, is asteroseismology. The asteroseismic estimation of the stellar inclination angle rests on our ability to resolve and extract signatures of rotation in the power spectra of non-radial modes of oscillation. Its applicability depends entirely on the stellar properties, namely, on the intrinsic ratio $\nu_{\rm s}/\Gamma$ between the rotational splitting and the full width at half maximum (or linewidth) of the oscillation mode profiles, and not on the planetary or orbital parameters, which is a clear advantage when measuring the obliquities of systems with small planets or long-period planets.

2. Spin-orbit geometry

In Fig. 1, z points toward the observer and the xy plane is the plane of the sky. Moreover, $\mathbf{n}_{\rm o}$ and $\mathbf{n}_{\rm s}$ denote the orbital and stellar angular momentum vectors, respectively.

For a transiting planet, the orbital inclination i_o is measurable via transit photometry, while the projected spin-orbit angle λ is measurable via the Rossiter–McLaughlin effect or the analysis of planetary transits over starspots. In addition, the stellar inclination i_s can in principle be constrained using asteroseismology. Nevertheless, we should keep in mind that the only angle of intrinsic physical significance is the true obliquity or spin-orbit angle, ψ, between $\mathbf{n}_o$ and $\mathbf{n}_s$, which is generally not directly measurable.

The various angles are related according to (Fabrycky & Winn 2009):

$$\cos\psi = \sin i_s \cos\lambda \sin i_o + \cos i_s \cos i_o \,. \tag{2.1}$$

For a transiting system, we would still expect to place mild constraints on ψ even when lacking a measurement of λ. In Fig. 2 we show the posterior probability distribution for ψ conditioned on i_s and i_o, $p(\psi|i_s,i_o)$, sampled by means of Monte Carlo simulations using Eq. (2.1). We have assumed an edge-on orbit ($i_o=90°$), having selected three error-free values for the stellar inclination ($i_s=30°$, $i_s=60°$, and $i_s=85°$). A uniform distribution in λ was used to express our ignorance with respect to this quantity. We have also derived a potentially useful analytical expression for $p(\psi|i_s,i_o)$ (T. L. Campante et al., submitted),

$$p(\psi|i_s,i_o) \propto \frac{\sin i_s \sin\psi}{\sqrt{\sin^2\psi \sin^2 i_o - (\cos\psi \cos i_o - \cos i_s)^2}}, \quad \text{for all } |i_o - i_s| < \psi < i_o + i_s, \tag{2.2}$$

that we plot in Fig. 2. The main conclusions to be drawn from this simple exercise follow immediately from an inspection of Fig. 2: For a transiting system, a small value of i_s implies a spin-orbit misalignment. The converse is not true, since a large value of i_s is consistent with, but does not necessarily imply, a spin-orbit alignment. For instance, if we consider $i_s=85°$, then any spin-orbit angle in the range $5°<\psi<175°$ has a nonnegligible probability. The interpretation of the spin-orbit alignment in terms of the measured i_s can thus be ambiguous. Therefore, in order to draw general inferences about spin-orbit alignment, a statistical analysis of an ensemble of such measurements is needed (Fabrycky & Winn 2009; Morton & Winn 2014).

3. Estimation of the stellar inclination angle via asteroseismology

Solar-like oscillations are predominantly acoustic global standing waves. Commonly called p modes, owing to the fact that pressure plays the role of the restoring force, these modes are intrinsically damped and stochastically excited by near-surface convection. Consequently, all stars cool enough to have an outer convective envelope may be expected to exhibit solar-like oscillations. The oscillation modes are characterized by the radial order n (related to the number of radial nodes), the spherical degree l (specifying the number of nodal surface lines), and the azimuthal order m (with $|m|$ specifying how many of the nodal surface lines cross the equator). Radial modes have $l=0$, whereas non-radial modes have $l>0$. Values of m range from $-l$ to l, meaning that there are $2l+1$ azimuthal components for a given multiplet of degree l. Observed oscillation modes are typically high-order modes of low spherical degree.

The asteroseismic estimation of i_s rests on our ability to resolve and extract signatures of rotation in the power spectra of non-radial modes of oscillation. Rotation lifts the degeneracy in the frequencies, ν_{nl}, of non-radial modes and introduces a dependence of the oscillation frequencies on m, with prograde (retrograde) modes (with $m>0$ and $m<0$, respectively) having frequencies slightly higher (lower) than the axisymmetric mode ($m=0$) in the observer's frame of reference. For the fairly modest values of the stellar angular velocity Ω that are typical of solar-like oscillators, the effect of rotation can

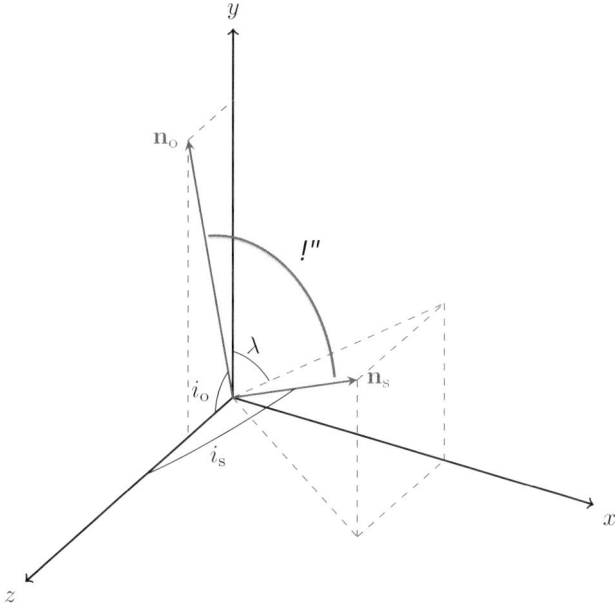

Figure 1. Observer-oriented coordinate system. Here the z axis points toward the observer, the x axis points along the line of nodes, the y axis completes a right-handed triad, and the xy plane is the plane of the sky. The unit vectors $\mathbf{n}_\mathrm{o}$ and $\mathbf{n}_\mathrm{s}$ denote the orbital and stellar angular momentum unit vectors, respectively. All depicted angles are introduced in the text.

be treated following a perturbative analysis and the star is generally assumed to rotate as a solid body (i.e., Ω=const.). In the limit of solid-body rotation, the frequency ν_{nlm} of a mode, as observed in an inertial frame, can be expressed to first order as (Ledoux 1951):

$$\nu_{nlm} = \nu_{nl0} + m\frac{\Omega}{2\pi}(1 - C_{nl}) \approx \nu_{nl0} + m\nu_\mathrm{s} \,. \tag{3.1}$$

The kinematic splitting $m\Omega/(2\pi)$ is corrected for the effect of the Coriolis force through the dimensionless Ledoux constant, C_{nl}. For high-order p modes, $C_{nl} \ll 1$, with the rotational splitting being dominated by advection and given approximately by the angular velocity, i.e., $\nu_\mathrm{s} \approx \Omega/(2\pi)$. To a second order of approximation, centrifugal effects that disrupt the equilibrium structure of the star can be taken into account through an additional frequency perturbation (e.g., Ballot 2010). This perturbation scales as the ratio of the centrifugal to the gravitational forces at the stellar surface, i.e., $\Omega_\mathrm{surf}^2 R_\mathrm{s}^3/(GM_\mathrm{s})$, where G is the gravitational constant, R_s is the stellar radius, and M_s is the stellar mass.

Assuming energy equipartition between multiplet components with different azimuthal order, the dependence of mode power on m is given by (e.g., Gizon & Solanki 2003):

$$\mathcal{E}_{lm}(i_\mathrm{s}) = \frac{(l-|m|)!}{(l+|m|)!} \left[P_l^{|m|}(\cos i_\mathrm{s}) \right]^2 \,, \tag{3.2}$$

where $P_l^m(x)$ are the associated Legendre functions and the sum over m of $\mathcal{E}_{lm}(i_\mathrm{s})$ has been normalized to unity. Measurement of the relative power of the azimuthal components in a non-radial multiplet thus provides a direct estimate of the stellar inclination angle. The above formalism further relies on the assumption that contributions to the observed intensity across the visible stellar disk depend only on the angular distance from the disk center, which is valid for photometric observations. According to Eq. (3.2), when the

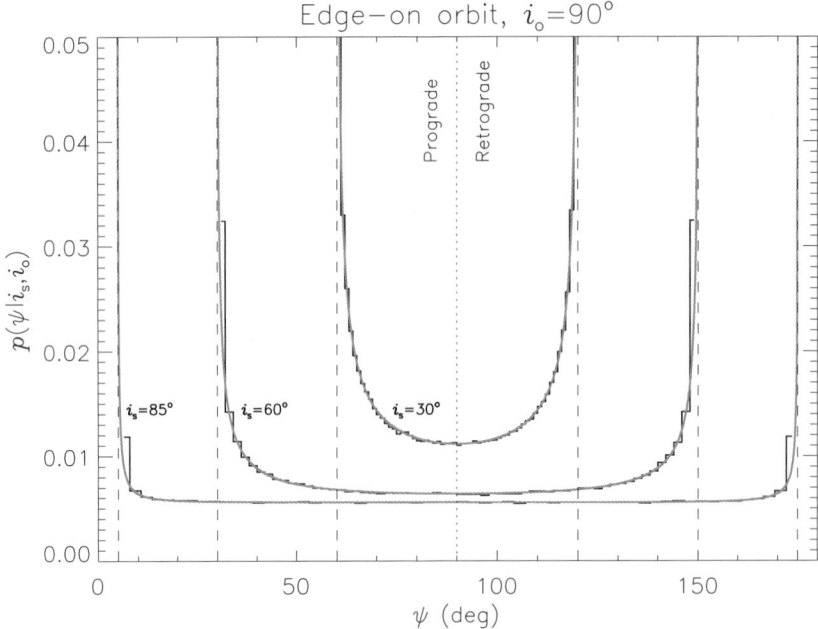

Figure 2. Posterior probability distribution for the spin-orbit angle ψ conditioned on $i_{\rm s}$ and $i_{\rm o}$, $p(\psi|i_{\rm s},i_{\rm o})$. The posterior distributions have been sampled by means of Monte Carlo simulations using Eq. (2.1) and are displayed as histograms. The analytical expression for $p(\psi|i_{\rm s},i_{\rm o})$ given in Eq. (2.2) is plotted (after normalization) in gray. The vertical dashed lines are placed at the asymptotes $\psi=|i_{\rm o}-i_{\rm s}|$ and $\psi=i_{\rm o}+i_{\rm s}$. The vertical dotted line at $\psi=\pi/2$ marks the transition between a prograde and a retrograde orbit.

stellar spin axis points toward the observer (pole-on configuration), only the axisymmetric mode is visible and no inference can thus be made about rotation. When the spin axis lies on the plane of the sky (edge-on configuration), as is approximately the case of the Sun, observations are essentially sensitive only to modes with even $|l-m|$. Figure 3 displays the theoretical profiles of dipole ($l=1$) modes as a function of the angle $i_{\rm s}$ for three different $\nu_{\rm s}/\Gamma$ ratios. Dipole modes are approximately three times more prominent in the power spectra of intensity observations than quadrupole ($l=2$) modes of similar frequency, and consequently it is the former modes that ultimately determine our ability to constrain $i_{\rm s}$. It is evident from Fig. 3 that, given sufficient frequency resolution and S/N, it will be the intrinsic ratio $\nu_{\rm s}/\Gamma$ which determines whether it is possible to resolve the azimuthal components (Ballot, García & Lambert 2006).

A detailed fitting of the modes of oscillation to extract signatures of rotation from the power spectrum is normally done in a Bayesian manner using an MCMC (Markov chain Monte Carlo; e.g., Handberg & Campante 2011) or similar sampler. Analytical approximations for the errors on $i_{\rm s}$ and $\nu_{\rm s}$ are given in Ballot et al. (2008). Bayesian inference, however, is capable of providing us with the full posterior probability distribution (ppd) of these parameters, from which credible regions can be conveniently defined. We note that even though $i_{\rm s}$ is independent of $\nu_{\rm s}$, their measured values are highly correlated. The ppd of $i_{\rm s}$ can then be used to sample that of the spin-orbit angle ψ via Eq. (2.1).

Broad mode profiles (characteristic of late F-type stars) hinder our ability to resolve and extract signatures of rotation in the power spectrum. When coupled to a moderate-to-low S/N in the p modes, this means that the effect of the correlated background noise on the mode profiles may bias the outcome of the detailed fitting analysis. A

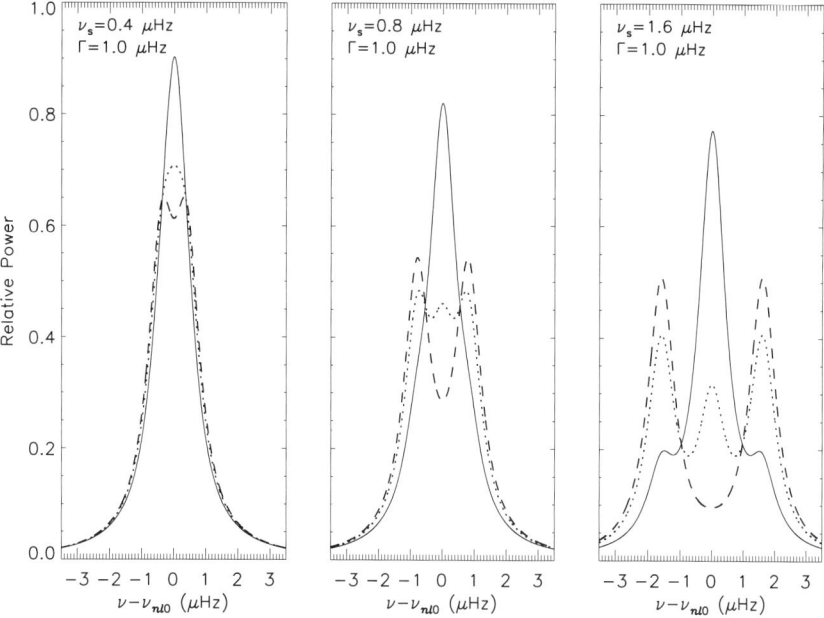

Figure 3. Theoretical profiles of dipole multiplets as a function of $i_{\rm s}$ for three different $\nu_{\rm s}/\Gamma$ ratios. The linewidths of the mode profiles of the azimuthal components have been assigned solar values ($\Gamma = 1.0\,\mu{\rm Hz}$). The rotational splitting ranges from solar ($\nu_{\rm s} = 0.4\,\mu{\rm Hz}$; left-hand panel) to twice (middle panel) and four times (right-hand panel) solar. The three stellar inclination angles are represented by different line styles: $i_{\rm s} = 30°$ (solid), $i_{\rm s} = 60°$ (dotted), and $i_{\rm s} = 85°$ (dashed).

reduced visibility of the multiplet components is an issue that is frequently encountered. In order to test the asteroseismic method, and in particular the robustness of the returned uncertainties on $i_{\rm s}$, we recommend that in such situations tests with artificial data be performed.

4. Applications

Following the application of the asteroseismic technique to a few host stars with single, non-transiting large planets discovered using the radial-velocity method (Wright et al. 2011; Gizon et al. 2013), the asteroseismic technique has recently been applied to several Sun-like hosts observed with NASA's *Kepler* space telescope (Chaplin et al. 2013; Benomar et al. 2014; Lund et al. 2014; Van Eylen et al. 2014). In addition, Huber et al. (2013) used asteroseismology to measure a large obliquity for Kepler-56, a red giant hosting two transiting coplanar planets, thus showing that spin-orbit misalignments are not confined to hot-Jupiter systems. Another instance of an asteroseismic obliquity measurement of an evolved host star is that of Kepler-432 (Quinn et al. 2015), for which the obliquity of one of its two long-period giant planets may have been shaped by the same process that realigns hot-Jupiter systems. Recently, the stellar inclination angles of the solar analogs 16 Cyg A and B were determined using asteroseismology by Davies et al. (2015). The B component hosts a Jovian planet in an eccentric, low-obliquity orbit, which is consistent with Kozai cycling driven by the A component. The first statistical analysis of an ensemble of asteroseismic obliquity measurements obtained for solar-type stars hosting transiting planets was recently conducted (T. L. Campante et al., submitted).

References

Albrecht, S., Winn, J. N., Johnson, J. A., et al. 2012, *ApJ*, 757, 18
Ballot, J., García, R. A., & Lambert, P. 2006, *MNRAS*, 369, 1281
Ballot, J., Appourchaux, T., Toutain, T., & Guittet, M. 2008, *A&A*, 486, 867
Ballot, J. 2010, *Astronomische Nachrichten*, 331, 933
Benomar, O., Masuda, K., Shibahashi, H., & Suto, Y. 2014, *PASJ*, 66, 94
Chaplin, W. J., Sanchis-Ojeda, R., Campante, T. L., et al. 2013, *ApJ*, 766, 101
Davies, G. R., Chaplin, W. J., Farr, W. M., et al. 2015, *MNRAS*, 446, 2959
Fabrycky, D. C. & Winn, J. N. 2009, *ApJ*, 696, 1230
Gizon, L. & Solanki, S. K. 2003, *ApJ*, 589, 1009
Gizon, L., Ballot, J., Michel, E., et al. 2013, *Proc. of the National Academy of Science*, 110, 13267
Handberg, R. & Campante, T. L. 2011, *A&A*, 527, A56
Huber, D., Carter, J. A., Barbieri, M., et al. 2013, *Science*, 342, 331
Ledoux, P. 1951, *ApJ*, 114, 373
Lund, M. N., Lundkvist, M., Silva Aguirre, V., et al. 2014, *A&A*, 570, A54
Morton, T. D. & Winn, J. N. 2014, *ApJ*, 796, 47
Quinn, S. N., White, T. R., Latham, D. W., et al. 2015, *ApJ*, 803, 49
Van Eylen, V., Lund, M. N., Silva Aguirre, V., et al. 2014, *ApJ*, 782, 14
Wright, D. J., Chené, A.-N., De Cat, P., et al. 2011, *ApJ (Letters)*, 728, L20

FM2:
Astronomical Heritage: Progressing the UNESCO-IAU Initiative

Focus Meeting 2, "Astronomical Heritage: Progressing the UNESCO–IAU Initiative" Introduction and overview

Clive Ruggles[1] & Anna Sidorenko[2]

[1] School of Archaeology and Ancient History
University of Leicester, Leicester LE1 7RH, United Kingdom
email: rug@le.ac.uk

[2] UNESCO World Heritage Centre
7, Place de Fontenoy, 75352 Paris CEDEX 07, France
email: a.sidorenko@unesco.org

Abstract. Marking seven years of formal cooperation between the IAU and the UNESCO World Heritage Centre to implement UNESCO's "Astronomy and World Heritage" Thematic Initiative, this Focus Meeting reviewed achievements, challenges, and progress on particular World Heritage List nomination projects.

Keywords. Astronomical heritage, World Heritage

1. Introduction

Since 2008 the International Astronomical Union has worked with the UNESCO World Heritage Centre to implement its "Astronomy and World Heritage" Thematic Initiative (whc.unesco.org/en/astronomy/). Through deliverables such as the *ICOMOS–IAU Thematic Study on the Heritage Sites of Astronomy and Archaeoastronomy* (Ruggles & Cotte 2010) and the Portal to the Heritage of Astronomy (www.astronomicalheritage.net), it has been influential in developing broad criteria for assessing the heritage values, and ultimately the potential Outstanding Universal Value (needed for inclusion on the World Heritage List), of cultural sites of all ages relating to astronomy. Since 2012, representatives of the IAU have begun to work directly with State Parties to help develop particular potential nominations for inscription onto the World Heritage List.

This Focus Meeting set out to review achievements to date, discuss some of the most challenging issues in the assessment of different types and categories of astronomical heritage, and evaluate progress in projects focusing on particular potential nominations.

Highlights included Teasel Muir Harmony and David DeVorkin's panel discussion on "The Development of Mauna Kea as an Astronomical Site"; an hour-long presentation on Polynesian archaeoastronomy by archaeologist Patrick V. Kirch; and two very productive sessions focusing on the preservation of dark skies in a World Heritage context, organised in collaboration with FM21 on "Mitigating Threats of Light Pollution and Radio Frequency Interference".

A taster event held at the Bishop Museum in Honolulu on August 9, two days before the start of the Focus Meeting itself, featured the announcement that the AURA Observatory in Chile has become the world's first IDA Dark Sky Sanctuary, and the launch of a revised edition of the book *Nā Inoa Hōkū*, long regarded as a definitive source of reference for anyone studying the use of astronomy in Polynesian voyaging, by Hawaiian authors Rubellite Johnson and John Mahelona working in collaboration with Clive Ruggles.

PDF versions of presentations in this Focus Meeting, as available, can be found on the FM2 page on the UNESCO–IAU Portal to the Heritage of Astronomy (www.astronomical heritage.net/index.php/community/news-events/focus-meeting-at-iau-general-assembly). Those in the joint sessions are also available on the FM21 pages on the NOAO website, www.noao.edu/education/IAUGA2015FM21.

2. Sessions and topics

2.1. *The implementation of the Astronomy and World Heritage Initiative: achievements, issues and prospects*

This session, intended as an introductory briefing for all participants, featured invited reviews by Anna Sidorenko, Clive Ruggles and Michel Cotte on behalf of UNESCO, the IAU, and UNESCO's advisory body ICOMOS, respectively.

2.2. *The potential for archaeoastronomical World Heritage sites*

Given increasing interest by various governments in nominating ancient sites connected with astronomy for inscription on the World Heritage List, the second session addressed relevant issues such as credibility (Clive Ruggles) and serial nomination (Juan Belmonte *et al.*). Morgan Saletta pointed out that the astronomical significance may derive from illumination hierophanies rather than observations of celestial targets. A second paper by Juan Belmonte (not included below) addressed the problem of dealing with popular "fringe" theories concerning astronomy that pollute sensible interpretations, even at existing World Heritage sites: this is a particular problem in the case of Egyptian pyramids.

2.3. *Recognizing the twentieth-century heritage of astronomy*

Following a provocative historical perspective by Virginia Trimble, papers by Christina Barboza and James Hesser *et al.* provided case studies of 20th-century observatories in Brazil and Canada respectively, and how they are being, and might be, dealt with as heritage sites. The session concluded with an hour-long panel session organised by David DeVorkin and Teasel Muir Harmony in which John Jefferies and former colleagues shared their recollections and views relating to the history of the establishment of Mauna Kea Observatory, which forms part of the "Windows to the Universe" project (see below).

2.4. *World heritage and the protection of working observatory sites*

Dark skies cannot of themselves be recognised under the World Heritage Convention. However, light pollution not only affects night sky quality but also affects the integrity of other resources and, indeed, whole ecosystems. It is also linked to the issue of energy waste through lighting. These factors affect the sustainable management of both cultural and natural sites, including existing and potential World Heritage Sites.

This—the first of two joint sessions with FM21 ("Mitigating Threats of Light Pollution and Radio Frequency Interference") dealing with preserving dark skies and protecting against light pollution in a World Heritage framework—focused upon working observatory sites and their dark skies.

The first half of the session featured presentations by several of the key participants in the "Windows to the Universe" project, one of the main nomination projects being advanced within the Astronomy and World Heritage Initiative. This is being led by Chile, but with potential partnership from Spain, the United States and France. In addition to the papers by M. Smith *et al.* and C. Smith *et al.* included below, there was an important presentation by Gabriel Rodríguez, from the Energy, Science & Technology and Innovation Direction, Ministry of Foreign Affairs of Chile, outlining Chile's plans to continue to attract and facilitate the installation of international radio and optical observation projects and to protect their exceptional skies. These plans include various initiatives connected with UNESCO's World Heritage programme.

A number of presentations described efforts to protect observatories from light pollution and/or radio frequency interference, in Chile (Pedro Sanhueza, not included below), Hawai'i (Richard Wainscoat, not included below), Arizona and the Canary Islands (Richard Green, not included below), and South Africa (Ramotholo Sefako).

Finally, Rémi Cabanac and Michel Cotte outlined a practical approach to the recognition of Dark Sky places as possible World Heritage sites using the Pic du Midi Observatory as a case study (not included below).

A fuller report on this session will be found in the FM21 pages in this volume.

2.5. *Preserving dark skies and protecting against light pollution in a World Heritage framework*

This, the second of the joint sessions with FM21, shifted the focus away from cultural sites (modern observatories) onto natural sites and landscapes with dark skies. Following a stunning presentation of nightscape photography by Babak Tafreshi (not included below), Michel Cotte described a way in which the "Windows to the Universe" concept can be elaborated in a heritage context. John Hearnshaw, Dan Duriscoe and Arkadiusz Berlicki then reported respectively on night sky preservation issues at the Aoraki-Mackenzie International Dark Sky Reserve in New Zealand, National Parks in the United States, and the Izera Dark Sky Park in Poland and the Czech Republic (none included below).

A fuller report on this session will be found in the FM21 pages in this volume.

2.6. *Observatories, observations and archives: scientific, historical and heritage issues*

This session focused upon the challenge of balancing different categories of astronomical heritage and in particular how moveable and intangible heritage can best be taken into account when considering the heritage value of fixed places. It also highlighted two other important potential nomination projects in progress: the "Route of astronomical observatories" project, which is concerned with classical observatories from the Renaissance to the rise of astrophysics (overview by Gudrun Wolfschmidt), and the "Odyssey of human creative genius" project, which is concerned with scientific and technological heritage related to space exploration (overview by Olga Dluzhnevskaya and Mikhail Marov). The remaining papers by Elizabeth Griffin and Areg Mickaelian *et al.* concerned the digitization of plate archives—moveable items of continuing scientific as well as heritage value whose adequate preservation presents a range of challenges.

2.7. *Hawaiian and Polynesian cultural heritage relating to astronomy*

The centrepiece of this session was a presentation by archaeologist Patrick V. Kirch (not included below) on recent applications of archaeoastronomy to the interpretation of prehistoric monuments in Polynesia. Drawing upon case studies of temple sites in Mangareva and Hawai'i, he demonstrated that Polynesianritual architecture frequently exhibits regular patterns of orientation, including alignments upon astronomical phenomena such as the solstices and the rising position of the Pleiades. He concluded that Polynesian temples were not only places of offering and sacrifice to the gods but also locations for formal astronomical observations that were crucial for keeping the lunar calendar synchronized with the solar year. Clive Ruggles *et al.* followed with a report on the publication of the revised edition of *Nā Inoa Hōkū* and its implications for the recognition and preservation of Hawaiian star knowledge. This incomparable intangible heritage relating to Polynesian navigation also informs and motivates living cultural traditions.

2.8. Dealing with movable and intangible heritage in a World Heritage framework

The final batch of case studies relating to moveable and intangible heritage featured an invited paper by Alejandro López on the potential pitfalls as well as the advantages of recognising living practices as "cultural heritage". The intangible heritage of living cultural practices also featured in a presentation on native D(L)akota skywatchers in the United States by Annette Lee (not included below), as well as in that on Gufa, a Nepalese cultural ritual, by Pritisha Shrestha, and in the broad comparative study by Sona Famanyan *et al.*

The concluding paper by Rosa Ros and Beatriz García describes how cultural practices relating to astronomy can form an integral part of astronomy outreach activities and suggested important ways of connecting education and outreach on the one hand and history and heritage on the other.

3. Outcomes

In view of several examples of astronomical heritage in danger highlighted at the meeting, such as the prehistoric dolmens of Jordan (presentation by Belmonte *et al.*), an important recommendation to emerge from the general discussions was that the new Commission C4 on World Heritage and Astronomy should—in addition to three Working Groups already proposed to advance the "Windows to the Universe", "Route of astronomical observatories" and "Odyssey of human creative genius" nomination projects—aim to establish a Working Group on Astronomical Heritage in Danger, which would (unlike UNESCO itself) be able to identify and publicise such cases whether or not the site(s) in question were already inscribed on the World Heritage List.

In view of the importance of preserving dark skies at places whose heritage value, and potential OUV, is largely or exclusively natural rather than cultural, it was noted in discussions that Commission C4 needs to establish firm links and seek cooperative projects with the IUCN in the same way as it does with ICOMOS.

The strong focus upon intangible heritage in the form of living cultural heritage, both in Hawai'i and elsewhere, led to the question of whether it might be appropriate to recognise some living cultural practices relating to astronomy under UNESCO's Convention for the Safeguarding of Intangible Cultural Heritage (www.unesco.org/new/en/santiago/culture/intangible-heritage/convention-intangible-cultural-heritage/). This would represent a new avenue for the AWHI, and Michel Cotte considered this in some detail at a follow-up meeting on Hawaiian, Oceanic and Global Cultural Astronomy held in Hilo in the week following the GA (www.astronomicalheritage.net/index.php/community/news-events/cultural-astronomy-meeting-big-island), suggesting Polynesian navigation as a possible case study. On the other hand, Alejandro López has raised a number of critical issues in the very conception of intangible heritage. His observation that, in various ways, the very concept of heritage is defined within, and constrained by, a Western mental framework presents a real issue if it is to be reconciled with the "universal" aspect of OUV (meaning that something should be recognised as valuable by all cultures).

The meeting concluded by identifying a number of important links between heritage, history, outreach and education activities within Division C.

References

Ruggles, C. & Cotte, M. (eds.) 2010, *Heritage Sites of Astronomy and Archaeoastronomy in the context of the UNESCO World Heritage Convention* (Paris: ICOMOS –IAU)

The UNESCO Thematic Initiative "Astronomy and World Heritage"

Anna Sidorenko

Coordinator, Thematic Initiative "Astronomy and World Heritage"
UNESCO World Heritage Centre
7 Place de Fontenoy, 75352 Paris CEDEX 07, France
email: a.sidorenko@unesco.org

Abstract. My presentation is divided into two parts: the first part retraces chronologically all the main achievements accomplished within the framework of this Thematic Initiative; the second provides key information regarding the nomination process.

Keywords. Astronomical heritage, World Heritage, UNESCO

Dear colleagues,
It is a great pleasure for me to take part in this Focus Meeting dedicated to astronomical heritage. I would like to extend my sincere gratitude to the International Astronomical Union for having organized this important meeting in close coordination with the World Heritage Centre.

1. Development and achievements of the Initiative

At the present time, outstanding properties related to astronomy are still under-represented on the World Heritage List.

Created in 2003 within the framework of the Global Strategy for a balanced, representative and credible World Heritage List, the Thematic Initiative on Astronomy and World Heritage aims to establish a link between Science and Culture towards recognition of monuments and sites connected with astronomical observations.

The Initiative is being implemented on the basis of the following orientations:
• Definition of major issues in the identification and nomination of sites relating to astronomy;
• Definition of the priorities for the safeguarding and promotion of these sites in order to open new pathways for cooperation and to mobilize extra-budgetary funds and new partner support;
• Development of pilot projects at selected sites;
• Development of cooperation agreements between the States Parties aiming to promote these sites, through the elaboration of serial nominations, as well as to develop conservation and research activities focusing on heritage of science and technology;
• Creation of partnerships between scientists and other stakeholders involved in the process of identification, conservation and management of this specific category of properties.

This Initiative is implemented in conformity with the decisions of the World Heritage Committee chronologically presented in this presentation. Information is also available on the web page of the Initiative: http://whc.unesco.org/en/astronomy.

Over the years, UNESCO has developed very close collaboration and partnership with numerous States Parties and International Organizations aiming to promote astronomical sites.

In October 2008, UNESCO signed a Memorandum of Understanding with the International Astronomical Union as a result of which the IAU has become integrally involved in the process of advancing the initiative. One of the first actions of the IAU was to set up a Working Group on Astronomy and World Heritage.

In order to facilitate the identification and nomination process of astronomical sites, a cycle of activities "Astronomy and World Heritage: across time and continents" was launched by the Director-General of UNESCO in 2009 during the opening ceremony of the International Year of Astronomy, "The Universe, Yours to Discover".

Among the most important references, I would like to mention the Kazan Resolution adopted by the participants during an International Conference organized in 2009 with support of the Russian Federation.

A proposal regarding the definition of categories of Space Technological sites was included in this Resolution. As proposed by the participants, groups of Space Technological Sites could fall into the following categories:

1. Launch Pads, such as the human-engineered Gagarin Launch Pad in Baikonur Space Centre, together with related auxiliary facilities and roads;

2. Clearly defined landscapes designed and created intentionally by people and intrinsically associated with a given Launch Pad and the development of related structures, e.g. the Baikonur area;

3. Prominent parts of space networks specially designed and constructed to support manned flights, such as Gagarin Star City in the Moscow region with its facilities for cosmonauts' pre-flight training, education, and the verification of space operations using spacecraft mock-ups;

4. Historical sites where the pioneering concepts of space flight and the design of original space vehicles were tested, such as Kaluga town and the K.E. Tsiolkovsky Museum of Cosmonautics, with their related natural landscapes.

More recently, in 2014, the participants of an Associated Event on technological heritage connected with space exploration organised during the 40th COSPAR Scientific Assembly (Moscow, Russian Federation, 6 August 2014) agreed that the main issues that should be raised in conjunction with the identification of potential technological sites connected with space exploration are:

• Recognition of technological sites connected with space exploration at the national level as cultural heritage sites and adoption of appropriate legislative, regulatory and contractual measures for their protection;

• Specificities of the legal status of use of the technological installations recognized as cultural heritage sites;

• Obligations related to the protection, management and sustainable use of such sites and the rules of application of these obligations;

• Conservation Master Plan, including requirements and deadlines for completion of restoration, renovation and others works;

• Definition of requirements regarding the conditions of visitor access to the site;

• Other requirements necessary for the protection of the site within the framework of the World Heritage Convention (UNESCO 1972) and the Operational Guidelines for its implementation (UNESCO 2015).

The participants launched an appeal to the international community to actively contribute to the development and implementation of studies and research on space technological heritage towards recognition of the most representative sites by their inscription on the World Heritage List.

During the "Starlight Initiative"'s meeting in 2009 the Working Group "Starlight Reserves and World Heritage" developed the concept of a "Starlight Reserve". The ICOMOS–International Astronomical Union Global Thematic Study on Astronomical Heritage (Ruggles & Cotte 2010) includes a study on a "Starlight Reserve" proposal.

In 2009, during an International Workshop "Starlight Reserves and World Heritage", it was agreed to cooperate towards a potential serial nomination to cover outstanding examples of astronomical heritage and observation sites.

However, neither Starlight Reserves, nor Dark Sky Parks can be recognized by the World Heritage Committee as specific types or categories of World Heritage cultural and natural properties since no criteria exist for considering them under the World Heritage Convention.

Moreover, taking into account the growing number of requests to UNESCO concerning the recognition of the value of the dark night sky and celestial objects, the World Heritage Centre made its official statement underlining that the sky or the dark night sky or celestial objects or starlight as such cannot be nominated to the World Heritage List within the framework of the Convention concerning the Protection of the World Cultural and Natural Heritage.

In 2011, at its 35th session, the World Heritage Committee (Decision 35 COM 9C, UNESCO, Paris) encouraged States Parties to take into account the recommendations provided by the Science and Technology Expert Working Group in the context of World Heritage Nominations (London, 2008), as well as recommendations developed within the framework of the Thematic Initiative "Astronomy and World Heritage" while preparing nominations to the World Heritage List.

At its 36th Session (St. Petersburg, 2012), the World Heritage Committee welcomed financial and technical support provided by States Parties and the International Astronomical Union for the Thematic Initiative "Astronomy and World Heritage" since 2003 and also encouraged cooperation between the UNESCO World Heritage Centre, specialized agencies and relevant interdisciplinary scientific initiatives towards the elaboration of a Global Thematic Study on Heritage of Science and Technology, including studies and research on technological heritage connected with space exploration.

The World Heritage Centre has informed all States Parties that one of the activities of this Initiative will be the development of an upstream project proposal for the nomination of serial properties composed of astronomical sites.

In 2013, UNESCO and the IAU signed a new Memorandum of Understanding and reconfirmed their commitment to promote astronomical sites, as well as to provide States Parties to the World Heritage Convention with expertise, in preparing nominations for inscription on the World Heritage List of exceptional sites that bear witness to major breakthroughs in the development of scientific knowledge.

Identification of potential sites could be based on the ICOMOS–IAU Thematic Study on Astronomical Heritage endorsed by the World Heritage Committee. It provides clear guidance to State Parties on sites that have the best potential for inscription but also facilitate serial and/or transnational nominations which might better reflect the different values and periods. This study constitutes the background for a comparative analysis that could be carried out to assess the Outstanding Universal Value of a specific site of the same type proposed for World Heritage listing.

It identifies the main characteristics and astronomical values of the generic type of heritage site from a World Heritage perspective, examines a select number of representative examples included or not included in the World Heritage List, determines possible gaps in the latter and, with reference to the Operational Guidelines (UNESCO 2015), indicates the criteria under which such sites might be nominated for inscription on the World Heritage List.

The role of experts and specialists in the development of serial nomination projects is crucial in making progress in safeguarding this specific heritage.

I would like to inform you that Austria, Brazil, Canada, Côte d'Ivoire, Ecuador, Arab Republic of Egypt, French Republic, Germany, Republic of Mauritius, Mexico, Russian Federation, Kingdom of Thailand, Ukraine, and United States of America have provided updated information regarding the institutions and focal points in charge of this initiative at the national level.

The National Focal Points are expected to organize and participate in national workshops/seminars in order to prepare, in collaboration with the World Heritage Centre, the Advisory Bodies, the AWH Working Groups and the relevant national authorities, a national strategy aimed at protecting and promoting scientific and technological heritage and to contribute to the upstream process seeking to identify sites relating to astronomy of potential Outstanding Universal Value, to the establishment of collaboration with Specialized Agencies (Space Agencies, Institutes etc), as well as to the development of a Global Thematic Study on Heritage of Science and Technology, including heritage connected with space exploration, as requested by the World Heritage Committee at its 36th session (St. Petersburg, 2012).

2. Questionnaires

The World Heritage Centre has developed a Questionnaire for National Focal Points in charge of the implementation of the UNESCO Thematic Initiative "Astronomy and World Heritage", in order to obtain a clearer view on the impact of the Initiative on astronomical heritage conservation and identification in the participating States Parties. Hence, the States Parties of Germany, Egypt, the Russian Federation and the United States of America have provided the World Heritage Centre with concrete information about the implementation of the Initiative, and more specifically on awareness raising, nomination procedures of astronomical heritage sites, and the difficulties and challenges that arise around the preservation of these sites. Furthermore, the States Parties provided suggestions on how to better enhance the Thematic Initiative. These include the creation of a newsletter to keep Focal Points and interested parties updated on activities and discussions, the organization of multinational workshops, the initiation of a similar thematic study on space heritage, and the establishment of a Board for scientific and technological heritage.

The expected outputs are:
1. Provide participants with current status of the implementation of the Thematic Initiative "Astronomy and World Heritage";
2. Review status of national strategies aimed at protecting and promoting scientific and technological heritage in the States Parties by the Focal Points;
3. Outline of a strategic plan, including a time table and definition of responsibilities,

for development of an upstream project proposal for the nomination of serial transnational property composed of sites of potential Outstanding Universal Value which demonstrate the transition from classical astronomy to modern astrophysics;

4. Discuss framework and funding for development of a Global Thematic Study on Heritage of Science and Technology, including studies and research on technological heritage connected with space exploration;

5. Discuss specific issues focusing on mobilization of voluntary contributions, extra-budgetary funds and partner support for the implementation of the Thematic Initiative.

Some proposed questions for discussion are:
- Do the existing national legal and regulatory frameworks correspond to the needs for protection and promotion of heritage of science and technology?
- Has your country developed a mechanism aimed at protecting and promoting scientific and technological heritage?
- What steps have been taken by your authorities to integrate the identification, protection and promotion of heritage of science and technology into these national programmes / annual work plan and budget provision / strategies?
- How to effectively ensure broader participation of the interested parties and raise awareness on identification, protection and promotion of heritage of science and technology?
- Which approach has been/could be developed to establish/enhance international collaboration and co-operation among the States Parties?
- How to best ensure transnational, regional and sub-regional cooperation and networks?
- What types of collaboration and cooperation could be developed with Space Agencies with a view of undertaking joint activities within the framework of this Initiative?

3. Nomination process

I would like to briefly present the main issues relevant to the nomination process, as well as an upstream approach developed to accompany States Parties and to provide assistance throughout the whole World Heritage nomination process.

The objective of the upstream process is to explore creative approaches and new forms of guidance that might be provided to State Parties in considering nominations before their preparation, as well as in relation to the nomination process.

A Power Point reference presentation which describes also the main stages of the nomination process will be available on the web site of the Initiative (http://whc.unesco.org/en/astronomy).

The preparatory stages prior to launch the development of a potential draft nomination by the State(s) Party(ies) concerned could be:
- gathering of all the available information (legal/management/protection/use);
- creation of the expert team involving the representatives of all stakeholders concerned;
- identify the potential Outstanding Universal Value of the site;
- develop analysis and study of all elements exhibiting value, as well as their relationship to each other, which will determine the nature of the management and protection plan for the property;
- ensure that this is justified through a comparative analysis involving relevant facilities;

- make sure adequate protection, conservation and management is provided by the relevant authorities in conformity with the existing international normative instruments;
- develop a shared understanding of the nominated property and shared responsibility for its future.

It is desirable to carry out initial preparatory work to establish that a property has the potential to justify Outstanding Universal Value, including integrity or authenticity, before the development of a full nomination dossier which could be expensive and time-consuming. Such preparatory work might include collection of available information on the property, thematic studies, scoping studies of the potential for demonstrating Outstanding Universal Value, including integrity or authenticity, or an initial comparative study of the property in its wider global or regional context, including an analysis in the context of the Gap Studies produced by the Advisory Bodies. Such work will help to establish the feasibility of a possible nomination at an early stage and avoid use of resources on nominations that may be unlikely to succeed.

States Parties are invited to contact the World Heritage Centre and the Advisory Bodies at the earliest opportunity in considering nominations to seek information and guidance (UNESCO 2015, paragraph 122).

Inscribing a site on the World Heritage List is not the end of the story. Site managers and local authorities continuously work towards managing, monitoring and preserving the World Heritage properties. States Parties have an obligation to regularly prepare reports about the state of conservation and the various protection measures put in place at their sites.

4. Conclusion

This Focus Meeting contributes to the exchange of information and the establishment of new partnerships, with the goal of promoting and protecting technological heritage connected with space exploration and developing all necessary mechanisms to safeguard our common heritage.

Allow me to underline the importance to implement the decisions of the World Heritage Committee and to co-ordinate international activities aiming to promote and protect astronomical heritage. We count on your support and contribution for the success of the Thematic Initiative "Astronomy and World Heritage".

I thank you for your attention.

References

Ruggles, C. & Cotte, M. (eds.) 2010, *Heritage Sites of Astronomy and Archaeoastronomy in the context of the UNESCO World Heritage Convention* (Paris: ICOMOS –IAU)

UNESCO 1972, *Convention concerning the protection of the World Cultural and Natural Heritage* (Paris: UNESCO) http://whc.unesco.org/en/conventiontext

UNESCO 2015, *The Operational Guidelines for the Implementation of the World Heritage Convention* (Paris: UNESCO) http://whc.unesco.org/en/guidelines

The IAU's involvement in the Astronomy and World Heritage Initiative: achievements and challenges

Clive Ruggles

School of Archaeology and Ancient History
University of Leicester, Leicester LE1 7RH, United Kingdom
email: rug@le.ac.uk

Abstract. Since 2008 the IAU has worked with UNESCO and its advisory bodies to help recognise, promote and protect all types of astronomical heritage and to encourage nominations for World Heritage Sites relating to astronomy. I review the main challenges and achievements so far, and indicate how the Astronomy and World Heritage Initiative is likely to develop in the future.

Keywords. Astronomical heritage, World Heritage

1. Introduction

Within weeks of a formal Memorandum of Understanding (MoU) being signed between UNESCO and the IAU in 2008, under which the two organisations would work together to implement UNESCO's "Astronomy and World Heritage" Thematic Initiative (AWHI), the IAU's new WG on Astronomy and World Heritage discovered that it could not simply create a list of what it considered to be the world's most important astronomical heritage sites that might constitute potential World Heritage. Only national governments (State Parties to the World Heritage Convention) can nominate properties (places or landscapes) for inscription on the World Heritage List, and only UNESCO, with the aid of its advisory bodies ICOMOS (for cultural sites) and the IUCN (for natural sites) can decide whether the site concerned really does demonstrate the outstanding universal value (OUV) required for inscription.

Since that time, the IAU has established three main ways in which it can work together with UNESCO and its advisory bodies to help member states prepare credible nomination dossiers for sites with a connection to astronomy:

- Thematic studies;
- The Portal to the Heritage of Astronomy; and
- Direct involvement with State Parties as part of the "upstream process".

The MoU was renewed in 2013.

2. First ICOMOS–IAU Thematic Study

ICOMOS Thematic Studies aim to develop an overall vision of some aspect of cultural heritage, aided by a wide range of case studies. In doing so, they help to establish robust general principles according to which World Heritage List nominations relating to the type of heritage in question can be judged. In 2009 the IAU started to work with ICOMOS to produce a joint Thematic Study on the Heritage Sites of Astronomy and Archaeoastronomy. This was published electronically in June 2010 (Ruggles & Cotte 2010) and presented at the 34th session of UNESCO's World Heritage Committee (34COM) in Brasilia. A printed version followed in 2011.

Its scope was very wide, ranging from early prehistory to modern astrophysics and space heritage, including working observatory sites and dark-sky places. In view of the publication of a report on classical observatories a year earlier by ICOMOS–Germany and the University of Hamburg (Wolfschmidt 2009), it was not considered necessary to give special emphasis to classical observatories from the renaissance to the mid-20th century, which were included as just one among 14 other cultural heritage themes.

Among the key issues identified in the first Thematic Study were:

- the importance of different categories of heritage—moveable (e.g. objects and artefacts) and intangible (e.g. knowledge and ideas)—in assessing the heritage value of the fixed property (site or landscape) to which they are linked;
- the fact that from a science heritage standpoint, as opposed to an architectural heritage standpoint, alteration with time is inevitable and inherent and tends to add value, therefore requiring different approaches to authenticity and integrity; and
- the need to find ways to recognise the value of the dark night sky, given that dark sky sites cannot, in themselves, be recognised under the World Heritage Convention.

3. Portal to the Heritage of Astronomy

The "Portal to the Heritage of Astronomy" (www.astronomicalheritage.net) is a dynamic, publicly accessible database, discussion forum, and document-repository on astronomical heritage sites throughout the world. All contributions are moderated by an editorial group comprising representatives of UNESCO, ICOMOS and the IAU. The Portal was formally launched in 2009 during the IAU GA in Beijing.

The Portal is connected to the UNESCO World Heritage Centre's own website at various levels, including many two-way links between relevant sites on UNESCO's World Heritage List site (whc.unesco.org/en/list/). Publishing a Case Study on the portal is widely perceived as a useful first step in the development of a potential nomination project. At the time of writing there are 69 case studies on the portal, with more gradually coming on-line all the time.

4. Extended case studies and the second ICOMOS–IAU Thematic Study

The second Thematic Study (Ruggles & Cotte 2015) continues the development of a common vision and robust general principles by presenting a selection of case studies in greater depth, structured as segments of draft dossiers, that raise and help explore key issues relating to astronomical heritage that had first been identified in the 2010 work. A list of these "extended case studies" and the issues they address is presented in Table 1. Among the most challenging issues for the IAU has been how to recognise and protect the value of the dark night sky at places connected with astronomy—whether ancient sites, indigenous cultural landscapes, or modern observatories—and more than half of the extended case studies address this issue in one way or another.

In most cases, the primary aim of the extended case studies is to explore how potential OUV in relation to astronomy might best be demonstrated. Specific extended case studies might well facilitate the eventual preparation of a full nomination dossier should a State Party decide to prepare one, but this process must involve a wide range of stakeholders and must cover a range of legal and management issues as well as the scientific and heritage ones.

Table 1. Case studies included in the second Thematic Study and issues addressed.

Property	State(s)	Main themes and issues
Seven-stone antas	Portugal, Spain	Potential for serial nomination of a group of prehistoric monuments whose astronomical significance is only evident from the group as a whole
Stonehenge World Heritage Property	United Kingdom	Management issues given due recognition of astronomical values
Chankillo	Peru	Values of specific site in relation to astronomy as against broader values of archaeological landscape and related sites
Royal Observatory, Cape of Good Hope	South Africa	Importance of movable and intangible heritage in strengthening value of fixed heritage
Observatoire de Paris	France	Relative strength of individual v. serial nomination of classical observatory sites
Baikonur Cosmodrome	Kazakhstan	Relationships between science heritage and technology heritage[1]
Astronomical timing of irrigation	Oman	Cultural practices explicitly dependent upon dark night skies
Pic du Midi de Bigorre Observatory	France	High-mountain observatories
AURA Observatory / Canarian Observatories / Mauna Kea Observatory, Hawai'i	Chile / Spain / USA	Leading optical observatories under direct threat from light pollution
Aoraki–Mackenzie International Dark Sky Reserve	New Zealand	Pristine dark-sky area with broad cultural connections to the sky
Eastern Alpine and Großmugl starlight areas	Austria	Relatively dark dark-sky areas with few or no direct cultural connections to the sky

Notes:
[1] Not fully developed as an extended case study.

Stonehenge, which was inscribed in 1986, is an exception. Here, a modified statement of OUV was accepted in 2011 which recognises the significance of the solstitial alignments. The statement of authenticity reads: "At Stonehenge several monuments have retained their alignment on the Solstice sunrise and sunset, including the Stone Circle, the Avenue, Woodhenge, and the Durrington Walls Southern Circle and its Avenue." In this case the extended case study deals mainly with the management issue of maintaining the integrity of the astronomical sightlines.

A preliminary version of the second Thematic Study was distributed to interested participants at the 39COM side-event (see below) and publication is imminent.

5. Potential nomination projects

We continue to progress two potential transnational, multi-site ("serial") nomination projects. In such cases, IAU members can play a vital role in facilitating the exchange of ideas between different governments in the early stages of development of a proposal. The projects concerned are:

• the "Route of astronomical observatories" project, which focuses on the possible serial nomination of a number of classical observatories in European countries and their former colonies; and

• the "Windows to the Universe" project, focusing on the world's leading observatory sites and their dark skies, which currently involves Chile, Spain (Canarian Observatories) and the USA (Mauna Kea). We are currently investigating the possible inclusion of other "High-mountain observatories" such as Pic du Midi in France and Mt Wilson and Palomar in California.

The "upstream process" provides a way for the IAU to provide more direct assistance to State Parties in the identification of properties of significance, and potentially of OUV, in relation to astronomy. It enables UNESCO's advisory bodies, and the IAU in the case of properties with a connection to astronomy, to provide advice and assistance to governments who are considering potential nominations. Since January 2015 it has been possible for State Parties to request advisory missions which could involve the IAU.

In order to encourage such approaches and make personal contacts, the IAU hosted its first lunchtime "side-event" at the 39th session of UNESCO's World Heritage Committee (39COM) in Bonn, Germany, in July 2015. Side-events at annual UNESCO World Heritage Committee meetings are very important for raising awareness of activities and issues among the members of National Commissions attending these meetings.

6. The AWHI into the future

Prior to 2010, the only World Heritage List inscription explicitly relating to the heritage of astronomy was the Struve Geodetic Arc (10 European countries), a triangulation network constructed in the 19th century to measure the precise size and shape of the Earth. (Rather than being an astronomical monument per se, it really marks a technological achievement applying astronomy in the service of geodesy.) The first truly astronomical inscriptions took place in 2010, with the recognition not only of the Jantar Mantar observatory at Jaipur (India), but also of Dengfeng observatory as part of the "Historic Monuments of Dengfeng in 'The Centre of Heaven and Earth' " (China). By 2013, several State Parties had placed astronomical heritage sites on their Tentative Lists, varying in time range from prehistoric (Chankillo astronomical complex, Peru) through classical (Mykolayiv Observatory, Ukraine) to 20th-century (Jodrell Bank Observatory, UK).

Now, in 2015, the IAU has a direct involvement in several nomination projects and Commission C4 and its Working Groups are available to advise interested parties on other proposals. The Portal to the Heritage of Astronomy is very active and will be maintained for the foreseeable future.

Work will continue with ICOMOS on further Thematic Studies, addressing broader issues relating to science and technology heritage. Work has already commenced on a third Thematic Study, which will concentrate on heritage associated with space exploration. Dark skies issues can be addressed from the viewpoint of both cultural and natural heritage and Commission C4 plans to work with the IUCN's Dark Skies Advisory Group to promote a much stronger awareness of dark skies issues within the IUCN.

A Steering Group of National Focal Points appointed by national governments is in place and will retain strategic oversight of the AWHI on behalf of interested State Parties.

References

Ruggles, C. & Cotte, M. (eds.) 2010, *Heritage Sites of Astronomy and Archaeoastronomy in the context of the UNESCO World Heritage Convention* (Paris: ICOMOS –IAU)

Ruggles, C. & Cotte, M. (eds.) 2015, *Heritage sites of Astronomy and Archaeoastronomy in the Context of the UNESCO World Heritage Convention: Volume II* (Bognor Regis: Ocarina Books), in press

Wolfschmidt, G. (ed.) 2009, *Astronomical Observatories: from Classical Astronomy to Modern Astrophysics* (Berlin: ICOMOS, Monuments and Sites XVIII)

What makes astronomical heritage valuable? Identifying potential Outstanding Universal Value in cultural properties relating to astronomy

Michel Cotte[1,2]

[1] World Heritage Unit, ICOMOS, Tournon, France

[2] Centre François Viète d'histoire des sciences et des techniques, University of Nantes, Nantes, France

email: cotte.michel@orange.fr

Abstract. This communication presents the situation regarding astronomical and archaeoastronomical heritage related to the World Heritage Convention through recent years up until today. Some parallel events and works were promoted strongly within the IAU–UNESCO Initiative during the International Year of Astronomy (2009). This was followed by a joint program by the IAU and ICOMOS—an official advisory body assisting the World Heritage Committee in the evaluation of nomination dossiers. The result of that work is an important publication by around 40 authors from 20 different countries all around the world: *Heritage Sites of Astronomy and Archaeoastronomy in the Context of the UNESCO World Heritage Convention* (Ruggles & Cotte 2010). A second volume is under preparation (2015). It was also accompanied by some initiatives such as the "Windows to the Universe" organisation and the parallel constitution of local "Starlight Reserves". Some regional meetings studying specific facets or regional heritage in the field giving significant knowledge progresses also accompanied the global trend for astronomical heritage.

WH assessment is defined by a relatively strict format and methodology. A key phrase is "demonstration of Outstanding Universal Value" to justify the WH Listing by the Committee. This communication first examines the requirements and evaluation practices about of demonstrating OUV for a given place in the context of astronomical or archaeoastronomical heritage. That means the examination of the tangible attributes, an inventory of the property in terms of immoveable and moveable components and an inventory of intangible issues related to the history (history of the place in the context of the history of astronomy and cultural history). This is also related to the application to the site of the concept of integrity and authenticity, as regards the place itself and in comparison with other similar places (WH sites already listed, sites on national WH Tentative Lists, or other similar places in the region). The second issue of the communication is to give a glimpse of today's WH List, including some difficulties with listing and occasional failures, and trends and promising approaches.

Keywords. Astronomical heritage, Archaeoastronomical heritage, Outstanding Universal Value (OUV)

1. Introduction: New themes for the World Heritage List?

Since the 1990s, the "Global strategy for a representative, balanced and credible World Heritage List" has extended the scope of the WH Convention beyond the classical fields studied since its inception (UNESCO 1972). The new fields include:

- Canals, corridors of transportation;
- Cultural roads;
- Industrial heritage; and
- Heritage of technology and sciences.

Examples of listed properties include the Cultural road of Camino Real (Mexico), the industrial site of saltpeter works at Humberstone and Santa Laura (Chile), and the mountain railway of India (serial nomination, India).

There are also two new trans-thematic methods for a broader implementation of the WH Convention:

- Cultural landscapes; and
- The possibility of serial nominations (with national or trans-national boundaries).

These have permitted, for example, the inscription of the Grand Canal (China), Rice terraces (Phillippines), and Qhapaq Ñan, the Andean road system (6 states).

Nevertheless, such enlargement must be done in a common format for all the past and new categories of the WH List, for both natural and cultural properties, as is stated in the official Guidelines (UNESCO 2015).

2. The ICOMOS–IAU joint Thematic Study

At the end of the 2000s, the joint Initiative for Astronomy by UNESCO and the International Astronomical Union underlined the importance of astronomical and archaeoastronomical heritage. It also demonstrated the need to develop case studies for tangible evidence and related intangible values, and the need for an interdisciplinary methodology for heritage analysis both in scientific and heritage terms.

The initial question was: "What are the best ways to support and encourage the recognition of the most outstanding examples of astronomical heritage?". The best answer could be the World Heritage List. The methodology and case studies must accord with the WH Guidelines and recommendations. The natural partners become ICOMOS for cultural heritage and the IUCN for natural heritage.

Regarding methodology, the subject of astronomical and archaeoastronomical heritage appears as a specific multidisciplinary field, merging architecture, archaeology, sociology, technology and science. This complex heritage needs global overviews of subfields (periods, regions?), well understood and well analyzed case studies, and inventories and comparisons. All this assumes a shift from enthusiasm to professionalism.

The ICOMOS–IAU joint *Thematic Study on the Heritage of Astronomy and Archaeoastronomy* (vol. 1) was achieved in 2010 (Ruggles & Cotte (2010)). The list of themes aims to provide a short and clear overview of the recognized evidence relating to astronomical heritage in general, from prehistoric times to 20th-century heritage, and for every part of the world through historical times.

Contributions involved more than 40 authors from some 20 different countries from all regions of the world. The result is 16 chapters following chronology and geography with a consistent introductioon and conclusion by the editors highlighting epistemological and methodological questions. This Thematic Study, a tool for a first valuable approach, is freely available on-line (www.astronomicalheritage.net/index.php/about/astronomy-and-world-heritage/icomos-iau-thematic-study).

3. WH recognition for astronomical and archaeoastronomical heritage

Few sites are recognized for their astronomical values alone, but some, such as the Struve Arc (serial nomination, Europe) and Jantar Mantar at Jaipur (India), are

Figure 1. The Thirteen Towers as seen from the Fortress. ©Iván Ghezzi

already inscribed on the WH List. Some important places are listed with their important associated value for astronomy or archaeoastronomy being recognized. These include Samarkand (central Asia), Denfeng (China), various places in Egypt and pre-Columbian Mesoamerica, Stonehenge (UK), etc.

Since 2010 we have pursued and deepened the Thematic Study with "Extended case Studies" published on the UNESCO–IAU Portal to the Heritage of Astronomy (www.astronomicalheritage.net). A number of sites have appeared on national "Tentative Lists". A second volume of the Thematic Study will be submitted to ICOMOS for peer review in order to be published as an official document of the WH Committee.

These include some rare but extremely important archaeoastronomical individual cases. A paradigmatic example is certainly Chankillo in Peru (Fig. 1), but there are others.

Frequently, the astronomical value of already WH-listed places is not really recognized; we have to encourage State Parties to refurbish it correctly during the process of OUV revision. A case in point is Pulkovo Observatory, St Petersburg, Russia.

Many examples of astronomical heritage are members of important subcategories along with similar sites. This encourages serial approaches and strict selection. One of the most promising possible series is the "High-mountain observatories" which merge various qualities: precise definition of the series, global repartition, merging with dark sky reserve initiative, outstanding contribution to the scientific understanding of the Universe, etc.

4. Conclusion: Astronomical and archaeoastronomical heritage in context

Astronomy can take many forms but it is never alone: it is always a part of a larger ensemble of attributes that characterize a human society at a given time. Astronomy was permanently not only "pure" knowledge but an important part of human symbolic representations (cosmology). Consequently, the study of astronomical and archaeoastronomical heritage is a multidisciplinary field and the efforts made in this manner by the Thematic Studies is promising.

Although the World Heritage Convention focuses on "tangible immoveable heritage", we have to present immovable as well as moveable, and intangible as well as tangible; only the global approach make sense and indicates the real value of the place. The core of scientific knowledge is mainly intangible. It is an intellectual framework of the human spirit using specialized languages (written language, mathematics, etc.) and images (drawings, maps, photographs, physical information such as spectra, and so on).

The distinction between property and moveable objects is important from the juridical and heritage perspective, but has no real significance for astronomers. Archives, collections and bibliographies are hugely important. These documents are the product of scientific activities in their cultural context. The only reasonable conclusion is that the dichotomy between "fixed" and "moveable" makes little sense as a classificatory criterion in astronomical heritage in particular, or in science or technology heritage in general.

Regarding cultural and natural heritage relating to astronomy, one importance of natural and cultural landscapes (e.g. the High-mountain observatories) is that they provide a possible way for the recognition of the "Dark Sky" quality of a place among other attiributes. Astronomy represents a rich and significant aspect of cultural and natural heritage. Recognizing this permits us to identify and to clarify astronomical value in the context of the World Heritage Convention.

References

Ruggles, C. & Cotte, M. (eds.) 2010, *Heritage Sites of Astronomy and Archaeoastronomy in the context of the UNESCO World Heritage Convention* (Paris: ICOMOS –IAU)

UNESCO 1972, *Convention concerning the protection of the World Cultural and Natural Heritage* (Paris: UNESCO) http://whc.unesco.org/en/conventiontext

UNESCO 2015, *The Operational Guidelines for the Implementation of the World Heritage Convention* (Paris: UNESCO) http://whc.unesco.org/en/guidelines

Establishing the credibility of archaeoastronomical sites

Clive Ruggles

School of Archaeology and Ancient History
University of Leicester, Leicester LE1 7RH, United Kingdom
email: rug@le.ac.uk

Abstract. In 2011, an attempt to nominate a prehistoric "observatory" site onto the World Heritage List proved unsuccessful because UNESCO rejected the interpretation as statistically and archaeologically unproven. The case highlights an issue at the heart of archaeoastronomical methodology and interpretation: the mere existence of astronomical alignments in ancient sites does not prove that they were important to those who constructed and used the sites, let alone giving us insights into their likely significance and meaning. The fact that more archaeoastronomical sites are now appearing on national tentative lists prior to their WHL nomination means that this is no longer just an academic issue; establishing the credibility of the archaeoastronomical interpretations is crucial to any assessment of their value in heritage terms.

Keywords. Astronomical heritage, Credibility, Chankillo

1. Introduction

In 2011, the Republic of Macedonia nominated Kokino, an "ancient observatory" site, for inscription on the World Heritage List. According to the nomination dossier, the site was used throughout the Bronze Age both a ritual area and for observations of the celestial bodies. It featured natural "stone seats" and "stone markers"—gaps and niches in the nearby rocks that ostensibly served as horizon foresights. The problem was that the case for astronomical observations was strongly based upon alignment evidence, and the number of possible observing points and putative astronomical markers could easily be explained away as chance occurrences emphasised by the judicious selection of data. This danger has been very well known in archaeoastronomy since its early development during the 1970s and 1980s. The nomination dossier was rejected.

2. Credibility as a heritage issue

The credibility of archaeoastronomical sites, then, is a major issue for astronomical heritage. It is linked to authenticity, since authenticity concerns the degree to which a 'fact' has been proven, or a particular interpretation can be supported, according to current standards in the discipline concerned—e.g. history, archaeology or archaeoastronomy (Ruggles & Cotte 2010, p. 267).

In a first qualitative attempt to characterise different degrees of archaeoastronomical credibility, the 2010 ICOMOS–IAU Thematic Study (*ibid.*, pp. 270–271) identified four categories as follows:

(1) *Generally accepted.* Examples include the alignment of the Neolithic passage grave at Newgrange in Ireland upon the rising sun at midwinter, and the solstitial alignment of Stonehenge in the United Kingdom. Both are already on the World Heritage List.

(2) *Debated among specialists.* An example here is the interpretation of the circular Caracol at Chichen Itza, Mexico, as an observatory for watching the risings and settings of the sun, planets and stars. Archaeoastronomers continue to debate this interpretation (e.g. Aveni 2006; Schaefer 2006; Ruggles 2014, pp. 363–364). Despite the uncertainties, the building is widely referred to as the "Observatory".

(3) *Unproven.* An example is the "equinox hierophany" at the Kukulcan pyramid, again at Chichen Itza. This phenomenon attracts many thousands of visitors each year, but there is no convincing historical or contextual evidence to support the conjecture that it was actually deliberate and intentional (Carlson 1999). Despite this, it is widely cited as the principal astronomical connection of the site.

(4) *Completely refuted.* An example is the interpretation of four rock-painted symbols at Chaco Canyon, USA as a depiction of the 1054 supernova. Cultural evidence strongly suggests that the symbols marked a sun-watching station for native priests (Ellis 1975). Yet the "supernova" was until recently, and may well remain, signposted at the site.

Archaeoastronomical sites are also particularly susceptible to overinterpretation driven by political factors, because they can too easily be seen as demonstrating the intellectual achievements of ancestors. "A sense of national pride or political expediency can sometimes be implicit in claims that, for example, a site is 'the oldest observatory' in a given region. The irony is that such attitudes are founded upon a mistaken belief in a single 'path of progress' towards modern science, something that actually places a negative value onother cultural perspectives and practices" (Ruggles & Cotte 2010, p. 271).

3. Chankillo: an extended case study

Clearly, we need more sophisticated methods to assess archaeoastronomical credibility, and they must reflect the current state of theory and practice in the discipline. Advances in archaeoastronomy over several decades have resulted in the development of a substantial body of theory and practice where the most favoured interpretations depend upon integrating methods from astronomy, anthropology and other disciplines (cf. Ruggles 2011; 2014). Nonetheless, individual cases can still engender significant methodological debate, as recently regarding Chankillo (Ghezzi & Ruggles 2011; Malville 2011), which is now on Peru's tentative list. It is also included as an extended case study in the second ICOMOS–IAU Thematic Study (Ruggles & Cotte 2015).

The thirteen towers of Chankillo, dating to c. 300BC, have been interpreted as a solar observation device by Ghezzi & Ruggles (2007), an interpretation that is widely accepted although it is disputed by Malville. It is not disputed that the overall plan of the various structures at Chankillo is laid out on a roughly solstitial axis, nor that they had much to do with solar-related cults and rituals. The methodological debate focuses particularly on three issues:

1. *The selection of observing points.* While it is clear archaeologically that the western observing point—an open doorway facing the towers and surrounded by a scatter of cult offerings—is a place of unique ritual and ceremonial significance, the same is not true of the supposed eastern observing point.

2. *The towers are not an exact fit to the sunrise arc.* Although December solstice sunrise as viewed from the western observing point occurs directly over the southernmost tower, June solstice sunrise occurs somewhat to the left of the northernmost tower, at the foot of a more distant hill rising to the left away from the towers.

3. *The towers are evenly spaced.* If the successive gaps between the towers were marking off equal intervals of time through the year, then one would expect the towers to be more closely spaced towards the ends of the line.

The resolution of such issues depends upon how we factor in archaeological evidence pertaining to the broader cultural context, and also upon a theoretical framework built upon a broad range of anthropological experience. This experience indicates the likely complexity of cultural practices, meaning that observation devices are unlikely to have operated with "western" consistency or exactitude. Cases where we can make a strong case for an astronomical association on the basis of alignment statistics alone—as is true for the seven-stone antas (see paper by Belmonte *et al.*)—are rare; more often, the strongest interpretation is one supported by multiple lines of evidence and convergent methodologies, often from several different disciplines (Ruggles 2011, pp. 12–13).

4. Conclusion

Significant progress has been made within the Astronomy and World Heritage Initiative towards establishing broadly acceptable measures of archaeoastronomical credibility that make sense in the context of the heritage evaluation process. Expert opinion may now be sought from members of the new IAU Commission C4 as part of UNESCO's upstream process, as described in other papers in this Focus Meeting.

At one end of the scale, a claim that an ancient site is strongly related to astronomy, if based on putative astronomical alignments alone, will nearly always be problematic because of data selection issues; a broader interpretative context, from archaeology or history, will generally be crucial in supporting such claims. Comparative studies of sites that raise similar methodological issues, even in very different cultural contexts, may also be important in demonstrating potential outstanding universal value (OUV).

On the other hand, interpretations will always be subject to modification in the light of new archaeological evidence or improved methods. A site such as Newgrange or Stonehenge can clearly be recognised as being of exceptional importance in relation to astronomy even if many of the details of that relationship in the minds of those who built and used it remain unclear, and are likely to remain so.

References

Aveni, A. F. 2006, in: T.W. Bostwick & B. Bates (eds), *Viewing the sky through past and present cultures* (Phoenix: Pueblo Grande Museum), pp. 57–70, 79–83

Carlson, J. B. 1999, *Archaeoastronomy*, 14(1), 136

Ellis, F. H. 1975, in: A.F. Aveni (ed.), *Archaeoastronomy in Precolumbian America* (Austin: University of Texas Press), pp. 59–87

Ghezzi, I. & Ruggles, C. L. N. 2007, *Science*, 315, 1239

Ghezzi, I. & Ruggles, C. L. N. 2011, in: C.L.N. Ruggles (ed.), *Archaeoastronomy and Ethnoastronomy: Building Bridges between Cultures* (Cambridge: Cambridge University Press), pp. 144–153

Malville, J. McK. 2011, in: C.L.N. Ruggles (ed.), *Archaeoastronomy and Ethnoastronomy: Building Bridges between Cultures* (Cambridge: Cambridge University Press), pp. 154–161

Ruggles, C. L. N. 2011, in: C.L.N. Ruggles (ed.), *Archaeoastronomy and Ethnoastronomy: Building Bridges between Cultures* (Cambridge: Cambridge University Press), pp. 1–18

Ruggles, C. L. N. (ed.) 2014, *Handbook of Archaeoastronomy and Ethnoastronomy* (3 vols.) (New York: Springer)

Ruggles, C. L. N. & Cotte, M. (eds.) 2010, *Heritage Sites of Astronomy and Archaeoastronomy in the Context of the UNESCO World Heritage Convention* (Paris: ICOMOS–IAU)

Ruggles, C. L. N. & Cotte, M. (eds.) 2015, *Heritage Sites of Astronomy and Archaeoastronomy in the Context of the UNESCO World Heritage Convention: Volume II* (Bognor Regis: Ocarina Books), in press

Schaefer, B. E. 2006, in: T.W. Bostwick & B. Bates (eds), *Viewing the sky through past and present cultures* (Phoenix: Pueblo Grande Museum), pp. 27–56, 71–77

Astronomy, Illumination and Heritage: the Arles-Fontvieille megalithic monuments and their implications for archaeoastronomy and world heritage

Morgan Saletta

History and Philosophy of Science, University of Melbourne
Victoria 3010, Australia
email: msaletta@unimelb.edu.au

The Arles-Fontvieille monuments, or hypogées, have long had a special place in megalithic studies. Their unique architecture, blending "Atlantic" megalithic construction with subterranean rock-cut architecture more commonly found in the Mediterranean, and their size, especially that of the Grotte de Cordes, place them among the most important monuments in France and Europe (Daniel 1960, Guilaine 1998, Sauzade 1999, Hoskin 2001, Saletta 2014). My discovery and interpretation of seasonal light and shadow hierophanies (Saletta 2011, 2014)) within the Arles-Fontvieille monuments has important implications for identifying astronomically related Outstanding Universal Value for late prehistoric European monuments.

Despite much discussion of illumination hierophonies, most archaeoastronomers investigating late prehistoric European monuments continue to think in terms of the "celestial targeting hypothesis" —assuming a monument's orientation is targeting a celestial body at some point in the sky, generally on or near the horizon (Saletta 2014). This, and the often implicit focus on astronomically important dates such as the equinoxes or solstices, etc., is an enduring influence of the now rejected Thom paradigm. In the context of the monumental landscape of Orkney, where, as in Europe more generally, many monuments show no such calendrically significant orientation, Downes & Richards 2005 suggest that seasonal illumination of chambered monuments at different times was creating an "ancestral time". Such "ritual time" (Bloch 1977) is often associated with ritual feasting, which has been proposed by many researchers as playing a part in the construction and use of megaliths and other monuments in late-prehistoric Europe (e.g. Renfrew 1973; Tilley 1996). Numerous cross-cultural ethnographic examples also suggest a link between astronomy, time-reckoning and ritual feasting (Hayden & Villeneuve 2011).

Furthermore, multiple lines of evidence suggest a cosmologically symbolic link between houses of the living and houses of the dead in later prehistoric Europe (Ruggles 2010). Solar orientation of domestic architecture for functional purposes (i.e. lighting and heat) has also been suggested as an explanation for the orientation patterns of Neolithic domestic structures (Topping 1996). Thus the solar orientation and illumination in monuments may have originated in a functional orientation of domestic architectures which took on symbolic significance and was then incorporated in the houses of the dead.

The similarity of the seasonal illumination of the Arles-Fontvieille monuments with the illumination events at Newgrange in Ireland and Maeshowe in Scotland, suggests that astronomical orientation and seasonal illumination of monuments were part of widely shared cosmological beliefs and practices related to monumentality in late prehistoric Europe (Midgley 2008). Illumination hierophanies in monuments with passages and chambers occur for a number of days and thus produced a ceremonial time period, rather than marking a precise date. The wide range of regional monumental orientations may have produced what Dwyer (1990) has termed social synchrony by distributing seasonal rituals spatially and temporally, thus helping to organize ceremonial and seasonal ecological activities. In interpreting these orientations, and assessing their astronomically related heritage value, archaeoastronomers need to move beyond the celestial targeting paradigm to a more anthropological approach to astronomical orientation and seasonal illumination within late prehistoric European monuments.

References

Bloch, M. 1977, *Man*, 12, 278

Daniel, G. E. 1960, *The Prehistoric Chamber Tombs of France: A Geographical, Morphological, and Chronological Survey* (London: Thames and Hudson)

Downes, J. & Richards, C. 2005, in: C. Richards (ed.), *Dwelling among the monuments: the Neolithic village of Barnhouse, Maeshowe passage grave and surrounding monuments at Stenness, Orkney* (Cambridge: McDonald Research Monograph), p. 57

Dwyer, P. D. 1990, *The Pigs That Ate the Garden: a Human Ecology from Papua New Guinea* (Ann Arbor: University of Michigan Press)

Guilaine, J. 1960, *Au temps des dolmens: mégalithes et vie quotidienne en France méditerranéenne il y a 5000 ans* (Toulouse: Ed. Privat)

Hayden, B. & Villeneuve, S. 2011, *A. Rev. Anthropol.*, 40, 433

Hoskin, M. 2001, *Tombs, Temples and their Orientations : a New Perspective on Mediterranean Prehistory* (Bognor Regis: Ocarina Books)

Midgley, M. S. 2008, *The Megaliths of Northern Europe* (London: Routledge)

Renfrew, A. C. 1973, *Before Civilization: The Radiocarbon Revolution and European Prehistory* (London: Jonathan Cape)

Ruggles, C. L. N. 2010, in: C.L.N. Ruggles & M. Cotte (eds.), *Heritage Sites of Astronomy and Archaeoastronomy in the context of the UNESCO World Heritage Convention* (Paris: ICOMOS –IAU), p. 28

Saletta, M. 2011, in: C.L.N. Ruggles (ed.), *Archaeoastronomy and Ethnoastronomy: Building Bridges between Cultures* (Cambridge: Cambridge University Press), p. 364

Saletta, M. 2014, in: X. Margarit *et al.* (eds.), *Les monuments mégalithiques d'Arles-Fontvieille, État des connaissances, contextes et nouvelles données* (Aix-en-Provence: DRAC-SRA de PACA, Lampea), p. 143

Sauzade, G. 1999, in: J. Guilaine (ed.), *Mégalithisme de l'Atlantique à l'Éthiopie* (Paris: Errance), p. 125

Tilley, C. 1996, *An Ethnography of the Neolithic: Early Prehistoric Societies in Southern Scandinavia* (Cambridge: Cambridge University Press)

Topping, P. 1996, in: T. Darvill, & J. Thomas (eds), *Neolithic Houses in Northwest Europe and Beyond* (Oxford: Oxbow Books), p. 157

Serial nominations for the AWH initiative: The paradigm of seven-stone antas and beyond

Juan Antonio Belmonte[1], César González García[2] and Michael Hoskin[3]

[1] Instituto de Astrofísica de Canarias, Spain
email: jba@iac.es

[2] Incipit, CSIC, Santiago de Compostela, Spain
email: cesar.gonzalez-garcia@incipit.csic.es

[3] Cambridge University, United Kingdom
email: michael.hoskin@ntlworld.com

Abstract. In this short report we examine the ideal status of the seven-stone antas (a type of very ancient megalithic monument in the southwest of the Iberian Peninsula) as an excellent candidate for a serial nomination within the Astronomy and World Heritage Initiative. This case will be compared with an extraordinary set of dolmens at the other side of the Mediterranean, within the Transjordan Plateau, worthy of being protected under the umbrella of the same initiative but which are in serious danger of 'extinction'.

1. Seven-stone antas

Seven-stone antas are a group of megalithic monuments that were built in the southwest of the Iberian Peninsula for a period of some one thousand years in the 4th millennium B.C. —and possibly earlier —for the burial of people presumably belonging to a pastoral culture. They can be described as standard corridor dolmens but their most impressive fact is that they were constructed with a surprisingly consistent architectural concept over an extended period of time (hence their common name) and with a pattern of orientations that certainly situate them among the oldest monuments on Earth with indisputable astronomical orientations (Hoskin 2001). They were built in an extended area of the present day regions of Alentejo, in Portugal, where most of the antas are located, and Extremadura, in Spain (see Fig. 1). The presence of such impressive, extremely old monuments at both sides of the Spanish-Portuguese frontier and their reliability as an extremely important Case Study for the UNESCO Astronomical Heritage initiative makes of this particular set of megalithic tombs an extraordinary opportunity as a paradigm for serial nominations (Belmonte *et al.* 2015).

The numbers of these monuments is around 200 but somewhat fewer are in a sufficiently good state of preservation that alignment studies can be performed. The area occupied by the antas is so huge that it is evident that a full protection of all the monuments would be an extremely difficult task. However, we strongly believe that seven-stone antas as a group, and especially a particularly well-preserved set of them in the areas of Valencia de Alcántara (Spain) and central Alentejo (Portugal), deserve recognition within the framework of the Astronomy and World Heritage Initiative. This is supported not only by the fact that they represent the oldest group of monuments in the world with an unmistakable astronomical interest in their orientation —and presumably an interest in

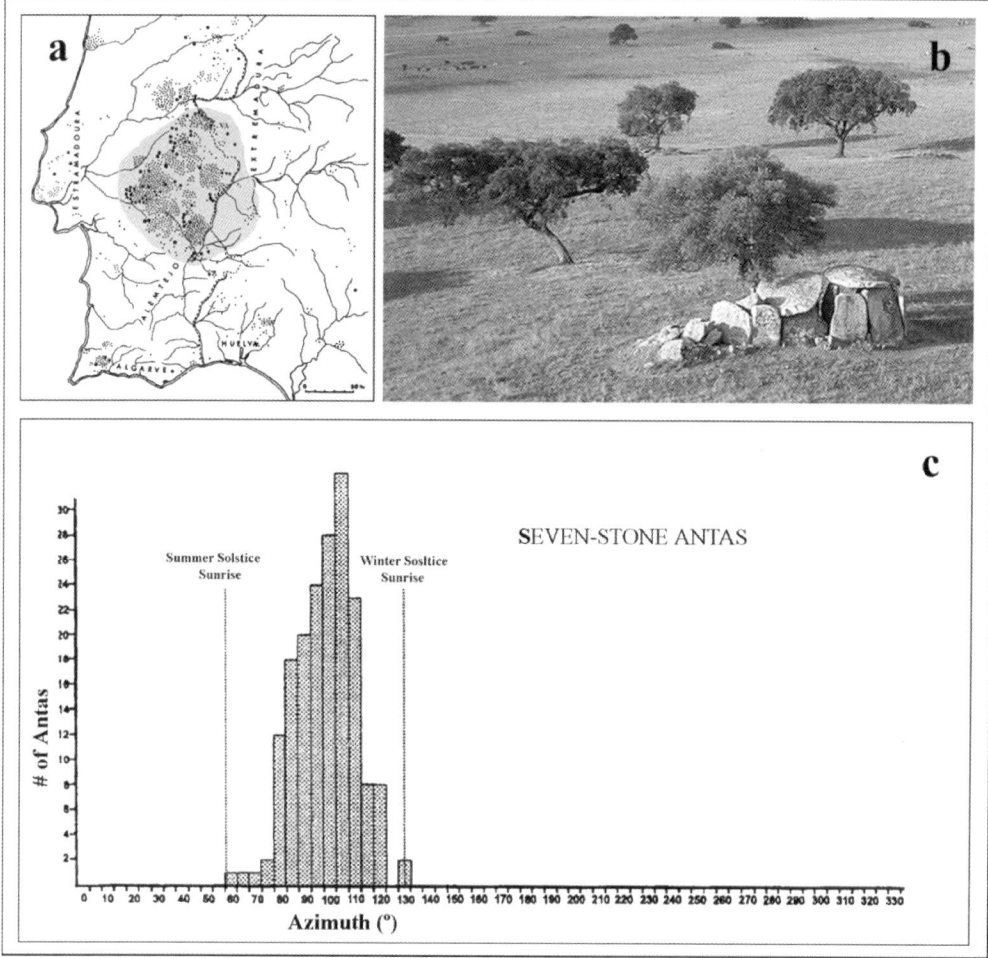

Figure 1. Seven-stone antas: area of distribution (a), together with a well-preserved example in the Portuguese Alentejo (b) and a histogram of orientation of a set of 172 measurable monuments. Diagram by the authors.

the sky as the motivation for this aspect of their construction —but also because they extend over a vast geographical area within both of the modern Iberian states: Spain and Portugal. This would certainly serve to guarantee the protection of the sites from any possible challenge in the future and for gaining a deserved recognition among modern society.

2. Dolmens in Jordan

Interestingly, at the other end of the Mediterranean, on the plateau to the east of the Jordan River valley, exist some of the largest and most attractive groups of dolmens of the Levant, which are fully comparable to the best exemplars to the west (see Fig. 2). The orientations of more than 180 dolmens have been measured in the field in Jordan, permitting the study of the astronomical aspects of the megalithic phenomenon in the region (Belmonte et al. 2013). The evidence shows that Early Bronze Age populations of the Transjordan plateau also orientated their tombs according to their celestial

Figure 2. Two examples of endangered megalithic necropolises in Jordan: a nicely preserved dolmen and another on the way to destruction at Ala Safat (a) and the huge necropolis at Al Murayghat (b) where a gigantic quarry is steadily approaching the site and may engulf it within a few years if nothing prevents it. Diagram by the authors.

connections (but also, to a minor extent, according to the terrestrial landscape), providing a second potential case for serial nomination. However, in contrast to the Iberian Peninsula, it is especially troubling that specialists in the Levant are now faced with the rapid destruction of the Hashemite Kingdom of Jordan's megalithic heritage owing to the huge quarries that are rapidly expanding and, more generally, to the fast urban development in the country (see Fig. 2). It is indeed necessary to raise awareness through UNESCO promotion of these beautiful and extremely interesting monuments —especially among the local people, perhaps the most difficult ones to persuade —so that they may be seriously protected as soon as possible by local authorities before it is too late. Otherwise, we might be facing the loss of a relevant component not only of archaeological, but also of astronomical, world heritage.

References

Belmonte, J. A., González García, A. C. & Polcaro, A. 2013, *J. Hist. Ast.*, 44, 429

Belmonte, J. A., Tirapicos, L., & Ruggles, C. 2015, in: C. Ruggles & M. Cotte (eds.), *Heritage sites of Astronomy and Archaeoastronomy in the Context of the UNESCO World Heritage Convention: Volume II* (Bognor Regis: Ocarina Books), in press

Hoskin, M. 2001, *Tombs, Temples and their Orientations : a New Perspective on Mediterranean Prehistory* (Bognor Regis: Ocarina Books)

As international as they would let us be

Virginia Trimble[1,2]

[1] Dept of Physics & Astronomy, U. Calfornia, Irvine CA 92697-4575 USA
email: vtrimble@astro.umd.edul

[2] Queen Jadwiga Observatory, Rzepiennik, Biscopi, Poland

Extended Abstract. Astronomers wanting to chart the whole sky, or even part of it, 24/7, must collaborate across nations, continents, and hemispheres. The Greeks were perhaps the first to do this, though Eratosthenes' measurement of the diameter of the earth was done when Alexandria and Syene were both part of Ptolemaic Egypt. The Golden Age of Arabic/Moslem astronomy coincided with times when there were very large caliphates and similar empire-like structures. The situation was very different for European astronomy at all times, with periods of successful international collaborations alternating with periods of disaster. Von Zach's "celestial police" agreed in 1800 at Lilienfeld to divide up the sky and look for the "missing" planet between Mars and Jupiter. The observatory was ransacked and papers burned by retreating Napoleonic soldiers in 1813. Skipping ahead most of a century, the Carte du Ciel, George Ellery Hale's Solar Union, Kapteyn's selected areas, and several other cooperative projects had scarcely started when the Great War broke out, and a German eclipse expedition under Erwin Freundlich, which had gone to the Crimea to look for bending of starlight, was imprisoned. All members eventually got home, but the equipment was never recovered.

While British eclipse expeditions organized by Eddington saw the bending of light from Sobral and Principe in 1919, the IAU, founded in the same year, explicitly excluded Germany, Austria-Hungary, Turkey, and the other central powers (even neutrals were not admitted for several years). Indeed Germany did not adhere until 1952. IAU membership now is generally driven by financial issues, but the solution of one nation, two adhering organizations for China was a major victory for international compromise. Coming down to the present, it is interesting that Europe, the site of the outbreak of both the World Wars, has led in the establishment of the European Southern Observatory, the European Space Research Organization (now ESA), and *Astronomy and Astrophysics*. Even the USA is gradually learning to cooperate with other astronomical nations (though we are still not brilliantly good at it). And the current projects for very large ground-based observing facilities, with acronyms like LOFAR, SKA, GMT, EVLT, and TMT, involve most of Europe, North America, China, Japan, India, and other parts of Asia and South America. A printed version of my FM2 presentation is in press for Vol. 43 (December 2015 issue) of the *Journal of American Association of Variable Star Observers* (roughly half of whose membership resides outside the US).

I thank Gudrun Wolfschmidt for a list of the astronomers who made up the "celestial police".

Twentieth-century astronomical heritage: the case of the Brazilian National Observatory

Christina Helena Barboza

Museum of Astronomy and Related Sciences
Rio de Janeiro, Brazil
email: christina@mast.br

Abstract. This paper aims at contributing to the UNESCO-IAU Astronomy and World Heritage Initiative's discussions by presenting the case study of a 20th-century observatory located in a South American country. In fact, the National Observatory of Brazil was created in the beginning of the 19th century, but its present facilities were inaugurated in 1921. Through this paper a brief description of the heritage associated with the Brazilian observatory is given, focused on its main historical instruments and the scientific and social roles it performed along its history. By way of conclusion, the paper suggests that the creation of the Museum of Astronomy and Related Sciences with its multidisciplinary team of academic specialists and technicians was decisive for the preservation of that expressive astronomical heritage.

Keywords. Heritage, history of astronomy, observatory

1. Introduction

The National Observatory (ON) is the oldest observatory in Brazil still in activity (Granato 2009). It was founded in 1827, soon after the political independence of that country, with the aim of transmitting practical astronomy to the students of the military and naval academies. From the mid-19th century to the early 20th century it was installed in the ruins of a Jesuit church located in the center of Rio de Janeiro, former capital of Brazil. In 1921 it was transferred to its present location, on the top of the São Januário hill. Since 1985 the wooded area of approx. 44km^2 where it is located is shared with the Museum of Astronomy and Related Sciences (MAST), a research institution originally created with the main purpose of studying and safeguarding the Observatory's old buildings and obsolete instruments (Barboza 1994).

2. Brief inventory of the tangible heritage at the ON-MAST site

The ON collection of moveable instruments, now under the safeguard of MAST, comprises approx. two thousand objects, which pertain to a wide range of scientific disciplines, such as astronomy, geodesy, topography, meteorology, magnetism and electricity, among others. They are evidence not only of the multiple activities developed in the Observatory at the turn of the 20th century, but also of its central role in the State building then in course, since the institution organized many demarcation expeditions and lent its instruments to other governmental agencies.

The immoveable heritage of the ON is formed by eight instruments sheltered in pavilions spread throughout the campus:

1) a 21cm Equatorial refractor by Gustav Heyde;

2) a 32cm Equatorial refractor by Cooke & Sons;
3) a 14cm Zenith telescope by Gustav Heyde;
4) a 46cm Equatorial refractor by Cooke & Sons (the only one not belonging to the MAST collection);
5) a 7cm Meridian circle by Askania;
6) a 8cm Meridian circle by Bamberg;
7) a 19cm Meridian circle by Gautier; and
8) a 10cm Photoheliograph by Zeiss.

The majority of these instruments were planed for the new area in São Januário hill, but two of them, the 32cm Equatorial and the Gautier circle, are 19th-century instruments. The Gautier was even installed in the old building of the ON. This obsolete instrument and its pavilion, in particular, recently underwent a deep restoration process, performed by the MAST team of museologists and technicians.

3. Cultural and symbolic dimensions of the Observatory activities

Between the end of the 19th century and the first two decades of the 20th century the ON became one of the most important scientific institutions in Brazil. Firstly, because under the direction of Emmanuel Liais (1871–1881) and his successor, Luiz Cruls (1881–1908), the links with the Military Academy were broken and it was given a variety of other tasks. Some of them had practical purposes, such as providing visual time signals useful to the ships and to the urban economic life in general, or establishing a meteorological network throughout the country, in order to understand the tropics and fight old climatic prejudices that were seen as the main threat to immigration. Other initiatives fit more strictly in what was then called "pure" science, such as the engagement in the international efforts to observe the Venus transit of 1882, or in the French "Carte du Ciel" photographic project. Anyway, it is possible to argue that during the last years of the Empire the ON rivalled with Cordoba and Cape of Good Hope Observatories in the discovery and investigation of celestial objects within the Southern skies. The Great Comet of 1882 (1882b), for example, was discovered independently and almost simultaneously in Cape Town and Rio de Janeiro. Cruls was even awarded the Valz Prize of the Paris Academy of Sciences for his pioneer chemical study of it (Videira 2001).

Secondly, after the coup d'État that set up the Republican regime, in November 1889, the ON became responsible for some decisive tasks in the new State building process, such as the geographical determination of the territory borders and of the area for the new capital, in the Central Plateau (which was built only in the 1950s). It was then, under the direction of Henrique Morize (1908–1930), that the Observatory finally succeeded in moving from its improvised house in the city center to a planned one in the (poor) suburbs, on a small hill still close enough to the port. It was also then that the greatest part of the instruments now deposited in MAST were acquired (Morize 1987).

4. Conclusion

The complex ON-MAST was listed by the Brazilian National Institute of Historic and Artistic Heritage in 1986, and by the Rio de Janeiro State Cultural Heritage Institute in 1987, as a result of the same mobilization of intellectual and scientific forces that is in the origin of MAST. The listed heritage comprises the buildings as well the collection of scientific instruments and other significant artifacts derived from the Observatory, including its original furniture.

Since its creation, the multidisciplinary team of MAST, which is formed by historians, anthropologists, museologists, archivists and educators, among other specialists, has developed academic research on the ON collections of objects and documents, in order to deepen the understanding of the past astronomical practices in Brazil. These researches are at the basis of the Museum exhibitions. Most importantly, they have supported the successful restoration of many instruments and the astronomical pavilions of the Observatory.

References

Barboza, C. H. 1994, *O encontro do Rei com Vênus; a trajetória do Observatório do Castelo no ocaso do Império* (Niterói: Universidade Federal Fluminense, Dissertação de Mestrado (História))

Granato, M. 2009, in: G. Wolfschmidt (ed.), *Astronomical Observatories from Classical Astronomy to Modern Astrophysics. Proceedings of the International ICOMOS Symposium in Hamburg, 2008* (Berlin: ICOMOS), p. 122

Morize, H. 1987, *Observatório Astronômico; um século de História (1827–1927)* (Rio de Janeiro: MAST, Salamandra)

Videira, A. A. 2001, in: A. Heizer & A.A. Videira (eds.), *Civilização e Império nos Trópicos* (Rio de Janeiro: Access), p. 123

Canada's Dominion Astrophysical Observatory and the rise of 20th Century Astrophysics and Technology

James E. Hesser, David Bohlender & Dennis Crabtree

Dominion Astrophysical Observatory, Herzberg Astronomy and Astrophysics,
National Research Council of Canada, Victoria, BC, Canada, V9E 2E7
email: jim.hesser@nrc-cnrc.gc.ca

Abstract. Construction of Canada's Dominion Astrophysical Observatory (DAO) commenced in 1914 with first light on 6 May 1918. As distinct from the contemporaneous development with private funding of major observatories in the western United States, DAO was (and remains) funded by the federal government. Canada's initial foray into 'big science', creation of DAO during the First World War was driven by Canada's desire to contribute significantly to the international rise of observational astrophysics enabled by photographic spectroscopy. In 2009 the Observatory was designated a National Historic Site. DAO's varied, rich contributions to the astronomical heritage of the 20th century continue in the 21st century, with particularly strong ties to Maunakea.

Keywords. General: history and philosophy of astronomy; Instrumentation: adaptive optics, spectrographs; Astronomical databases: miscellaneous

1. Introduction

Principles and practices behind DAO's high scientific and technological impact trace back to the founding Director, John Stanley Plaskett (1865-1941). He worked in close collaboration with the Warner and Swasey Company and the Brashear Optical Company to design and test the 1.82 m telescope today bearing his name. Indeed, perusal of *Publications of the Dominion Astrophysical Observatory Volume 1* suggests that in modern parlance Plaskett acted as the "systems engineer," "project scientist" and "program manager" for the Observatory's development, commissioning and scientific operations until his 1935 retirement. The design was copied six times around the globe into the 1960s, reflecting high respect for the facility whose development Plaskett led.

2. Scientific Programs

To ensure that Canada's new observatory would have immediate, as well as lasting, scientific impact, Plaskett consulted widely regarding the critically important challenges and opportunities emerging in stellar astrophysics. In the first two decades or so of DAO operation the tiny staff embarked upon ambitious programs that today would be characterized as 'Key Projects'. The study of spectroscopic binaries became an Observatory hallmark and continued through the 1980s and 1990s not only for traditional binaries, but also for identification and characterization of newly-discovered X-ray sources. Those early efforts included amassing over 15 years radial velocities of O and B stars that, interpreted by Plaskett and J.A. Pearce through Oort's ideas of differential rotation, determined the most reliable estimate of the size and mass of the Milky Way available until radio astronomical techniques emerged in the 1950s. The first organic molecule, CH, in interstellar space was discovered by DAO astronomer Andrew McKellar, who also deduced from interstellar CN observations the first, puzzling estimate of ∼3 K for the temperature of interstellar space that presaged the discovery of the cosmic microwave background radiation. In 1962 DAO's spectroscopic facilities were significantly augmented by the commissioning of the 1.2 m aperture telescope feeding its coudé spectrograph (subsequently named the McKellar Spectrograph), which E.H. Richardson's brilliant optical innovations made competitive with high-dispersion instruments of much larger telescopes.

3. Institutional Evolution

Throughout its history DAO has welcomed scientists from throughout Canada and around the world to use its facilities for their research. While there have been enormous changes to the operational environment since 1918, Plaskett's high standards and vision for scientific and technological excellence and relevance continue.

In 1970 Canada's Parliament charged the National Research Council (NRC) to "operate and administer any astronomical observatories established or maintained by the Government of Canada." All Federal astronomical facilities, including DAO, were brought into the NRC, which created the Herzberg Institute of Astrophysics to carry out the new mandate. In 1974 Canada undertook its first formal international agreement to build the CFHT, which was followed later by the JCMT, Gemini Observatory, ALMA, TMT and the development of the SKA, as well as leading roles in the development of space astronomy facilities funded by the Canadian Space Agency. Today DAO hosts the headquarters of the diverse NRC Herzberg organization that functions as a national astronomy laboratory supporting Canadian participation in those international facilities and their use by university astronomers and students. In addition to scientific research, NRC Herzberg activities comprise a) oversight of Canadian telescope allocation committees, international agreements, industrial partnerships, etc.; b) the Astronomy Technology Program (ground, space instrumentation); c) the Radio Astronomy Program; the Optical Astronomy Program (which operates the DAO telescopes; d) the Canadian Gemini Office; the Canadian Astronomy Data Centre (CADC); and e) public outreach (now in collaboration with community organizations).

As in Plaskett's day, ancillary instrumentation is critical to scientific breakthroughs. His pioneering efforts on spectrograph design continue through DAO's contributions to the CFHT and Gemini spectrographs under the leadership of David Crampton and the instrumentation team whose development he led (1980s-2000s). Their efforts included the introduction of multi-object capabilities on workhorse CFHT (MOS/SIS, 1992) and Gemini (GMOS 2001, 2002) spectrographs which enabled Canadian astronomers and their international partners to become leaders in observational cosmology beginning in the 1990s. In the same era the instrumentation team began its highly successful efforts on adaptive optics (AO), e.g., CFHT's 1988 experimental High Resolution Camera, that enabled the first detection of Virgo Cluster Cepheids used to determine the Hubble Constant; the MOS/SIS spectrograph (CFHT 1992); PUEO (CFHT, 1996); the ALTAIR AO camera for Gemini North (2002); and the Gemini Planet Imager 2013. These efforts have involved experts from Canadian universities and international partners.

'Data archiving' in Plaskett's day meant careful storage of the photographic plates on which spectra were taken prior to the 1980s. He likely would be astonished by the CADC (established in 1986) which today hosts and distributes data via the internet from diverse ground- and space-based telescopes to some 7,000 users worldwide. Among CADC's holdings are spectra selected from DAO's >100K photographic spectral archive that Elizabeth Griffin, David Balam and David Bohlender are digitizing. Twentieth century digitized photographic spectra are being used in research from ozone in the Earth's atmosphere to the properties of eclipsing spectroscopic binaries such as ϵ Aurigae observed over its 27 year orbit since the 1950s.

4. Current Perspective

Another important heritage dating to DAO's 1918 opening is sharing staff astronomical research with the public. Throughout its history the Observatory's outreach has functioned in synergy with members (predominantly amateur astronomers) of the Victoria Centre of the Royal Astronomical Society of Canada (RASC). In 2001 NRC opened a dedicated visitor facility, the *Centre of the Universe* (CU), to offer public outreach plus programming delivered by professional informal educators that was aligned with the school curriculum to complement and enrich the formal educational system. A change in NRC management philosophy led to CU closure in mid 2013. Today the Observatory is working with community organizations towards restoration of CU programming while individual staff continue to respond to frequent requests from community groups and schools to share their knowledge.

As prescient as Plaskett was, he could not have foreseen that a century later much of DAO's scientific, engineering and technical impact and heritage would have arisen from opportunities provided to international astronomy, including Canada, through access to Maunakea, for which Canadian astronomers are extremely grateful. Mahalo!

The Development of Mauna Kea as an Astronomical Site
Panelists: John Jefferies, Ann Boesgaard, Alan Stockton, Eric Becklin, and Alan Tokunaga

Teasel Muir Harmony[1] and David DeVorkin[2] (moderators)

[1]Center for History of Physics, American Institute of Physics, College Park, MD 20740, USA
email: tmuir@aip.org

[3]National Air and Space Museum, Smithsonian Institution, Washington, DC 20560, USA
email: devorkind@si.edu

Abstract. On August 11 we held a panel discussion at the 2015 IAU General Assembly, within the three-day Focus Meeting FM2, "Astronomical Heritage: Progressing the UNESCO–IAU Initiative". Our purpose was to both honor and explore the contributions of John Jefferies to the creation and development of Mauna Kea as an astronomical site.

1. Overview

The panel format (see Fig. 1) provided a platform wherein these astronomers, and the audience, shared their recollections and views relating to the history of the establishment of Mauna Kea and the Institute for Astronomy and Jefferies' role in bringing both into existence. Organizing the panel also provided the opportunity for identifying and assessing the state of preservation of the historical record covering the early history of modern astronomy in Hawaii.

DeVorkin opened the session with a short illustrated overview (see the pdf file on the FM2 website www.astronomicalheritage.net/index.php/community/news-events/focus-meeting-at-iau-general-assembly, which also contains Becklin's presentation) and then invited Jefferies to provide a short account of what attracted him to Hawaii. His remarks were then followed by remarks and queries by those noted in the order above, followed by responses by Jefferies. Finally Muir-Harmony posed general questions to the panel and moderated audience responses.

Principal questions posed to, and by, each of the panelists during the session, and in more depth during oral history sessions held during the meetings, are:

1. What were the chief opportunities, challenges and then hurdles to overcome in the identification and then implementation of Mauna Kea as an ideal site?
2. How did the University of Hawaii become the host institution for its establishment?
3. What was learned from tracking sites established on Haleakela by the U. S. Air Force and the Smithsonian that focused attention on Mauna Kea?
4. How and why did NASA become interested in establishing a large observatory there?
5. And once the site was established and its qualities fully appreciated, how did astronomical institutions from all over the world become attracted to the site?

Figure 1. The panel. From left to right: John Jefferies, Ann Boesgaard, Alan Stockton, Eric Becklin, Alan Tokunaga, and Teasel Muir Harmony

The panel discussion was recorded as were all the interview sessions, which are now being processed by the American Institute of Physics. All of the panelists, as well as two other participants in the early years, were interviewed privately during the first and second weeks of the meeting. The resulting 25 hours of interviews as well as findings during an inspection tour of the archival vault maintained by the Institute for Astronomy will, we hope, stimulate increased attention by historians to the history of modern astronomy in Hawaii.

The AURA Observatory in Chile—part of the IAU/UNESCO Extended Case Study

Malcolm G. Smith[1], R. Chris Smith[2] and Pedro Sanhueza[3]

[1] NOAO/CTIO,
Avenida Juan Cisternas 1600, La Serena, Chile
email: msmith@ctio.noao.edu
[2] AURA Observatory in Chile,
Avenida Juan Cisternas 1600, La Serena, Chile
email: csmith@aura-o.aura-astronomy.org
[3] OPCC,
Avenida Juan Cisternas 1606, La Serena, Chile
email: psanhueza2007@gmail.com

Brief report. The Extended Case Study for AURA-O as a "Window to the Universe" (http://www2.astronomicalheritage.net/index.php/show-entity?identity=000059&idsub-entity=005) was prepared in the context of supporting the desire to preserve humanity's scientific/cultural heritage of outstanding, high-mountain, ground-based, observatory sites developed over the period 1870–2000.

AURA–O includes the Cerro Tololo InterAmerican Observatory (CTIO) established in 1962 as the first of the major international observatories to be installed in Chile. The future of AURA–O now includes the Large Synoptic Survey Telescope (LSST). This Extended Case Study has provided the context for the development of possible initiatives to protect a variety of sites in Chile (e.g. Tololo, Pachón, La Silla, Las Campanas, Paranal, Armazones and Chajnantor) for their historical and scientific value to humanity. The dark skies and ideal weather patterns of northern Chile, along with its location in the southern hemisphere, have made this area of the world a major centre for astronomical facilities.

While this talk touched on the importance of dark skies as part of the Windows to the Universe concept, others will discuss the current status and future plans (of the Chilean Government and the observatories) for protecting the dark skies of northern Chile.

Site Protection Efforts at the AURA Observatory in Chile

R. Chris Smith[1], Pedro Sanhueza[2] and Malcolm G. Smith[3]

[1]AURA Observatory in Chile,
Avenida Juan Cisternas 1600, La Serena, Chile
email: csmith@aura-o.aura-astronomy.org

[2]OPCC,
Avenida Juan Cisternas 1606, La Serena, Chile
email: psanhueza2007@gmail.com

[3]NOAO/CTIO,
Avenida Juan Cisternas 1600, La Serena, Chile
email: msmith@ctio.noao.edu

Abstract. The AURA Observatory site in northern Chile, which includes Cerro Tololo and Cerro Pachon, has been operational for over 50 years now, facing a variety of challenges to its long-term future. The site now hosts over 20 operational telescopes, ranging from small projects with 0.4m telescopes to the Blanco 4m, the SOAR 4.1m, and the 8m Gemini-South telescopes. In addition, we have recently begun the construction of the Large Synoptic Survey Telescope (LSST) on the summit of Cerro Pachon. We summarize our efforts over the past 20-30 years to highlight the importance of site protection through education and public outreach as well as through more recent promotion of IDA certifications in the region and support for the World Heritage initiatives described by others in this conference.

Keywords. atmospheric effects, site testing

1. Overview

The AURA Observatory in Chile (AURA-O) was the first of the major international observatories to be established in northern Chile to exploit the optimal astronomical conditions available there. The site was originally established in 1962 to host the Cerro Tololo Inter-American Observatory (CTIO). It now hosts more than 20 operational telescopes, including some of the leading U.S. and international astronomical facilities in the southern hemisphere, such as the 4m Blanco telescope on Cerro Tololo and the 4m SOAR and 8m Gemini-South telescopes on Cerro Pachon. Construction of the next generation facility, the Large Synoptic Survey Telescope (LSST), has recently begun on Cerro Pachn, while additional smaller telescopes continue to be added to the complement of facilities on Cerro Tololo.

While the site has become a major platform for international astronomical facilities over the last 50 years, development in the region has led to an ever-increasing threat of light pollution around the site. AURA-O has worked closely with local, regional, and national authorities and institutions (in particular with the Chilean Ministries of Environment and Foreign Relations) in an effort to protect the site from threats so that future generations of telescopes, as well as future generations of Chileans, can benefit from the dark skies in the region. In the following we briefly describe our site protection efforts, starting with the protectin of the site from mining interests and moving to protection from light pollution.

2. Outstanding Sites & Evolution of Protection

Northern Chile hosts some of the best sites in the world for astronomical ground-based facilities. The combination of high mountains, one of the driest deserts in the world, and the smooth air flow coming off the Pacific provides excellent conditions for optical, infra-red, and radio observations. In addition to these natural conditions, the solid infrastructure and stable economic and political conditions of Chile allow international observatories to operate efficiently and effectively plan for long-term investments.

In the late 1950s, Dr. Frederico Rutllant, then Director of the Chilean National Observatory, noted the development of the next generation of major observatories in the Northern hemisphere and recognized the opportunity that the conditions in northern Chile provided for astronomy. He traveled to the United States and visited key figures in the development of new astronomical facilities there, inviting them to consider establishing one or more astronomical observatories in Chile. This initiative led to the selection of Cerro Tololo as the site for the first major international observatory in Chile in November of 1962. The observatory was named the Cerro Tololo Inter-American Observatory, or CTIO, in recognition of the collaborative nature of the initiative.

Although atmospheric conditions over northern Chile provide outstanding conditions for astronomical observations, the geology has provided a much more tangible benefit for the Chilean economy. Northern Chile is rich in copper, gold, and other valuable minerals, and the mining industry has been the main economic engine for Chile for many decades. As such, mining poses a significant threat to the astronomical facilities sited in northern Chile. The Chilean government recognized this threat, and developed a mechanism for protecting the sensitive areas around the many observatories that have come to Chile since the 1960s. The sites have been declared "areas of scientific interest", a designation that requires approval by the President of Chile for any mining activity within the areas.

The rapid economic development during the 1980s and 1990s in northern Chile brought a new threat to the forefront, that of the lights of the cities in the region around the observatories. In the late 1990s, AURA-O and the other international observatories worked with the Chilean government to highlight the threat and look for ways to mitigate it while not interfering with the economic development in the region. The result was the first "norma", or regulation, on light pollution, published in 1998 as Decreto Supremo 686. In addition, the "Oficina de Protecion de la Calidad de los Cielos del Norte de Chile", or OPCC, was established to support the implementation of the norma through education and support of municipalities in the selection of appropriate public lighting.

3. Education & Public Outreach initiatives

Following in the pattern developed by OPCC of stressing education and support of the community, as opposed to an emphasis on enforcement, AURA-O has developed a multi-faceted approach to raising the awareness of the impacts of light pollution and the importance of protecting the dark skies of northern Chile. Through its divisions of CTIO, Gemini, and more recently LSST, the Observatory has developed a variety of educational and public outreach (EPO) programs to carry the messages into the community.

NOAO/CTIO: The EPO program of CTIO consists of a diverse set of activities in the community, with an emphasis on training teachers and other community educators, both in formal and informal settings. The strategy is to take advantage of the multiplicative effect that one educator may have, touching hundreds if not thousands of students and/or public. Together with this training, NOAO's EPO experts have developed a variety of educational kits for the classroom and other events, including the Galileoscopes (for the

International Year of Astronomy), Dark Sky Awareness kits, and most recently a Quality of Light kit for the International Year of Light.

Gemini: The Public Information Office (PIO) of Gemini also supports a variety of community outreach, emphasizing sharing the wonders of the Universe through major public events. Gemini sponsors two large events annually, AstroDay and Viaje al Universo, touching thousands of public participants with each event. Messages of appreciation of dark skies and the impacts of light pollution are woven into these events to sensitize the community about these important issues.

LSST: The LSST is the newest of AURA's initiatives in Chile, and with it comes a new perspective on outreach, one of public participation in the science. LSST is developing a strong "Citizen Science" program in which the public can browse through the petabytes of images that LSST will take each year and help classify galaxies, stars, and other celestial objects and events. Through these programs, AURA and LSST hope to make the Universe more accessible to the public, and in doing so also help them understand the importance of protecting this special window to the Universe that Chile provides.

In addition to these focused EPO programs, AURA has supported initiatives to create municipal and touristic observatories in the region. What started as an idea for the first municipal observatory in the nearby community of Vicuna has developed into a growing industry of "astro-tourism" in northern Chile! The industry now includes not only touristic observatories, but has spread to many of the nearby hotels (providing telescopes and nighttime programs) as well as nighttime activities such as guided horseback rides with astronomical observations included.

4. Path Forward

Through collaborative efforts with the Chilean government at local, regional, and national levels, we have made substantial progress in protecting the astronomical sites in northern Chile from both mining and the ever-growing issue of light pollution. In 2013 the Chilean government issued updated regulations for outdoor lighting in the region, and those regulations are beginning to be implemented as of this writing.

Beyond these regulations, the AURA-O and the other international observatories, including the European Southern Observatory, Carnegie Observatory, and the Giant Magellan Telescope, continue to strengthen their ties with the Chilean community through educational programs and outreach activities, in an effort to both get the public excited about the science and the wonders of the Universe while also sensitize them to the fragile condition of the dark skies and the threats those skies face. Most recently the AURA-O site was recognized as the first international Dark Sky Sanctuary by the International Dark-Sky Association, in recognition of the importance of the dark skies in northern Chile and the efforts to protect them. These efforts of international recognition are being furthered by the Chilean government through its initiative to propose key astronomical sites in northern Chile as UNESCO World Heritage sites. We all hope that through these initiatives and collaborations, we can ensure that the skies of northern Chile remain dark for decades, if not centuries, to come for the enjoyment and wonder of future generations of astronomers as well as children of all ages.

Protection of SAAO observing site against light and dust pollution

Ramotholo Sefako[1] and Petri Väisänen[1,2]

[1]South African Astronomical Observatory, PO Box 9, Observatory, 7935, South Africa
[2]Southern African Large Telescope, PO Box 9, Observatory, 7935, South Africa
email: rrs@saao.ac.za

Abstract. The South African Astronomical Observatory (SAAO) observing station near Sutherland, Northern Cape in South Africa, is one of the darkest sites in the world for optical and IR astronomy. The SAAO hosts and operates several facilities, including the Southern African Large Telescope (SALT) and a number of international robotic telescopes. To ensure that the conditions remain optimal for astronomy, legislation called the Astronomy Geographic Advantage (AGA) Act, of 2007, was enacted. The Act empowers the Department of Science and Technology (DST) to regulate issues that pose a threat to optical and/or radio astronomy in areas declared Astronomy Advantage Areas in South Africa. For optical astronomy, the main challenges are those posed by light and dust pollution as result of wind energy developments, and petroleum gas and oil exploration/exploitation in the area. We give an update of possible threats to the quality of the night skies at SAAO, and the challenges relating to the AGA Act implementation and enforcement. We discuss measures that are put in place to protect the Observatory, including a study to quantify the threat by a planned wind energy facility.

Keywords. Protection, telescopes, sites, light pollution, AGA Act.

1. Introduction and Background

The SAAO observing site at Sutherland is among the best astronomical sites in the world for optical and Infrared astronomy, and is recognised as one of the few regions in the Southern Hemisphere that are suitable for large 10-m class telescopes. Indeed, Sutherland hosts SALT, the largest single telescope in the Southern Hemisphere. The site is very dark at night, and relatively dust-free due to its location in the arid Karoo, far away from major cities and industrial activities. Measurements done at SAAO indicate a V-band zenith sky-brightness, at Solar minimum, to be close to 22.0 mag arcsec^{-2}, i.e. at the level of the darkest observatories on the planet (e.g. Benn & Ellison 1998). It is mostly clear and atmospherically stable, with 70% observable nights in the year. As well as SALT and other SAAO telescopes, a number of robotic international telescopes are hosted at Sutherland†, including LCOGT (USA), KMTNet (South Korea), MASTER (Russia), KELT (USA), SuperWASP (UK), Solaris (Poland), Monet (Germany) and IRSF (Japan).

These wonderful conditions for optical astronomy, that attracted so many international facilities, are at risk due to developments geared towards addressing energy problems in South Africa (see e.g. Sefako 2012). The planned developments include solar energy, wind farms and petroleum gas and oil exploration, some of which may impact negatively on the high quality observing site due to light and dust pollution that will be generated either during their construction or operation.

† Details on SAAO and hosted telescope facilities: http://www.saao.ac.za/science/facilities/telescopes/

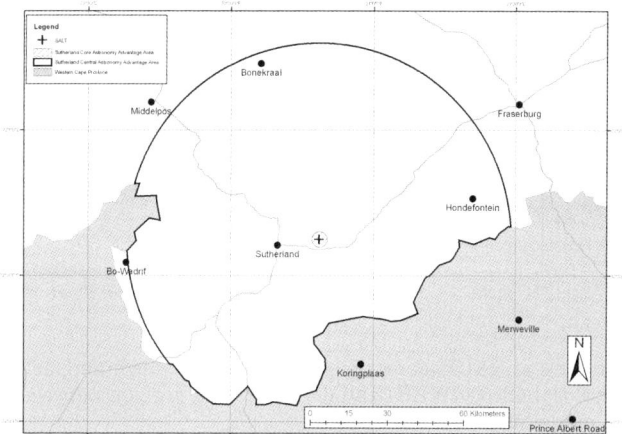

Figure 1. The Astronomy Advantage Areas for optical astronomy around the SAAO site.

To address the challenges, the South African government enacted the AGA Act No. 21 of 2007, which makes provision for the minister of DST to declare Astronomy Advantage Areas (AAAs) within which activities detrimental to astronomy may be controlled or restricted. The DST's Astronomy Management Authority (AMA) is charged with monitoring and ensuring that the AAAs are complied with. The AAA for optical astronomy is divided into to three areas with differing levels of protection (Figure 1). The **Core** is the physical area of the observatory covering about 3 km radius from SALT; the **Central AAA** extends to 75 km from SALT in the Northern Cape surrounding the Core AAA (Figure 1). The **Coordinated Area**, which has the least regulated restrictions, surrounds the Central Area and extends to 250 km radius from SALT.

2. Current status and challenges

A number of proposals around the Sutherland area are currently considered, including shale gas extraction, wind energy and solar power. Some or all of these energy developments are likely to cause light or dust pollution during construction or operation. There is on-going discussions between the SAAO, AMA and energy developers on ways in which light and dust pollution can be controlled or reduced during construction and/or during operation of the facilities.

The Civil Aviation Authority (CAA) has also been engaged given their requirement that structures taller than 45m must have aircraft warning lights (The Aviation Act of 1962). The CAA recognise both the need for renewable energy and importance of astronomy in the country, and has considered helpful options, such as Pilot Activated Lighting (PAL; lights on a specific wind farm would be activated by the aircraft pilot when flying lower than specified height above the facility) and Automatic Primary Radar Sensors, capable of detecting aircraft and activating the lighting (at the wind farm) for the period of the aircraft operating within the sensor range. Either of these methods will efficiently reduce the light pollution compared to having permanent lighting structures on wind turbines within the 75 km of the Observatory.

Other current activities include light pollution modelling for one of the proposed wind energy facilities (WEF) called Gunsfontein. The results show that wind turbines that do *not* have direct lines of sight to the telescopes 20 to 25 km away will not have significant effect to the general night sky brightness, as seen if Figure 2 showing modelling by the

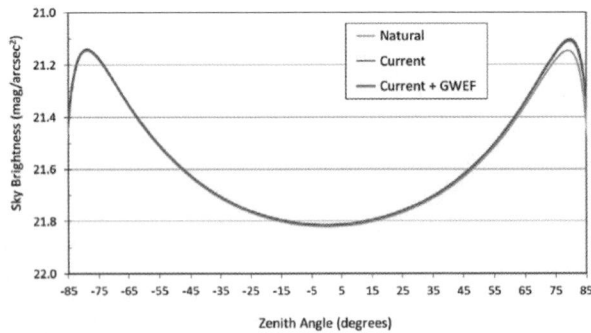

Figure 2. Sky brightness at SALT along the great semicircle beginning at the horizon opposite the Gunsfontein WEF (labelled GWEF) and ending at the horizon between SALT and GWEF, resulting from DSP modelling.

Dark Sky Partners (DSP), LLC (from Arizona), using modified Garstang Models (e.g. Garstang 1991, and references therein).

However, it is not clear whether light pollution will be the only serious problem from the energy facilities. Construction phases require mining quarries for roads and other development, generating substantial amounts of air-borne dust. For hydraulic fracturing (aka fracking), this is an issue during operations as well as during construction. For wind farms, this is likely to be a challenge during construction and erecting of wind turbines. Turbulence (seeing), and even potential changes in weather patterns may be expected when wind farms are operating. All these are issues which would need careful consideration and modelling with limited resources, while multiple plans are being put forward. It is also not clear what levels of compliance from developers we will have once facilities are up and running.

3. Summary

Sutherland Observatory is one of the darkest astronomical sites in the world, and the South African Government has enacted legislation to help protect the skies. Nevertheless, there are multiple large facilities being proposed in the region catering for the growing energy needs of the country, which makes for a challenging decade ahead for the Observatory. Collaboration between different stakeholders is needed to address the issues of protection of the Observatory's research facilities against light and dust pollution, and any other detrimental activity to optical astronomy, while acknowledging energy developments addressing the energy crisis in South Africa.

References

The Astronomy Geographic Advantage Act, 2007 (Act No. 21 of 2007), *Government Gazette* No. 31157, 17 June 2007
Aviation Act, 1962 (Act. No. 74 of 1962), *Part 139.01.33, the 13th Amendment of the Civil Aviation Regulation*, 1997 (or SA-CATS AH 139.01.33)
Benn, C. R. & Ellison, S. L., 1998, *New Astr. Rev.*, 42, 503
Garstang, R. H. 1991, *Pub. Astron. Soc. Pacific*, 103, 1109.
Sefako, R. R. 2012, *African Skies*, 16, 38

How can UNESCO World Heritage Criteria be applied to the "Windows to the Universe" Sites?

Michel Cotte[1,2]

[1]World Heritage Unit, ICOMOS, Tournon, France

[2]Centre François Viète d'histoire des sciences et des techniques,
University of Nantes, Nantes, France
email: cotte.michel@orange.fr

Abstract. This communication proposes a methodical approach trying to link the concept of "Windows to the Universe" to the uses of the Criteria defined by the World Heritage Convention (UNESCO 1972). The first issue is well advanced today after more than 10 years of active studies and preservation projects such as "Starlight Reserves" by specialists of astronomy, archaeoastronomy and environmental sciences. The second issue is related to a UNESCO Convention ruled by the WH Committee that has led to the recognition of around 1000 World Heritage sites over 40 years. The official booklet *Operational Guidelines for the Implementation of the World Heritage Convention* (latest edition 2015) (UNESCO 2015) summarizes conceptual ideas and methodological recommendations for WH nominations. In practice the WH Committee's decisions rely on the scientific and professional evaluation of each site by UNESCO's advisory bodies: ICOMOS for cultural heritage and IUCN for natural heritage.

The first goal of this presentation is to establish appropriate understanding of a very specific conceptual approach (Windows to the Universe) in the context of a very large UN Convention (the World Heritage List) related both to cultural and natural heritage in general. The second goal is to give a readable understanding of the WH requirements coming from the strict evaluation of the "Outstanding Universal Value" (OUV) of a given place, including the choice of WH Criteria expressing OUV with respect to the format of the *Guidelines*. Furthermore, and due to concepts coming from two very different fields, the communication aims to present a practical methodology in the case of a possible WH nomination: how to understand relationships between different classes of value and how to demonstrate OUV and justify the choice of Criteria for the place. Beyond potential WH projects, obviously limited in number, the communication tries to propose an efficient and general methodology for assessing the value and creating understanding of places having a "Windows to the Universe" facet.

Keywords. Astronomical heritage, Outstanding Universal Value (OUV), Windows to the Universe

---

1. Introduction

Within the terms of reference issued from the UNESCO World Heritage Convention (UNESCO 2015), the key sentence for WH recognition is the "Justification of Outstanding Universal Value" (OUV) of a given place, which means a "place" or a "site" within the international convention. The OUV of the site is expressed by the WH criteria, 6 of which express possible cultural value and 4 express possible natural value.

The six WH cultural criteria are:
I. represent a masterpiece of human creative genius;

II. exhibit an important interchange of human values;

III. bear a unique or at least exceptional testimony to a cultural tradition;

IV. be an outstanding example of a type of building, architectural or technological ensemble or landscape which illustrates (a) significant stage(s) in human history;

V. be an outstanding example of a traditional human settlement, land-use, or sea-use which is representative of a culture (or cultures), or human interaction with the environment;

VI. be directly or tangibly associated with events or living traditions, with ideas, or with beliefs, with artistic and literary works of outstanding universal significance.

The four WH natural criteria are:

VII. contain superlative natural phenomena or areas of exceptional natural beauty and aesthetic importance;

VIII. be outstanding examples representing major stages of earth's history;

IX. be outstanding examples representing significant ongoing ecological and biological processes in the evolution of plants and animals;

X. *in-situ* conservation of biological diversity.

There are also other important associated issues for the demonstration of OUV in WH, in particular the integrity and authenticity of the place, and the comparative analysis with similar sites.

2. The "Windows to the Universe" concept in the heritage view

The question now is: how can the UNESCO World Heritage Criteria be applied to the "Windows to the Universe" sites? The basic features shaping the "Windows to the Universe" concept are:

• The sky itself (object of the observation);

• The site as a property in local permanent context (geography, atmosphere, architecture, landscape, nature ...); and

• Humankind using the observation place, eventually with artifacts/instruments.

Of course, these three basic issues of the sky observation are intimately related together. But they do not have the same sense in the WH view and difficulties arise as a result. The sky itself can not be seen as a "property" in the same human/juridical sense as for all the other WH sites (since this would imply anthropocentrism). It follows that the "Dark Sky" cannot be taken alone in consideration for a WH nomination. The thing most spontaneously adapted to the goals of the WH Convention is the place itself, as a local tangible property. The human presence of the observers and astronomers gives meaning and life to the place, supporting important additional intangible value by the history of knowledge and human representations.

On the other hand, dark sky quality can be considered as a local environmental attribute. Studying a place in the WH perspective starts with an inventory of its tangible attributes: generally speaking, a given place has a series of natural attributes and cultural attributes supporting and expressing its value. Clearly, intrinsic dark sky quality is a natural attribute of the place among others. Assessing its importance will involve an analysis of its properties (purity as minimum of physical constraints) and consideration of the visible relationships between attributes that give a landscape value.

Dark Sky quality is one among a cluster of natural attributes. It contributes to the global natural context of a given place, and it belongs to a larger group of natural attributes of the site, forming its natural environment components. One of the best ways to use Dark Sky value is among a set of other remarkable attributes that characterize a

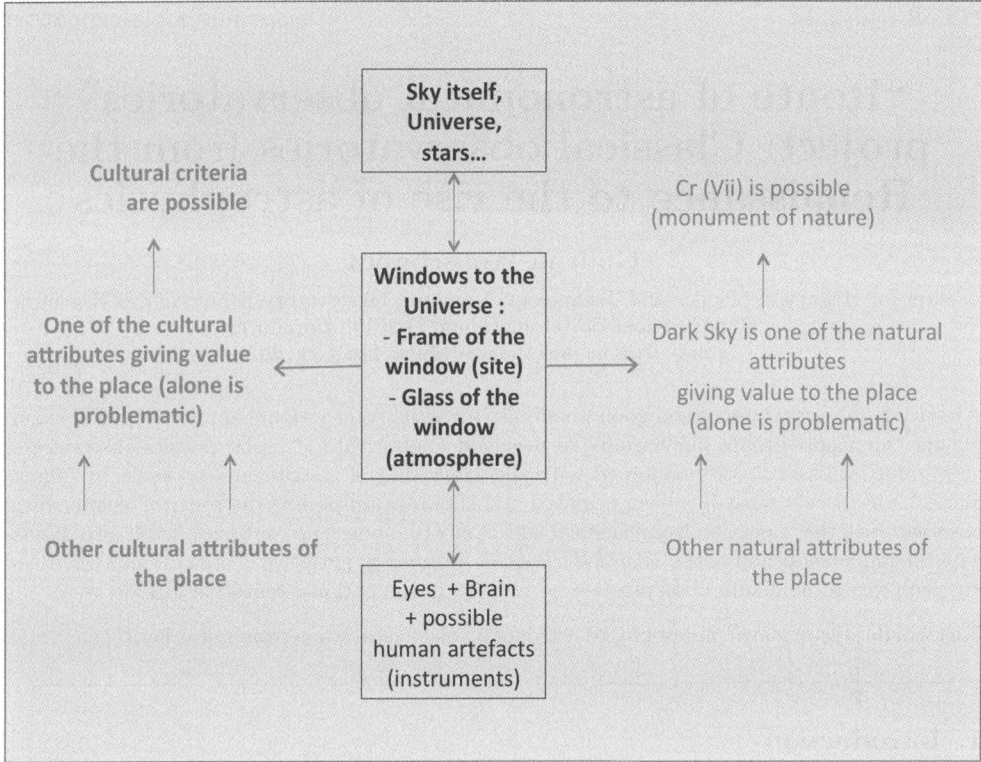

Figure 1. Basic features shaping the "Windows to the Universe" concept and ways to use Dark Sky value among other natural and/or cultural attributes. Diagram by the author

generally remarkable landscape both by night and by day. Potential OUV results from the combination of a set of remarkable attributes that contribute to the significance and beauty of the whole.

Dark skies can also be recognized among a cluster of cultural attributes or mixed attributes. Dark sky quality could be considered as a cultural attribute in the context of the history of the place as an observatory: thus, the remarkable atmospheric quality of a given place explains its identification as a "Window to the Universe" and its choice for a High-Mountain Observatory. The other good way to use "dark sky" value is to link it to a set of remarkable cultural attributes or—equally often—a mixture of cultural and natural features.

These considerations are summarized in Fig. 1.

A related issue is Dark Sky management, both for scientific efficiency and for heritage preservation. The international success and importance of Dark Sky Reserve initiatives has presented a strong challenge to control human light pollution. The maintenance of dark sky quality could certainly, in itself, be a strong management goal for an astronomical site and region. This has implications relating to the WH Buffer Zone concept.

References

UNESCO 2015, *The Operational Guidelines for the Implementation of the World Heritage Convention* (Paris: UNESCO) http://whc.unesco.org/en/guidelines

"Route of astronomical observatories" project: Classical observatories from the Renaissance to the rise of astrophysics

Gudrun Wolfschmidt

Center for History of Science and Technology, Hamburg Observatory, University of Hamburg
Bundesstrasse 55 Geomatikum, D-20146 Hamburg,
email: Gudrun.Wolfschmidt@uni-hamburg.de

Abstract. Observatories offer a good possibility for serial transnational applications. For example one can choose groups like baroque or neoclassical observatories, solar physics observatories or a group of observatories equipped with the same kind of instruments or made by famous firms. I will discuss what has been achieved and show examples, like the route of astronomical observatories, the transition from classical astronomy to modern astrophysics. I will also discuss why the implementation of the *World Heritage & Astronomy* initiative is difficult and why there are problems to nominate observatories for election in the national tentative lists.

Keywords. history and philosophy of astronomy, observatories, astronomical heritage

1. Introduction

Cultural Heritage of observatories contains (monuments, groups of buildings) sites with historical, aesthetic, archaeological, scientific, etc. value. From the ten criteria for the assessment of *Outstanding Universal Value* (OUV) mainly three (UNESCO criteria ii, iv and vi) are well suitable for classical observatories from Renaissance to 20th century.

We find astronomical observatories not only in Europe but all over the world, cf. Ruggles & Cotte (2011) and here: *Commission C4 on World Heritage and Astronomy* (http://www2.astronomicalheritage.org/). But I will not include here the Islamic observatories like Samarkand – important in the Middle Ages – or the five Indian Observatories in Delhi, Ujjain, Mathura, Varanasi and Jaipur (Maharaja Jai Singh II), the last of which is already on the UNESCO World Heritage List since 2010. My emphasis in this contribution is about occidental astronomical heritage but there are not only observatories involved! There are also astronomical clocks, Baroque frescos like Michelangelo's "Creation of Sun and Moon" in the Sixtin Chapel or ceiling frescos in the *Mathematical Hall* in Prague. One should also mention predecessors of the planetarium like the *Gottorf globe* (Schleswig castle, now in St. Petersburg) or *Eise Eisinga Planetarium* in Franeker. But here I would like to present four groups of examples of observatories from Renaissance to the 20th century, which offer a good possibilities for serial transnational applications. A well-known example for a thematic programme is the Struve arc, already recognized as World Heritage in 2005.

2. Observatories for the Calendar Reform

Nicolaus Copernicus (1473–1543) measured the length of the solar year in the castle in Allenstein/Olsztyn, Poland. Cathedrals were also used as solar observatories – *"beauty and utility"* (Heilbronn (1999)). Here some examples for cathedrals with meridian lines: 67-m-meridian line in San Petronio in Bologna, 1655, made by Jean Dominique Cassini

(1625–1712), Santa Maria del Fiore in Florence, St. Sulpice in Paris, Santa Maria degli Angeli in Rome, *Tower of the Winds* with Meridian Room (1578–1580) in Vatican City for the *Gregorian Calendar Reform* (1582). Meridian lines were also integrated in observatories in Italy and in Breslau/Wrozław.

In the Renaissance and early Baroque time the first real observatories were built. I skip also these observatory buildings of famous astronomers which are no longer existing like Tycho' or Hevelius' observatory on Island Hven 1576 to 1599) and in Danzig/Gdansk (1649 to 1679) respectively.

3. Baroque Roof and Tower Observatories, 17th/18th century

In the 17th and 18th century for the Baroque observatories step by step a special architecture was developed, I show some examples (cf. Müller (1978)). The Round Tower (Rundetårn) Copenhagen (1642) was used by the University of Copenhagen until 1861. The prominent Royal observatories of the 17th century like Paris (Claude Perrault, 1667), Greenwich (Sir Christopher Wren, 1675) and Stockholm (Carl Hårleman, 1753) are well-known.

Some examples for roof observatories are the University Observatory (1711–1803) in Altdorf near Nuremberg, the observatory in St. Petersburg on the roof of the Royal Academy (1725), the observatory of Palermo (1791) using the Norman tower of the Royal Palace or the *Mathematical Tower* (1811) – the impressive Breslau (Wrozław) University Observatory.

Many Baroque observatories are tower observing posts, old buildings are re-used and changed for astronomical purpose: *La Specola* in Bologna (1721–1725) – first University Observatory in Italy, built on a fortification tower, and *La Specola*, Osservatorio Astronomico di Padova (1761).

A remarkable group form the observatories in Jesuit colleges (cf. Udías (2003)); they can be found all over the world, for example Beijing (1669), India, Philippines, Middle and South America, and Georgetown Observatory in Washington D. C. (1841–1844) in the USA as well in Europe: Prague the Jesuit College *Clementinum* (1722–1725) with the *Astronomical Tower*. In Mannheim the Jesuit tower observatory (1772/74) was used for astronomy and meteorology; the *Societas Meteorologica Palatina* (1780) had set up the *Mannheim Hours*, a standard until today. Vilnius Observatory, Lithuania, two towers erected on the top of the three-storey university building, is a spectacular Jesuit observatory (1753).

Concerning the group of Baroque monastery observatories I would like to mention the *Csillagász* (observatory) in Eger, Hungary, (1785) with meridian line and with a *Camera Obscura* (1779), for observing the Sun and (today) the town. The most impressive building, the "sky scraper" of the 18th century is the *Specula Cremifanensis*, the *Mathematical Tower* in Kremsmünster (1758).

4. Neoclassical Observatories around 1800 in the shape of the Greek cross – Invention of the Dome

The dome, a new feature – which is now regarded as typical for observatories – came up before 1800; the dome symbolizes the sky. Early examples are Zach's influential *Seeberg Observatory* Gotha (1788), no longer existing, the Dunsink Observatory of Trinity College in Dublin (1783–1785), Armagh Observatory (1790, restored), and Göttingen Observatory (1802–1816, Georg Heinrich Borheck), designed for Carl Friedrich Gauß (1777–1855), preserved in an excellent way in original condition.

The typical architecture of observatories around 1800 is characterized by observatories in the shape of a greek cross (four wings of equal length) in neoclassical style with a central rotating dome (sometimes still in a cylindrical or conic shape). Many famous architects were involved: Madrid (1790, Juan de Villanueva), Tartu (1810), Naples (1820), Warsaw (1820–1824), Cape Town (1828), Christiania/Oslo (1831, Chr. H. Grosch), Turku/Åbo (1819, Carl Ludwig Engel), Berlin (1835, Karl Friedrich Schinkel, demolished 1915), Bonn (1844, Karl Friedrich Schinkel), Athens (1846, Theophil Hansen) and later the ETH Observatory Zürich (1861–1864, Gottfried Semper). Very striking due to the six "domes" is the *Observatorio Astronomico* in Quito (1873), Ecuador, inspired by Bonn and with the help of the German Father Juan Bautista Menten, built in Victorian style; Quito is the oldest observatory in South America. A final example is the Vienna University Observatory (1883, Ferdinand Fellner and Hermann Helmer), the world's largest observatory, a "Theatre for Stars" as the director Karl Littrow (1811–1877) called it and he wanted to reside like a prince of science (cf. Schnell (2009), p. 142-149). Very impressive is also the large staircase in the entrance hall like in a theatre or a palace; the hall is covered with a glass roof. But the building was not extremely suitable for astronomical observations.

The architect Carl Ludwig Engel (1778–1840), a friend of Schinkel, designed the Helsinki Observatory (1834); it was among the most modern observatories of its time besides Kazan old observatory (1837, Mikhail P. Korinfsky). The new idea of the 1830s was the three domes facade (in the beginning still cylindrical domes), because more instruments were in use (refractor, heliometer, meridian circle and transit instrument). The good design principles of Helsinki Observatory were adopted in Russia for the design of the Pulkovo Central Observatory in St. Petersburg (1839, Alexander Brüllow), the leading observatory in that field, an outstanding example for classical astronomy; thus Pulkovo became a prototype for the 19th century observatory architecture, cf. Lissabon (1861), Yerkes Observatory, USA (1897), Georgetown College Washington, D. C. (1841–1844) and even the *Astrophysical Observatory* (APO) Potsdam (1874/79) as well as the *Deutsches Museum* München (1925) had this characteristic front.

5. Observatories around 1900 – From classical astronomy to the rise of modern astrophysics

Important features for this serial transnational group of observatories are the following:
- Change of the research field – from classical astronomy to modern astrophysics
- Impressive architecture (well preserved and renovated)
- Ensemble of buildings in a park or a mountain observatory, cf. *Observatoire de la Côte d'Azur*, Nice, France (1888), cf. Le Guet Tully (2009)
- Change of the instrumentation – old: meridian circle, refractor, heliometer – new: astrograph, reflector, Schmidt telescope and several astrophysical instruments.

Classical astronomy includes time keeping – providing the official time signal for the city and for navigation; around 160 time-balls existed, about 60 are still existing, many in English speaking countries (cf. Howse (1997)), e. g. time balls in Greenwich (1933), Cape of Good Hope (1836), Washington, D. C. (1845), Liverpool (1845), Edinburgh (1852), San Francisco (1852), Sydney (1858), Lyttelton in New Zealand (1876), New York (1877), and Hong Kong (1885), Danzig/Gdansk (1894), Hamburg (1876 to 1934) and other places in the world; this could be an interesting serial transnational group of observatories.

The other task of classical astronomy is determining stellar positions with meridian circles and compiling star catalogues with the coordinates. Examples for this serial transnational group are the French observatories including Nice and Algier with French instruments and Strasbourg, Munich, the Ukraine observatories Nikolaev (Mykolayiv)

(1821–1829) and Kiev (1845), Rio de Janeiro (1827, new 1913 to 1920) and Hamburg-Bergedorf (1802, 1825 and 1912) equipped with Repsold instruments.

The rise of astrophysics, the "new astronomy", as Simon Newcomb (1835–1909) called it in 1888, started around 1900: *"that the age of great discoveries in any branch of science had passed by, yet so far as astronomy is concerned, it must be confessed that we do appear to be fast reaching the limits of our knowledge."* (Newcomb (1888))

The new topics in astrophysics were photometry, astrophotography, spectroscopy / spectralanalysis and solar physics (for a detailed discussion see Wolfschmidt (2009)). Karl Friedrich Zöllner (1834–1882) coined the expression "astrophysics" in 1865 and presented a first instrument for measuring the magnitudes, the wide spread Zöllner photometer (1860). In astrophotography first portrait objectives were used (Max Wolf in Heidelberg, Edward Emerson Barnard in Lick and Chicago observatories). The Frères Henry had an excellent idea to set up the *Carte du Ciel* project (Paris, 1887), this cooperation for 18 zones of the sky with a standard astrograph (34cm, 3,40m, this ratio of 1:10 offered for measuring the plates that 1mm on the plate corresponds to $1'$ in the sky). But this project turned out to be unsuccessful in the end. In 1964 finally the catalogue was published, the photographic chart of the sky was never completed. But soon around 1900 in Hamburg and in the USA astrographs were used for astrophysics, for spectral surveys of the sky, using an objective lens prism. And with the invention of the Schmidt telescope this photographic map of the sky could be done much quicker and also in different colours, more useful for astrophysics (the "Big Schmidt" did it in seven years with blue and red sensitive plates). The next important step for astrophysics was the introduction of the glass reflecting telescope, important for spectroscopy; in Europe the leading firm became Zeiss of Jena (70-cm-Waltz Reflector Heidelberg 1907 and 1-m-reflector Hamburg 1911). The large reflectors were built in the USA, cf. in Mt. Wilson the $60'' = 1.5$-m-reflector (1904) and the $100'' = 2.5$-m-Hooker-Telescope (1917).

Twelve observatories (Algiers, La Plata, Rio de Janeiro, Greenwich, Tartu, Paris-Meudon, Nice, Hamburg, Kodaikanal, Lisbon, Pulkovo, US Naval Obs. Washington D.C.) were chosen in 2008 in discussion with the Conservation Department of Hamburg and with Prof. Dr. Michael Petzet, former president of ICOMOS. One should add the Cape Observatory (cf. Glass (2015)) and the Canadian *Dominion Observatory* (1902 to 1970) in Ottawa for classical astronomy in combination with the *Dominion Astrophysical Observatory* (1918) Victoria, B.C., especially added for astrophysical research. These important observatories contributed remarkably to astronomical science of that time, to our modern view of the world, as well as Mt. Wilson, Einstein Tower, Meudon and Pic du Midi in the field of solar physics.

6. Conclusion

It is not difficult to develop ideas for cooperations, for a serial transnational application. But the realization is very difficult. The observatories have to be acknowledged in the states where the monument protection people mainly come from history of art and prefer castles and churches as cultural heritage. And if you succeed with this, you have to convince the politicians who change every four years. And then you have to restart. And the partner observatories have the same problems.

But in general observatories offer a good possibility for serial transnational applications. A well-known example for a thematic programme is the Struve arc, already recognized as World Heritage. I have shown groups like Renaissance, Baroque or neoclassical observatories, which can be chosen for serial transnational applications. There are other possibilities like solar physics observatories (Wolfschmidt (2005)) or a group of observatories equipped with the same kind of instruments (meridian circles) or made by famous firms. The group of colonial observatories of Spain, Portugal, France (Algiers (1890), cf. Irbah et al. (2001)) and the Netherlands needs to be studied much more.

I have discussed what has been achieved in the route of astronomical observatories project, this group of observatories around 1900 showing the transition from classical astronomy to modern astrophysics (cf. La Plata, Hamburg, Nice, etc.), visible in the architecture, the choice of instruments, and the arrangement of the observatory buildings in an astronomy park. This corresponds to the main categories according to which the *"outstanding universal value"* (UNESCO criteria ii, iv and vi) of the observatories have been evaluated: historic, scientific, and aesthetic. This proposal is based on the criteria of a comparability of the observatories in terms of the urbanistic complex and the architecture, the scientific orientation, equipment of instruments, authenticity and integrity of the preserved state, as well as in terms of historic scientific relations and scientific contributions. This is perhaps the most promising group for a serial transnational UNESCO application.

References

Glass, Ian S. 2015, *The Royal Observatory at the Cape of Good Hope – History and Heritage.* Cape Town: Mons Mensa Publishing.

Heilbronn, John L. 1999, *The Sun in the Church: Cathedrals as Solar Observatories.* Harvard University Press.

Howse, Derek 1997, *Greenwich Time and the Longitude.* London: Philip Wilson 1997.

Irbah, Abdenour; Abdelatif, Toufik & Sadsaoud Hamid 2001, *Astronomy in Algeria: past and present developments.*, *PASP* In: Astronomy for developing countries. Edited by Alan H. Batten. San Francisco: Astronomical Society of the Pacific, p. 171–178.

Le Guet Tully, Françoise 2009, Raphaël Bischoffsheim und die Gründung eines astronomischen Observatoriums an der französischen Riviera. In: Wolfschmidt 2009, p. 111–129.

Müller, Peter 1978, *Sternwarten. Architektur und Geschichte der astronomischen Observatorien.* Bern, Frankfurt am Main: Peter Lang Verlag (Europäische Hochschulschriften. R. 32, 1) 1975, 2nd edition 1978.

Newcomb, Simon 1888, The Place of Astronomy among Sciences. In: *Sideral Messanger* 7, p. 14–20, 65–73.

Udías, Agustín 2003, *Searching the Heavens and the Earth: The History of Jesuit Observatories.* Berlin: Springer Science & Business Media.

Ruggles, Clive & Michel Cotte (ed.): *Heritage Sites of Astronomy and Archaeoastronomy in the context at the UNESCO World Heritage Convention. A Thematic Study.* Paris: International Council on Monuments and Sites, Monuments and Sites (ICOMOS) and International Astronomical Union (IAU) 2011.

Schnell, Anneliese 2009, The University Observatory Vienna. In: Wolfschmidt 2009, p. 142–149.

Wolfschmidt, Gudrun 2005, Development of Solar Tower Observatories. In: *Development of Solar Research. Entwicklung der Sonnenforschung.* Edited by Axel Wittmann, Gudrun Wolfschmidt und Hilmar Duerbeck. Frankfurt am Main: Harri Deutsch (Acta Historica Astronomiae Vol. 25), p. 169–198.

Wolfschmidt, Gudrun (ed.) 2009, *Cultural Heritage: Astronomical Observatories (around 1900) – From Classical Astronomy to Modern Astrophysics.* Proceedings of International ICOMOS Symposium, Hamburg, October 15–17, 2008. Berlin: Hendrik Bäßler (Monuments and Sites; Nr. XVIII).

Are Historical Observations "Ancient" or "Modern"?

R. Elizabeth Griffin

Dominion Asrophysical Observatory, 5071 West Saanich Road, Victoria, V9E 2E7, Canada
email: elizabeth.griffin@nrc-cnrc.gc.ca

Extended Abstract. The demarcation between "old", "historic" and "heritage" is fuzzy. To a large degree it depends upon purpose and usefulness, and it will always be subjective. At what point does the intrinsic value of an historic item outpace the mystique associated just with its age? When, for instance, does an "old" car become a "vintage" car? When do archived astronomical records contribute something of quantitative value to science? When can they be extricated from the realms of the museum and placed in the context of modern research?

Celestial objects vary. Some do so explosively, often irreversibly; many vary periodically over time-scales from a hour or less to a century or more. Furthermore, all celestial objects change as they evolve, mostly so slowly as to be practically imperceptible, but while the general time-scale of that evolution is millions of years there are a few stages (such as the collapse from AGB towards planetary nebula and white dwarf) which happen rather suddenly, and invaluable examples of "before–after" can be found in some plate stores. Astrophysics has a comprehensive need to investigate the nature and time-scales of all types of change, especially ones which *only* access to its "heritage" data can describe. Surely in this day and age we have enough tools, capacity and technologies to fulfil such a basic requirement?

The frustrating answer is that we do have some of the necessary tools, and most of the technologies, but as a community we lack "capacity" if that means manpower and funds. The problem is a technical one of accessing the older data in useable formats; it was generated by the universal change in detector technology from photography to electronic device, an exciting development in efficiency and scope that heralded a new era of research capability and data management, archiving and sharing, but it left pre-digital photographic data right out of the picture. Developments of that nature should have made research *more* inclusive, instead of the seriously exclusive picture that is currently seen. The longer the situation prevails, the greater the inertia and scepticism to be overcome.

Fortunately, some of the challenges *are* being tackled successfully, the most productive to date being the DASCH project (dasch.rc.fas.harvard.edu) at Harvard College Observatory to digitize and share all the images and objective-prism spectra from its collection (the world's biggest) of over 0.5M large plates. The DAO has commenced a programme to digitize its collection of > 16,000 high-dispersion spectra (~70% are good enough to scan and convert), and to scan plates from its larger but older Cassegrain collection of > 90,000 spectra upon request. The instrument for this Herculean task is its own PDS, now suitably upgraded to meet the demands of speed and accuracy; the DAO has also acquired and upgraded a second PDS, with which it plans to share the load. Some smaller observatories in Europe are trying with less sophisticated equipment, but the rest have not the resources to give such data transformation any priority. Despite the unquestionable advantages, it is still necessary to convince colleagues that the medium is not the message, and that the scientific need comes before technological expedience.

BAO Plate Archive digitization, creation of electronic database and its scientific usage

A. M. Mickaelian, H. V. Abrahamyan, H. R. Andreasyan,
N. M. Azatyan, S. V. Farmanyan, K. S. Gigoyan,
M. V. Gyulzadyan, K. G. Khachatryan, A. V. Knyazyan,
G. R. Kostandyan, G. A. Mikayelyan, E. H. Nikoghosyan,
G. M. Paronyan and A. V. Vardanyan

Byurakan Astrophysical Observatory (BAO), Byurakan 0213, Aragatzotn Province, Armenia
email: aregmick@yahoo.com

Abstract. Astronomical plate archives created on the basis of numerous observations at many observatories are important part of the astronomical heritage. Byurakan Astrophysical Observatory (BAO) plate archive consists of 37,000 photographic plates and films, obtained at 2.6m telescope, 1m and 0.5m Schmidt telescopes and other smaller ones during 1947–1991. In 2002-2005, the famous Markarian Survey (First Byurakan Survey, FBS) 2000 plates were digitized and the Digitized FBS (DFBS) was created. New science projects have been conducted based on these low-dispersion spectroscopic material. In 2015, we have started a project on the whole BAO Plate Archive digitization, creation of electronic database and its scientific usage. The project will run during 3 years in 2015–2017 and the final result will be an electronic database and online interactive sky map to be used for further research projects.

Keywords. telescopes, instrumentation: miscellaneous, methods: data analysis, techniques: image processing, techniques: spectroscopic, standards, astronomical data bases: miscellaneous

1. Introduction

A project on **Digitization of Byurakan Astrophysical Observatory (BAO) Plate Archive and creation of BAO Interactive Astronomical Database** (shortly BAO Plate Archive project, BAO PAP) has started in 2015. It is aimed at preservation of BAO valuable observational material accumulated during 1947-1991, creation of full Database of all BAO observations, creation of BAO Interactive Sky Map with visualization of all observations and quick access to the data, development and accomplishment of new research projects based on the existing observational material, and integration of BAO observations into the international databases. A number of BAO young astronomers are involved in this project and it will last 3 years.

BAO Plate Archive is one of the largest astronomical archives in the world and is considered to be BAO main observational treasure. Today BAO archive holds some 37,000 astronomical plates, films or other carriers of observational data.

2. BAO telescopes and observing programmes

BAO observers worked with a number of BAO telescopes during 1947-1991 and obtained several dozens of thousands plates, films and other products. The table gives general understanding on observations of 10 BAO telescopes that worked on photographic photometry, electrophotometry, slit and objective prism spectroscopy, and polarimetry of many thousands astronomical objects. We give in Table 1 an overview of BAO telescopes

Table 1. Overview of BAO telescopes and produced observational material.

Telescope	Sizes [cm]	Years	Observing methods	Plates
5" double–astrograph	13	1947–1950	photometry	3000
6"	15	1947–1950	photometry	3000
8" Schmidt	20/20/31	1949–1968	photometry	4500
20" Cassegrain	51/800	1952–1991	electrophotometry	
10" telescope–spectrograph	25	1953–19??	spectra	
Nebular spectrograph		1954–19??	spectra	
16" Cassegrain	41/400	1955–1991	electrophotometry	
21" Schmidt	53/53/183	1955–1991	photometry	12000
40" Schmidt (AZT–10)	102/132/213	1960–1991	photometry, spectra	7500
ZTA–2.6m	264/1016	1975–1991	photometry, spectra	7000
All telescopes		1947–1991		37000

and produced observational material. Telescope "Sizes" are given for the mirror and focal length for classical telescopes and for the correcting lens, mirror and focal length for Schmidt type telescopes.

Main observational projects run at three most important BAO telescopes were: **21" (0.5m) Schmidt** (Polarization of cometary nebula NGC 2261, Nuclei of nearby Sa and Sb galaxies, Nuclei of nearby Sc galaxies, Search for flare stars in Pleiades, Orion, NGC 7000 (Cygnus), Praesepe and Taurus Dark Clouds (TDC), Variability of Markarian galaxies, Monitoring of extragalactic supernovae in certain areas, etc.), **40" (1m) Schmidt:** (Detailed colorimetry of bright galaxies, First Byurakan Survey (FBS, Markarian survey; Markarian 1989), Search for flare stars in Pleiades, Orion, NGC 7000 (Cygnus), Praesepe and Taurus Dark Clouds (TDC), Second Byurakan Survey (SBS; Stepanian 2005), Extension of the FBS in the Galactic Plane, etc.) and **ZTA-2.6m telescope:** (Morphological study of Markarian galaxies, Investigation of star clusters, Investigation of groups and clusters of galaxies, Spectroscopy FBS blue stellar objects, FBS late–type stars, SBS galaxies and stellar objects (BAO/SAO), T Tauri and flare stars, Byurakan-IRAS Galaxies (BIG objects) and ROSAT AGN candidates (BAO/HS/OHP/INAOE), and Direct images of the central regions of Markarian galaxies).

Especially efficient were Byurakan surveys accomplished by Markarian and colleagues.

3. BAO Plate Archive Project

The digitization of astronomical plates and films pursues not only the maintenance task, but also it will serve as a source for new scientific research and discoveries, if only the digitized material runs according to modern standards and, due to its accessibility, it will become an active archive. The project is aimed at compilation, accounting, digitization of BAO observational archive photographic plates and films, as well as their incorporation in databases with modern standards and methods, providing access for all observational material and development of new scientific programs based on this material.

Scientific Programs Board (SPB) is created to evaluate the existing observational material, to select sets of priorities to be scanned first and to propose new research projects. It consists of BAO Director and most experienced BAO observers, as well as researchers from NAS RA Institute of Informatics and Automation Problems (IIAP) are involved for their experience in computer science related to databases and computational methods. Project Executing Team (PET) consists of 14 members. The scanning will be carried out with 2 EPSON Perfection V750 Pro scanners.

The project consists of the following works: Creation of the Project Database, Project Webpage and User Interface, Scanning of photographic plates and films, Astrometric solution, Extraction of images and spectra, Wavelength calibration (for spectra), Density

Figure 1. BAO Plate Archive Project webpage.

and flux calibration, Making up template low-dispersion spectra, Numerical classification of spectra, Visualization of BAO observations on sky map, Creation of electronic interactive sky map and search system, Scientific analysis of existing observational material and providing new research possibilities, and Proposing and discussing new research projects.

BAO PAP webpage (http://www.aras.am/PlateArchive/) was recently open and contains information on BAO observations, previous digitization projects, present Project details, teams, follow-up research projects, deliverables and related links (many items will be filled in during the next months). However, the main products will be **"Data Access"** and **"Interactive Sky Map"**. The first one will contain BAO Observational Database, Search by any parameter, Data Visualization and Download of the digitized plates, films, part of them or individual objects images or spectra. "Interactive Sky Map" will visualize the observed by BAO telescopes sky and will give possibility to check observed areas for a given observational project, given telescope, observer, observing method, limiting magnitude, etc. There will be possibility to check individual fields for presence and number of plates to propose further research projects.

4. Previous digitization programmes at BAO and research projects

A number of digitization projects have been accomplished at BAO, including the most important one, **Digitized First Byurakan Survey (DFBS**; Mickaelian *et al.* 2007) based on the digitization of the famous Markarian Survey (Markarian *et al.* 1989). Pixel size is 15.875μm or 1.542", each plate is 9601×9601 pixels and each plate is 180 MB file. Each low-dispersion spectrum is 107×5 pixels (1700μm length). Astrometric solution has ∼1" rms. Average dispersion is 33Å/pix (22–60 from blue to red part). Photometric accuracy is ∼0.3^m. In total there are 1874 digitized plates and the database is ∼400GB. There are 40,000,000 spectra for 20,000,000 objects. DFBS plate database is available in Vizier, Strasbourg (Mickaelian *et al.* 2005). The spectra extraction and analysis software is described in Mickaelian *et al.*(2010) and Knyazyan *et al.* (2011).

The Second Byurakan Survey (SBS; Stepanian 2005) plates are also subject for digitization, as they are hypersensitized and their emulsion is more sensitive for deterioration. 180 plates have been digitized so far. Due to SBS smaller photographic grains, 2400 dpi (10μm pixel size) is being used and 512 MB files are being obtained for each plate.

Photographic spectra of the FBS blue stellar objects (BSOs) have been obtained using 2.6m telescope and UAGS spectrograph on photographic films. ∼700 such spectra have been scanned with 1600 dpi, 16 bit and 650×21 pix sizes images were obtained (FBS

BSOs; Mickaelian 2008 & late-type stars; Gigoyan & Mickaelian 2012). All spectra were put in a standard format, so that automatic reduction was possible. 101 FBS blue stellar objects were published and a number of planetary nebulae, white dwarfs, hot subdwarfs and HBB stars have been revealed (Sinamyan & Mickaelian 2009).

Another project was the study of long-term variability of ON 231, which appeared in the Coma field, where photographic chains for discovery of flare stars were carried out. In total 189 plates with a total number of more than 1200 exposures in 1969–1976 with the Byurakan 21" and 40" Schmidt telescopes were obtained. This was a valuable material for study of ON231 long-term variability (Erastova 2004).

5. Summary

BAO Plate Archive Project will lead to preservation of BAO valuable observational material obtained during 1947-1991. However, our goal is not only to create a passive archive of scanned plates and films, but also to make use of especially those fields, where more studies are possible. Proper motion and variability studies are most important, as time domain material is contained in historical plates.

There are a number of further possible research projects that will be conducted having the plates digitized, such as Correction of ephemerides of known asteroids and search for new asteroids (Berthier *et al.* 2009), Discovery and study of variable stars, Revealing high proper motion stars, Study of variability of known blazars and discovery of new blazars, Revealing Novae and Supernovae progenitors, Discovery of new QSOs, new white dwarfs, new late-type stars (Gigoyan *et al.* 2010) and optical sources of gamma-ray bursts, Optical identifications of X-ray, IR and radio sources (Mickaelian & Sargsyan 2004; Mickaelian & Gigoyan 2006; Mickaelian *et al.* 2006; Hovhannisyan *et al.* 2009).

Armenian Virtual Observatory (ArVO) was created in 2005 to maintain and actively use DFBS database, and BAO Plate Archive Project will enrich it with many more images and spectra that are subject for further studies together with other available multi-wavelength and multi-time-domain data (Mickaelian *et al.* 2010).

References

Berthier J., Sarkissian A., Mickaelian A., & Thuillot W. 2009, *EPSC*, 4, 526
Erastova, L. K. 2004, *A&AT*, 23, 209
Gigoyan K. S. & Mickaelian A. M. 2012, *MNRAS*, 419, 3346
Gigoyan K. S., Sinamyan P. K., Engels D., & Mickaelian A. M. 2010, *Astrophysics*, 53, 123
Hovhannisyan L. R., Weedman D. W., Mickaelian A. M., *et al.* 2009, *AJ*, 138, 251
Knyazyan A., Mickaelian A. M., & Astsatryan H. 2011, *Int. J. "Inf. Theories and Appl."*, 18, 243
Markarian, B. E., Lipovetsky, V. A., Stepanian, J. A., *et al.* 1989, *Comm. SAO*, 62, 5
Mickaelian, A. M. 2008, *AJ*, 136, 946
Mickaelian, A. M. & Gigoyan, K. S. 2006, *A&A*, 455, 765
Mickaelian, A. M., Hagen, H.-J., Sargsyan, L. A., & Mikayelyan, G. A. 2005, *Cat. VI/116 at CDS*
Mickaelian, A. M., Hovhannisyan, L. R., Engels, D., *et al.* 2006, *A&A*, 449, 425
Mickaelian, A. M., Nesci, R., Rossi C., *et al.* 2007, *A&A*, 464, 1177
Mickaelian, A. M. & Sargsyan, L. A. 2004, *Astrophysics*, 47, 213
Mickaelian, A. M., Sargsyan, L. A., & Mikayelyan, G. A. 2010, *Proc. of Science*, 30
Mickaelian, A. M., Sargsyan, L. A., Nesci, R., *et al.* 2010, *ASP CS, Vol. 434: ADASS XIX*, p. 325
Sinamyan, P. K. & Mickaelian, A. M. 2009, *Astrophysics*, 52, 76
Stepanian, J. A. 2005, *RMxAA*, 41, 155

Odyssey of Human Creative Genius: From Astronomical Heritage to Space Technology Heritage

Olga Dluzhnevskaya[1] and Mikhail Marov[2]

[1]Institute for Astronomy, Russian Academy of Sciences,
National Focal Point for the Russian Federation, Astronomy and World Heritage Initiative
email: olgad@inasan.ru

[2]Vernadsky Institute, Russian Academy of Sciences,
Nominated Chair of WG on Space Technology, UNESCO WH Committee
email: marovmail@yandex.ru

1. Introduction and Historical Highlights

Astronomy was one of the most important sciences in the ancient world. It was rooted in naked eye observations and primitive stone instruments for astrometric measurements to determine the positions of the Sun, Moon, planets and some stars that had both practical and sacred meaning. That is why the majority of archaeoastronomical monuments are simultaneously observatories and sanctuaries, with burials and altars.

Interest in the investigation of ancient monuments as instruments for astronomical observations has grown significantly in recent decades. Since 1981, the "Oxford" international conferences on archaeoastronomy have been held every 3–5 years, organized since 1995 by the International Society for Archaeoastronomy and Astronomy in Culture (ISAAC) which was founded in the USA in that year. The European Society for Astronomy in Culture (SEAC) also deals with the problems of archaeoastronomy and holds scientific conferences annually. In Russia, the first conference on archaeoastronomy was held in 1996 in Moscow, followed by the SEAC conference "Astronomy of ancient civilizations" during the JENAM meeting in 2000, and the International symposium on "Astronomy 2005—modern state and prospects" in 2005.

A new milestone started with the UNESCO Initiative on "Astronomy and World Heritage". Throughout the years, UNESCO has been working hard to preserve humankind's achievements, astronomy included, as World Heritage. The breathtaking monuments of ancients civilizations such as the Decorated Grottoes of The Vézère Valley (France), Stonehenge (Great Britain), the Lines of Nasca (Peru), the Pyramids of Giza (Egypt), the Temple of Heaven (China), and Ulugh Beg's Observatory in Samarqand (Uzbekistan), to mention a few, have been recognized. These also bear an invaluable educational mission.

In 2003, the UNESCO World Heritage Centre (WHC) set up a new project to reveal and preserve the objects of archaeoastronomy of historical and cultural value all over the world. The First International Meeting of experts "Archaeoastronomical objects and observatories", organized by the WHC and the Regional European Bureau on Science, was held in 2004 in Venice, Italy. The strategy of the thematic program "Astronomy and World Heritage" and the general criteria for the selection of archaeoastronomical sites/observatories were considered, including (i) Objects situated or related to celestial objects or astronomical events; (ii) Images of the sky and/or of celestial objects and astronomical events; (iii) Observatories and instruments; and (iv) Objects closely connected with the history of astronomy.

2. Space Exploration: Baseline and Venue

Space exploration manifested a new great milestone in the development of human civilization. It made possible observations in every wavelength with incremental precision and, also, direct *in situ* measurements on other worlds. The Astronomy and World Heritage Initiative should encompass various facilities related to space exploration. Basically, great breakthroughs of modern astronomy were achieved thanks to space-born instruments and planetary space missions. Indeed, space astronomy ensured very significant progress in astrophysics, gaining invaluable knowledge about space objects and the Universe as a whole, thus broadening human horizons tremendously. It is therefore essential to include heritage related to Space Astronomy as an important segment of astronomical World Heritage.

The project on technological heritage connected with space exploration arises as a logical extension of the Astronomy and World Heritage Initiative because it is intrinsically related with the most important breakthroughs in space science and additionally, it is rooted in space technology. The idea was put forward at the Astronomy and World Heritage meeting held in Kazan, Russia in August, 2009 (ASTROKAZAN 2009), subject to further discussions and clarifications. A first step towards the goal has been undertaken in the contributions by David DeVorkin and Mikhail Marov on Space Achievements as World Heritage that form Ch. 15 of the first ICOMOS-IAU Thematic Study on the Heritage Sites of Astronomy and Archaeoastronomy (Ruggles & Cotte (2010)). It is generally understood that the proposed segment on Space Astronomy/Technology World Heritage should have an international significance in terms of human beings' tight relationships with the sky. Obviously, an international team of specialists should be further involved in the process of collecting respective materials/documentation and writing proposals to accommodate the World Heritage Convention.

A synopsis of the most important events along the way is as follows:
• Round-table discussion in the Russian Academy of Sciences aiming to define a core of the Thematic Initiative on the heritage of science and technology (2005);
• 50th anniversary of the Sputnik: *"Fenêtres sur le Cosmos: Spoutnik et l'Aube de l'Age Spatial"* organized at the French Senate by ESA/CNES, Paris, France;
• Thematic research proposal "Odyssey of human creative genius: towards the protection of space technological heritage connected with space exploration" (2007);
• International Conference on "Astronomy and World Heritage: Across Time and Continents", Kazan, Republic of Tatarstan, Russia (2009);
• ICOMOS/IAU Thematic Study on the Heritage of Astronomy, Cairo, Egypt (2010);
• International Seminar on the Heritage of Astronomy in the Institute of Astrophysics, Paris, France (2011); and
• UNESCO World Heritage Committee meeting in St. Petersburg, Russia (2012).

At the Kazan conference, a preliminary definition was given for the first time of types of technological sites and facilities connected with space exploration, and it was proposed how to identify the clearly defined landscape designed and created intentionally by man and intrinsically associated with the Launch Pad and the development of related structures. The focus was upon the most prominent parts of space networks specially designed and built for manned flights and the historical sites where the concepts of spaceflight were pioneered and original space vehicle designs were tested. At the Cairo meeting, space heritage was additionally specified as heritage related to the process of carrying out science in space, heritage related to manned space flight/exploration, and human cultural heritage that remains off the surface of planet Earth. In turn, at the Paris meeting the tentative proposal of fixed sites and facilities pertaining specifically to space astronomy and/or generally to space science—in particular, ground space facilities and launch pads (cosmodromes)—was discussed.

The UNESCO World Heritage Committee in St. Petersburg in 2012 preliminarily endorsed Space Astronomy as a segment of Astronomical Heritage and, basically, as a segment of Space Technology Heritage. It was recommended to set up an International Working Group under the UNESCO umbrella in order to discuss the main issue and develop proposals on how to progress the Space Technology Initiative and to accommodate the World Heritage Convention. In Decision 36COM5D adopted at the 36th Session of the World Heritage Committee in 2012 it is stated: "The World Heritage Committee ... Also welcomes financial and technical support provided by States Parties and the International Astronomical Union for [the] Thematic Initiative 'Astronomy and World Heritage' since 2003 and also encourages cooperation between the UNESCO World Heritage Centre, specialized agencies and relevant interdisciplinary scientific initiatives towards the elaboration of a Global Thematic Study on [the] Heritage of Science and Technology, including studies and research on technological heritage connected with space exploration; [and] Further encourage States Parties, international organizations and other donors to contribute to the thematic programmes and initiatives and also requests an updated report on Thematic Programmes to the World Heritage Committee at its 38th session in 2014".

Following this resolution, the Director of the WHC, Mr. Kishore Rao, in a letter to interested parties on June 30, 2012, emphasized the point that "In the light of this decision, the World Heritage Centre would like to identify all main actors concerned in order to enhance international cooperation and to define new partnerships. Two working groups were already created within the framework of this initiative—an International Working Group on Astronomy and World Heritage chaired by Prof. Clive Ruggles and a first expert Working Group on technological heritage connected with space exploration chaired by Prof. Mikhail Marov, Academician of the Russian Academy of Science." In addition, the UNESCO Associated Director-General for Culture, Mr. Francesco Bandarin, in his letter to Prof. Mikhail Marov dated July 20, 2012, underlined that "a number of decisions regarding development of innovative strategic approaches in the implementation of the World Heritage Convection were taken by the World Heritage Committee at the 36th session. Among them, the Committee's first decision on technological heritage connected with space exploration is indeed highly symbolic. I would like to underline the very active role the high-level Russian specialists play in the promotion and support of this new initiative ... I am confident that under your chairmanship and with the involvement of all specialized agencies, institutions and centres concerned, this study on technological heritage connected with space exploration will be the start to our common flagship activities."

It is worth noting that the International Astronomical Union has supported this new UNESCO Initiative since the very beginning. Quoting from the MOU adopted during the 28th IAU General Assembly (Beijing, China, August 24, 2012): "The IAU expresses its continuing support to the UNESCO Thematic Initiative 'Astronomy and World Heritage' and in response to the UNESCO promotion, is willing to further extend this Initiative over Space Science and Technology with the main focus placed on Space Astronomy and relevant facilities."

3. Space Astronomy and Space Technology: A Synergy

There is a basic synergy between Space Astronomy and Space Technology, the latter serving as a driver to progress with astronomy. The Space Heritage Initiative can be further extended to many historically important achievements in space science and technology spin-offs. Robotic and manned flights aiming eventually to establish a permanent human habitat in space and to follow up by extending this through the Solar system serve as brilliant examples of Space Science and Technology Synergy. However, many important questions remain unsolved and are not yet clearly defined. The key question is how to recognize and/or commemorate space-related objects and distinguish between tangible and intangible entities. In other words, how to commemorate material artifacts in space that are, after launch, rather more virtual than tangible objects.

The bottom line is how to select, alongside the valuable sites, monuments, observatories and instruments, outstanding objects operating in space and how they could be listed as significant material artifacts among historically important astronomical facilities. Needless to say, Astronomical Space Observatories and Lunar and Planetary Space Vehicles have been of outstanding value to astronomy. Space observatories such as the Hubble Space Telescope, Quant, Chandra, WMAP, Spitzer, Planck, Kepler, etc. literally revolutionized our world attitude and tremendously advanced physics astrophysics and astronomy. In turn, planetary orbiters and landers such as Luna, Venera, Viking, Voyager, Galileo, Cassini-Huygens, Rosetta, etc opened up to our eyes neighbor space and allowed us to have close-up views of virtually all major members of the Solar system family.

The goal is therefore to find a consistent approach to the legacy of various human artifacts and activities in this particular field of astronomy which is intimately related with the progress of space technology. Mock-ups of several generations of world-recognized astronomical and planetary spacecraft preserved in space facilities around the world could serve as replicas of the virtual space objects and after acceptance by the international bodies (IAU, ICOMOS, etc.) could possibly be assigned as UNESCO Space World Heritage. In the short list of such objects one might suggest Yu. Gagarin's orbital flight capsule VOSTOK and Gagarin's space-suit; the Apollo 11 lander EAGLE and N. Armstrong's space-suit. Also in the list could be the first orbital stations Salyut, Skylab and MIR, which paved the way to the International Space Station (ISS), and several generations of space launchers including the Space Shuttle and Buran, as well as a few robotic spacecraft.

4. Extraterrestrial Material/Lunar Soil and Proceedings/ Manuscripts of Space Pioneers

These entities are of special importance in terms of Space World Heritage. Samples of extraterrestrial origin delivered by APOLLO astronauts and LUNA space vehicles back to Earth promoted unique opportunities to gain insight into the formation of the Earth-Moon system and manifested the first steps undertaken by humans towards the in-depth study of pristine matter. They encapsulated unique information on the solar system's origin and early evolution. Currently, the main bulk of the samples are preserved in the Johnson Space Flight Center, Houston, USA, and in the Vernadsky Institute, Moscow, Russia. These samples represent real accomplishments of human culture and they deserve to be recognized as UNESCO World Heritage.

Among the important topics selected as space legacy, the Proceedings and/or Manuscripts of space pioneers could be considered. This particular aspect of Astronomical/Space

World Heritage is addressed with a caveat to meet the criteria of the World Heritage Convention. Among space pioneers, one might suggest listing Konstantin Tsiolkovsky, Hermann Oberth, Robert Goddard, Yuri Kondratyuk, Serge Korolev, and Verner von Braun. The respective archives are pertinent to selecting necessary documents, publications, etc. in an appropriate format. There should be rather strict regulations while soliciting UNESCO patronage over these documents.

5. Space Facilities and Launch Pads (Cosmodromes)

This particular segment of technological heritage connected with space exploration must be specially addressed and could be considered as the first step towards the implementation of the new UNESCO Initiative. Indeed, Space Facilities where recognized spacecraft were designed and manufactured and Launch pads (cosmodromes) are regarded as an important part of the overall space infrastructure. They ensured the development and launch of spacecraft and thus are to be regarded as historical cornerstones of space exploration. Examples are: OKB-1 (RSC "Energiya"), NPO–Lavochkin, the Jet Propulsion Laboratory (JPL), Johnson Space Flight Center (JSFC), Toulouse Space Facility, Bayconour Space Center, Cape Canaveral Space Center, and Koru Launch Pad. One should bear in mind, however, that the existing formal restrictions are subject to rather elaborative negotiations before one could announce space ventures/cosmodromes as potential UNESCO World Heritage. Nonetheless, we are keen to anticipate solid progress towards understanding the principal concepts underlying such a complex topic which could be accomplished with a step-by-step approach.

Baykonour Launch pad could be suggested as a Case Study. Historically it was the first site linking mankind with the skies and it has a priority as the site holding the historical facilities that provided the launch of the first artificial satellite and the Gagarin flight, which manifested a great breakthrough in human civilization. It is one of the most advanced possessions of the space era and occupies a historically important position in human culture. Its recognized achievements and credibility would surely satisfy the Operational Guidelines of the World Heritage Convention (UNESCO 2015), which state in Article 49 that "Outstanding Universal Value means cultural and/or natural significance which is so exceptional as to transcend national boundaries and to be of common importance for present and future generations of all humanity". They also, surely, "represent a masterpiece of human creative genius" (*ibid.*, Article 77(i)).

We are aware that the Case Study must satisfy the basic standards concerning the tangible evidence of sites and/or objects in order to be selected as cultural entities of globally recognized value. Unfortunately, until recently no special guidelines have been proposed by interested countries in order to satisfy the requirements and criteria for the selection of space science and technology sites/artifacts in order to fully accommodate UNESCO WH standards. Obviously, alongside the inventory and description of the historical development, management issues must be addressed as most critical in terms of the justification for inscription. They include comparative analysis, integrity-authenticity, criteria under which inscription might be proposed, suggested statement of OUV, etc., which are vital in the study.

Another important point one should bear in mind is that Baykonour is a huge area with numerous launch pads and facilities in its infrastructure; it has been dramatically extended since those historical events. Thus, in order to satisfy integrity-authenticity requirements, UNESCO patronage could be extended over only a small part of the overall Baykonour site/infrastructure—e.g., the famous "Launch pad No. 2" known as "Gagarin's start site". In addition to Gagarin's start site, two small cottages "Gagarin" and "Korolev", where the first cosmonaut and Chief Designer stayed overnight before the historical flight, could be suggested. Following the WH Convention (1972), such a selection is based upon tangible evidence. Gagarin's launch pad is in place and operational. The start's facility is integrated in terms of the capable space technologies used for the testing and launch of manned craft. Many cosmonauts and crews have been launched since the Gagarin flight. Also, in terms of integrity, it is important that no removal or partial demolition has occurred since Gagarin's time.

The authenticity is undoubted and well documented, including available historical archives. Commemoration of the historical events related to the launch pad is inscribed on the stone plate erected *in situ* beside the pad. While the original launch pad and close environment were partially reconstructed incorporating modern equipment this does not influence the authenticity. The Gagarin and Korolev cottages have been repaired but did not experience any changes: they include the former furniture and personal belongings. Thus Gagarin's start site and cottages satisfy the criteria under which inscription might be proposed, as well as contributing to the suggested statement of OUV. The places fully preserve the tangible attributes of their history and could be commemorated as UNESCO WH. This, it seems, will not impose political obstacles despite the fact that Baykonour is under international Russia-Kazakhstan joint jurisdiction. Some important issues are to be negotiated with ROSCOSMOS including managerial aspects of the implementation of the Initiative.

6. Conclusion

Space Astronomy and Planetary Exploration have ensured very significant progress in gaining knowledge about the Solar system and the Universe, which has tremendously broadened human horizons. There is a basic synergy between Space Astronomy and Space Technology, the latter serving a driver to progress with astronomy. It is therefore essential to include Space Astronomy/Technology, as represented by astronomical spacecraft and probes, as an important segment of Astronomical World Heritage. It is proposed to develop a consistent approach to the legacy of space facilities and human artifacts so that they could be recognized as Space Technology Heritage. Launch pads and ground-based space facilities could be selected as a starting point for accepting the proposed Space Heritage Initiative. The Initiative could be further expanded to other historically important achievements in space science and technology involving robotic and manned flights. An International Expert Working Group was set up under the UNESCO umbrella in 2012 to discuss the main issues and to develop proposals on how to progress the Space Technology Initiative and to accommodate the criteria of the World Heritage Convention.

References

Ruggles, C. & Cotte, M. (eds.) 2010, *Heritage Sites of Astronomy and Archaeoastronomy in the context of the UNESCO World Heritage Convention* (Paris: ICOMOS –IAU)

UNESCO 2015, *The Operational Guidelines for the Implementation of the World Heritage Convention* (Paris: UNESCO) http://whc.unesco.org/en/guidelines

Nā Inoa Hōkū: Hawaiian and Polynesian star names

Clive Ruggles[1], Rubellite Kawena Johnson[2] and John Kaipo Mahelona[3]

[1] School of Archaeology and Ancient History
University of Leicester, Leicester LE1 7RH, United Kingdom
email: rug@le.ac.uk

[2] Center for Hawaiian Studies, University of Hawai'i at Manoa, Honolulu, Hawai'i, U.S.A.

[3] Academia Language School, Honolulu, Hawai'i, U.S.A.

Abstract. In this paper we report on a 15-year project to construct a comprehensive catalogue of Hawaiian starnames documented in historical sources, published during the IAU General Assembly. Hawaiian star knowledge represents incomparable intangible heritage relating to Polynesian navigation in the Pacific. It both informs and motivates living cultural traditions aiming to reconstruct and build upon such knowledge.

Keywords. Astronomical heritage, Intangible heritage, navigation, cultural astronomy

1. Introduction

An extensively revised edition of *Nā Inoa Hōkū*, a catalogue of Hawaiian and Polynesian star names published by two of the authors 40 years ago (Johnson & Mahelona 1975), has been published during this IAU General Assembly (Johnson, Mahelona & Ruggles 2015). The first edition of *Nā Inoa Hōkū* contained many first-hand translations of primary sources researched in archives during the 1950s to 1970s. Since that time, a number of new primary sources have been identified (e.g. Chun 2004), and these and other primary sources have been translated or re-translated as part of the project.

One of the major original goals was "To document what remains of a once flourishing mastery of celestial navigation by accumulating the star lore which has managed to survive centuries of meager regard". During the last four decades the reality of Polynesian and Micronesian navigation skills has become widely acknowledged, supported by a wealth of ethnohistorical, archaeological, and linguistic evidence (e.g. Kirch 2000; 2012; Kirch & Green 2001).

It is no longer doubted that the sky occupied a prominent place in the rich cultural and historical legacy of the Hawaiian people—at many levels, and in many different ways. The new project, which has taken 15 years to complete, aims to provide a resource that will help to do justice to this undeniable fact.

2. Issues and complexities

While a number of Hawaiian star names are well known, a major challenge is to separate reliable first-hand information, mostly in Hawaiian-language archival sources dating back to the mid-19th century, from later commentaries and interpretations, many of which have introduced assumptions and errors that have become embedded in the literature.

Some new star names have also been introduced recently, in the traditional style, as part of the living tradition of Hawaiian and Polynesian voyaging.

The sources, often fragmentary, reveal much more than just the use of star observations for navigation and wayfinding, hugely important as this was. There was no single tradition but a complex and dynamic body of astronomical knowledge. Particular star names are not always consistently applied to the same stars. Accounts of physical characteristics such as the dates and times of appearance and disappearance of particular stars do not necessarily make sense in a Western, objective sense: they may, for example, represent times when the asterisms in question became important for divinatory purposes. In short, there is no simple 1-to-1 correspondence between Hawaiian and Western star names.

Added to this, there is much confusion in the historical accounts and dictionaries owing to transcription and printing errors, ethnographer errors, and the fact that early Hawaiian texts lack diacritical markings that would have helped to clarify the intended meaning. Some translators added unsourced information of their own, and many Western "identifications" of indigenous star names have entered the literature in this way.

Such challenges make it all the more important to construct a resource that is as reliable as possible for future scholars, not only within Hawaiian cultural studies but also for comparative analyses with star names in other parts of Polynesia, which have the potential to shed important new light on beliefs and practices brought to the Hawaiian Islands by the earliest Polynesian settlers. The new edition of *Nā Inoa Hōkū* aims to do this.

3. Wider significance

The star knowledge and navigation skills of Hawaiians and other Polynesians represent incomparable intangible heritage. The Polynesian Voyaging Society (www.hokulea.com), who are responsible for the many successful voyages of the sailing canoe *Hōkūle'a*, has done immense service in helping to reconstruct and revive these traditional concepts and practices. Our project aims to complement this by clarifying what is evident from ethnohistory. The historical and living traditions cannot be separated from each other. Knowledge and awareness of past traditions helps to generate respect for, and hence to protect and preserve, the living tradition. Looking backward provides a solid basis for looking forward, while looking forward gives the point to looking backward. As it is said in Hawaiian: *I ka wā mā mua, ka wā mā hope* ("From the way of the past comes the way forward").

References

Chun, M. N. 2004, *The History of Kanalu: Mo'okū'auhau 'Elua, by Benjamin K. Nāmakaokeahi* (Honolulu: First People's Productions)

Kirch, P. V. 2000, *On the Road of the Winds: an Archaeological History of the Pacific Islands before European Contact* (Berkeley: University of California Press)

Kirch, P. V. 2012, *A Shark Going Inland is my Chief: the Island Civilization of Ancient Hawai'i* (Berkeley: University of California Press)

Kirch, P. V. & Green R. C. 2001, *Hawaiki, Ancestral Polynesia: an Essay in Historical Anthropology* (Cambridge: Cambridge University Press)

Johnson, R. K. & Mahelona, J. K. 1975, *Nā Inoa Hōkū: a Catalogue of Hawaiian and Pacific Star Names* (Honolulu: Topgallant Books)

Johnson, R. K., Mahelona, J. K., & Ruggles, C. L. N. 2015, *Nā Inoa Hōkū: Hawaiian and Pacific Star Names (revised edition)* (Bognor Regis: Ocarina Books)

Astronomical Heritage and Aboriginal People: Conflicts and Possibilities

Alejandro Martín López[1,2]

[1] CONICET-Universidad de Buenos Aires (UBA), Argentina
[2] Sociedad Interamericana de Astronomía en la Cultura (SIAC)
email: `astroamlopez@hotmail.com`

Abstract. In this presentation we address issues relating to the astronomical heritage of contemporary aboriginal groups and other minorities. We deal specially with intangible astronomical heritage and its particularities. Also, we study (from ethnographic experience with Aboriginal groups, Creoles and Europeans in the Argentine Chaco) the conflicts referring to the different ways in which the natives' knowledge and practice are categorized by the natives themselves, by scientists, state politicians, professional artists and NGOs. Furthermore, we address several cases that illustrate these kinds of conflicts. We aim to analyze the complexities of patrimonial policies when they are applied to practices and representations of contemporary communities involved in power relations with national states and the global system. The essentialization of identities, the folklorization of representations and practices, and the fossilization of aboriginal peoples are some of the risks of applying the label "cultural heritage" without a careful consideration of each specific case.

In particular we suggest possible ways in which the international scientific community could collaborate to improve the agenda of national states instead of reproducing colonial prejudices. In this way, we aim to contribute to the promotion of respect for ethnic and religious minorities.

Keywords. Intangible astronomical heritage, aboriginal people, conflicts

1. Heritage as a language

Today, heritage is an increasingly broad concept. It has a great impact in many crucial areas: public politics, NGOs' politics, public opinion, and aboriginal communities' strategies. The focus of the heritage concept is the idea of "culture" as a value to protect. In particular, at present we can see an increasing valorization of non-western achievements. But the concept of heritage has strong links with a specific western juridical language and property conceptions. For this reason, some of their key characteristics are: the demand for "authenticity"; the necessity for a clear "definition" of the boundaries of every specific heritage item; and the "preservation" of the integrity of the heritage. The use of the heritage concept has a tendency to privilege tangible aspects, the spectacularity, and the uniqueness of the proposed heritage. This "western" bias has the consequence of and implicit hierarchization of the different conceptions of humankind involved. The Western concepts have a very strong tendency to prevail in the international definition of what is heritage and what is not. Also, we can see a strong tendency to use the concept of "culture" to refer to the diversity of human forms of life but hiding the power relations between different societies, while making claims of political "neutrality".

At the present time, claims about world heritage are, in many cases, claims about ownership and rights, but in the case of aboriginal communities the conflicts involved are also conflicts about different ways of thinking about definitions of things, people and humankind; the idea of territory; ownership; history, change, and identity. Heritage—as ecology—is now a new language or arena for the display of the complex conflicts between societies, specially nation-states and the minorities within them.

2. Essentialization and folklorization

The ideas of "traditional" and "authentic" are conceptions frequently applied to aboriginal populations. Usually they are grounded in the idea that those kinds of society do not change (and if they change they lose their authenticity). They are thought of as societies that only enter history and change after the impact of the colonization processes. This implies the conception of ethnic identity as linked to some well defined group of features such as dances, clothes, specific ceremonies, or to well defined cosmological systems. This does not fit very well with the ways in which oral, or predominantly oral, societies actually function. An example of this is the need to understand the crucial role in present aboriginal communities in South America of their own forms of Christianity, developed from the complex relations with western missionaries. Many western experts involved in world heritage initiatives are looking for "real aboriginal life" and do not pay attention to crucial cultural manifestations, with deep roots in the aboriginal cultures, because these manifestations are in the contexts of aboriginal forms of Christianity. In many cases these practices are not part of an "acculturation" process: they are not a "mix". They are real cultural creations of these groups, in the peculiar historical situation that they face. They are truly reinventions of Christianity in terms of aboriginal logics, and are fundamental ways to legitimate, in the context of the national society, important cultural forms, leadership mechanisms, social organization, and conceptions about the relations between human and non-human beings (Altman, 2015).

Another very common idea is that aboriginal people lose their identity if they adopt western technology. But this is not necessarily the case. For example, in the Chaco region in South America, cellphones and computers make possible new versions of the oral culture of past centuries, reinforce old mechanisms of making marriages, and expand the production of texts in aboriginal languages without the control of western teachers or missionaries. In each case it is necessary to study these elements in context.

3. Folklorization and bureaucratization

The incorporation of some aboriginal cultural traditions into the agendas of nation-states or international agendas implies in most cases the bureaucratization of these practices. They are incorporated, for example, to scholarship or state ceremonies that are under the control of white people. This situation tends to result in the enforced unification of practices that have a very broad spectrum of variation, according to the uncentralized character of the societies involved. This usually leads to attempts to define clothes, movements, instruments, meanings, etc. The displeasure with this by non-centralized societies can result in the rejection of "world heritage" nominations, as in the case of the Nguillatún ceremony of the Mapuche people in the southern part of Argentina and Chile (Carlos Massota, personal communication).

As another example of this we should mention that in many parts of South America local and national governments are starting to "recognize" the "new year" celebrations of different aboriginal groups. This process is usually governed by some key factors: the assimilation of the festivities of religious minorities; the imposition of a Western calendrical form of definition; and a time definition based on a single day and a single "sign" in place of local definitions based on a ranking of days and a complex group of signs (birds, flowers, stars, rains). In some cases, such as the promotion of the new year of the Avá-Guaraní people by the national government of Bolivia (Pereira Quiroga, 2015), we can see the use of this festivity in order to control and domesticate political tensions.

4. Right to Free Prior and Informed Consent (FPIC)

The Right to Free Prior and Informed Consent (FPIC) is a key principle for the relations between aboriginal groups and national governments or international organizations. It is an obligation—for many nations, in fact, a legal obligation—not a gift. This principle is supported by the United Nations Declaration on the Rights of Indigenous Peoples (2006) and the International Labour Organization Convention 169 (1989) ratified by 20 countries. Some of the most important characteristics of the FPIC are that:
- it must be prior to the decisions to be made;
- it must be conducted through institutions representative of indigenous communities, and indigenous people should control the process by which representatives are determined; and
- it must be free of pressures and manipulations—for example pressures using the promises of potential economical and touristic benefits.

In societies of low stratification the decision-making processes usually involve the creation of a consensus, and strong discrepancies between different communities and leaders. In many occasions the western agendas are not minded to tolerate these processes and their timescales. The complex problems concerning the installation of great telescopes on the top of mountains that aboriginal populations consider sacred are good examples of the relevance of these issues.

5. An example of multiple interests in dispute

In Chaco province, Argentina, there exists a very important dispersion of nickel-iron meteorites: Campo del Cielo. This dispersion is very important for the cosmological ideas of Moqoit aboriginal people, and also has strong roots in the history and culture of the local Creole population from colonial times (Giménez Benítez et al., 2004). For aboriginal people the manipulation of these objects is linked with the relations between humans and celestial beings. For these reasons, through texts (Martínez, 2006) and public actions, Juan Carlos Martínez and other young Moqoit leaders have demonstrated the connection that the Moqoit see between their notions about the cosmos and their land and cultural claims (López, 2011). The Moqoit's successful opposition to the attempt to transfer the largest of those meteorites ("El Chaco", with a weight of 37 tonnes) to Germany for an artistic exhibition (dOCUMENTA13, at Kassel), promoted by two artists from Buenos Aires, should be understood in this context. Around this event a very strong public debate arose, with the centrality of the sky icons at its fore. Our research group participated on this debate and many members of the academic astronomical community gave their support to the opposition to this "artistic project", which was eventually cancelled. All this revealed very deep conflicts between the different definitions of the meteorites' function and importance (López, 2015).

6. Final remarks

Heritage, as ecology, has become ia new arena and a new language for very complex conflicts. For this reason heritage is a key issue for the relations between indigenous minorities and national states or international organizations. International recognition has the potential to empower aboriginal groups and to protect them against national and regional abuses. But the western origin of the implicit notions involved in that recognition—identity, authenticity, change, relevance and definition—is a source of risk for aboriginal communities. Aboriginal people are not relics of the past, without history and agency. The symbolic struggle for the definition of the meaning of objects, places, times, ceremonies, etc. is not politically neutral. This struggle is marked by the force of colonial relations. In this context, the right to Free Prior and Informed Consent (FPIC) is an obligation when aboriginal communities are involved; not a gift. Aboriginal people are usually misrepresented or underrepresented in national governments and agencies. Colonial logics are inscribed in official bodies and practices, and very strong epistemological vigilance is needed in order to avoid the risk of reproducing colonial looting in the name of science and culture. The example of the conflicts about the installation of great telescopes, as the case of Mauna Kea (Hawai'i), demonstrate the relevance of these issues for the academic astronomical community.

To recognize "Astronomical World heritage" is insufficient. We also need to recognize "Astronomical World heritage in danger" in order to induce local governments to take responsibility in such cases. Maybe the IAU, and especially Division C, could make a list of sites in this situation. Cultural astronomy must play a key role in the articulation of astronomical heritage initiatives involving aboriginal people. We need to be involved with people if we work with people. Science, culture and art must point the way for governmental logics and not vice versa.

References

Altman, Agustina 2015, "Sky travelers: cosmological experiences among evangelical indians from the argentinean Chaco", in *Stars and Stones: Voyages in Proceedings of the SEAC 2011 Conference* & Pimenta, et. al (eds.) (Archaeopress, BAR International Series 2720, Oxford), pp. 148–153

Giménez Benítez, S., López, A., & Granada, A. 2004, "Suerte, riqueza y poder. Fragmentos meteóricos y la presencia de lo celeste entre los mocovíes del Chaco", in *Etno y arqueoastronomía en las Américas. Memorias del Simposio ARQ-13: Etno y arqueoastronomía en las Américas, 51 ICA*, Boccas, et al. (eds.)

López, Alejandro 2011, "New words for old skies: recent forms of cosmological discourse among aboriginal people of the Argentinean Chaco", in *Archaeoastronomy and Ethnoastronomy: Building Bridges between Cultures*, Proceedings of IAU Symposium 278, ed. C. L. N. Ruggles (Cambridge University Press, Cambridge) pp. 74–83

López, Alejandro 2015, "Astronomy in Chaco region, Argentina", in *Handbook of Archaeoastronomy and Ethnoastronomy*, ed. C.L.N. Ruggles (Springer, New York), pp. 987–995

Martínez, Juan Carlos 2015, *El secreto de la tierra*, Revista La Educación en nuestras manos, Diciembre (77), http://www.suteba.org.ar/revista-la-educacin-en-nuestras-manos-n-77-voces-yluchas-de-los-pueblos-originarios-2523.html

Pereira Quiroga, Gonzalo 2015, "Chiriguano Astronomy—Venus and a Guarani New Year," in *Handbook of Archaeoastronomy and Ethnoastronomy*, ed. C.L.N. Ruggles (Springer, New York), pp. 967–973

Gufa, a Unique Cultural Ritual— a Tale of a Forbidden Sun and a Girl

Pritisha Shrestha

Central Department of English, Tribhuvan University, Kathmandu, Nepal
email: prishafun@gmail.com

Abstract. Gufa, one of the traditional rituals, has been performed in Nepal since time immemorial by indigenous Newar people. In Gufa, a young girl who just had her first period is hidden in a sunless room for twelve consecutive days. This paper expounds the importance of ritual and its nexus with astronomy especially while interpreting how the daily motions of celestial objects have influenced the establishment and devolvement of a deep-rooted custom of Gufa.

Keywords. Cultural Astronomy, Sun, Nepal, Newar, Gufa, Ritual and Performance

1. The ritual performance

Gufa is part of the intangible cultural heritage of the Newars, an indigenous group of people in Nepal (Levy 1990, Nepali 2015). Gufa—also known as *Surya Darshan* or "sun's observation"—is a unique ritual. It is a puberty ceremony of a girl and her life-long connection with the sun. During the twelve-day long ceremony, which starts when she gets her first period, she is prohibited from observing the sun and therefore stays in a room heavily curtailed to keep out the sun's rays. "Great care is taken to ensure that no ray of sunshine can enter the room since an explicit aim of the rite is to ensure that the girls are seen neither by men nor by the sun. The first three days are said to be especially dangerous because this is the ritually prescribed period of bleeding throughout the Hindu world" (Allen 1986, p. 14). The room becomes her own sanctuary where she learns her own body. From her female relatives, she learns the etiquette of being a woman and receives education about living in a society. The community celebrates her ritual as she is taken as a life bearer.

On the twelfth day, she is carefully brought out of the room and symbolically married to the sun, in a ritual. She is cleansed and dressed in new red attire, usually a *sari*, adorned with red bangles and golden jewelries to prepare for the long awaited reunion with the sun. She is cautiously taken to the rooftop of her house, with her head covered with a shawl. This step is believed to have removed all the remaining menarche potency in the girl. Her mother turns her to face the direction of the sun then removes the cover. Initially, she looks at the sun through its reflection in the basin of water put in front of her. Then she crosses her fingers together, in a unique fashion, and takes a glimpse of the sun by peeking through her fingers. After this point, her priest instructs her to sit on the floor where an array of worshipping items are placed, at the center of which is the image or an idol of sun itself. She pours a purifying liquid onto the image of the sun (*ibid.*, p. 117) and she is ceremoniously and symbolically married to the sun. The logic behind the union after a pure restriction is to protect her sensitive young body and to ward off any harm to her reproductive parts from the sun's rays.

The French philosopher Van Gennep (2011) marks three distinct stages in ritual performances of coming of age: (a) separation, (b) liminal period, and (c) re-assimilation.

At the onset of the ceremony, the Gufa girl is separated from her formerly possessed social status. Then, in the liminal phase, she gets a new identity of someone who is performing the ritual. Her former status from pre-ritual phase no longer holds and she has yet to gain her new identity from the ritual. When she completes her ritual, she is once again reintroduced into society with a new set of responsibilities and a new status. Her re-assimilation is marked at a grand feast where both male and female members of her community join to commemorate the event.

2. The cosmic connection

Although the sun is seen as the cosmic energy that energizes and revitalizes lives on earth, during menarche it is stringently forbidden to the Gufa girl. According to old Hindu scriptures, namely Vedas and Ayurveda, there is a constant build-up of energy in a woman's body in the days prior to her period, as the body is geared towards becoming pregnant. When there is no pregnancy, menstruation kicks off, releasing all the energy downward, and during this time, a woman is susceptible to absorb any energy floating in her surrounding environment. Maya Tiwari, an Ayurveda teacher, describes the innate psychology and collective consciousness that there is of cosmic memory of food. This memory, she writes, "[which] is derived only from plant life according to the Vedas ... is a rising energy flowing up from the earth towards the sun and the sky". On the contrary, there is a downward flow of energy during the period in which bodily air is pulled down from the body by the magnetic forces of the earth (Tiwari (2011), p. 67).

In many cultures the sun is revered as a male god. Dr. Tejeswar Babu Gonga, a culture expert and the president of the Nepal-Russia Literary Association, argues that the twelve days of the Gufa ritual symbolize the twelve-month cycle of the sun and the subsequent seasons it brings (personal interview, 17 Mar 2015). Ancient people were certainly aware of the alterations in the sun's temperature and various seasons, which made them conclude that its rays could be harmful to young girls who had just hit puberty. Thus, the Gufa ritual became a traditional and celebratory means to shield their developing reproductive organs from the energy of the sun.

3. Possibilities and Conclusion

There are many cultural practices dealing with unique traditions and beliefs spread throughout the world. At a time when astronomy and space exploration is moving forward apace, it is also important to look back into our various cultural practices in which cosmic forces are both feared and revered. Hidden meanings in ancient cultures like Gufa can hold key information about the cosmos as perceived by people centuries ago. Even today, it is joyfully celebrated in Newar households of Nepal and abroad. Studies and research in cultural astronomy can reveal a plethora of information. The value of such practices needs to be respected and protected as global intangible heritage.

References

Allen, M. 1986, *The Cult of Kumari* (Kathmandu: Mandala Book Point)
Levy, R. 1990, *Mesocosm: Hinduism and the Organization of a Traditional Newar City in Nepal* (Berkeley: University of California Press)
Nepali, G. S. 2015, *The Newars* (Kathmandu: Mandala Book Point)
Tiwari, M. 2011, *Women's Power to Heal: Through Inner Medicine* (New York: Mother Om Media)
Van Gennep, A. 2011, *The Rites of Passage* (Chicago: University of Chicago Press)

Astronomical Knowledge from Holy Books

Sona V. Farmanyan[1,2], Vardan G. Devrikyan[1] and Areg M. Mickaelian[2]

[1] NAS RA, M. Abeghian Institute of Literature, Armenia
email: sona.farmanyan@mail.ru

[2] NAS RA Byurakan Astrophysical Observatory (BAO), Armenia

Abstract. We investigate religious myths related to astronomy from different cultures in an attempt to identify common subjects and characteristics. The paper focuses on astronomy in religion. The initial review covers records from Holy books about sky related superstitious beliefs and cosmological understanding. The purpose of this study is to introduce sky related religious and national traditions (particularly based on different calendars; Solar or Lunar). We carried out a comparative study of astronomical issues contained in a number of Holy books. We come to the conclusion that the perception of celestial objects varies from culture to culture, and from religion to religion and preastronomical views had a significant impact on humankind, particularly on religious diversities. We prove that Astronomy is the basis of cultures, and that national identity and mythology and religion were formed due to the special understanding of celestial objects.

Keywords. Religion, Mythology, Religious Astronomy, Preastronomy

1. Introduction

The earliest visions of the Universe are mostly reflected in the holy books: Ancient Egyptian Religion (Pyramid Texts), Zoroastrianism (Avesta), Hinduism (Vedas), Buddhism (Tipitaka), Confucianism (Five Classics), Sikhism (Guru Granth Sahib), Christianity (Bible), Islam (Quran), Druidism (Mabinogion) and Maya Religion (Popol Vuh) and in the interpretations of those books. These books include various information on the creation of the Universe, Sun and Moon, the age of the Universe, Cosmic sizes, understanding about the planets, stars, Milky Way and description of the Heavens in different religions. The Holy Books can be understood either literally or contextually.

Literalism suggests:

- The holy books are the inerrant word of God
- The holy books are literally true

Contextualism suggests:

- The meaning of the text depends on its context
- Text should be examined rationally
- The holy books contain metaphors and symbols
- Religious Cosmology

2. Religious Cosmology

Religious Cosmology (also mythological cosmology) is a way of explaining the origin, the history and the evolution of the Cosmos or Universe based on the religious mythology of a specific tradition. Religious cosmologies usually include an act or process of creation by a creator deity or a larger pantheon.

In the Bible Universe of the ancient Jews was comprised of a flat disc–shaped Earth floating on water, heaven above, underworld below. Humans inhabited Earth during life and the underworld after death, and the underworld was morally neutral. In this period, the older three-level cosmology was widely replaced by the Greek concept of a spherical Earth suspended in Space at the centre of a number of concentric heavens. We can encounter this three–level structured Universe also in ancient Armenian Cosmology. The early Armenians understanding of the structure of the Universe is preserved in the ancient "Vahagns Birth folk-song, recorded by the 5th century historian Moses of Khoren. According to this song the Universe has tripartite structure: Sky–Earth–Sea. The Earth's creation, according to Mormon scripture, was not ex nihilo, but organized from existing matter. The faith teaches that this Earth is just one of many inhabited worlds, including a planet or star Kolob which is said to be nearest the throne of God. In Tipitaka, the Universe comes into existence dependent upon the actions of its inhabitants. Buddhists posit neither an ultimate beginning nor final end to the Universe, but see the Universe as something in flux, passing in and out of existence, parallel to an infinite number of other Universes doing the same thing. Quran teaches that God created the Universe, including Earth's physical environment and human beings (M. Iqbal 2007). The highest goal is to visualize the Cosmos as a book of symbols for meditation and contemplation for spiritual upliftment or as a prison from which the human soul must escape to attain true freedom in the spiritual journey to God. In Hindu cosmology it is believed that everything takes its birth in the Universe (Mitcham 2005). And the Universe is considered to constantly expand since creation and disappear into a thin haze after billions of years. Jain texts describe the shape of the Universe as similar to a man standing with legs apart and arm resting on his waist. This Universe, according to Jainism, is broad at the top, narrow at the middle and once again becomes broad at the bottom (Soni 1998).

3. Discussion

As Nobel Prize Winner Freeman Dyson said: "Science and religion are two windows that people look through, trying to understand the big Universe outside, trying to understand why we are here." Thus, Science and Religion are complementary. Both science and religion are not unchanging, timeless, or static because both are complex social and cultural endeavours that have changed through time across languages and cultures (Stenmark 2004). By discussing religious cosmologies, we come to the conclusion that Astronomy recognizes reason, observation, and proof, while religions incorporate disclosure, confidence and consecration whilst additionally recognizing Philosophical and Metaphysical clarifications with respect to the investigation of the Universe. We can claim that Astronomy is the basis of cultures, and that national identity and mythology and religion were formed due to the special understanding of celestial objects.

References

Mitcham, Carl 2005, *Encyclopedia of Science, Technology, and Ethics*, Macmillan Reference USA. p. 917

Muzaffar, Iqbal 2007, *Science & Islam*, Greenwood Press

Soni, Jayandra; Craig, E. (Eds.) 1998, *"Jain Philosophy"*. *Routledge Encyclopedia of Philosophy*, London: Routledge

Stenmark, Mikael 2004, *How to Relate Science and Religion: A Multidimensional Model*, Grand Rapids, Mich.: W.B. Eerdmans Pub. Co.

Astronomy in the City for Astronomy Education

Rosa M. Ros[1] and Beatriz García[2]

[1] NASE president,
Department of Applied Mathematics 4, Technical University of Catalonia,
Jordi Girona 1-3, 08034 Barcelona, Spain
email: ros@ma4.upc.edu

[2] ITeDA Mendoza (CNAE,CONCET,UNSAM) & National Technological University-FRM
Azopardo 313, Godoy Cruz, Mendoza, Argentina
email: beatriz.garcia@iteda.cnea.gov.ar

Abstract. Astronomy is part of our culture. Astronomy cannot be isolated in a classroom, it has to be integrated in the normal life of teachers and students. "Astronomy in the city" is an important part of NASE (Network for Astronomy School Education) (Ros & Hemenway 2012). In each NASE course we introduce a "working group session" chaired by a local expert in cultural astronomy. The chair introduces several examples of astronomy in their city and after that, the participants have the opportunity to discuss and mention several similar examples. After this session all participants visit one or two sites proposed and introduced by the chair.

After more than 5 years using this method we visited and discovered several examples of astronomy in the city:
- **Astronomy in ancient typical clothes.**
- **Archaeological temples oriented according to the sunrise or set.**
- **Petroglyphs with astronomical meaning.**
- **Astronomy in monuments.**
- **Sundials.**
- **Oriented Colonial churches.**
- **Astronomy in Souvenirs.**

In any case, teachers and students discover that Astronomy is part of their everyday life. They can take into account the Sun's path when they park their car or when they take a bus "what is the best part in order to be seat in the shadow during the journey?" The result is motivation to go with "open eyes" when they are in the street and they try to get more and more information about their surroundings.

In summary, one of the main activities is to introduce local cultural aspects in NASE astronomy courses. The participants can discover a new approach to local culture from an astronomical point of view.

Keywords. Cultural Astronomy, Astronomy Education

Note. The full version of this article appears in the supplementary on-line materials at http://dx.doi.org/10.1017/S1743921316002660.

1. Introduction

Astronomy is part of our life and our culture. It is important to introduce this idea in schools if we want humanity to recognise their past from an astronomical point of view and also to discover the relation between this past and the current kind of life. We want to offer to secondary and primary school teachers several examples that they can use in their classes with their students. The main goal is to introduce the proposal that teachers with their students try to discover different evidence of cultural astronomy in their cities, and the most important thing is that they enjoy doing it as a "detective investigation".

We will propose several examples in a very open approach. In any case we need a serious contribution from local astronomers and scientist in general that can give support to teachers in order to work in an appropriate way.

The Network for Astronomy School Education (NASE) is a Working Group of IAU Commission 1, created at the 2009 IAU General Assembly. NASE organises courses for secondary and primary school teachers and creates a team of teachers—a Local Working Group (LWG)—who will work with us during the visit by NASE members and will give continuity to the astronomy activities when the visitors leave the country. All NASE activities are organised in the language of the country. As of August 2015 NASE has 28 Local Working Groups in 20 different countries, involving more than 290 volunteers who help to organise courses, prepare materials, organise visits and create guided tours (see www.naseprogram.org).

2. Astronomical Visits

After more than 5 years using this method, we visited several places and discovered several ways to teach "Astronomy in the City". During these years we organized 70 courses mainly in Central and South America and the main number of examples are in these locations. The following examples show different approaches to this activity.

- *Archaeological temples oriented with sunrise/sunset.* In 2010 during a course organized in Lima, Peru, with the Universidad de San Marcos, we visited Cieneguilla and we had the opportunity to visit the Huaycan site with the archaeologist who works on this site. In one of the walls of the site we could see a set of 12 circles that are considered a Sun-Moon calendar and we observed a sunset not far away from the December solstice. For more information see http://www.arqueologiadelperu.com.ar/huaycan.htm and http://www.cultura.gob.pe/es/tags/huaycan-de-cieneguilla.
- *Petroglyph with astronomical meaning.* We visited "Sitio Polanco" in Capira, one of several little-studied petroglyphs not far from Panamá city. Several drawings on this big stone may be a rain calendar connected with the Moon; we hope that the University of Panamá can give more details about this in the near future.
- *Astronomy in pottery.* In Ecuador, for several courses, we visited the Archaeological Museum of "Museo del Banco Central" guided by a well-known local archaeologist from San Francisco University, Quito, and had the opportunity to observe several different objects. To give one example, on several ceramics from the "Tutamonos" culture, Orion appears as a monkey constellation. Stars and monkeys appear in the sky for inhabitants of the jungle. See http://revistas.arqueo-ecuatoriana.ec/es/apachita/apachita-19/270-de-estrellas-y-monos-en-la-cultura-pasto.
- *Astronomy in monuments.* In Ecuador, there is a monument to recognise the Equator line named "Mitad del Mundo". On its top appears a "Parallel Earth", accurately placed, that offers the visitor the opportunity to see the day-night line or sun-shadow line on the sphere of the monument as it would appear on the real Earth. This phenomenon gives a similar vision to that which could be seen by an astronaut from outside the Earth (see Fig. 1).
- *Oriented colonial churches in Central America.* In Tegucigalpa, Honduras, we organise a visit to three churches in the colonial area of the city. Two of them—San Francisco, the oldest, a small church whose construction began in 1592 and the nearby Cathedral named of San Miguel Archangel, constructed in 1765—are oriented in the East/West direction as was common in the colonial period, while the third, Los Dolores, has a

Figure 1. The "Parallel Earth" on top of the "Mitad del Mundo" monument in Ecuador. The picture was taken on the equinox and the day-night line is seen to be crossing the South Pole.

different structure, different decorations, and—according to the tradition prevailing at the end of 17th century—was the church for native habitants of ancient Tegucigalpa. This native church is oriented in the North/South direction.

- *Native People's Solar Observatory.* In Brazil, the government moved the "Yetore" tribe to a new site, named Ekerua, some years ago. They recreated their ancient village and in particular they reproduced the Solar Observatory that gives them the position of the Sun during the year (especially sunrise and sunset at the solstices and equinoxes. We have the possibility of visiting this tribe and listening to their explanations about the new observatory that they have now. There are other similar observatories in different places and all of them are used to recognise the seasons and to give calendrical information.
- *Sundials in the Forbidden City.* During the NASE course in Beijing with the Planetarium of Beijing we did not visit the "Forbidden City" because this would have needed an additional full day. But in fact, our participants knew the monument and we took the opportunity to teach them about the different kinds of sundials found there. The participants felt excited to understand something of the monuments that had not been understandable when they had visited them in the past.
- *North/South- and East/West-oriented streets in Mendoza.* A visitor arriving in Mendoza, Argentina, may be surprised to find a cardinally oriented street grid with streets numbered from the centre to the North, South, East or West. This structure originated when the Spanish arrived and created the Plaza de Armas in the form of a rectangle with its sides oriented North/South and East/West. All the streets in the foundational area are oriented cardinally—we used the compass to verify this at several locations—but not so the newer part of the city, constructed after the big earthquake of 1861. It is structured as a grid but there is an error in the orientation.
- *Zero point in Managua city.* In Managua, Nicaragua, we visited the old part of the city that was planned at the beginning of the 19th century. This area is structured as a grid with the streets oriented more or less North/South and East/West, each with names and numbers. We decide to visit this zone and try to locate the "zero point" in this structure. The majority of buildings in this area were destroyed in the earthquakes of 1931 and 1972. We found a circle of stones at the location of the "zero point" but did not discover its purpose (Fig. 2).

Figure 2. A group of teachers at the circle of stones found to mark the "zero point" of the street grid in Managua, Nicaragua.

- *Mistakes in Sundials.* In several courses we visited sundials constructed in public squares of cities. We found a lot of them with mistakes. This suggested a very interesting exercise: to try to discover is the sundial was well made or not. It is a good opportunity to test if the course participants really understand the information that they have been given. Two examples of sundials with mistakes are the Horizontal Sundial in Argentina, in a park in Santa Fe, whose North-South direction deviates significantly from the 12-hour line, and the Equatorial Sundial of Paraguay, in the main entrance of the Instituto Geogrfico Militar de Ecuador in Quito, whose plane is significantly inclined to the ground, which it should not be for a location close to zero degrees latitude. A legend on the sundial explains why: it was a present to the Paraguayan Army and so, of course, was calculated for very different latitude.

3. Conclusions

After the working group and the astronomical visit, teachers understand that Astronomy is part of their everyday life and think that it could be possible to discover some astronomical aspects in the city with their students. Our main goal is to motivate people to walk with open eyes when they are in the street so that they try to get more and more information about their surroundings. They can also use knowledge of astronomy in their current activities, for example taking into account the Sun's path when choosing a shady place to park their car or a seat that will remain shady during a bus journey.

We know that teachers cannot learn archaeology in one hour. We are interested in offering a new approach to astronomy by means of cultural heritage—ancient cultures or oriented buildings or monuments. Astronomy is part of people's lives and we would like them to discoverer it. The most interesting part of the activities consists of promoting discussion and the exchange of information between participants: the chair and guide of the session is only a promoter of this exchange of ideas.

References

Ros, R. M. & Hemenway, M. K. (Eds.) 2012, *14 steps to the Universe*, Network for Astronomy School Education, IAU, 2012, pp 151.

FM3:
Scholarly Publication in Astronomy: Evolution or Revolution?

IM3

Evolution of Scholarly Publishing and Library Services in Astronomy
Its Impact, Challenges, and Opportunities

Hema Wesley[1] and Geetha Sheshadri[2]

[1] Consultant Editor, Indian Academy of Sciences, Bangalore, India
email: ruthwes09@gmail.com

[2] Assistant Librarian, Raman Research Institute, Bangalore, India
email: geetha.sheshadri@gmail.com

Abstract. Scholarly publishing and its procedures have evolved rapidly, forcefully, and incredibly. Technical advances in the production and promotion of science content have dramatically augmented the visibility and reach, deepened the impact and intensified the thrust of science journal content. These changes range from checking text on perforated tapes to pit stop; from hot metal types to CTP; and from Gutenberg to colour digital printers. Intrinsic and inextricable to this revolutionary aspect of evolution in scholarly publishing is the evolution of library services in astronomy which catapulted library resources from preprints on shelves to customised digital repositories and from communicating observational data through postal telegrams to Tablets. What impact does this unique blend of revolutionary advances have on science and society, what are the consequent challenges, and what are the opportunities that can metamorphose from challenges inherent in the power and potential of the 'published word'?

The perspectives expressed in this paper stem from learning experiences of the authors at the Indian Academy of Sciences, publishers of ten science journals including the Journal of Astrophysics and Astronomy, and at the Raman Research Institute Library (in which Astronomy is one of the core subjects for research)

Keywords. Publishing, library services, evolution

---

1. Introduction

Publishing – the process of production and dissemination of information – is one of the metrics of scientific careers. 350 or so years ago the scientific journal was itself an innovation. This was in large part enabled by the emergence of the printing press: the Gutenberg Letter Press invented in the 15th century. While electronic publishing methods have resulted in rapid publication and wider dissemination, it, nevertheless, is also true that with speed and ease of automation, quality is to be ferociously guarded. Could traditional methods, when juxtaposed with the continuous onslaught of newer technologies in our publishing procedures, be the answer? Are people still the quintessential element in technology?

Why do we publish? The phrase "Publish or Perish" has itself, interestingly, evolved in its meaning. "Publish or perish" is a phrase coined to describe the pressure in academia to rapidly and continually publish academic work to sustain or further one's career. I see evolution in this very phrase which first appeared in a non-academic context in 1932 and became the buzz word for researchers about 30 years ago. It is now a software program that retrieves and analyses academic citations.

2. Impact of evolutionary developments

How have publishing procedures evolved through the years; what is the impact of these developments; what are the challenges and opportunities for scholarly publishers in this digital age? How do we cope with and pre-empt the abuse and distortion of internet data, and with unethical practices that threaten to cripple the peer review system, and jeopardize publishing ethics and academic integrity? How can publishers fulfil their responsibility to publish content that is explicitly original?

Juxtaposed to this is the scenario in library services with increasing demand for e-versions of journals and other e-resources and the gradual loss of value for printed journals. Librarians are developing and maintaining electronic libraries and vast digital resources combined with additional features of e-publishing, such as tools, voice, graphic arts, films, videos, CDs, DVDs, etc.

How can we prepare for the future of publishing or is it already here with the popularity of pre-print servers? How have publishers responded to the Open Access model while being aware that "the price of keeping something free comes with a cost"? Would this cost entail ensuring quality of the presentation and reliability of information? Are potential authors confused identifying traditional peer reviewed content, predatory journals, fake reviewers, depositing research in institutional repositories, copyrights and licences? Are libraries impacted by these developments?

3. Publishing procedures

Publishing procedures have evolved from traditional methods involving manual operations to advanced software (Fig. 1).

4. Challenges and opportunities

Reviewers play an essential part in research and scholarly publishing. Most academic publishing companies rely on effective peer-review processes to uphold the quality and validity of individual articles and the overall integrity of the journals. Editors are looking for original and innovative research. The peer review practice involves reviewers who are qualified to assess the significance of the research and comment on the content, methodology of the experiment, and the results. Publishers now use online submission and review management software through which authors can submit and track the processing of their article submissions. The software enables authors, reviewers, editorial board members, chief editors, and the editorial staff to interact within their subscribed domains to achieve the double blind peer review process.

There is, however, a flipside to this advanced technology — we are often confronted with duplicate submissions of MSS, multiple submissions to the same journal with different combinations of co-authors, withdrawal of papers midstream in the review cycle, co-authors submitting the same MS to another journal, and additions, deletions, and even changes in the order of author names. There are also authors who continue to commit fraudulent submissions even after publishers have imposed a ban on them.

The major forms of scientific misconduct are captured by three words: *fabrication, falsification and plagiarism.* Plagiarism jeopardizes careers and is unfair to the authors whose work they are stealing. The online tool Cross Check, endorsed by publishers worldwide, compares text with a database of millions of articles. If an article that includes plagiarized content has been published it has to be retracted. However, the article will still be shown on the website along with the reason for retraction. This ensures the

EVOLUTION OF PUBLISHING PROCEDURES

Then — **Now**

 Manuscript Submission →

 Peer Review →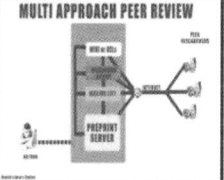

 Copy Editing →

 Formatting →

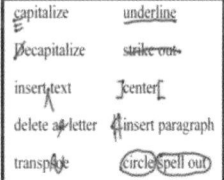 Proof Reading →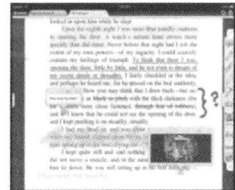

 Cover Designing →

 Quality Control →

Figure 1.

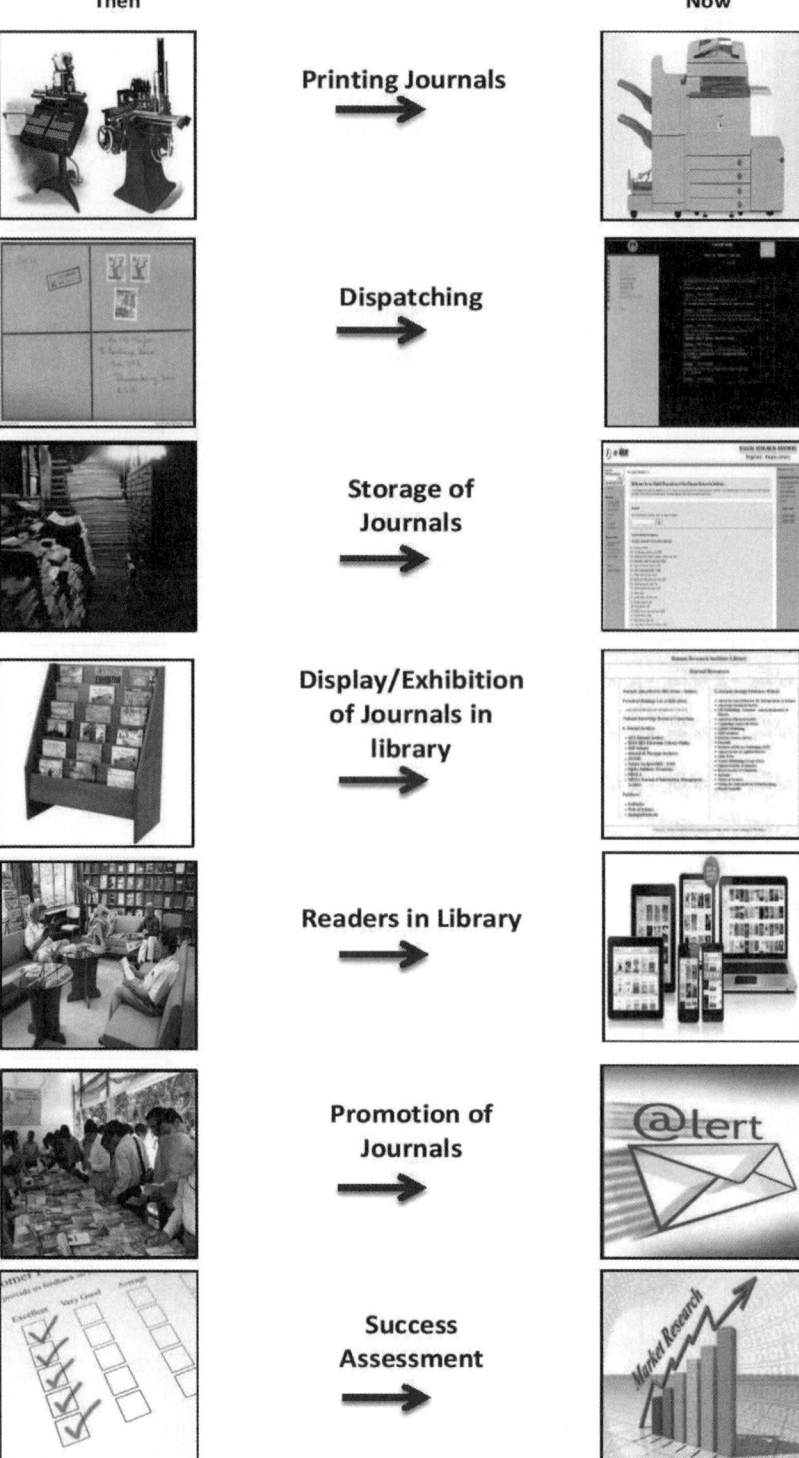

Figure 1. cont.

plagiarism and researchers responsible are highlighted. Publishers cope with misconduct through steps to detect plagiarism, deal with established cases, and find ways to preempt such cases.

Among the activities undertaken by the Indian Academy of Sciences to promote and disseminate science, publications have occupied pride of place. The first issue of the Academy's earliest publication titled 'Academy Proceedings', a monthly, appeared in July 1934, the same month that the Academy was founded by Sir C. V. Raman. Publication of science journals has since remained an important activity. Today the Academy publishes 10 science journals in various disciplines. Since 2007 the Academy journals are co-published with Springer. Submissions to JAA and other Academy journals are processed through an online submission and review management system. Papers undergo a blind peer review procedure. Springer provides access to our journals? content worldwide in an online full text database on Springerlink. Free access to all Academy journals worldwide is provided via the website of the Indian Academy of Sciences.

Today's technology has placed our journals' content on the world stage and we now deal with factors and features such as downloads, impact factor, citations, online subscribers, consortia packages, online deals, citation analysis, market surveys, sales incentives, marketing strategies, and the need to secure data. Our journals have reached new and previously unimagined geographical areas. Downloads from our archives date back to 1948!! Nevertheless, it is true that with speed, ease and automation of processes, quality must be ferociously guarded. This can be achieved by adopting technology, adapting it to our requirements and, most importantly, not compromising on the importance of trained and knowledgeable human interface which we believe is still the quintessential element in all technology.

The effectiveness of a personalized reminder email to reviewers and authors, in a particular instance perhaps, would outweigh an impersonal, automated letter. A dedicated proof reader's eye sees subtle inconsistencies missed by the machine and detect new mistakes in an otherwise correct piece of text unwittingly caused during conversion of files. A capable copyeditor's instinct is to catch the nuances of language expression. The indomitable spirit of the editorial staff ensures the accuracy of the most minute discrepancies such as position of an accent (tilde) on a letter, the distinguishing characteristic between an italic lower case 'v' and the Greek 'nu', the letters "fi" as a ligature, the minute difference between a minus sign and an en dash, the inbuilt space on either side of mathematical symbols, detect display equations in either TeX or Times Maths, or to look at a final proof and check by scanning it with an experienced eye for the finer aspects of formatting – symmetry, alignment, spacing, alliteration, hyphenation, layout, elegance, and clarity – before the page is passed for printing. An individual's skills vis-à-vis quality control, we have learned, can complement the limitations of automated processes and is therefore, to us, the imperative to achieve error free and well – produced publications – in content, format and style.

5. The future of publishing

What then are the challenges and opportunities for publishers in the digital future? How has the evolution of scholarly publishing impacted us and how has this evolution demonstrated itself? Here we enumerate the developmental changes impacting publishing; only time will provide the answers to these challenges.

• The 'serials crises' and demand for open access to government funded research; university libraries challenging publishers for raising journal prices (the 'cost of knowledge' debate, journal boycotts); the open access movement; birth and development of the ideas of gold and green OA publishing

- Publishers' responses to APC, OA and embargoes; allowing final accepted versions in author's Institutional Repositories; new OA journals by commercial publishers with APC as revenue models; fee-supported open access to papers in non-OA journals for authors who are willing to pay
- Journal pricing schemes; growth of online subscriptions; site licensing; journal bundling; archive packages; back volume pricing; issues concerning perpetual license; concessional pricing for developing countries; consortium pricing; price negotiations
- Evolution of the definition of OA itself; copyright and licenses for use of published papers and data; Creative Commons licenses; publisher-allowed authors' rights; mandates imposed by research funding agencies to modify copyright transfer agreements; movement towards discouraging APC: the matter of 'affordability of OA'
- Publishing negative data on research to help researchers understand what NOT to do - will Editorial Boards accept this decision
- Identifying and investigating the trap of predatory journals
- Will books be subsumed by eBooks? "Books are important but they are not as good as reading" contrasted with "Libraries are the future of reading"
- The threat of self-publishing by authors
- Science societies losing control (to the commercialisation of research) of the editorial and quality processes and production of their journals
- 'Accepted Manuscript' and 'Version of Record'— which one should the author cite
- Transparency within the traditional and confidential peer review system
- There is need to incentivise good reviewers
- Technology keeping pace with threats to integrity and assisting readers and researchers with DOI, Crosscheck, Fundref, CrossMark, Retraction Watch and data repositories for data mining
- Is the future of publishing already here with the popularity of preprint servers — new services that disrupt traditional publishing models?

How can publishers deal with the issues that confront them? Publishers, perhaps, will need to focus on adding value to content, such as explore newer formats, maintain a strong connect with authors, devise newer revenue models or develop multi-media platforms.

6. Library Services

How have library services evolved through the years and progressed to meet the requirements and demands of advanced technology? The focussed collection in any special library is at the core of its services and activities. Library service is the activity involved in information retrieval, delivery, maintenance, and preservation of its collection. The collection involves activities from acquisition of books, e-books and non-book materials; subscriptions to journals, e-journals, databases and maintaining perpetual access to open access books and journals to processing, maintaining and making information available to library users using both traditional methods (lending of books) as well as new emerging tools such as library web page and smart phones. Libraries are shifting focus from building local print collections to providing remotely accessed online resources to students and researchers and guiding their access needs through 'Federated search'.

The Raman Research Institute (RRI) Library has grown from the personal collection of the Nobel laureate Sir C. V. Raman into a modern library with print and electronic resources, an institutional repository, multimedia library and a partnership with National Knowledge Resource Consortium (NKRC) to access e-journals while affording internship training to students of library science from several universities. Astronomy is one of the core subjects for research at RRI; our dedicated focus is therefore on advancing methods and in proactive communication with astronomers, through forums such as LIS-FORUM, Forum for Resource sharing in Astronomy (FORSA) - custom made for astronomy libraries in India, NKRC, Astrolib and SLA-PAM.

LIBSYS - library management software used for library automation since 1994 has been updated with Libsys7 in 2015, which is a new web-based version with interactive features, customized federated search etc. The portals available on the RRI Library web page (http://www.rri.res.in/library.html) include: RRI Digital Repository, E-Books, E-Journal Portal, Imprints Collection, Multimedia Library, Library OPAC (Online Public Access Catalogue) and Open Access Portal.

7. How library services have evolved at RRI

Libraries have made constant transitions to become information centers that handle all types of knowledge, from paper to electronic media (CDs DVDs etc.), to networked resource (Figure 2).

8. How have libraries responded to the demands of the scientific community?

In the early 1980s, due to proliferation of information, library professionals working in the Institutes where astronomy was one of the main thrusts of research felt the need to come together and initiate a forum, which could act as a springboard for sharing and exchange of information. The first meeting of Astronomy librarians was held on July 29, 1981 at Raman Research Institute, Bangalore and informally launched Forum for Resource Sharing in Astronomy and Astrophysics (FORSA) with a vision and mission to share resources held in each library. Since 1989, FORSA members have met every year, in conjunction with the Annual Meeting of the Astronomical Society of India. At present, there are eleven institute members viz., ARIES, Bose Institute, CASA-OU, HRI, IIA, IUCAA, NCRA (TIFR), PRL, RRI, SINP and TIFR. The activities of FORSA are to facilitate access to journals/books and merged databases, access to websites of member institutions; actively participate in Resource Sharing; digitization of archival materials of the institute, and develop an open access Institutional Repository. Today, representatives of FORSA member institutions still maintain active informal interactions.

The NKRC, a national gateway of Science & Technology On-line resources, established in 2009, is a network of libraries and information centres of 40 Council of Scientific and Industrial Research (CSIR) institutions and 28 Department of Science and Technology (DST) institutes. Today, NKRC facilitates access to 5,000+ e-journals of all major publishers, patents, standards, citation and bibliographic databases. Apart from licensed resources, NKRC is also a single point entity that provides users with access to a multitude of open access resources. Another extremely important objective of NKRC is to build the Institutional Repository platform of all research organizations based on a common standard using D-Space open source software and access from a single national window (http://nkrc.niscair.res.in/indexpage.php).

Library and Information Services in Astronomy (LISA) conferences held since 1988 include participants from the all parts of the world. LISA builds astronomy librarians? network and influences the professional life of astronomy librarians. Library staff of FORSA libraries especially RRI, IIA, IUCAA, NCRA – Pune, PRL, and TIFR have attended and contributed to LISA. LISA VI was organized by FORSA and co-hosted by

EVOLUTION OF LIBRARY SERVICES

THEN	NOW
• Closed Access	→ Open Access
• **Library Building:** Library sections in different rooms of the Main Building of RRI	→ Separate Library Building : Basement, Ground Floor and Mezzanine
• **Collection Development:** Acquisition of Books, Journals and magazines in print form; Non-Book Materials - Slides, Microfiche & Films, CDs & DVDs, etc.	→ e-resources/ e-books/ e-journals
• **Circulation :** Lending of Books, slides and non-book materials – Entry in Registers/ Two-Card system; Barcoding books; Stock Verification	→ Issue/Return through LIBSYS - Library Management Software & Facilitating e-books via RRI homepage; Online Access /online Subscription
• **Periodical section:** Journal/magazine Subscription (Print) –Registers/ Kardex Entry	→ Entry through LIBSYS Software Access to E-journals from RRI webpage/ FORSA/NKRC Consortium
• **Reference Service:** Assistance in the use of the library including location of materials, use of the catalogs, Databases and the use of basic reference sources viz. almanacs, atlases	→ Web OPAC facility to access library collection from anywhere anytime (24X7)
• **Inter library loan:** Document delivery by post/manual: books/magazines /journals Exchanged/borrowed between the groups of libraries in Bangalore	→ Attending online queries` In house and ILL- article Requests (soft copy) delivered via e-mail
• **Notice boards** :Preprint services /Science News /Recent Additions of Journals and Books, etc.	→ Digital Display Board
• **Photocopy services :** Kores-stat and Xerox copies (Manual copiers); Reprography machines used	→ Facilitated with networked printers; Photocopying, scanning and converting to image/pdf and sent via email.
• **Preservation & Binding:** Binding Journals; Reports, thesis; Repair of Books and Packing	→ Purchase of Online Archives; Binding Journals diminished; Reports ,thesis; Repair of Books and other materials and Packing continues
• **SDI services :**	
✓ Compilation of Current awareness List & Bibliographies	→ Create lists using Online Databases -ADS & WOS
✓ Astronomy lectures, talks etc. in Audio/Video Cassettes, CDs and DVDs	→ Lectures, talks etc., access via Internet/ Repositories
✓ Preprints on shelves, memoirs of RRI – publications of RRI (Print), Annual Reports etc.	→ Placing preprints/articles online on arXive, Institutional Repositories, etc.
✓ Palomar sky survey charts were available as photographic prints till 1994, then as 102 CD	→ Palomar sky survey charts are available Online, since 2001
✓ IAU Circulars received by post	→ IAU Circulars available On-line as IAUCs

Figure 2.

Inter-University Centre for Astronomy & Astrophysics (IUCAA) Library and the National Centre for Radio Physics (NCRA) Library. Librarians, Astronomers, Computer Scientists, and Publishers met to discuss information maintenance and retrieval, new techniques and technologies, and the future of astronomy libraries and librarianship.

9. Challenges and opportunities for libraries in this digital age

Traditionally, libraries have focused on published information but new challenges also encompass non-explicitly published materials such as research data sets, interactive programs, complex visualizations, lab articles, diagrams, as well as web-based exchanges such as blogging. These materials have to be integrated into library resources in 'discovery tools'. This task requires new skills by librarians and demands they stay current on new and innovative approaches.

Information science instruction should occur throughout a Librarian's career and therefore be a part of their professional development. A variety of approaches, including online tutorials for basic skills, one-time classes, in-depth classes/workshops on strategies and tools for specific disciplines, and classes focused on information policy issues, such as intellectual property can be instituted by Information Science schools. Librarians also need to broaden their own concept of their role in the design of curriculum and provide outreach to faculty to help them understand how librarians can enhance the education of students.

Using new forms of technology such as mobile devices, You Tube, Internet of Things, 3D-printing etc. will enhance accessibility to library services and resources. The library's role in helping faculty and other researchers understand new ways to measure and evaluate research can be through the study of metrics.

10. The future of library services

With the gradual depletion of print astronomical journals and subscriptions to more and more e-journals, display space previously utilized for astronomy journals is being vacated. These 'maker spaces 'could be utilized for readers with Wi-Fi facility; providing e-book readers; meeting space for study and discussion; intellectual interactions with past and present students; or commons place for continuing education.

Change is an on-going evolution of library services. New technologies, new trends, and new ideas will emerge with new services. Are professional library skills useful in today's digital information world? Will libraries cease to exist? Will they become tourist attractions? Will they adapt to changing scenarios and shift focus to serve other related purposes?

11. Concluding thoughts

Notwithstanding all the issues, we need to remind ourselves that publishing is an agile, robust industry. Today the challenge is to discover, to create new ways of disseminating science research – through communication that's internet enabled and develop methods that will serve and justify a data-rich digital age. Intrinsic and inextricable to this very challenge is that libraries need to reinvent themselves and stay relevant by embracing the physical and the virtual, reinventing that defining feature of libraries as communities, as being more about "connections"than of "collections", and to stay committed to ensuring that knowledge is made accessible for the future – because "libraries are in the forever business".

References

Grothkopf, U. 2012, *Organizations, People and Strategies in Astronomy I (OPSA I)*, 227-244 (Ed. A. Heck, https://www.eso.org/sci/libraries/articles/opsa_grothkopf.pdf)

Isaksson, E., Lagerstrom, J., Holl, A., & Bawdekar, N. 2010, *ASP Conference Series*, 433 (Raman Research Institute Annual Report, 2014-2015)

Johnson, L., Adams Becker, S., Estrada, V., & Freeman, A. 2015, *NMC Horizon Report: 2015 Library Edition* (Austin, Texas: The New Media Consortium.)

Behind the Spam: A "Spectral Analysis" of Predatory Publishers

Jeffrey Beall[1]

[1] Auraria Library, University of Colorado Denver
1100 Lawrence St. Denver, Colorado, 80204 USA
email: jeffrey.beall@ucdenver.edu

Abstract. Most researchers today are bombarded with spam email solicitations from questionable scholarly publishers. These emails solicit article manuscripts, editorial board service, and even ad hoc peer reviews. These "predatory" publishers exploit the scholarly publishing process, patterning themselves after legitimate scholarly publishers yet performing little or no peer review and quickly accepting submitted manuscripts and collecting fees from submitting authors. These counterfeit publishers and journals have published much junk science ? especially in the field of cosmology ? threatening the integrity of the academic record. This paper examines the current state of predatory publishing and advises researchers how to navigate scholarly publishing to best avoid predatory publishers and other scholarly publishing-related perils.

Keywords. Scholarly publishing; Scholarly journals,; Predatory publishers; Open-access; Cosmology

1. Introduction

Perhaps no scientific fields have been hit as hard by predatory journals as astronomy and cosmology. These two fields, along with astrophysics, seem to attract many unqualified scholarly authors, and the predatory journals have given these people a publishing platform for their pseudo-scientific ideas that, to the general public, may appear authentic. The result has been a profusion of open-access journals happy and eager to accommodate publication of numerous pseudo-cosmology articles that purport to resolve some of the most important unanswered questions in the field.

2. Publishing Models for Scholarly Journals

We are all familiar with the traditional model of scholarly journals, also called the subscription model. In this model, libraries subscribe to journals and make the content available to their users or individuals subscribe to the journals to access and read them. The internet brought a new innovation to scholarly publishing ? the ability to purchase access to individual articles, sparing one the cost of having to pay the full subscription cost.

The subscription model has multiple strengths, the primary being that submitting authors are generally not charged for their submissions; publishing is free. While some non-profit, scholarly society publishers have imposed modest page charges on authors to defray publishing costs, these have been and remain the exception. Another advantage is the built in validation function of the subscription model. If a journal underperforms or otherwise becomes unsuitable for the subscriber, the subscriber cancels the subscription. Journals want to avoid cancelations, so they are keen to meet the needs of their subscribers and provide high quality content at a reasonable cost to preclude subscription cancelations. Next, the subscription model spreads out the cost of scholarly publishing among all subscribers. This enables publishers to create and employ many value added features to their published articles; enhancements benefitting both readers and authors. The subscription model also creates an economy of scale. Finally, subscription journals typically limit the number of articles they publish in each issue, performing a filtering function so only the best articles are published from among the many received.

I recognize three publishing models for scholarly open-access journals: gold, platinum, and green. In the gold open-access model, the publishing costs are financed by payments charged to authors upon acceptance of their manuscripts. This model has many major weaknesses. One weakness is that the more papers a journal accepts the more money it makes, thus creating a conflict-of-interest for the publisher. Good journals typically reject many article submissions resulting from recommendations made in the peer review process. However, for-profit journals aim to increase their revenue, a goal that conflicts with the rejection of manuscripts and the revenue they provide. Another weakness is that authors have to pay to publish, yet many lack funds for this, especially in fields where grant funding is uncommon. Finally, unlike subscription journals that spread the costs of scholarly publishing across subscribers, gold open-access journals focus the costs on the authors of each issue.

Another open-access publishing model for scholarly journals is the platinum model. In this model, publishing is free for authors and accessing the articles is free for readers ? there are no author fees or subscription charges. The publishing costs are funded benevolently, so usually the sponsors of platinum open-access journals are associations or institutes or universities. Very often, these publishers operate on a limited budget and are not able to offer all the value added benefits to scholarly publishing that larger subscription publishers offer.

For example, many low-budget open-access journals do not offer digital object identifiers (DOIs) and some do not follow best practices in digital preservation, meaning the content is at risk of being lost. In contrast, large publishers offer platforms with many value added features, such as direct importing of citations into citation management software, platforms that add great value to the published content and benefit authors by increasing the visibility of their work. Many academic libraries provide links to these platforms, and they serve as research portals for scholars.

Note that many open-access advocates do not differentiate between platinum open access and gold open access, lumping them together as gold open-access. But because the gold model involves payments from authors and the platinum model does not, the distinction between the two is important and merits distinct terminology.

Finally, the green open-access model refers to authors self-archiving their published works in open-access repositories. This model allows authors to benefit from publishing in high-quality subscription journals while also making a version of their article open-access via a repository. These include institutional repositories, such as those established and managed by academic libraries, and disciplinary repositories, those managed cooperatively by researchers in a particular field of study.

Despite these advantages, green open-access has weaknesses. There is low uptake on this open-access model. Once authors have published their works in subscription journals, few are motivated enough to self-archive them in a repository, a process that can require significant additional effort. Because the author has transferred copyright to the publisher, he or she is subject to conditions the publisher imposes, such as embargo periods before the papers can be mounted, often a year or more. Moreover, only the authors' post-prints can typically be archived, not the publishers' PDFs. The post-print, also called the author's accepted manuscript, is the author's final version of the article (usually a Word document, or in astronomy and physics, a PDF of a LaTex document) he or she submits to the journal.

3. Predatory Journals and Cosmology

In the five years plus I have been studying predatory publishers, I have noticed what I believe to be a disproportionate number of junk science articles published in the field of cosmology, articles the predatory publishers are happy to accommodate. Cosmology attracts amateur theorists and theorists from other disciplines who are inspired to dabble in cosmology.

Figure 1 shows a screenshot of one such article. It was published in a questionable journal in 2013. Its title is "Combating Climate Change with Neutrinos." I wrote a blog post about it right after it was published and the publisher removed the article almost immediately. It is no longer a published article, except for a copy on my blog. The publisher did not issue a formal retraction statement when it removed the article, the standard practice in such cases. This non-adherence to established standards is typical of low-quality, open-access journals. Regarding bogus cosmology articles, they are sometimes written in such a way that they cannot be proven or disproven. This is because the articles are speculative and not based on data. Though it is clear the articles are pseudo-science, it's difficult to disprove their theories and assertions because it's more difficult and more time consuming to prove a negative statement (this theory is false) than it is to prove a positive one (this theory is correct).

I have noticed a pattern of articles that purport to "correct" the findings of Einstein. Also, I've observed that in predatory journals, the nature of dark energy and dark matter has been "discovered" many times over. It seems there are many who want to be the hero who discovers the nature of dark energy and dark matter, so the author invents some explanation hoping he might get lucky and stumble on the actual discovery or that some may believe their theories. These authors want to create a shortcut to fame and achievement for themselves.

Environmental Sciences, Vol. 1, 2013, no. 2, 79 - 82
HIKARI Ltd, www.m-hikari.com

Combating Climate Change with Neutrinos

J. A. de Wet

Box 514, Plettenberg Bay, 6600, South Africa
jadew@global.co.za

Copyright © 2013 J. A. de Wet. This is an open access article distributed under the Creative Commons Attribution License, which permits unrestricted use, distribution, and reproduction in any medium, provided the original work is properly cited.

Figure 1. A screenshot of the title page of a now-removed article entitled "Combating climate change with neutrinos." The figure is copied under the terms of the Creative Commons Attribution License. The article was removed from the publisher's website soon after I published a blog post about it.

Interestingly, I have observed that many older men from fields other than cosmology tend to be the ones that attempt to answer cosmology's biggest questions.

As an example there is an article called "The Dark Side Revealed: A Complete Relativity Theory Predicts the Content of the Universe." It's written by Ramzi Suleiman (2013), a professor at the University of Haifa in Israel. I learned of his work because he emailed me recently asking for a recommendation.

He stated he found it easy to publish in the journals on my predatory publisher list, but that the legitimate cosmology journals had all rejected his article submissions. He wrote to ask for advice on an open-access book publisher. I responded to him that he should take the advice of the peer reviewers in cosmology and abandon his work in the field. His response was not favorable.

This university professor is a social psychologist by training, yet he thinks he has the solutions to cosmology's and physics' greatest mysteries; he is not the only one. It seems there are many established professors in fields outside cosmology who write nonsense articles about cosmology. What is the etiology of this pathology? I am at a loss to understand it.

In the abstract, the professor claims that his theory, "yields natural definitions of dark energy and dark matter and predicts the content of the universe with high accuracy" (Suleiman 2013, p. 34). Not surprisingly, his work is completely ignored by mainstream cosmology researchers. I have learned that it takes a Ph.D. to recognize good science in a particular field, but recognizing junk science requires only common sense.

A further example is the article entitled "Mathematical Prediction of Ying's Twin Universes" (Davvaz *et al.* (2014)). It was published in the journal American Journal of Modern Physics. I invite readers to access the article and judge its science themselves. I think it is pseudo-science, and the authors' choice of publishing venue – a low-quality journal – adds weight to this belief.

The journal's publisher is Science Publishing Group. I would call it a vanity press, but I think that would be offensive to all the real vanity presses. This publisher will publish anything for money and appears to be a favorite among pseudoscientists. It publishes over one hundred journals.

I have been unable to determine where this publisher is based or who is behind it. It claims to be based at 548 Fashion Avenue in New York City, but that's the address of a mail forwarding service.

These examples show the whole notion of selectivity in scholarly communication is disappearing. Many open-access journals will publish anything for money. Science is not something that should be democratized. It is not something that should be decided by popular vote. Unfortunately, predatory journals are selling the imprimatur of science to anyone with a manuscript and two or three hundred dollars.

4. The Damage to Science

Predatory journals negatively affect science, the communication of science, and scientists in many ways. First, because predatory journals often perform a fake peer review or skip it altogether (despite claiming to do it properly), there is an increased number of scholarly articles being published that contain one or more violations of scholarly publishing ethics. It is easy to find instances of plagiarism, self-plagiarism, salami slicing, and duplicate submission in predatory journals.

Next, as already mentioned, there is much pseudo-science being published in predatory journals, a result of their lax or non-existent peer review combined with their strategy to earn money through payments from authors. There are a number of scholarly indexes that, in their aim to be comprehensive and cover most scholarly journals, include predatory and low-quality journals. When these journals are indexed, the academic databases also index the junk science that is included in predatory journals. Students who use these indexes but who are unequipped to differentiate sound science from junk science may use junk science articles from search results, treating the papers as real.

Moreover, because research is cumulative, it is possible that some scholarly authors will, when researching and writing literature reviews, mistakenly include articles published in predatory journals, polluting the scholarly record. Citing research published in predatory journals, whether intentional or not, can stigmatize researchers and corrupt the cumulative nature of the progression of science. It may also provide unintended legitimacy to the junk science.

In addition to publishing journals, many questionable publishers are also cashing in on scholarly conferences. They do this by organizing conferences and spamming researchers, inviting them to present their research at the conferences. The venues are frequently in resort cities. Most submitted papers are accepted, and the registration fees are often high. Some conference organizers hold several conferences at the same hotel at the same time, maximizing profits. Some charge an additional fee for publishing abstracts or full papers in the conference proceedings, while others charge a separate fee for publishing the conference presentation as an article in one of their open-access journals.

In the biomedical sciences, some researchers are developing compounds such as nutraceuticals or medicines they hope to sell to a drug company or directly to the public. Before a drug can be marketed, however, medicines need, among other things, research proving their efficacy. Again, because of the fake peer review, predatory journals are the perfect place for an individual wanting to make an ineffectual compound appear effective. One can basically write an article showing 'efficacy' of a particular compound and submit it to a predatory journal, where it will be accepted and published upon payment of the author fee. The published article can then be used to demonstrate the effectiveness of the medicine to companies, investors, and the general public. Those being scammed may be unable to differentiate between predatory journals and authentic publications and are fooled into believing the research is real.

Finally, broadly, predatory open-access journals hurt science by preventing some of the science from being published. The gold (author pays) open-access model effectively silences authors who lack the funds to pay author fees. These authors may include those in developing nations, those in middle-income nations, and retired and emeritus faculty of all nations. Moreover, increasingly, the more respected an open-access journal is, the higher its author fee. Subscription journals, on the other hand, generally do not charge authors and seek to publish the best quality articles

It's no coincidence that the advent of predatory journals has occurred at the same time many are questioning the future of scholarly publishing, peer review, and the scholarly journal itself. Predatory publishers have polluted scholarly communication and threaten the very communication of science itself. Now anyone with a bizarre theory about cosmology, astrophysics, or astronomy can publish in a journal that appears legitimate and scholarly, potentially misleading many into believing that the work represents vetted science.

All reputable researchers need to develop a "scholarly publishing literacy" skill set (Zhao 2014) that enables them to recognize and avoid the increasing number of scholarly publishing scams that continue to appear. Academic librarians can assist with vetting scholarly journals. The future of scholarly communication is at stake, and all researchers must protect themselves from becoming victims of predatory publishers.

References

Davvaz, B., Santilli, R. M., & Vougiouklis, T. *Mathematical prediction of Ying's twin universes.* 2014, *American Journal of Modern Physics*, 4(3), 5-9

Suleiman, R. *The dark side revealed: A complete relativity theory predicts the content of the universe.* 2013, *Progress in Physics*, 9(4), 34-40

Zaho, L. *Riding the wave of open access: Providing library research support for scholarly publishing literacy.* 2014, *Australian Academic & Research Libraries*, 45(1), 3-18. doi:10.1080/00048623.2014.882873

The data sharing advantage in astrophysics

Bertil F. Dorch, Thea M. Drachen and Ole Ellegaard

University Library of Southern Denmark, University of Southern Denmark,
Campusvej 55, DK-5230, Odense, Denmark

Abstract. We present here evidence for the existence of a citation advantage within astrophysics for papers that link to data. Using simple measures based on publication data from NASA Astrophysics Data System we find a citation advantage for papers with links to data receiving on the average significantly more citations per paper than papers without links to data. Furthermore, using INSPEC and Web of Science databases we investigate whether either papers of an experimental or theoretical nature display different citation behavior.

Keywords. astronomical data bases, methods, sociology of astronomy, statistical

1. Introduction

Research funders increasingly require data management plans prior to applications for funding. Similarly, infrastructures and policies are arising regarding both archiving, documentation and sharing research data. While scientists are increasingly being evaluated and funded according to quantitative measures, e.g. by citations, it is relevant to ask whether a citation advantage exists that is related to the activity of sharing data, similar to the debated citation advantage related to Open Access (e.g. Kurtz *et al.* 2005 and Kurtz *et al.* 2007).

We present here a simple study of astrophysical publications to investigate a possible increased citation impact resulting from linking to data, using the NASA Astrophysics Data System, henceforth ADS (cf. Kurtz *et al.* 2000). This work is an extension of the initial study concerning data links in papers published in the journal *ApJ* during 2000–2010, presented in an unpublished working paper by Dorch (2012).

2. Publication data and method

The ADS, launched by NASA in 1992, is hosted by the Harvard-Smithsonian Center for Astrophysics. ADS is an online publication database of millions of astronomy and physics papers receiving abstracts or tables of contents from hundreds of journal sources. The ADS also lists citations for each paper. The ADS search engine is tailor-made for searching astronomical abstracts and can be queried for author names, astronomical object names, title words, abstract text, and results can be filtered according to a number of criteria (cf. Eichhorn *et al.* 2000).

For each publication record in ADS, a number of links are possible, including data links to online data, e.g. at external data centers. Links of this type are abbreviated "D" aka. D links (cf. Accomazzi & Eichhorn 2004 and Eichhorn *et al.* 2007). Therefore, it is possible to limit ADS to publications with or without D links.

In part of the work presented here, we invoke also a secondary source of publication data, the INSPEC database from the Institute of Engineering and Technology (formerly the IEE) and a source of citation data, and the Web of Science (WoS) science citation index from Thomson Reuters. Like WoS, but unlike ADS, INSPEC is a commercial major indexing database of scientific and technical literature.

Table 1. Data for the four journals *ApJ*, *A&A*, *MNRAS* and *AJ*: Journal Impact Factor 2013 (JF), the average number of papers published per year during 2000–2014, $\langle N \rangle^{\text{papers}}$, the average fraction of papers with D links $\langle n \rangle_{\text{D}}^{\text{papers}}$, the average fraction of citations resulting from papers with D links $\langle n \rangle_{\text{D}}^{\text{cite}}$, and the average D link citation advantage $\langle P_{\text{D}} \rangle$ during 2000–2014.

Journal	JIF 2013	$\langle N \rangle^{\text{papers}}$	$\langle n \rangle_{\text{D}}^{\text{papers}}$	$\langle n \rangle_{\text{D}}^{\text{cite}}$	$\langle P_{\text{D}} \rangle$
ApJ incl. *ApJL* and *ApJS*	6.280	3137	0.303	0.358	1.286
A&A	4.479	1941	0.386	0.441	1.302
AJ	4.052	-	-	-	409
MNRAS	5.226	1725	0.239	0.247	1.055

In this study, we perform two analyses:

(a) We investigate the number of papers and citations for papers with or without D links during the period 2000–2014 for *ApJ*, *A&A* and *MNRAS* using NASA ADS.

(b) We investigate the number of papers and citations for experimental and theoretical papers respectively during 2010 for *ApJ*, *A&A* and *AJ* using INSPEC and NASA ADS.

Firstly, (a) we limit the study to papers published in major astrophysical journals during the 15-year period in the current millennium 2000–2014, cf. Table 1. Furthermore, we define the citation advantage of papers that link to data P_{D} as the ratio of the number of citations per year to papers with links to data, and the number of citations per year to papers without such links. Publication data and derivatives for *ApJ* are illustrated in Fig. 1 left and right.

Secondly, (b) it is relevant to investigate whether we introduce a bias in selecting articles with data-links, e.g. whether experimental papers more often link to data, and whether experimental papers are cited more than theoretical papers.

To test this possibility, we apply the feature *treatment type* that the database INSPEC assigns to all indexed papers:

- *Theoretical or mathematical* is assigned when the subject matter is generally of a theoretical or mathematical nature.
- *Experimental* is used for documents describing an experimental method, observation or result. Includes apparatus for use in experimental work and calculations on experimental results.

Articles from the three journals *ApJ*, *A&A* and *AJ* are downloaded into the reference handling program Endnote in order to extract DOIs for further processing. The relevant DOIs are then entered into INSPEC and the articles are separated into the two tiers: either classified as theoretical or experimental work. The few articles classified as both experimental and theoretical are discarded from the analysis. Finally, we apply, in this case, WoS in order to extract the number of citations because DOIs are not searchable in ADS.

ApJ as registered by ADS includes letters as well as the supplement series but the articles published in those latter categories are not fully included in WoS and we discard them from the present analysis. The number of articles with or without data links (as well as citation data from WoS) is then downloaded directly from ADS.

A statistical analysis was performed as appropriate to test for any significance in mean citation counts between articles with and without datalinks as well as between theoretical and experimental articles. F tests were used to test for equal variance; two tailed t-tests were then run for unequal and equal variances as appropriate to test for significant

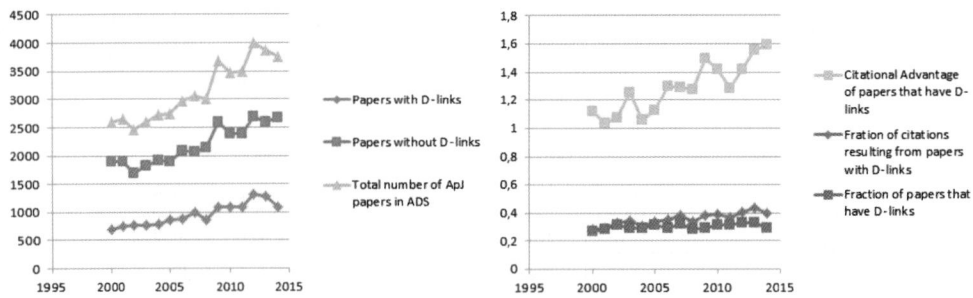

Figure 1. Left: The number of papers in ApJ 2000–2014 as a function of the year of publication as registered in ADS. Upper curve (green): Total number of papers. Middle curve (red): Papers without links to data. Lower curve (blue): Papers with links to data. Right: Upper curve (blue): The citation advantage P_D as a function of the year of publication as registered in ADS. Middle curve (blue): The fraction of the total number of citations that result from papers with links to data n_D^{cite}. Lower curve (red): The fraction of papers that actually have links to data n_D^{papers}.

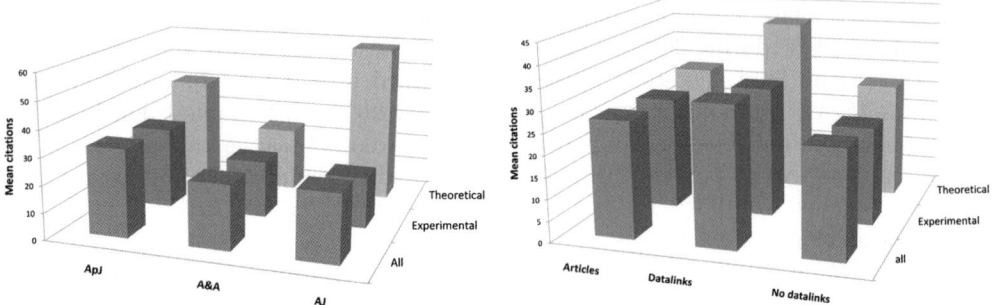

Figure 2. Left: Histogram of the total mean number of citations for ApJ, $A\&A$ and AJ papers with D links in 2010 and the corresponding contributions from experimental and theoretical papers. Right: Histogram of the total mean number of citations for ApJ in 2010 and for papers with and without D links (blue columns), and the differentiation for experimental papers (red columns) and theoretical papers (green columns).

difference between mean total citations per paper. Our focus is only on articles published in 2010. This ensures time to accumulate a sufficiently large number of citations.

3. Results and discussion

The papers with D links received, in total, fewer citations per year on average relative to the papers without D links (by approximately a factor of two). However, there being fewer papers with links to data, it turns out that these papers on the average received more citations per paper i.e. during the examined period the D link papers in ApJ on average receive 28% more citations per paper per year, than the papers without D links. Since 2009 that fraction is higher and in the case of ApJ more like 50% more citations, cf. Fig. 1 (right).

Next, we look at the journals and papers in term of their experimental or theoretical content. In case of papers published in ApJ the number of experimental papers is only slightly above the number of theoretical ones. The difference between the mean numbers of citations obtained by the two groups is small as well.

The situation is different when considering papers with or without data links. In case of D link papers, the number of experimental papers is much larger than the number of theoretical papers, while the latter has the largest mean number of citations. On the other hand, the number of theoretical non-link papers is above the similar number of experimental papers, but still the theoretical articles obtain the most citations.

The same pattern is observed in case of the two other journals *A&A* and *AJ*. The theoretical papers with data links obtain the highest number of citations. The difference is most pronounced in case of papers published in *AJ*, but this conclusion is based on rather few papers in the data. We have examined the statistical confidence level of our conclusions: In case of *ApJ* and *A&A* it is evident, although only evident at the 5% significance level in case of *ApJ* ($p < 0.05$), that papers with D links obtain the largest numbers of citations. In case of *AJ* a p value well above 0.05 indicates that the citation advantage is not statistically well founded. In a similar fashion a significant advantage for obtaining citations has been observed for theoretical D link papers compared to experimental D link papers in case of all three journals. On the other hand, it can only be proven at the $p > 0.05$ confidence level partly due to a low number of papers and scatter in citations data.

Our simple study indicates a clear tendency for papers with links to data to receive more citations per year on average, than papers that do not link to data. However, there are several biases that could be studied further, e.g. whether longer papers, papers with more authors etc. display generically different citation patterns. Also of potential importance is whether some subjects that "naturally" link to data have a higher citation impact than other fields, e.g. papers based on space missions or telescope data.

Henneken & Accomazzi (2011) performed an analysis restricting publication data using a set of 50 keywords: looking at cumulative citations to papers after a 10 year period. The report demonstrated a 20% increase in citation count for papers with D links, compared to those without. Alas, evidence is mounting that linking to data enabling sharing does indeed merit those who do so. This evidence thereby also supports initiatives furthering the development of data infrastructure.

A more comprehensive account of the study presented in these proceedings will be published by Dorch *et al.* (2016).

Acknowledgements

This research has made use of NASA's Astrophysics Data System Bibliographic Services.

References

Accomazzi, A. & Eichhorn, G. 2004, *ASP Conference Proceedings*, 314, 181
Dorch, S. B. F. 2012, https://halshs.archives-ouvertes.fr/hprints-00714715
Dorch, S. B. F., Drachen, T. M., Ellegaard, O., & Larsen, A. V. 2016, *LIBER Quarterly*, forthcoming
Eichhorn, G., Kurtz, M. J., Accomazzi, A., Grant, C. S., & Murray, S. S. 2000, *A&AS*, 143, 61
Eichhorn, G., Accomazzi, A., Grant, C. S., Kurtz, M. J., Thompson, D. M., & Murray, S. S. 2007, *Bull. Astron. Soc. India*, 35, 717
Henneken, E. A. & Accomazzi, A. 2011, http://arxiv.org/abs/1111.3618
Kurtz, M. J., Eichhorn, G., Accomazzi, A., Grant, C. S., Murray, S. S., & Watson, J. M. 2000, *A&AS*, 143, 41
Kurtz, M. J., Eichhorn, G., Accomazzi, A., Grant, C., Demleitner, M., Henneken, E. *et al.* 2005, *Information Processing & Management*, 41(6), 1395
Kurtz, M. J. & Henneken, E. A. 2007, http://arxiv.org/abs/0709.0896

That over-used and much abused 4-letter word: DATA

Elizabeth Griffin

Dominion Astrophysical Observatory, 5072 West Saanich Road,
Victoria, BC, V9E 2E7, Canada
email: `Elizabeth.Griffin@nrc-cnrc.gc.a`

Abstract. In its prime state, DATA is a Latin word meaning "[things] given", a plural noun derived from the verb "To Give". Its singular form is DATUM. Modern conversation equates DATA with "Information", while modern philosophies on information management are getting entwined with parallel philosophies on knowledge management. In some ways that is a positive development, and is greatly assisted by Open Access and Internet policies, but in others it is more detrimental, by threatening to blur the essential distinction between objectivity and subjectivity in our science. We examine that essential distinction from the view-points of observers, authors (and publishers), and database managers, and suggest where, when and how the distinctiveness of their fundamental contributions to the communication and validation of research results should be respected and upheld.

Keywords. Astronomical data bases:miscellaneous; catalogues

1. The Facts

The word "Data" is a plural Latin noun meaning "[things] given". It derives from the verb "to give"; its singular form is "datum". However, the way that modern conversation tends to equate "data" with "information" is actually a central contributor to the *apparent* problems surrounding the variety and complexity of 'data in science' or 'data in publishing'.

To start with, "data" is emphatically *not* the same as "information". Science needs a word to describe raw, or just preliminarily processed, observations or records – the unmodified signals which have been captured as an image or a spectrum – and "data" is the traditional word for that. Raw observations are objective: untouched, and distinct from any interpretation of what they show. They are the base reference; regardless of what laws we want to apply, or of what evidence for some theory we hope to extract, those original data constitute the same pristine observation or record for each and every researcher to study.

Admittedly we need to qualify the description "raw", since an observation will unavoidably bear the signature of the detector itself, or (in the case of a spectrum) of the spectrograph – its designed characteristics of limiting resolution and point profile, and its selected ones of wavelength region and focus. It will also have been pre-processed by (for example) a CCD readout, a telemetric-bit conversion, or a photographic development. But the observation stands by itself, and does not rely upon any introduced concept or interpretation like classification type or temperature to explain it, nor is any justification required for what it shows.

2. The Problem

Some of the confusion between "data" and "information" has sneaked in through somewhat sloppy journal editing, or lack of it, and the consequences of what has been happening are still not widely appreciated. Many of today's authors who do not have English as their first language tend to copy the way their English-born peers use their own language, and so one sees a number of new terms, new spellings and even new words gradually creeping into our papers, including this critical lack of appreciation of when, and when not, to use the word "data". Information embraces what is known about some object, process, or whatever; it includes what can be deduced from the basic observation by applying laws, theories and measurements, and represents an end-product of what an analysis of the original observation has produced *on this occasion*. The analysis may be different when carried out by different people, and is therefore fully subjective; the thinking behind each analyses is likely to evolve with time, so the information is neither static nor conclusive. Enough information gleaned from a suitable range of appropriate sources ultimately contributes to *knowledge*, which sounds as though it should be more stable and more comprehensive than the separate strands of information which fed into it. But these stages are a long way removed from the original Data from which they stemmed.

It is therefore essential to respect the differences between Data and Information, and to ensure that that respect prevails throughout our publications, our libraries, our archives and our databases. Unless we do, researchers will be presented with materials that already confuse issues and render it unclear just what has been measured, and what has been deduced and therefore has a temporal quality. One good example of this is the *Bright Star Catalogue*, which includes not only the fundamental positions of the stars and recorded *measurements* of radial velocities, photometry, proper motions and parallaxes, but mixes in the classification type and an opinion as to whether the object has a variable velocity or not, and (from that) whether it is likely to be single or multiple. Many of the velocity observations in the literature of hot stars, in particular, were made manually on photographic spectra, and when the object was rotating so that its (already rather few) lines were broad, it was difficult to measure line-positions (from which the velocity was then derived) with very high precision. As a consequence, the velocities tabulated in the literature show scatter, and rather than suspect that the scatter was caused by low measuring precision the *Catalogue* suggests that all the objects thus affected have variable velocity. It takes a very long time and resources to prove that something once labelled as 'variable' does not in fact vary. A recent study of 12 such systems, for which the *BSC* gave the verdict of variable velocity for 11 of them, finally showed – after numerous observations spanning 5 years – that *all but one* have constant velocity; the one exception proved to be a previously-unrecognized spotted star, whose rapid rotation gave rise to line-profile changes.

3. A Solution?

Despite the scientist's need to restrict the term "data" to a very special aspect of the discipline, the way that the same word is also used very loosely in everyday conversation as a synonym for "facts", "characteristics" or "parameters" is spilling over into science. The development of Open Access and the concomitant involvement of an increasing variety of relevant expertise in information management is now another factor that is adding to the confusion of our descriptive language. That development cannot and should not be checked, but our science is suffering in important ways. How often do young authors commence a paper with "These stars are known to be ...", or "It is well known that ...", when the truth of the matter is that someone once proposed a hypothesis which then got printed in a paper, and once it was printed rather than just discussed orally it immediately gained an undue credibility: the simple act of publishing the hypothesis conferred upon it a level of proof and acceptability that it did not [yet] deserve. Insisting on a more clear and rigorously maintained distinction between Data and Information will teach the need to honour the fundamental difference between the objective and the subjective. But Open Access has its own momentum, and any attempt to clean up our conversation so as to respect that basic scientific distinction is as futile as trying to stem a breached dyke-wall with one finger. Perhaps science should invent a new word of its own?

Evolution of Scholarly Publishing and Library Services in Astronomy
Its Impact, Challenges, and Opportunities

Hema Wesley[1] and Geetha Sheshadri[2]

[1]1Consultant Editor, Indian Academy of Sciences, Bangalore, India
email: ruthwes09@gmail.com

[2]2Assistant Librarian, Raman Research Institute, Bangalore, India
email: geetha.sheshadri@gmail.com

Abstract. Scholarly publishing and its procedures have evolved rapidly, forcefully, and incredibly. Technical advances in the production and promotion of science content have dramatically augmented the visibility and reach, deepened the impact and intensified the thrust of science journal content. These changes range from checking text on perforated tapes to pit stop; from hot metal types to CTP; and from Gutenberg to colour digital printers. Intrinsic and inextricable to this revolutionary aspect of evolution in scholarly publishing is the evolution of library services in astronomy which catapulted library resources from preprints on shelves to customised digital repositories and from communicating observational data through postal telegrams to Tablets. What impact does this unique blend of revolutionary advances have on science and society, what are the consequent challenges, and what are the opportunities that can metamorphose from challenges inherent in the power and potential of the 'published word'?

The perspectives expressed in this paper stem from learning experiences of the authors at the Indian Academy of Sciences, publishers of ten science journals including the Journal of Astrophysics and Astronomy, and at the Raman Research Institute Library (in which Astronomy is one of the core subjects for research)

Keywords. Publishing, library services, evolution

1. Introduction

Publishing – the process of production and dissemination of information – is one of the metrics of scientific careers. 350 or so years ago the scientific journal was itself an innovation. This was in large part enabled by the emergence of the printing press: the Gutenberg Letter Press invented in the 15th century. While electronic publishing methods have resulted in rapid publication and wider dissemination, it, nevertheless, is also true that with speed and ease of automation, quality is to be ferociously guarded. Could traditional methods, when juxtaposed with the continuous onslaught of newer technologies in our publishing procedures, be the answer? Are people still the quintessential element in technology? Why do we publish? The phrase "Publish or Perish"has itself, interestingly, evolved in its meaning. ?Publish or perish? is a phrase coined to describe the pressure in academia to rapidly and continually publish academic work to sustain or further one?s career. I see evolution in this very phrase which first appeared in a non-academic context in 1932 and became the buzz word for researchers about 30 years ago. It is now a software program that retrieves and analyses academic citations.

Table 1. Overview of current knowledge on circum-stellar condensate grains in meteorites.

Mineral	Size [μm] abund. [ppm][1]	Isotopic Signatures	Stellar Sources	Contribution[2]
diamond	0.0026 1500	Kr-H, Xe-HL, Te-H	supernovae	?
silicon carbide	0.1 − 10 30	enhanced ^{13}C, ^{14}N, ^{22}Ne, s-process elem. low ^{12}C/^{13}C, often enh. ^{15}N enhanced ^{12}C, ^{15}N, ^{28}Si; extinct ^{26}Al, ^{44}Ti low ^{12}C/^{13}C, low ^{14}N/^{15}N	AGB stars J-type C-stars (?) Supernovae novae	> 90 % < 5 % 1 % 0.1 %
graphite	0.1 − 10 10	enh. ^{12}C, ^{15}N, ^{28}Si; extinct ^{26}Al, ^{41}Ca, ^{44}Ti s-process elements low ^{12}C/^{13}C low ^{12}C/^{13}C; Ne-E(L)	SN (WR?) AGB stars J-type C-stars (?) novae	< 80 % > 10 % < 10 % 2 %
corundum/spinel/hibonite	0.1 − 5 50	enhanced ^{17}O, moderately depl. ^{18}O enhanced ^{17}O, strongly depl. ^{18}O enhanced ^{16}O	RGB / AGB AGB stars supernovae	> 70 % 20 % 1 %
silicates	0.1 − 1 140	similar to oxides above		
silicon nitride	1 0.002	enhanced ^{12}C, ^{15}N, ^{28}Si; extinct ^{26}Al	supernovae	100 %

Notes:
[1] For the abund. (in wt. ppm) the reported maximum values from different meteorites are given.
[2] Note uncertainty about actual fraction of diamonds that are pre-solar and for fraction of graphite attributed to SN and AGB stars (see discussion in text).

2. Impact of evolutionary developments

How have publishing procedures evolved through the years; what is the impact of these developments; what are the challenges and opportunities for scholarly publishers in this digital age? How do we cope with and pre-empt the abuse and distortion of internet data, and with unethical practices that threaten to cripple the peer review system, and jeopardize publishing ethics and academic integrity? How can publishers fulfil their responsibility to publish content that is explicitly original? Juxtaposed to this is the scenario in library services with increasing demand for e-versions of journals and other e-resources and the gradual loss of value for printed journals. Librarians are developing and maintaining electronic libraries and vast digital resources combined with additional features of e-publishing, such as tools, voice, graphic arts, films, videos, CDs, DVDs, etc. How can we prepare for the future of publishing or is it already here with the popularity of pre-print servers? How have publishers responded to the Open Access model while being aware that ?the price of keeping something free comes with a cost?? Would this cost entail ensuring quality of the presentation and reliability of information? Are potential authors confused identifying traditional peer reviewed content, predatory journals, fake reviewers, depositing research in institutional repositories, copyrights and licences? Are libraries impacted by these developments?

Silicon carbide. All SiC grains in primitive meteorites are of pre-solar origin, and they are the best characterized. This has been helped by their comparably high contents of minor and trace elements. Characteristic for most grains are enhanced ^{13}C, ^{14}N, former presence of ^{26}Al as indicated by overabundances of its daughter ^{26}Mg, neon that is almost pure ^{22}Ne [Ne-E(H)] and heavy elements showing the characteristic isotopic signatures of the s-process. These 'mainstream grains' quite obviously are condensates out of the winds of AGB stars (see contribution by Lugaro & Höfner, this volume). Only a percent or so have a clearly different origin tied to supernovae (the 'X-grains'). They are characterized by high ^{12}C, ^{28}Si, very high former abundances of ^{26}Al as well as ^{44}Ti and not fully understood signature in the heavy trace elements (see contribution by Amari & Lodders, this volume).

Oxides and silicates. Besides diamonds (see below) silicates - not unexpectedly - are the most abundant of the pre-solar grains that have been found. The most characteristic features of oxides and silicates are contained in the oxygen isotopic composition that can be used for assigning each grain to one of four groups (Nittler *et al.* 1997). Grains without evidence for the former presence of ^{26}Al are assumed to originate from RGB stars, those with ^{26}Al from AGB stars.

Graphite and silicon nitride. The characteristics of most grains (see Tab. 1) have traditionally led to assume a SN origin (e.g., Zinner 1998; Hoppe & Zinner 2000). However, this percentage may have been overestimated as most high-density graphite grains, although showing enhanced ^{12}C abundances, contain s-process signatures and so are more likely to originate from AGB stars (Croat *et al.* 2005). The rare Si_3N_4 grains show isotopic signatures similar to SiC-X and SN graphite grains and derive probably from supernovae as well.

Nanodiamonds. In several ways these are the most enigmatic. Although discovered first, their pre-solar credentials are based solely on trace elements Te and noble gases that they carry. They are too small for individual analysis each consisting of some 1000 carbon atoms only on average and the carbon isotopic composition of 'bulk samples' (i.e., many diamond grains) is within the range of Solar System materials. What fraction of the diamonds is truly pre-solar is an as yet open question.

3. Implications

Isotopic structures and nucleosynthesis. As isotopic structures are the key for establishing the grains as pre-solar, isotope studies are at the core of investigations that have been performed. Results from isotopic studies in turn are also those that bear strongest on astrophysics. For one, they allow us to pinpoint the grains' stellar sources. In addition, given the precision of the laboratory isotopic analyses, which far exceed whatever can be hoped for in remote analyses, they allow conclusions with regard to details of nucleosynthesis and mixing in the parent stars as well as Galactic Chemical Evolution. They have borne strong on, e.g. the need for an extra mixing process (cool bottom processing) in Red Giants and provide detailed constraints on the operation of the s-process in AGB stars (e.g., Busso *et al.* 1999). A non-standard neutron capture process ('neutron burst') may be implied by the SiC-X grains from supernovae (Meyer *et al.* 2000) and possibly the trace Xe in the diamonds (e.g., Ott 2002). The progress in analytical techniques promises more important results in the near future.

Grain formation. Chemical composition, sizes, and microstructures of grains constrain conditions during condensation in stellar winds and supernova ejecta. Condensation of SiC apparently occurred under close to the equilibrium conditions (e.g., Lodders & Fegley 1998). Additional constraints are imposed by trace element contents both on average (Yin *et al.* 2006) as well as in individual grains (Amari *et al.* 1995). An important relevant observation is the occurrence of subgrains of primarily TiC within graphite (Croat *et al.* 2005).

The lifecycle of pre-solar grains (and maybe interstellar grains in general). Interstellar grains are expected to be processed and eventually destroyed by sputtering or astration (e.g., Draine 2003), with an as yet unidentified process needed to account for the balance between formation and destruction. Pre-solar grains preserved in meteorites carry, in principle, a record of conditions they have been exposed to, which, however, is difficult to read. Determining an absolute age using long-lived radioisotopes is virtually ruled out by the fact that these systems use decay of rare constituents (e.g., K, Sr, Re, U) decaying into other rare elements with uncertain non-radiogenic composition. However, appearance and microstructures of pristine (i.e. not chemically processed) SiC show little evidence for being processed, indicating either that they were surprisingly young when entering the forming Solar System or that they were protected (Bernatowicz et al. 2003); a similar situation is indicated by the lack of detectable spallation Xe produced by exposure to cosmic rays during residence in the ISM (Ott et al. 2005). The distribution, finally, among various types of meteorites, provides a measure of processing in the early Solar System.

References

Amari, S., Hoppe, P., Zinner, E., & Lewis R. S. 1995, *Meteoritics*, 30, 490
Anders, E. & Zinner, E. 1993, *Meteoritics*, 28, 490
Bernatowicz, T. J., Messenger, S., Pravdivtseva, O., Swan, P., & Walker, R. M. 2003, *Geochim. Cosmochim. Acta*, 67, 4679
Busso, M., Gallino, R., & Wasserburg, G. J. 1999, *ARAA*, 37, 239
Croat, T. K., Stadermann, F. J., & Bernatowicz, T. J. 2005, *ApJ*, 631, 976
Draine, B. T. 2003, *ARAA*, 41, 241
Hoppe, P. & Zinner, E. 2000, *J. Geophys. Res.*, A105, 10371
Hoppe, P., Ott, U., & Lugmair, G. W. 2004, *New Astron. Revs*, 48, 171
Lodders, K. & Fegley, B. 1998, *Meteorit. Planet. Sci.*, 33, 871
Meyer, B. S., Clayton, D. D., & The, L.-S. 2000, *ApJ* (Letters), 540, L49
Nittler, L. R. 2003, *Earth Planet. Sci. Lett.*, 209, 259
Nittler, L. R., Alexander, C. M. O.'D., Gao, X., Walker, R. M., & Zinner, E. 1997, *ApJ*, 483, 475
Ott, U. 1993, *Nature*, 364, 25
Ott, U. 2002, *New Astron. Revs* 46, 513
Ott, U., Altmaier, M., Herpers, U., Kuhnhenn, J., Merchel, S., Michel, R., & Mohapatra, R. K. 2005, *Meteorit. Planet. Sci.*, 40, 1635
Yin, Q.-Z., Lee, C.-T. A., & Ott, U. 2006, *ApJ*, 647, 676
Zinner, E. 1998, *Ann. Rev. Earth Planet. Sci.*, 26, 147
Zinner, E. 2004, in: K.K. Turekian, H.D. Holland & A.M. Davis (eds.), *Treatise in Geochemistry 1* (Oxford and San Diego: Elsevier), p. 17

Discussion

MASSEY: Im wondering if you have considered the expected intrinsic dispersion in absolute magnitude of WRs – if you consider the (large) mass range that becomes an early WN or late WC according to the evolutionary models, wouldnt you expect a large dispersion in M_v?

VAN DER HUCHT: Indeed, we will be always left with some intrinsic scatter in M_v due to mass differences within the same spectral subtype. But in my opinion, the current large dispersion is for a large fraction due to uncertainties of the adopted distances of open clusters and OB associations.

WALBORN: I think that the scatter in WNL absolute magnitudes is dominated by intrinsic spread rather than errors. In the LMC, one finds a range of 5 to nearly 8. This in turn likely reflects different formation channels: mass-transfer binaries, post-RSG, and extremely massive stars in giant H II regions.

VAN DER HUCHT: As said above, there is likely to be intrinsic scatter. But, I wonder whether a scatter of 3 magnitudes perhaps reflects undetected multiplicity.

MAÍZ-APELLÁNIZ: I could not agree more with your comment on the need for an updated catalogue of O-type stars (as a follow up of that of Garmany *et al.* 1982). We are currently working on precisely that (see our poster, these Proceedings) and we will soon make it available.

VAN DER HUCHT: Wonderful.

KOENIGSBERGER: Is the ratio WR/O-stars in clusters similar or different from this ratio for the field stars?

VAN DER HUCHT: I think it is different because incompleteness among field stars is even larger than that among cluster stars. But perhaps it should also be different because WR stars are older and could have drifted away from clusters, more than O-type stars.

GIES: How many of the WR stars in your catalogue might be low mass objects?

WALBORN: Comment: PN central stars in the WR sample would be only [WC].

VAN DER HUCHT: Among the WR stars in our VIIth Catalogue we doubt only one: WR109 (V617 Sgr), which is a peculiar object (not even a [WR] central star of a PN). All other stars in our catalogue are true massive Population I WR stars, and properly classified as such. We have not listed known Population II [WC] objects, as we did separately in our VIth Catalogue (van der Hucht *et al.* 1981). [WN] objects are not known to exist, see the comment by Nolan.

ZINNECKER: Are all Galactic WR stars in open clusters and OB association or are there many WR stars in the field?

VAN DER HUCHT: See the VIIth WR catalogue (van der Hucht 2001): of the listed 227 Galactic WR stars, only 53 are in open culsters and OB associations, or believed to be. The other 184 are supposedly field stars.

FM 8:
Statistics and Exoplanets

Introduction

Suzanne Aigrain[1] and Eric Feigelson[2]

[1] Department of Physics, University of Oxford, Keble Road, Oxford, OX3 9UU, UK
email: suzanne.aigrain@astro.ox.ac.uk

[2] TBD
email: Department of Astronomy & Astrophysics and Center for Astrostatistics, Penn State University, 525 Davey Laboratory, University Park PA 16802 USA

Abstract. The IAU's *Statistics and Exoplanets* Focus Meeting brings together observers, modelers and methodologists to discuss the intricate challenges of extracting and interpreting faint planetary signals from dominant starlight. Initiated by the IAU's new groups concentrating on astroinformatics and astrostatistics, the meeting stimulated the wider exoplanetary community as well as experts in data and science analysis. This proceedings presented selected papers from the Focus Meeting.

Keywords. methods: data analysis, methods: statistical, (stars:) planetary systems

1. Motivation

The discovery and characterization of exoplanets requires both superbly accurate instrumentation and sophisticated statistical methods, to extract weak planetary signals from dominant starlight, very large samples and noisy datasets. While numerous exoplanet conferences take place each year, these mostly focus on observational results, their physical implications, and current or future instrument developments; the statistical aspects of the papers presented, while important, are not usually central to the program. We felt it would be timely to hold a meeting dedicated specifically to the statistical work underpinning much of exoplanet science – from the detection of tiny signals buried in correlated noise to the robust inference of planet demographics from diverse, incomplete and often biased surveys. Encouraged by the IAU's Working Group on Astrostatistics and Astroinformatics (now evolved into the Commission B.3 on Astroinformatics and Astrostatistics), we therefore proposed *Statistics and Exoplanets* as a Focus Meeting for the 2015 General Assembly of the IAU. Hot on the heels of the first astro-statistics IAU Symposium – *Statistical Challenges in 21st Century Cosmology*, IAU Symposium #306, Lisbon, 2014 – this would, we hoped, help engage a fruitful discussing between astronomers and statisticians, disseminate and foster best practice, and lead to new collaborations and novel applications of state-of-the-art statistical methods to exoplanetary data and science.

2. Overview

Perhaps the most important intention for this Focus Meeting was to bring astronomers working on exoplanets face to face with mathematical and computational statistics experts. This enables the astronomers to discuss and address key problems they encounter when analysing their datasets, on the one hand, and relevant statistical methods, on the other, with a view to making definite progress on some key challenges that are at the forefront of the field today, such as the robust detection of exoplanets signals in time-series data affected by correlated noise, and the estimation of the incidence of Earth-sized planets in the habitable zones of Sun-like stars, known as η-Earth. We sought to achieve this by inviting a number of eminent professional statisticians, whom we knew to be interested in engaging with data from a variety of disciplines, but would bring a distinct perspective to the problems regularly faced by exoplanet astronomers.

The speakers at the Focus Meeting oral sessions are listed in the table below with their titles. They include astrostatisticians with an established track record of work on exoplanets, as well as observational astronomers and modelers of planetary systems. They provided overviews of the principal statistical challenges associated with various subfields of exoplanetary research. Together with a selection of contributed talks, the talks were arranged over six sessions covering:
- statistician's perspective
- planet demographics and η-Earth
- planetary signals in sparse datasets
- planetary signals in continuous light curves
- high contrast imaging
- characterisation of exoplanet atmospheres

The final session included a panel discussion with audience participation to reflect on the highlights, lessons learned and ideas expressed during the meeting.

The talks were complemented by a well-populated poster session of $\sim$75 contributed papers. These often presented fascinating results using the latest methodological techniques. The Focus Meeting was capped by a 'Hack Day' splinter session where the authors of software packages designed for the analysis of exoplanet datasets. These informal presentations presented important new methods to potential users, who then had the opportunity to try them out on the spot.

The Focus Meeting was extremely well attended, with from 180 to 130 IAU members at the oral sessions. This is a testament both to the buoyant nature of exoplanets as a field of study and to the timely nature of a meeting concentrating on statistical methods. Methodology developed for one application is often relevant to others, and thus the meeting appealed to a wide range of IAU astronomers. The standard of the talks was uniformly high and the feedback received by the Scientific Organizing Committee very positive. The talks were filmed and archival video posted on the meeting website, www.exostats.org, where the interested reader will also find some of the slides of the talks presented during the Focus Meeting and Hack Day.

3. Panel discussion

The meeting concluded with a panel discussion between Wesley Traub (Jet Propulsion Laboratory, US), Thomas Loredo (Cornell University, US), Shay Zucker (Tel Aviv University, IL) and Suzanne Aigrain (Oxford University, UK), chaired by Andrew Collier Cameron (University of St. Andrews, UK). In his opening remarks, the chair highlighted some 'meta-issues' which arose several times during the meeting:
- exoplanet studies often work close to the detection limit, making the need to understand selection functions all the more pressing,
- astronomers seeking sophisticated statistical methods have a tendency to 'reinvent the wheel'. This partly arises from a difficulty with statistical jargon; how do you search for a solution to a problem if you do not know how to describe your problem in the language of those who might be able to solve it?
- reproducibility of reducing and modeling observational data is important. We need systematic and standardised means of providing intermediate results alongside formal publications.

The discussion then focussed on a few topics that were recurrent throughout the meeting such as: (a) the measurement of η-Earth and the associated problem of defining what we mean by an Earth-like planet; and (b) the detection of an unknown number of planetary signals in sparse (radial velocity) datasets 'decorated' with correlated noise. All panelists recognised that the tremendous progress in the last few years depends critically on the use of powerful statistical procedures that are becoming standard practice among exoplanetary researchers. But the panelists also agreed that we have much more to learn from the statistical community, as well as from the literature in other fields ranging from electrical engineering to econometrics.

4. Conclusion and thanks

We feel this was a very successful and enjoyable Focus Meeting and hope that it will be the first in continuing cross-disciplinary interchanges statistical aspects of exoplanet studies. The management and archival of the meeting website www.exostats.org is being taken over by Penn State University, and the intention is that it should be used by future meetings on similar topics too.

The Scientific Organizing Committee was excellent in structuring the meeting and selecting speakers. Committee members were Suzanne Aigrain (University of Oxford, UK, co-chair), Andrew Collier-Cameron (University of St Andrews, UK), Laurent Eyer (Observatoire de Geneve, CH), Eric Feigelson (Penn State University, US, co-chair), Philip Gregory (University of British Columbia, CA), Chris Koen (University of Western Cape, ZA), Michael Liu (University of Hawaii, US), Oleg Malkov (Russian Academy of Science, RU), Claire Moutou (University of Marseille-Provence and CFHT, FR), Sascha Quanz (ETH Zurich, CH), and James Berger (Duke University, US).

We thank the IAU Working Group on Astrostatistics and Astroinformatics, under the leadership of Eric Feigelson and Prajval Shastri (Indian Institute for Astrophysics, IN), for stimulating the organization of this meeting. Joseph Hilbe (Arizona State University US) represented the International Astrostatistics Association affiliated with the International Statistical Institute (sister organization to the IAU) that cosponsored the meeting. James Berger, G. Jogesh Babu (Penn State University, US), Jessi Cisewski (Yale University, US) and Joseph Hilbe are professors of statistics who attended and stimulated the meeting. Andrew Collier-Cameron moderated the panel discussion during the final session. Andrew Howard (University of Hawaii US) and Michael Liu assisted as the Local Organizing Committee. Paul Wilson (Institut d'Astrophysique FR) developed the superb exostats.org Web site and arranged the recording, broadcasting, and archiving of oral talks.

FM8 *Statistics and Exoplanets* Speakers

Title	Presenter
Transit Timing Variations as a Tool for the Bayesian Characterization of Exoplanets	Eric B. Ford
Overview of modern Bayesian statistical methods	James Berger
Planet Demographics from Transits	Andrew Howard
Kepler Reliability Metrics and Their Use in Occurrence Rate Calculations	Steve Bryson
The Occurrence of Earth-Like Planets Around Other Stars	Will M. Farr
Hierarchical inference for exoplanet populations	Daniel Foreman-Mackey
Bayesian planet searches in radial velocity data	Phil Gregory
Wide Giant Planets are Rare: Planet Demographics from Direct Imaging	Beth Biller
The Various Challenges of Subtracting Speckles and Planet Detection/Characterization in High Contrast Imaging	Christian Marois
Planet Frequency beyond the Snow Line from MOA-II Microlensing Survey	Daisuke Suzuki
Astrometric exoplanet surveys in practice: challenges, opportunities, and results	Johannes Sahlmann
Dealing with activity in RV planet searchers	Isabelle Boisse
Estimations of uncertainties of frequencies	Laurent Eyer
Significance of noisy signals in periodograms	Maria Süveges
Advances in the Kepler Transit Search Engine and Automated Approaches to Identifying Likely Planet Candidates in Transit Surveys	Jon M. Jenkins
Combining Transit and Radial Velocity Data to Infer the Planet Mass-Radius-Flux Distribution	Leslie A. Rogers
BART: A Probabilistic and automated tool for the vetting of transits	Olivier Demangeon
Validation of transting planet candidates: a Bayesian view	Ridrigo F. Diaz
Probabilistic Mass-Radius Relationship for Sub-Neptune-Sized Planets	Angie Wolfgang
Probabilistic stellar rotation periods with Gaussian processes	Ruth Angus
A population-based Habitable Zone perspective	Andras Zsom
Reliable extraction of transmission and emission spectra using deterministic and stochastic systematics models	Neale Gibson
Overcoming Degeneracies in Exoplanet Spectra	Björn Benneke
Measuring Transmission Spectra from the Ground	Andres Jordan
Approximate Bayesian Computation	Jessi Cisewski
Analyzing Complex and Structured Data via Unsupervised Learning Techniques	Kai L. Polsterer

Stellar rotation period inference with Gaussian processes

Ruth Angus[1], Susanne Aigrain[1] and Daniel Foreman-Mackey[2]

[1] Subdepartment of Astrophysics, University of Oxford, Oxford, UK
email: ruth.angus@astro.ox.ac.uk, suzanne.aigrain@astro.ox.ac.uk

[2] Sagan Fellow, Department of Astronomy, University of Washington, Seattle
email: danfm@uw.edu

The light curves of spotted, rotating stars are often non-sinusoidal and Quasi-Periodic (QP) and a strictly periodic sinusoid is therefore not a representative generative model. Ideally, a physical model of the stellar surface would be conditioned on the data, however the parameters of such models can be highly degenerate. Instead, we use an appropriate *effective* model: a Gaussian Process (GP) with a QP covariance kernel function,

$$k_{i,j} = A \exp\left[-\frac{\sin^2(\pi(x_i - x_j)/P)}{2g^2} - \frac{(x_i - x_j)^2}{2l^2}\right]. \qquad (0.1)$$

By modelling the covariance matrix of the light curve with a QP GP, we remain agnostic about model choice, whilst sampling directly from the posterior probability distribution function of the periodic parameter and marginalising over the other kernel hyperparameters.

We simulated 300 light curves with a range of rotation periods and spot lifetimes and attempted to recover the rotation periods using three methods: our GP method, a sine-fitting periodogram method and an AutoCorrelation Function (ACF) method (McQuillan *et al.* 2014). Results are shown in Figure 1. The posterior probability distribution of the rotation period parameter was sampled using the affine invariant ensemble MCMC sampler, `emcee` (Foreman-Mackey *et al.*, 2013) and the GP operations were performed using the `george` python package (Foreman-Mackey 2015). This method produces rotation periods that are more precise than the periodogram and both more accurate and precise than the ACF method. Furthermore, the improvement is expected to be even more dramatic when applied to real, noisy *Kepler* light curves, since the GP method is well suited to modelling rotation signals and correlated noise simultaneously.

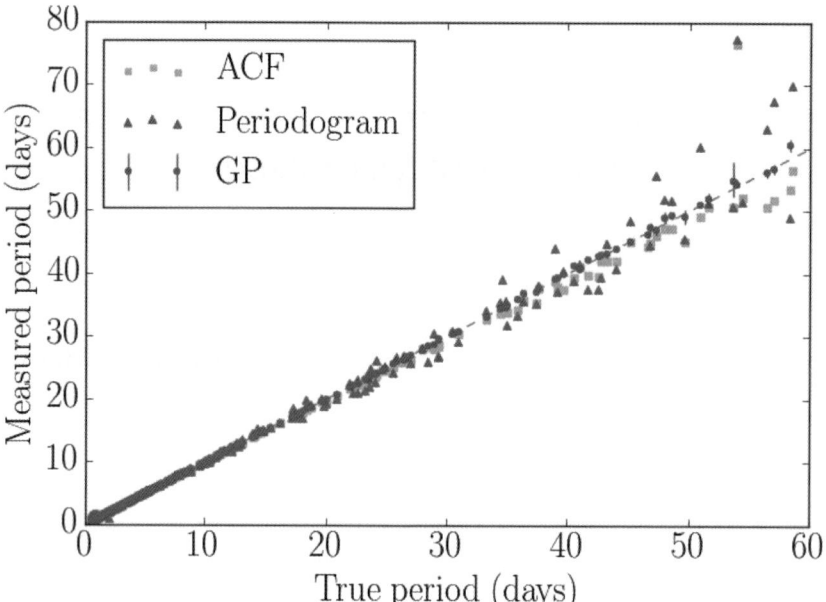

Figure 1. Measured vs true rotation periods for 300 simulations of light curves from spotted, rotating stars. Three different methods were tested: the ACF method, a Lomb-Scargle periodogram (sine-fitting) method and our new GP method. The GP method measures the most precise and accurate rotation periods and is expected to perform even better on real data.

References

McQuillan, A., Mazeh, T., & Aigrain, S., 2014, *ApJ*, 211, 24
Foreman-Mackey, D., Hogg, D. W., Lang, D., & Goodman, J., 2013, *PASP*, 125, 306
Foreman-Mackey, D. 2015, https://github.com/dfm/george

Dealing with activity in RV planet searches

Isabelle Boisse

Aix Marseille Université, CNRS, LAM (Laboratoire d'Astrophysique de Marseille) UMR 7326, 13388, Marseille, France —- isabelle.boisse@lam.fr

Abstract. Precise radial velocity measurements of a star allow to search for planets. But this method has to face with irregularly time series. Stellar variabilities: pulsation, granulation, stellar activity on a short and long timescale, also modify the measure of the radial velocities. There is indeed a growing literature of controversies on how a signal is interpreted as a planet or due to stellar activity. I present how the star variations change the measured RVs, which techniques and indices are used by several teams to disentangle activity and planets, and the future options that are being studied.

Keywords. Stellar activity, Spectroscopy,

1. Introduction

I focus this short review on the impact of stellar activity on radial velocity (RV) measurements. I omit most of the work done on its impact on planetary parameters from transit observations, and do not attempt an exhaustive review of the literature.

Precise RVs are derived from the measure of the Doppler shift of thousands of spectral lines. Dwarfs stars of spectral type F, G and K are the core of RV planet searches with several thousand of lines available in the visible wavelength range. Contrary to a Doppler effect du to an orbiting companion that shifts the line, stellar variability affects the line shape. Different timescales exist:

(a) pulsations induce variability of 0.1 to 4m/s on a few min (5 min for the Sun),

(b) flares have a signature of few m/s that least hours,

(c) granulation impacts RV on 0.1 to 2 m/s on hours,

(d) stellar activity, i.e. the rotation of spots and plages with the line of sight, leads to m/s to km/s variability with periods of days to weeks, and

(e) the long term variability, often related to magnetic cycle have a years basis with amplitudes of m/s to 20 m/s.

These numbers are indicative and are not always well constrained. They are moreover dependent, at least, on the spectral type.

With the increase precision of RV spectrographs, the literature now has increasing controversies on the data analysis of RV surveys, depending on the way the data are analysed and how researchers take into account the stellar variability. Periodicities found in the RV could be interpreted as due to the star or to planets. Examples include CoRoT-7 (Haywood *et al.* 2014), HD 41248 (Santos *et al.* 2014, Jenkins & Tuomi 2014), on GJ 581 (Robertson *et al.* 2015a, Anglada-Escude & Tuomi 2015), and GJ 191 (Anglada-Escude *et al.* 2015, Robertson *et al.* 2015b).

2. A clever observational strategy

When the frequency of the noise (here the stellar variability) is different from the periodic one of the planet that we search, the noise should be averaged. For that, the way the measurements are treated is very important. Pulsation is averaged thanks to a 15 min exposure. Several measurements per night (2 or 3 depending on the spectral type) allow the average of the granulation noise. For a planet with a year-period, the stellar activity should be averaged thanks to several measurements per rotational period (Dumusque *et al.* 2011).

On the other hand, to analyse the RV amplitude induced by CoRoT-7b, Hatzes *et al.* (2011) proposed to analyse the data without modelling the activity signal. Even if the star is strongly active, since the planet has a period of 0.85 day, one can fit the planet periodic signal considering a free offset between each night. Since the rotational period of the star is ∼20 days, one can consider that the stellar activity does not change much during one night.

A careful selection of the targets and a clever observational program is the first strategy of RV surveys to disentangle activity and orbiting companions.

3. Observational diagnostics

Since RV observations are coming from high-resolution spectra, several diagnostics are available to monitor the degree of activity of the star: line bisector variations (e.g. Vspan, BIS), active lines (CaII H&K, Hα, HeI, NaI), and also simultaneous observations can be done in photometry, polarimetry or spectroscopy in the near infrared.

There was few cases of simultaneous high resolution spectroscopy and high precision photometry observations. HD189733 was monitored with SOHIE@OHP and the satellite MOST. It shows that the star is dominated by dark spots (Boisse *et al.* 2009, Lanza *et al.* 2011, Aigrain *et al.* 2012, Dumusque *et al.* 2014). The observations of CoRoT-7 with CoRoT and HARPS@LaSilla allow to refine the number and the masses of the planets in the system (Haywood *et al.* 2014, Barros *et al.* 2014). Spectroscopic observations of stars simultaneously observed by K2 are under study. These observations will help to understand how degenerate is the photometric signal and how it could be physically related to the spectroscopic one, but also its dependance on the stellar parameters.

There are several programs under analysis of simultaneous spectroscopic and polarimetric observations (Hébrard *et al.*, subm.). The idea is to observe how the large scale magnetic field structure could help to understand the RV. The programs focus on different targets: on M dwarfs in the preparation of the SPIRou instrument (Moutou *et al.* 2015), on G and K dwarfs, and on young stars, cTTs and wTTs stars.

Soon, a new generation of nIR spectrographs will be on the sky (CARMENES, SPIRou, HPF, IRD). The impact of a dark spot is less important in the nIR than in the visible (e.g. Figueira *et al.* 2010) and allow to disentangle between the signature of a planet and stellar activity.

4. Simulation tools

Several tools have been developed to simulate the RV and the photometry of dark spots and bright plages. Among them, SOAP 2.0 is a code available to the community† (Dumusque *et al.* 2014). It uses different mean spectral lines for spots, plages and quiet photosphere, that allows the modelling of convective blueshift and its inhibition in active regions. This is an evolving code that will be improved (differential rotation, evolution of the spot with time, simulated spectrum,...). The code StarsSim (Herrero *et al.*, subm.) model the spectra with BT-Settl with Phoenix code. Spots and plages are simulated by

† http://www.astro.up.pt/resources/soap2/

stars with different temperature. An important challenge of thiese codes is to manage the number of free parameters when fitting the data.

5. The Sun as a star

The Sun is our closest star. The high spatial resolution available could be used to model the stellar activity impact (e.g. Meunier *et al.* 2010a,b, Lagrange *et al.* 2010, Borgniet *et al.* 2015, Marchwinski *et al.* 2015). Meunier & Delfosse (2011) found that the correlation between the Hα and the CaII activity indices depends on the activity cycle.

The Sun has almost never been observed in the same way as the other stars are measured in RV, i.e. without the spatial resolution on the complete visible bandwidth. Recently observations were get with HARPS of Sunlight reflected on the VESTA asteroid (Haywood *et al.* subm.). We got a program accepted to observe reflected light on the Moon with HARPS (PI: P. Figueira), and a new instrument, the solar telescope is installed on HARPS-N@TNG. These observations could be compare to the high-spatial resolution observations obtained at the same time. This will allow to relate the RV, line width and active index variations to the appearance and disappearance of spots, plages and filaments.

6. Conclusion

In a few words to conclude: 1) the observational strategy has to be though anyway, 2) all the available diagnostics should be scrutinised, 3) we should be aware that the diagnostics may behave differently depending on spectral type, 4) in the modelling of observations, the red noise should be implemented.

References

Aigrain, S., Pont, F., & Zucker, S. 2012, *MNRAS*, 419, 3147
Anglada-Escud, G., Tuomi, M., Arriagada, P. *et al.* 2015, arXiv: 1506.09072
Barros, S., Almenara, J., Deleuil, M. *et al.* 2014, *A&A*, 569, 74
Boisse, I., Moutou, C., Vidal-Madjar, A. *et al.* 2009, *A&A*, 495, 959
Boisse, I., Bouchy, F., Hébrard, G. *et al.* 2011, *A&A*, 528, A4
Borgniet, S., & Meunier, N. and Lagrange, A.-M. 2015, *A&A*, 581, 133
Dumusque, X., Lovis, C., Ségransan, D. *et al.* 2011, *A&A*, 535, 55D
Dumusque, X., Boisse, I., & and Santos, N. C. 2014, *ApJ*, 796, 132
Figueira, P., Marmier, M., Bonfils, X. *et al.* 2010, *A&A*, 513, 8
Hatzes, A., Fridlund, M., Nachmani, G. *et al.* 2011, *ApJ*, 743, 75
Haywood, R., Collier Cameron, A., Queloz, D. *et al.* 2014, *MNRAS*, 443, 2517
Herrero, E., Ribas, I., Jordi, C. *et al.* 2015, hsa, 494
Jenkins, J. & Tuomi, M. 2014, *ApJ*, 794, 10
Lagrange, A.-M., Desort, M., & Meunier, N. 2010, *A&A*, 512, 38
Lanza *et al.* 2011, *A&A*, 533, A44
Marchwinski,R., Mahadevan, S., Robertson, P., Ramsey, L., & Harder, J. 2015, *ApJ*, 798, 63
Meunier, N. *et al.* 2010a, *A&A*, 512, 39
Meunier, N. *et al.* 2010b, *A&A*, 512, 39
Meunier, N. & Delfosse, X. 2011, *A&A*, 532, 18
Moutou, C., Boisse, I., Hébrard, G. *et al.* 2015, sf2a, arXiv:1510.01368
Robertson, P., Mahadevan, S., Endl, M., & Roy, A. 2015a, *Science*, 347, 1080
Robertson, P., Roy, A., & Mahadevan, S. 2015b, *ApJ*, 805, 22
Santos, N. C., Mortier, A., Faria, J. *et al.* 2014, *A&A*, 566, 35

An artificial Kepler dichotomy? Implications for the coplanarity of planetary systems

Timothy Bovaird[1] and Charles H. Lineweaver[1,2]

[1] Research School of Astronomy and Astrophysics, Australian National University, Canberra, ACT 2611, Australia

[2] Research School of Earth Sciences, Australian National University

Abstract. We challenge the assumptions present in previous efforts to model the ensemble of detected Kepler systems, which require a dichotomous stellar population of 'fertile' and 'sterile' planet producing stars. We remove the assumption of Rayleigh distributed mutual inclinations between planets and show that the need for two distinct stellar populations disappears when the inner part of planetary disks are assumed to be flat, rather than flared.

Keywords. Kepler, multiple-planet systems, coplanarity, protoplanetary disks

The Kepler Dichotomy

The Kepler dichotomy is an inference drawn from modeling the number of Kepler detected planetary systems. While the simulated populations are well matched by the detected sample, the number of systems with a single transiting planet is underestimated by a factor of ~ 3 (Lissauer et al. (2011), Johansen et al. (2012), and Ballard & Johnson (2014)). These studies propose that the Kepler stellar sample is dichotomous in its planet producing capability, separated into stars which produce many planets, and stars which largely produce a single planet.

Model Dependence

An assumption in the above studies is that the mutual inclinations between the orbital plane's of planets are Rayleigh distributed. This is equivalent to a flared disk, where a planet's vertical distance from the invariable plane is proportional to its semi-major axis. In our Solar System, this is only seen exterior to the orbit of Jupiter, and the vertical height above the invariable plane is approximately constant for the inner planets. We propose using a new disk model for our simulated Kepler systems, a 'flat' disk model where a planet's height above the invariable plane is independent of semi-major axis.

Another assumption in the simulated systems is that the number of planets in a given system can be drawn from a Poisson distribution. This implies that planet formation within a given system in an independent process. For our simulated Kepler systems, we trial a planet formation process where planets in the inner disk are formation sequentially Chatterjee & Tan (2014). This results in an exponential multiplicity distribution.

Under these new assumptions, our simulations show that the Kepler stellar sample can be represented by a single stellar population with respect to planet formation. The requirement of a dichotomous stellar population with distinct fertile and sterile planetary formation properties is either significantly reduced, or removed entirely.

References

Lissauer, J. J. et al. 2011, *ApJS*, 197, 8
Johansen, A., Davies, M. B., Church, R. P., & Holmelin, V. 2012, *ApJ*, 758, 39
Ballard, S. & Johnson, J. A. 2014, *ApJ*, submitted
Chatterjee, S., & Tan, J. C., 2014, *ApJ* 780, 53

Kepler Reliability and Occurrence Rates

Steve Bryson[1] and The Kepler Team

[1] NASA Ames Research Center   email: steve.bryson@nasa.gov

The *Kepler* mission has produced tables of exoplanet candidates ("KOI table"), as well as tables of transit detections ("TCE table"), hosted at the Exoplanet Archive (http://exoplanetarchive.ipac.caltech.edu). Transit detections in the TCE table that are plausibly due to a transiting object are selected for inclusion in the KOI table. KOI table entries that have not been identified as false positives (FPs) or false alarms (FAs) are classified as planet candidates (PCs, Mullally *et al.* 2015). A subset of PCs have been confirmed as planetary transits with greater than 99% probability, but most PCs have < 99% probability of being true planets. The fraction of PCs that are true transiting planets is the PC reliability rate. The overall PC population is believed to have a reliability rate > 90% (Morton & Johnson 2011).

The PC reliability rate is not homogeneous across the PC population: low S/N transiting objects, particularly Earth-size planets at 365-day orbital periods, may have reliability rates substantially lower than 90%. Astrophysical false positives, such as grazing or background eclipsing binaries, are identified through pixel analysis that determines the location of the transit source, or through light curve shape analysis that determines the nature of the transiting object. When the transit signal is very shallow both pixel and light curve analysis can fail, resulting in a detected transit that cannot be localized to the target star and whose nature cannot be determined. Such objects are classified as planet candidates, but have a lower reliability than higher S/N objects that can be localized and whose light curve shapes can be analyzed.

A variety of instrumental phenomena can generate spurious detections not caused by a transiting object. Such FAs can be very difficult or impossible to distinguish from shallow planetary transits. A particularly important class of FAs are thermal systematics that lead to a large excess of shallow transit detections near the *Kepler* 372-day orbital period. Other FA sources include temporary pixel sensitivity changes due to cosmic rays, and stellar variability.

Exoplanet occurrence rate calculations, particularly in the $\eta_\oplus$ regime, require understanding the rate of the above FPs and FAs relative to the rate of true planets in the PC population. Studies by the *Kepler* mission are currently underway that should significantly improve our understanding of reliability. These studies include transit searches of inverted and shuffled data. Real planets will not be detected in either study, so they will provide insight into FA distributions.

Once our understanding of reliability has improved, occurrence rate estimates such as that in Burke *et al.* (2015) will need to be modified to use that understanding. The best way to use knowledge of PC reliability in occurrence rate calculations is currently an open problem.

References

Burke, C. J., *et al.* 2015 *ApJ* 809, 1
Morton, T. D. & Johnson, J. A. 2011, *ApJ* 738, 170
Mullally, F., *et al.* 2015, *ApJS* 21,l 31

A probabilistic and automated tool for the vetting of transit candidates

Olivier Demangeon[1] and Pascal Bordé[2]

[1] Aix Marseille Université, CNRS, LAM (Laboratoire d'Astrophysique de Marseille) UMR 7326
13388, Marseille, France
email: olivier.demangeon@lam.fr

[2] Université de Bordeaux, Observatoire Aquitain des Sciences de l'Univers, BP 89,
33271 Floirac Cedex, France
e-mail: pascal.borde@u-bordeaux.fr

Abstract. We developed, based on the CoRoT experience, an automated tool called BART for Bayesian Analysis for the Ranking of Transit in order to perform a homogeneous and automated ranking of planetary candidates. We applied it to the candidates detected in the campaign 1 of K2.

Keywords. planets and satellites: general, methods: statistical, techniques: photometric

1. Introduction: Need for automated vetting of planetary candidates

CoRoT (Baglin *et al.* 2006), Kepler (Koch *et al.* 2010) and now K2 (Howell *et al.* 2014) have to deal with tens of thousands of transit events. Such events are well known to be treacherous as they can reveal as easily planets and eclipsing binaries. CoRoT, Kepler and K2 planet harvests are thus hiding a tremendous effort in order to vet, follow-up and finally reveal the planetary nature of some of these transits (for eg. Léger *et al.* 2009). If up to now this process as been mainly done "manually" by vetting and follow-up teams, the future TESS and Plato missions will require automation. Follow-up observations are very difficult to automatize, but the vetting process is much more suitable for that. Recently, automated tools to vet the Kepler Objects of Interest have appeared, see Morton 2012 and Coughlin *et al.* 2015. We are going to present a tool, developed on the experience acquired with the CoRoT, that we applied to the candidates found in K2 campaign 1.

2. Method: The bart software

BART for Bayesian Analysis for the Ranking of Planets is an automated software to rank planetary candidates based on the probability for a given transit to be due to a planet orbiting the target star. This probability is computed in the framework of Bayesian model comparison. The set of models is : planetary system (P.S., a planet orbiting the target star), contaminating planetary system (C.P.S., a planet orbiting another star which contaminates the light curve of the target star), eclipsing binaries (E.B., a star orbiting the target star) and contaminating eclipsing binaries (C.E.B., the eclipsing binary doesn't involve the target star). The models E.B. and C.E.B. are actually divided into two separated models depending on whether the real orbital period of the eclipsing binary is the detected one (E.B.P.) or if it's the double of it (E.B.2.P.) due to similar depths of the primary and secondary eclipses. Finally, it's important to mention that hierarchical triple system model is included in C.P.S. or C.E.B.P., C.E.B.2.P. models depending on the

Table 1. Fragment of the ranking for K2 campaign 1

Rank	Candidate	Nature (M+15)	Nature (BART)	Probability of the different models [%]					
				P.S.	C.P.S.	E.B.P.	E.B.2.P.	C.E.B.P.	C.E.B.2.P.
1	201569483.01	FP	PS	99.7	0.17	0.0	0.0	0.13	0.0
2	201779067.01	FP	PS	71.6	28.4	0.0	0.0	0.0	0.0
3	201565013.01	Candidate	PS	52.9	42.9	0.15	0.0	4.0	0.05
4	201367065.01	Planet	PS	51.64	47.4	0.07	0.04	0.85	0.05

nature of the eclipsing system. For all these models the set of parameters are: orbital period, radius ratio, temperature ratio, impact parameter, density proxy $\rho' = \frac{M_{prim}+M_{sec}}{4/3\pi R_{prim}^3}$, and the contamination of the light-curve.

All those models are confronted to the detected transits, but contrary to Morton 2012, in order to reduce the computational time, we don't use the folded transit light-curve directly, but a set of observables measured on this light-curve : detected period of the transit, odd and even depths, secondary depth, inner and outer durations.

In order to automatize the computation, the exploration of the parameter space is done thanks to an adaptation of Gregory 2005 Automated Metropolis-Hasting MCMC algorithm. The adaptation consists in including an estimate of the autocorrelation of the trace of each free-parameter into the loop which adapt the size of the proposal distribution. This exploration allows to do parameters inference for each of the models and then to define a narrowed parameter space for the computation of the un-normalized posterior probability of each model thought Monte Carlo Integration. Finally the probability of each model is normalized by the sum of all the considered models. Then all the studied transits are ranked according to the posterior probability of the P.S. model.

3. Results and discussion

We applied BART to the candidate list announced in Montet *et al.* 2015 thereafter M+15. Due to space limits, we only present in table 1 the first 4 candidates. The first two candidates in our ranking are mentioned as false positive M+15. But this is easily explained by our non-detection of the secondary eclipses due to our circular orbit hypothesis. This is a known weakness which will be corrected soon. The third candidate is also still a candidate in M+15 and the fourth turns out to be one of the K2-3 planets Crossfield *et al.* 2015. Those results are thus very promising. We will thus apply BART to all the candidates detected by K2 in the near future. Such an automated analysis will be of particular interest for the estimate of the reliability and completeness of planet detection with K2.

References

Baglin, A., Auvergne, M., Boisnard, L. et al. 2006, *36th COSPAR Scientific Assembly*, 36, 3749
Coughlin J. F. et al. 2015, *Submitted to ApJ*
Gregory 2005, *Cambridge University Press*
Howell S. B. et al. 2014, *PASP*, 126, 398H
Koch D. G. et al. 2010, *ApJ*, 713, 79
Léger A. et al. 2010, *A&A*, 506, 287
Montet, B., Morton T. D., Foreman-Mackey, D., et al. 2015, *ApJ*, 809, 25
Morton, T. D. 2012, *ApJ*, 761, 6

Validation of transting planet candidates: a Bayesian view

Rodrigo F. Díaz[1] and Jose Manuel Almenara[2] and Alexandre Santerne[3]

[1] Observatoire Astronomique de l'Université de Genève, Versoix, Switzerland.

[2] Université Grenoble Alpes, IPAG, Grenoble, France.

[3] Instituto de Astrofísica e Ciencias do Espaço, Universidade do Porto, Portugal.

Abstract. Transiting candidate validation is essentially a Bayesian model comparison problem: different models, all explaining the observations comparably well, compete for the support of the available data. The basic characteristics of the planet validation problem are discussed and the different approaches taken to tackle its difficulties are reviewed.

Keywords. planetary systems, techniques: photometric, methods: statistical

---

Transiting candidate validation consists in securing the planetary nature of a transiting candidate not by a dynamical measurement of its mass, but by accumulating enough evidence for the planetary hypothesis with respect to all possible false positive (FP) scenarios. This technique is particularly relevant for faint transit host stars, such as those from the CoRoT and Kepler surveys.

Planet validation is therefore essentially a Bayesian model comparison problem, and consists in computing the odds ratio between the planetary hypothesis and $H_{\rm FP}$, the probabilistic union of all FP scenarios†. Three particularities of planet validation render the task very complex and difficult to tackle: i) the relevant data sets are of diverse nature (transit light curves, broad band photometry, radial velocity measurements, etc.), ii) the prior information is non-trivial, and involves knowledge of the subjects such as stellar formation rate and evolution, and galactic structure, and iii) the computation of the models representing each hypothesis is in some cases very time-consuming.

Despite its clear Bayesian nature, the planet validation problem has received in the past mainly a frequentist treatment (Torres *et al.* 2011). The VESPA code (Morton 2012) computes Bayesian odds ratios but employs simplistic models that increase speed but only partially exploit the available datasets. The Planet Analysis and Small Transit Investigation Software (Diaz *et al.* 2014) was developed keeping the characteristics of the problem in mind. It computes the Bayesian evidence for a full set of FP scenarios and the planet hypothesis, modelling all the available data self-consistently to produce rigorous Bayes factors. Its object-oriented architecture allows constructing a vast set of FP models easily. An early PASTIS result on a CoRoT Neptune-size candidate is presented by Moutou *et al.* (2014).

References

Díaz, R. F., Almenara, J. M., Santerne, A., *et al.* 2014, *MNRAS*, 441, 983
Morton, T. D. 2012, *ApJ*, 761, 6
Moutou, C., Almenara, J. M., Díaz, R. F., *et al.* 2014, *MNRAS*, 444, 2783
Torres, G., Fressin, F., Batalha, N. M., *et al.* 2011, *ApJ*, 727, 24

† See ancillary material for a graphical representation of all FP scenarios.

Estimations of uncertainties of frequencies

Laurent Eyer[1], Jean-Marc Nicoletti[2] and Stephan Morgenthaler[3]

[1] Department of Astronomy, University of Geneva, CH-1290, Sauverny, Switzerland
email: `Laurent.Eyer@unige.ch`
[2] Swiss Statistics Federal Office, CH-2010 Neuchâtel, Switzerland
[3] Swiss Federal Institute of Technology, EPFL, CH-1015 Lausanne, Switzerland

Abstract. Diverse variable phenomena in the Universe are periodic. Astonishingly many of the periodic signals present in stars have timescales coinciding with human ones (from minutes to years). The periods of signals often have to be deduced from time series which are irregularly sampled and sparse, furthermore correlations between the brightness measurements and their estimated uncertainties are common. The uncertainty on the frequency estimation is reviewed. We explore the astronomical and statistical literature, in both cases of regular and irregular samplings. The frequency uncertainty is depending on signal to noise ratio, the frequency, the observational timespan. The shape of the light curve should also intervene, since sharp features such as exoplanet transits, stellar eclipses, raising branches of pulsation stars give stringent constraints. We propose several procedures (parametric and nonparametric) to estimate the uncertainty on the frequency which are subsequently tested against simulated data to assess their performances.

Keywords. methods: statistical, data analysis, stars: variables, planetary systems

1. Summary

The estimation of uncertainties of frequency has been studied from the statistics side (e.g. Whittle 1952, Walker 1971, Walker 1973) and astronomy side (e.g. Kovacs 1981, Cuypers 1987). For a sinusoidal curve, the uncertainty on the frequency is inversely proportional to the signal to noise ratio, the time span of the observations, and the square root of the number of observations. Hall *et al.* (2000) took into account the light curve shape for this estimation, it is with IID (Independent and Identically Distributed) errors and an irregular sampling. A generalisation was made by Nicoletti (2012), where the hypothesis of IID is relaxed. An application to a box signal (for exoplanet transits) has been studied and asymptotic behaviour are derived. As for the signal sinusoidal the frequency uncertainty is inversely proportional to the signal to noise ratio (depth of transit/measurement error), the time span and differently from the sine-shape the number of measurements. The next steps of our studies are: (1) From simulations, to determine the boundaries of the validity of the asymptotic behaviour, (2) We will derive, if possible, ad-hoc corrections when the signal has a small number of measurements.

References

Cuypers, J. 1987, *Acad. Analecta*, 49, 3
Nicoletti, J.-M. 2012, *PhD Thesis 5296 of EPFL*
Kovacs, G. 1981, *Ap&SS*, 78, 175
Hall, P., Reimann, J., & Rice, J. 2000, *Biometrika*, 87, 545
Walker, A. M. 1971, *Biometrika*, 58, 21
Walker, A. M. 1973, *Advances in Applied Probability*, 5, 217
Whittle, P. 1952, *Biometrika*, 39, 309

Reliable inference of light curve parameters in the presence of systematics

Neale P. Gibson[1,2]

[1] European Southern Observatory, Karl-Schwarzschild-Str. 2, 85748 Garching bei München, Germany
[2] Astrophysics Research Centre, School of Mathematics and Physics, Queens University Belfast, Belfast BT7 1NN, UK
email: n.gibson@qub.ac.uk

Abstract. Time-series photometry and spectroscopy of transiting exoplanets allow us to study their atmospheres. Unfortunately, the required precision to extract atmospheric information surpasses the design specifications of most general purpose instrumentation. This results in instrumental systematics in the light curves that are typically larger than the target precision. Systematics must therefore be modelled, leaving the inference of light-curve parameters conditioned on the subjective choice of systematics models and model-selection criteria. Here, I briefly review the use of systematics models commonly used for transmission and emission spectroscopy, including model selection, marginalisation over models, and stochastic processes. These form a hierarchy of models with increasing degree of objectivity. I argue that marginalisation over many systematics models is a minimal requirement for robust inference. Stochastic models provide even more flexibility and objectivity, and therefore produce the most reliable results. However, no systematics models are perfect, and the best strategy is to compare multiple methods and repeat observations where possible.

Keywords. methods: data analysis, techniques: spectroscopic, planetary systems:

1. Introduction

Transiting exoplanets allow us to study their compositions and atmospheres via wavelength dependent variations in their light curves using transmission and emission spectroscopy, and significant progress has been made since the first detection of an exoplanet's atmosphere (Charbonneau *et al.* 2002). These techniques require exquisite temporal stability in the light curves in order to measure $\sim 10^{-4}$ or smaller variations in the transit or eclipse depth as a function of wavelength (e.g. Seager & Sasselov 2000; Brown 2001). Despite the enormous challenges and huge scientific reward, there are still no dedicated instruments for the study of transiting exoplanet atmospheres. Consequently, we have been using common-user instruments for many years that were not designed with this level of precision in mind. While we can easily collect sufficient photons to reach $\sim 10^{-4}$ precision, instrumental systematics limit the achievable precision in all of our observations. Our ability to reliably model these systematics is therefore of fundamental importance to our understanding of exoplanet atmospheres, and we are yet to reach a consensus in the community on how to address this.

Typically, systematics are modelled as a function of auxiliary inputs. These are measurements obtained simultaneously with the exoplanet time-series that are (thought to be) related to the underlying cause of the systematics, e.g. position of the image/spectral trace on the detector, width of the instrumental profile, or specifically for the *Hubble Space Telescope (HST)*, its orbital phase. There are multiple approaches to constructing a systematics model. This short review aims to place the most commonly-used models in context, comparing the use of deterministic and stochastic models, and outlining the best practices for modelling exoplanet light curves in the presence of systematics. It is largely based on Gibson (2014), to which the reader is referred for more details. Here, I only focus on general systematics models, and do not consider instrument specific cases.

2. Modelling instrumental systematics

Systematics models fall into two main classes: deterministic and stochastic processes. A deterministic model is where the systematics are defined as a simple function of inputs, ranging from the simple case of a polynomial of time, to more complex functions of multiple inputs. Stochastic processes allow us to use probability distributions over functions, and offer much more flexibility. Gibson (2014) compares these models and tests their applicability on simulated transit light curves. Here, I summarise the models and results.

2.1. Arbitrary systematics models

Deterministic models are the simplest to construct, and the default choice is a linear model of the auxiliary inputs. With enough inputs (or higher order terms), these often appear to accurately model the systematics in the light curves, leaving the residuals adequately 'whitened'. However, many different systematics models can appear to account for instrumental systematics, yet produce significantly different results. Gibson *et al.* (2011) discussed this in the context of *HST* transmission spectra. In short, such methods produce spurious results that depend on arbitrary human choices, and should be avoided.

2.2. Model selection

Where many different systematics can be constructed, Bayesian model selection should allow us to select the best model, and remove human bias in this process. Indeed, this is the approach most widely used in the exoplanet community today (e.g. Sing *et al.* 2011, Nikolov *et al.* 2014). In general, we should calculate the Bayesian evidence of each model, however, this is often difficult to compute, and simple model selection criteria are used to estimate it. The most popular of these is the Bayesian Information Criterion (BIC), which is based on a simple addition of a term to the maximum likelihood value (or equivalently minimised χ^2) which penalises the complexity of the model (via the number of free parameters).

Given its widespread use, Gibson (2014) tested the validity of this approach, by simulating tens of thousands of transit light curves with injected systematics, and attempted to recover the transit depth with uncertainties using a family of systematics models, *including the correct model*. This found that simple model selection routinely results in underestimated uncertainties in the extracted light curve parameters (in the case of the BIC, typically 100%!). Full calculation of the Bayesian evidence fared better, but still did not provide reliable statistics. This means that *model selection does not work in practice*, and we should be careful when interpreting results extracted in this way.

2.3. Marginalisation over many models

The above result might seem surprising. However, when performing model selection, we are neglecting that there is *uncertainty associated with the choice of systematics model*, as well as the parameters of a specific model. In order to account for this, we can marginalise over many systematics models rather than force ourselves into selecting a single one. This is straightforward in practice, and requires minimal additional computation on top of model selection. In a nutshell, the evidence or information criteria for each model can be converted into a probability for each model (generally assuming uniform priors on the models), and the light curve parameters can be determined from a weighted average (based on these probabilities) of the marginalised posterior distribution for each individual model. Gibson (2014) shows that this provides much more reliable statistics of the recovered light curve parameters than model selection. Where the evidence clearly prefers one model over the others, this process reduces to model selection.

2.4. *Stochastic models*

Marginalisation over many models (and model selection) assumes that one of the systematics models tested is the correct one. Stochastic processes offer an even more flexible alternative to modelling instrumental systematics, and largely avoids this problem. These were introduced by Gibson (2012) in the form of Gaussian processes (GPs). Rather than defining our model as a specific function of the inputs (i.e. a deterministic process), a GP allows us to place a probability distribution over a class of functions. For example, using the 'squared exponential' GP defines a distribution over any smooth function of the inputs, without needing to specify the exact form of the function itself. GPs are also intrinsically Bayesian, therefore penalise complexity in much the same way as model selection/marginalisation, although do so over an effectively infinite number of models. GPs can therefore marginalise out our ignorance of the unknown form of the systematics, and therefore provides a robust measurement of light curve parameters. Gibson (2014) tested the use of GPs on tens of thousands of simulated light curves (injected with deterministic and stochastic forms of systematics), and confirmed that they are the most reliable method to model systematics among those discussed here.

3. Conclusions

The method used to model instrumental systematics in exoplanet light curves is of central importance to our understanding of exoplanet atmospheres, and great care is needed to construct useful models. However, the choice of systematics model will always contribute to the error budget, and is usually neglected. This must be taken into account by integrating out our ignorance of the form of the systematics model. I recommend the use of marginalisation over many models when using deterministic models; however, stochastic processes will over an even greater degree of flexibility and objectivity, and it is even possible to marginalise over many stochastic models for very complex problems.

Of course, no systematics models are perfect, and a combination of the above methods is the best approach, tested alongside instrument-specific models and preferably with repeated observations. However, the quality of extracted spectra will always be limited by the systematics, where uncertainty in model parameters *and* model selection always contribute to the error budget. To make the next leap in our understanding of exoplanets, dedicated instrumentation is needed to minimise the systematics in our observations.

References

Brown, T. M. 2001, *ApJ*, 553, 1006
Charbonneau, D., Brown, T. M., Noyes, R. W., & Gilliland, R. L. 2002, *ApJ*, 568, 377
Gibson, N. P., Pont, F., & Aigrain, S. 2011, *MNRAS*, 411, 2199
Gibson, N. P., Aigrain, S., Roberts, S. *et al.* 2012, *MNRAS*, 419, 2683
Gibson, N. P. 2014, *MNRAS*, 445, 3401
Nikolov, N. *et al.* 2014, *MNRAS*, 437, 46
Seager, S. & Sasselov, D. D. 2000, *ApJ*, 537, 916
Sing, D. K. *et al.* 2011, *MNRAS*, 416, 1443

Bayesian Planet Searches for the 10 cm/s Radial Velocity Era

Philip C. Gregory

Physics and Astronomy, University of British Columbia, 6224 Agricultural Rd, Vancouver, British Columbia V6T 1Z1, Canada
email: gregory@phas.ubc.ca

Abstract. A new apodized Keplerian model is proposed for the analysis of precision radial velocity (RV) data to model both planetary and stellar activity (SA) induced RV signals. A symmetrical Gaussian apodization function with unknown width and center can distinguish planetary signals from SA signals on the basis of the width of the apodization function. The general model for m apodized Keplerian signals also includes a linear regression term between RV and the stellar activity diagnostic $\ln(R'hk)$, as well as an extra Gaussian noise term with unknown standard deviation. The model parameters are explored using a Bayesian fusion MCMC code. A differential version of the Generalized Lomb-Scargle periodogram provides an additional way of distinguishing SA signals and helps guide the choice of new periods. Sample results are reported for a recent international RV blind challenge which included multiple state of the art simulated data sets supported by a variety of stellar activity diagnostics.

Keywords. stars: planetary systems; methods: statistical; methods: data analysis; techniques: radial velocities.

At the current m/s RV precision level, intrinsic stellar activity (SA) has become the main limiting factor. New spectrographs are under development like ESPRESSO and EXPRES that aim to improve RV precision by a factor of approximately 100 over the current best spectrographs, HARPS and HARPS-N. Clearly, the success of these developments hinges on our ability to distinguish true planetary signals from SA induced signals. At the 'Towards Other Earths II" meeting held in Porto Portugal in September 2014, Xavier Dumusque challenged the community to a large scale blind test using simulated RV data at the 0.7 m/s level of precision, to understand the limitations of present solutions to deal with SA signals and to select the best approach. This paper describes a new approach using apodized Keplerian models which was tested on the first five of the RV Challenge data sets.

For the apodized Keplerian (AK) models, the semi-amplitude of the Kepler RV model is multiplied by a symmetrical Gaussian of unknown width, τ, and with an unknown center of the apodizing window, t_a. Since a true planetary signal spans the duration of the data τ will be large while SA induced RV signals generally vary on shorter time scales. The general model for m apodized Keplerian signals also included a linear regression term between RV and the stellar activity diagnostic $\ln(R'hk)$, as well as an extra Gaussian noise term with unknown standard deviation. The correlation term was particularly useful in removing long term SA signals associated with the stars magnetic cycle.

In addition to the RV measurements, the challenge data sets includes simultaneous observations of three stellar activity diagnostics. Two of these come from additional information on the spectral line shape that are extracted from the cross correlation function (CCF), the average shape of all spectral lines of the star. These two shape parameters are the CCF width (FWHM) and bisector span (BIS). The third diagnostic, $\ln(R'hk)$, is based on the Ca II H & K line flux that is sensitive to active regions on the stellar surface. A preliminary analysis of the first 5 data sets indicated a strong correlation between RV and the $\ln(R'hk)$ diagnostic and slighlty reduced correlation with the FWHM diagnostic.

The AK models were explored using an automated fusion MCMC algorithm (FMCMC; Gregory 2013), a general purpose tool for nonlinear model fitting and regression analysis. The AK models combined with the FMCMC algorithm† constitute a multi-signal apodized Keplerian periodogram.

The primary role of the AK models is to distinguish planetary signal candidates from SA signals. Suppose the results indicate that k of the signals are planetary and $m - k$ are SA signals. Final model parameter estimates and model comparisons are based on subsequent runs using a model of k Keplerians and $m - k$ apodized Keplerians.

The methodology is partially illustrated for the second challenge data set. The raw RV 2 data had a standard deviation of 8.58 m/s. After removing the best linear regression fit to $\ln(R'hk)$ as the independent variable, the standard deviation was reduced to 3.95 m/s. The top two rows of Figure 1 show the RV data and FWHM after removing the $\ln(R'hk)$ diagnostics together with their GLS periodogram on the right. In this analysis the FWHM (rhk corrected) is treated as a control.

It proved useful to construct a differential form of the Zechmeister & Kürster (2009) Generalized Lomb-Scargle (GLS) periodogram of selected period regions for the RV residuals. Two examples of this are shown in the bottom two panels. The black trace is from the upper right GLS periodogram in Figure 1. The dark gray trace is the negative of the GLS periodogram of the FWHM control (middle right panel) and the light gray trace shows the difference, the black trace plus the gray trace. The strongest signals at $P = 3.77$ and 10.63 d have no significant counterpart in the control. Strong yearly aliases signal are also evident. For both periods the light gray trace and the black trace coincide closely near the positive peaks, consistent with a planetary signal or its alias. Signals in common to both black and dark gray traces (e.g., near $P = 12.5$ d) indicates SA.

Figure 2 shows sample FMCMC results for an eight AK signal model. It is clear from this that there are three planetary candidates with periods of 3.77, 10.63, and 75.56 d whose apodization windows span the duration of the data. The apodization time constant of ~ 800 d for the 10.63 d signal make it a borderline P candidate. Also, the presence of SA activity at a period of 11.06 d in the close vicinity of the 10.63 d signal called into question a planetary interpretation of the 10.63 d signal. For the competition, the 10.63 d signal was reported as a probable planetary signal. The true planetary signals injected into the RV 2 data set included three with $K \geqslant 1$ m/s at $P = 3.77, 10.64, 75.28$ and two with $K < 1$ m/s at $P = 5.79, 20.16$ d. Starting from the raw data (standard deviation of 8.58 m/s), which was dominated by SA, we have been able to achieve residuals of 1.42 m/s which is a factor of six lower but still a factor of two higher than the mean measurement uncertainty of 0.7 m/s. Further studies including consideration of other apodization functions (e.g., an asymmetrical Gaussian) are warranted.

---

† A more detailed description of FMCMC is available in Chapter 1 of "Supplement to 'Bayesian Logical Data Analysis for the Physical Sciences'," a free supplement available in the resources section of the Cambridge University Press website for my Textbook "Bayesian Logical Data Analysis for the Physical Sciences: A Comparative Approach with *Mathematica* Support." A *Mathematica* implementation of fusion MCMC is also available from the resource section. The supplement includes a detailed discussion of the priors adopted by this author for exoplanet RV analysis. Chapter 1 also includes a comparison of three marginal likelihood estimators used for Bayesian model comparison and concludes in favor of the Nested Restricted Monte Carlo (NRMC) estimator which is is used in this work.

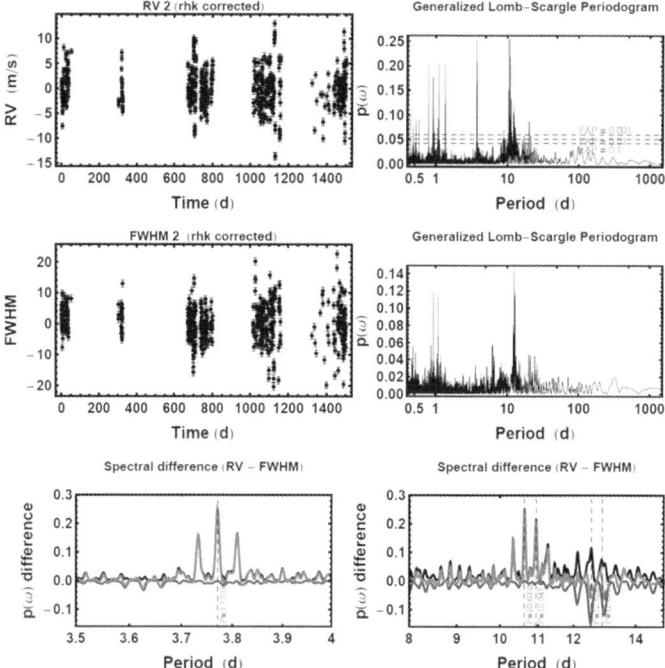

Figure 1. The RV 2 data and FWHM (control) after removing the $\ln(R'hk)$ diagnostics (rhk corrected) together with their GLS periodograms on the right. The differential GLS periodogram for two selected period ranges is shown in the bottom two panels.

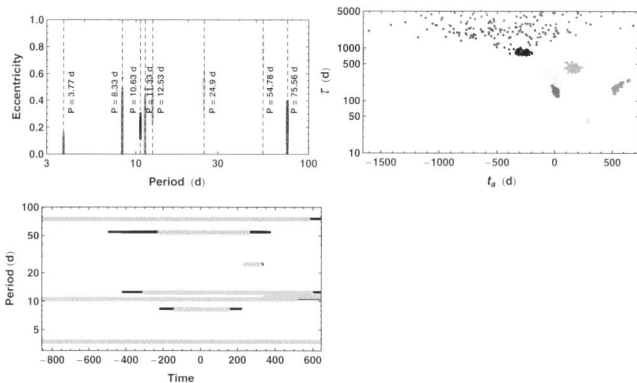

Figure 2. The upper right panel shows the eccentricity versus period parameters for sample MCMC iterations for the 8 signal apodized Kepler periodogram of the RV 2 data. The upper left panel is a plot of the apodization time constant, τ, versus apodization window center time, t_a. The lower panel shows the apodization interval for each signal (gray trace for MAP values of τ and t_a, black for a representative set of samples which is mainly hidden below the gray).

References

Gregory, P. C., 2007, *MNRAS*, 381, 1607-1616
Gregory, P. C., 2013, in *Astrostatistical Challenges for the New astronomy*, Joseph M. Hilbe (ed.), Springer Series in Astrostatistics 1,121
Zechmeister, M. & Kürster, M., 2009, *A&A*, 496, 577

Measuring the mass of Kepler-78b using nonparametric Gaussian process estimation

Samuel K. Grunblatt[1], Andrew W. Howard[1] and Raphaëlle D. Haywood[2]

[1] Institute for Astronomy, University of Hawaii,
2680 Woodlawn Drive, Honolulu, HI 96822
email: skg@ifa.hawaii.edu

[2] SUPA School of Physics and Astronomy, University of St Andrews,
North Haugh, St Andrews, Fife KY16 9SS, United Kingdom

Abstract. Measuring the masses of rocky planets is quite difficult, as the relevant signal produced by such planets is often dwarfed by stellar activity by an order of magnitude or more. Developing a more robust way to isolate the stellar activity in these measurements is crucial to the search for Earth-like planets. We estimate the mass of Earth-size planet Kepler-78b using a Gaussian process estimator to describe the stellar activity in both photometric and radial velocity (RV) data, confirming previous results with a more robust technique that can be extended toward Earth analogues.

Keywords. eta-Earth, Gaussian process, radial velocity, etc.

Kepler-78b is a transiting planet with a radius of 1.16 ± 0.16 $R_\oplus$ that orbits a young, active K dwarf in 8 hours (Sanchis-Ojeda *et al.* (2013)). The mass of Kepler-78b was reported from RV measurements using the HIRES and HARPS-N spectrographs. Even in quiet stars, stellar RV variations are an order of magnitude larger than the RV signal of Earth, and this effect is magnified in young stars (Hillenbrand *et al.* (2015)). Howard *et al.* (2013) and Pepe *et al.* (2013) independently modeled the RV signal of Kepler-78 parametrically to remove stellar activity while measuring the planetary signal.

In this study, we model the RVs using a quasiperiodic GP estimator trained on Kepler photometry. Using the package `emcee`, we determine parameter distributions with a Monte Carlo Markov Chain to simultaneously model the quasiperiodic stellar activity and Keplerian signal, allowing us to measure a mass of 1.87 ± 0.27 $M_\oplus$.

We find that despite the stellar activity producing radial velocity shifts ten times the size of the planetary signal, we can robustly isolate the planetary signal. This technique is dependent on using the more frequently sampled photometric signal to constrain the stellar rotation period, thereby restricting the Gaussian process estimator to a period distinct from the planetary orbital period. This technique is most valuable for obtaining parameters for small, rocky planets or planets at large orbital distances with orbital periods distinct from the stellar rotation. For more details, see Grunblatt *et al.* (2015).

References

Grunblatt, S., Howard, A., & Haywood, R. 2015 *ApJ*, 808,127
Hillenbrand, L., Isaacson, H., Marcy, G. *et al.* 2014, *Cool Stars*, 18, 759
Howard, A., *et al.* 2013, *Nature*, 503, 381
Pepe, F., *et al.* 2013, *Nature*, 503, 377
Sanchis-Ojeda, R., Rappaport, S., Winn, J., *et al.* 2013, *ApJ*, 774, 54

Improving Accuracy of Quasars' Photometric Redshift Estimation by Integration of KNN and SVM

Bo Han[1], Hongpeng Ding[1], Yanxia Zhang[2] and Yongheng Zhao[2]

[1]International School of Software, Wuhan University, Wuhan, P.R.China
email: bhan@whu.edu.cn; 2010282160014@whu.edu.cn
[2]Key Laboratory of Optical Astronomy, National Astronomical Observatories, Chinese Academy of Sciences, 20A Datun Road, Chaoyang District, 100012, Beijing, P.R.China

Abstract. Catastrophic failure is an unsolved problem existing in the most photometric redshift estimation approaches for a long history. In this study, we propose a novel approach by integration of k-nearest-neighbor (KNN) and support vector machine (SVM) methods together. Experiments based on the quasar sample from SDSS show that the fusion approach can significantly mitigate catastrophic failure and improve the accuracy of photometric redshift estimation.

Keywords. Photometric redshifts, K-Nearest-Neighbour, Support Vector Machine, SDSS

Methodology and Results

We analyze the reasons of catastrophic failure of photometric redshift estimation for quasars and point out that the outlier points are resulted by non-linearly separation in Euclidean feature space of input pattern. Therefore we propose a new approach by integration of SVM and KNN methods. By Gaussian kernel function in SVM, we map input pattern from an original Euclidean space into a high dimensional feature space. In this way, many outlier points can be identified by the hyperplane and then corrected. As shown in Figure 1, the experimental results based on quasars from SDSS database indicate that the integration approach can significantly mitigate catastrophic failure and improve the photometric redshift estimation accuracy (RMS of Δz is reduced from 0.258 to 0.246, and the percent when $|\Delta z| <0.3$ is improved from 87.02% to 90.03%, for $|\Delta z| <0.2$ from 85.68% to 88.78%, for $|\Delta z| <0.1$ from 78.70% to 81.70%). While other researchers mitigate catastrophic failure by cross-match of data from multiple surveys, our approach achieves the similar objective only from a single survey and therefore can be applied to much larger data avoiding cross-match efforts.

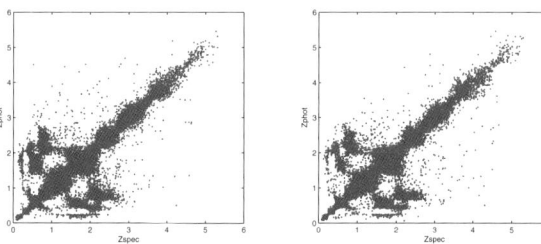

Figure 1. Left: Estimation by KNN; right: Estimation by KNN+SVM

Advances in the *Kepler* Transit Search Engine

Jon M. Jenkins

NASA Ames Research Center,
M/S 244-30, Moffett Field, CA 94035 U.S.A.
email: jon.jenkins@nasa.gov

Abstract. Twenty years ago, no planets were known outside our own solar system. Since then, the discoveries of ∼1500 exoplanets have radically altered our views of planets and planetary systems. This revolution is due in no small part to the *Kepler Mission*, which has discovered >1000 of these planets and >4000 planet candidates. While *Kepler* has shown that small rocky planets and planetary systems are quite common, the quest to find Earth's closest cousins and characterize their atmospheres presses forward with missions such as NASA Explorer Program's Transiting Exoplanet Survey Satellite (TESS) slated for launch in 2017 and ESA's PLATO mission scheduled for launch in 2024.

These future missions pose daunting data processing challenges in terms of the number of stars, the amount of data, and the difficulties in detecting weak signatures of transiting small planets against a roaring background. These complications include instrument noise and systematic effects as well as the intrinsic stellar variability of the subjects under scrutiny. In this paper we review recent developments in the *Kepler* transit search pipeline improving both the yield and reliability of detected transit signatures.

Many of the phenomena in light curves that represent noise can also trigger transit detection algorithms. The *Kepler Mission* has expended great effort in suppressing false positives from its planetary candidate catalogs. Over 18,000 transit-like signatures can be identified for a search across 4 years of data. Most of these signatures are artifacts, not planets. Vetting all such signatures historically takes several months' effort by many individuals. We describe the application of machine learning approaches for the automated vetting and production of planet candidate catalogs. These algorithms can improve the efficiency of the human vetting effort as well as quantifying the likelihood that each candidate is truly a planet. This information is crucial for obtaining valid planet occurrence rates. Machine learning approaches may prove to be critical to the success of future missions such as TESS and PLATO.

Keywords. statisics, machine learning, transit photometry, exoplanets

1. Transit Detection and Additional Statistical Tests

Jenkins (2002) introduced an adaptive, wavelet-based matched filter for detecting transiting planet signatures against time-varying, non-white observation noise. The transit detection is conducted in a joint-frequency domain through the wavelet transform, but can more simply be represented in the time domain as a simple matched filter where the whitened flux time series, $\tilde{x}$, is projected onto the whitened transit pulse waveform, $\tilde{s}$:

$$Z = \frac{\tilde{x} \cdot \tilde{s}}{\sqrt{\tilde{s} \cdot \tilde{s}}} = \frac{\sum_{i=1}^{p} \tilde{x}_i \cdot \tilde{s}_i}{\sqrt{\sum_{i=1}^{p} \tilde{s}_i \cdot \tilde{s}_i}}, \quad (1.1)$$

where Z is the detection statistic (thresholded at 7.1 σ for *Kepler* transit searches), and the rightmost term is written in terms of the individual transits, of which we assume there are p. Fig. 1 shows the light curve for one of *Kepler's* stars along with the whitened light curve. Fig. 2 shows a montage of the impulse response of the adaptive whitener,

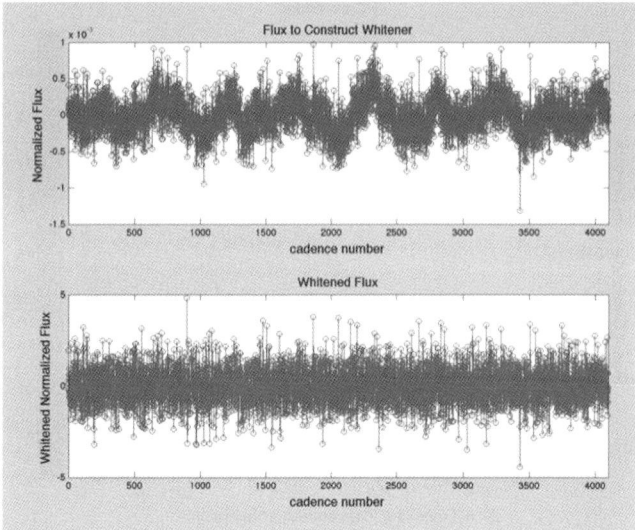

Figure 1. Scatter plot of a section of a flux time series for a *Kepler* star (top) and the whitened flux time series for this same star (bottom).

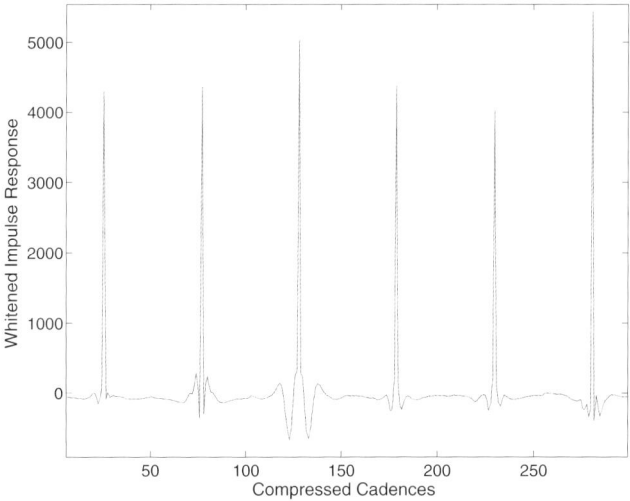

Figure 2. Montage of time-varying impulse response of the whitener applied in Fig. 1.

illustrating the fact that the whitener adapts to non-stationary noise. Eq. 1.1 explicitly takes the non-stationary noise into account when formulating the detection statistic.

To help control the false alarm rate due to structured noise in the flux light curves, Seader *et al.*(2013) added a robust version of the detection statistic, Z_{robust} based on an iterative kernel-based weighting of the individual in-transit data points as per Holland & Welsch (1977). Z_{robust} is thresholded typically at 6.8 σ. Seader *et al.*(2013) also formulated two different χ^2 metrics that examine the degree to which the scatter of the residual in-transits points matches expectations after the transit model has been subtracted from the data. $\chi^2_{(2)}$ pursues this question at the level of the individual transits and has $p-1$ degrees of freedom, while χ^2_{GOF} handles this at the level of the in-transit points, and has $n_{transit}$ degrees of freedom, where $n_{transit}$ is the number of points in transit.

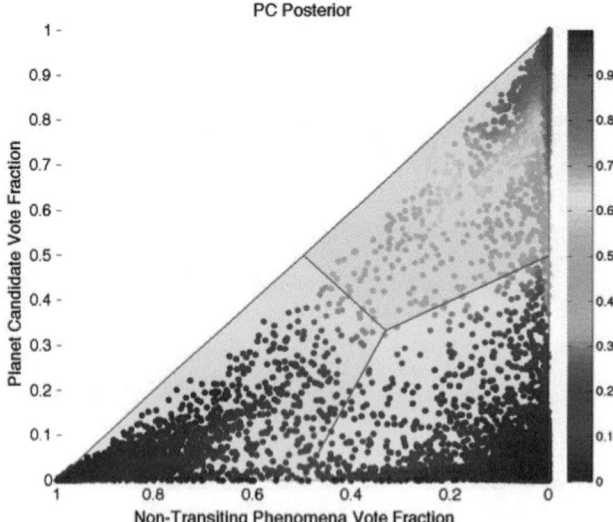

Figure 3. Ternary plot of the posterior probability of the classification of 'planet candidate' for ∼16,000 TCEs based on the vote fractions of a random forest. The color bar indicates the posterior probability for each TCE.

If a transit-like feature exceeds the thresholds for all four tests, it is registered as a Threshold Crossing Event (TCE) and subjected to a suite of diagnostic tests and physical model fitting, and passed on the the Threshold Crossing Event Review Team (TCERT) for dispositioning as a planet candidate, an astrophysical false positive, or an artifact.

2. Random Forests for Automated Classification of Transit-Like Features

We are exploring supervised machine learning approaches to dispositioning TCEs that leverage the extensive history of the TCERT in vetting previous sets of TCEs. Figure 3 shows the result of training a random forest on the TCEs resulting from a search through the first 16 quarters of *Kepler* data and overlaying a Bayesian classification scheme on top of the feature space defined by the vote fractions of the random forest for each of three classes: planet candidate, astrophysical false positive, and non-transiting phenomena. This allows us to construct posterior probabilities for each class based on the behavior of the training sets. These posteriors can be used in occurrence rate calculations to reduce the sensitivity to which objects are or are not included.

References

Breiman, L. 2001 *Machine Learning*, 45, 1
Jenkins, J. M. 2002, *ApJ* 575, 493
Holland, P. W.,& R. E. Welsch, Communications in Statistics: Theory and Methods, A6 , 1977, 813–827.
Seader, S., Tenenbaum, P., Jenkins, J. M., & Burke, C. J. 2013, *ApJS*, 206, 25

Low-contrast pre-coronagraph for extra contrast of dark-hole

Jun Nishikawa[1,2,3] Masahito Oya[4,1] Naoshi Murakami[5]
Motohide Tamura[6,1,3] Takashi Kurokawa[1,7] Yosuke Tanaka[7] and
Takayuki Kotani[3,1]

[1] National Astronomical Observatory of Japan, Extrasolar Planet Detection Project Office,
2-21-1 Osawa, Mitaka, Tokyo, Japan, 181-8588
email: jun.nishikawa@nao.ac.jp

[2] SOKENDAI (Graduate University for Advanced Studies), Faculty of Physical Sciences,
2-21-1 Osawa, Mitaka, Tokyo, Japan, 181-8588

[3] National Institute of Natural Sciences, Astrobiology Center,
2-21-1 Osawa, Mitaka, Tokyo, Japan, 181-8588

[4] Nihon University, Graduate School of Physics,
Surugadai 1-8-14, Chiyoda, Tokyo, Japan, 101-8308

[5] Hokkaido University, Fuculty of Engineering,
Kita 13 Nishi 8, Kita-Ku, Sapporo, Hokkaido, Japan, 060-8628

[6] The University of Tokyo, Graduate School of Science, Department of Astronomy,
7-3-1 Hongo, Bunkyo-ku, Tokyo, Japan, 113-0033

[7] Tokyo University of Agriculture and Technology, Graduate School of Engineering,
Koganei, Tokyo, Japan, 184-8588

Abstract. We propose a low-contrast pre-coronagraph that can provide additional dark-hole contrast to a main coronagraph.

Keywords. instrumentation: adaptive optics, techniques: interferometric.

1. Pre-coronagraph under dark-hole control

The low-contrast pre-coronagraph (LPC) is a new style of the unbalanced nulling interferometer (UNI) which was developed for precise wavefront control (Nishikawa et al. 2008). The LPC is used in the four-stage coronagraph system: the first deformable mirror (DM), the LPC, the second DM, and the main coronagraph, to obtain an additional contrast to the main coronagraph. Originally a wavefront sensor was used around the UNI (LPC) and we characterized it using a four-quadrant phase mask coronagraph (Kobayashi, et al. 2012). Recently we have found that the two deformable mirrors in the system can be controlled by the dark-hole algorithm with a final focal-plane detector if we use two steps. First, the control is made by the first DM with a normal mask at the pre-coronagraph and without a mask at the main coronagraph. Second, the control is made by both two DMs to produce the circular dark hole with a low-contrast mask exchanged at the pre-coronagraph and with a normal mask at the main coronagraph.

References

Nishikawa, J., Abe, L., Murakami, M., & Kotani, T. 2008, *A&A*, 489, 1389
Kobayashi, T., Nishikawa, J., Tanaka, Y., Kurokawa, T., Kashiwagi, K., Murakmai, N., Baba, N., & Hashimoto, N. 2012, *Proc. SPIE*, 8421, 84213D

Combining Transit and Radial Velocity Data

Leslie A. Rogers[1,2,3]

[1] Hubble Fellow, Cahill Center for Astronomy & Astrophysics, Caltech
[2] Sagan Fellow, Department of Earth and Planetary Sciences, University of California Berkeley
501 Campbell Hall #3411, Berkeley, CA 94720, USA
[3] Department of Astronomy & Astrophysics, University of Chicago
5640 S Ellis Ave, Chicago, IL 60637, USA
email: larogers@uchicago.edu

Abstract. The *Kepler* Mission, combined with ground based radial velocity (RV) follow-up, has revolutionized the observational constraints on sub-Neptune-size planet compositions. *Kepler*'s unprecedentedly large and homogeneous samples of planets with both mass and radius constraints open the possibility of statistical studies of the underlying planet composition distribution. This presentation describes the application of hierarchical Bayesian models to constrain the underlying planet composition distribution from a sample of noisy mass-radius measurements. This approach represents a promising avenue toward a quantitative measurement of the amount of physical scatter in small planet compositions, the identification of planet subpopulations that may be tied to distinct formation pathways, and empirical constraints on the dominant compositional trends in the planet sample. Both the transit and radial velocity techniques are subject to selection effects, and approaches to mitigate the resulting biases will be addressed. In addition to distilling composition-distribution insights from the current sample of Kepler planets with RV masses, this framework may be used to optimize the target selection for future transiting planet RV follow-up surveys.

Keywords. methods: data analysis, methods: statistical, planetary systems, planets and satellites: composition, techniques: photometric, techniques: radial velocities

---

1. Introduction

In terms of constraining planet bulk compositions, the subset of planets detected both with transits and with radial velocity (RV) measurements or transit timing variations (TTVs) are most valuable. The planet radius measured from the transit depth and the planet mass measured from the planet's gravitational influence (on their host star or other planets in the system) together combine to give the planet's mean density and some handle on the planet's bulk composition. Planet bulk compositions in turn provide insights into the a planet's formation and evolution. The presence of a H/He envelope surrounding a planet constrains the timing of the planet's formation relative to the protoplanetary disk lifetime, $t_{\rm disk} \sim 3$ Myr. The water-to-rock mass ratio of a planet is a tracer of the formation location relative to the snow line in the protoplanetary disk.

In this paper, we focus on how to combine insights from transit and RV and/or TTV measurements to constrain the range of possible compositions for a single planet (Section 2), and to derive insights into the composition distribution of planets from a statistical sample of noisy mass-radius measurements (Section 3).

2. Single Planets

The main planet properties that can be observationally constrained by transit and RV data are mass (M_p), radius (R_p), and incident flux (F_p). Inferring planet compositions from mass-radius data can be cast in a Bayesian framework as first formulated by Rogers & Seager (2010a),

$$p(M_p, \{x_i\} \mid \mathcal{D}) \propto \left\{ \int p(\mathcal{D} \mid M_p, R_p) \, p(R_p \mid M_p, \{x_i\}) \, dR_p \right\} p(M_p, \{x_i\}). \quad (2.1)$$

We denote by $\{x_i\}$ the composition of the planet (with x_i representing the mass fraction of the planet in the ith component). The likelihood function, $p(\mathcal{D} \mid M_p, R_p)$ represents the likelihood of the transit and RV data (denoted by $\mathcal{D}$) given true planet M_p and R_p. The next factor, $p(R_p \mid M_p, \{x_i\})$, encapsulates insights from planet interior structure and evolution models, which provide a mapping $f : [M_p, \{x_i\}] \mapsto R_p$. Most analyses to date have effectively placed absolute trust in the models, treating this term as a Dirac-delta function $p(R_p \mid M_p, \{x_i\}) = \delta(R_p - f(M_p, \{x_i\}))$. Finally, $p(M_p, \{x_i\})$ denotes the priors on the mass-composition distribution given the assumptions and information available before any data is collected. Equation 2.1 above is distilled for clarity; it could readily be generalized *i)* to be cast in terms of the radial velocity amplitude, transit depth and host star properties, and/or *ii)* to explicitly include additional dependencies (e.g., on age, F_p, host star metallicity, host star chemical abundances etc.).

Marginalizing over the planet mass gives the posterior pdf of the planet compositions,

$$p(\{x_i\} \mid \mathcal{D}) = \int p(M_p, \{x_i\} \mid \mathcal{D}) \, dM_p. \quad (2.2)$$

Model degeneracies complicate exoplanet compositional inferences (e.g, Valencia *et al.* 2006, Adams *et al.* 2008, Rogers & Seager 2010ab). The mapping $\tilde{f} : [M_p, \{x_i\}] \mapsto [M_p, R_p]$ is not one-to-one; planets with very different compositions can have identical masses and radii, especially if > 2 distinct chemical components are considered (e.g., H/He, ices, rock). Consequently, single planet compositional inferences are intrinsically in an under-constrained regime where the prior assumptions strongly affect the results (posteriors) obtained. It is thus crucial to ask the questions that can be most robustly answered. For instance, the mass-radius relations of rocky planets are most sensitive to the total iron mass fraction of the planets and are less sensitive to the degree of differentiation or oxidation within the planets (e.g., Elkins-Tanton & Seager 2008, Rogers *et al.* 2011, Zeng & Sasselov 2013). Similarly the mass-radius relations for planets with gas envelopes are most sensitive to the H_2/He gas mass fraction, but less sensitive to the ice-to-rock mass ratio (e.g., Rogers *et al.* 2011, Valencia *et al.* 2013). As a result, the total iron mass fraction of rocky planets and the H/He mass fraction of gas-laden planets are more readily constrained than other aspects of the planets' compositions. Examples of Bayesian compositional inferences for individual planets include Kepler-36 b, c (Carter *et al.* 2012) and PH3 b, c, d (Schmitt *et al.* 2014).

3. Planet Populations

The accumulating census of exoplanets with observational constraints on both their mass and radius offers the opportunity to move beyond case studies of individual objects to studies of the overall population. Hierarchical models provide a natural framework to constrain the underlying composition distribution of planets from a noisy sample of mass-radius measurements. Hierarchical models open the priors on the planet compositions to modeling. Given observations of N planets and a model for the joint planet mass-composition distribution that depends on some hyperparameters ϕ, we may generalize Equation 2.1,

$$p(\phi, \{M_{pj}\}, \{x_{ij}\}|\mathcal{D}) \propto \prod_{j=1}^{N}\left[\left\{\int p(\mathcal{D}|M_{pj}, R_{pj})p(R_{pj}|M_{pj}, \{x_{ij}\})\mathrm{d}R_{pj}\right\}\right.$$
$$\left. \times p(M_{pj}, \{x_{ij}\}|\phi)\right] p(\phi) \qquad (3.1)$$

Equation 3.1 is simplified to treat each jth planet as independent; it can, however, be readily generalized to account for correlations between multiple planets within the same system (from corellated measurements and/or from physical correlations between the true planet properties). So far, hierarchical models have been applied to the sample of $\mathcal{O}$ (50) *Kepler* planets having Keck HIRES RV follow-up to constrain, as a function of planet size, the fraction of planets that are sufficiently dense to be rocky (Rogers 2015), and to fit a power-law mass-radius relationship allowing for intrinsic scatter in the planet masses at a given radius (Wolfgang *et al.* 2015). Once the sample of small transiting planets with RV mass constraints grows to $\mathcal{O}$ (200), analyses focused on the joint mass-composition distribution of planets (as in Equation 3.1) could plausibly constrain the typical mass scale of rocky planets, the core mass-envelope mass relationship of gas-laden planets, and extent of physical scatter in exoplanet compositions (Rogers, in prep.).

A future with $\mathcal{O}$ (200) small planet mass-radius measurements may be only a couple years away. There is a suite of space-based transit surveys on the horizon (K2, TESS, CHEOPS and PLATO) that will find planets transiting bright stars that are more amenable to ground-based RV follow-up than typical *Kepler* planets. This combined with ongoing investment in the technological development of ground-based spectrographs (e.g., MAROON-X, Carmenes, HPF, SHREK, SPIRou, ESPRESSO, MINERVA) means the pace of exoplanet discovery and characterization is poised to continue accelerating.

We conclude with a note of caution. To ensure that future transit-search missions and RV follow-up instruments are leveraged to maximum advantage in characterizing the composition distribution of small exoplanets, care must be taken to characterize and to mitigate selection effects and biases. *Kepler* mass-radius populations studies (e.g., Rogers 2015, Wolfgang *et al.* 2015) have so far focused on mass distributions conditioned on planet radius to factor out the dominant selection effects. The next major advances in the study of the planet composition distribution will require a more sophisticated treatment of selection effects. Transiting planet radial velocity follow-up surveys directed at population level inferences (as opposed to case studies of individual systems) should *i)* aim to follow an algorithmic approach to selecting their targets and *ii)* report all measurements in the survey (including non-detections and RV upper-limits).

References

Adams, E. R., Seager, S., & Elkins-Tanton, L. 2008, *ApJ*, 673, 1160
Carter, J. A., Agol, E., Chaplin, W. J., *et al.* 2012, *Science*, 337, 556
Elkins-Tanton, L. T. & Seager, S. 2008, *ApJ*, 688, 628
Rogers, L. A. 2015, *ApJ*, 801, 41
Rogers, L. A., Bodenheimer, P., Lissauer, J. J., & Seager, S. 2011, *ApJ*, 738, 59
Rogers, L. A. & Seager, S. 2010a, *ApJ*, 712, 974
—. 2010b, *ApJ*, 716, 1208
Schmitt, J. R., Agol, E., Deck, K. M., *et al.* 2014, *ApJ*, 795, 167
Valencia, D., O'Connell, R. J., & Sasselov, D. 2006, *Icarus*, 181, 545
Valencia, D., Guillot, T., Parmentier, V., & Freedman, R. S. 2013, *ApJ*, 775, 10
Wolfgang, A., Rogers, L. A., & Ford, E. B. 2015, ArXiv e-prints
Zeng, L. & Sasselov, D. 2013, *PASP*, 125, 227

Astrometric exoplanet surveys in practice

Johannes Sahlmann[1,2]

[1] European Space Agency, ESAC, P.O. Box 78, Villanueva de la Cañada, 28691 Madrid, Spain

[2] European Space Agency, STScI, 3700 San Martin Drive, Baltimore, MD 21218, USA
email: Johannes.Sahlmann@esa.int

Abstract. Conversely to the transit photometry and radial velocity methods, the astrometric discovery of exoplanets is still limited by the sensitivity of available instruments. Ground-based surveys are now sensitive to giant planets in orbit around nearby low-mass stars and brown dwarfs. In 2014, ESA's Gaia mission began its survey, which is expected to discover thousands of giant exoplanets by detecting the astrometric orbital motions of the host stars.

Keywords. Astrometry, planetary systems, binaries, telescopes, space vehicles

1. Introduction

In comparison to other observational techniques, the contributions of astrometry to the discovery of exoplanets have so far been limited (Fischer *et al.* 2014). However, promising results were obtained with past (Muterspaugh *et al.* 2010) and ongoing surveys.

2. Ongoing surveys

Most present-day observational programs employ optical/infrared cameras on intermediate and large telescopes (e.g. Boss *et al.* 2009; Lurie *et al.* 2014; Sahlmann *et al.* 2014). Typical targets are low-mass stars and brown dwarf within tens of parsec of the Sun and the planet detection sensitivities reach Jupiter-mass for year-long periods (e.g. Weinberger *et al.* 2014; Sahlmann & Lazorenko 2015).

3. Gaia

The combination of our current knowledge of giant planet occurrence with the astrometric precision, sampling, and 5-year duration of the all-sky Gaia mission (de Bruijne 2012) translates into a number of expected exoplanet discoveries in excess of several thousand (Casertano *et al.* 2008; Sozzetti *et al.* 2014; Perryman *et al.* 2014; Sahlmann *et al.* 2015). This major step in instrumental capabilities will have to be matched by improved algorithms that optimally exploit the data. Using very precise ground-based astrometry we employed genetic and MCMC algorithms and proved them to be efficient in constraining all astrometric parameters of a low-mass binary system (Sahlmann *et al.* 2013). Similar algorithms will be applied to some of the Gaia exoplanet data (cf. Sozzetti 2013), which makes them important tools for harvesting the results of the first major astrometric exoplanet survey.

References

Boss, A. P., Weinberger, A. J., Anglada-Escudé, G., *et al.* 2009, *PASP*, 121, 1218
Casertano, S., Lattanzi, M. G., Sozzetti, A., *et al.* 2008, *A&A*, 482, 699

de Bruijne, J. H. J. 2012, *Ap&SS*, 341, 31
Fischer, D. A., Howard, A. W., Laughlin, G. P., et al. 2014, *Protostars and Planets VI*
Lurie, J. C., Henry, T. J., Jao, W.-C., et al. 2014, *AJ*, 148, 91
Muterspaugh, M. W., Lane, B. F., Kulkarni, S. R., et al. 2010, *AJ*, 140, 1657
Perryman, M., Hartman, J., Bakos, G. Á., & Lindegren, L. 2014, *ApJ*, 797, 14
Sahlmann, J. & Lazorenko, P. F. 2015, *MNRAS*, 453, L103
Sahlmann, J., Lazorenko, P. F., Ségransan, D., et al. 2013, *A&A*, 556, A133
Sahlmann, J., Lazorenko, P. F., Ségransan, D., et al. 2014, *A&A*, 565, A20
Sahlmann, J., Triaud, A. H. M. J., & Martin, D. V. 2015, *MNRAS*, 447, 287
Sozzetti, A. 2013, *European Physical Journal Web of Conferences*, 47, 15005
Sozzetti, A., Giacobbe, P., Lattanzi, M. G., et al. 2014, *MNRAS*, 437, 497
Weinberger, A. J., Boss, A. P., & Anglada-Escudé, G. 2014, *IAU Symposium*, 299, 230–231

Significance of periodogram peaks

Maria Süveges[1], Leanne Guy,[1] Shay Zucker,[2] and the Gaia CU7 team[1]

[1] Dept. of Astronomy, University of Geneva, Switzerland
[2] Dept. of Geosciences, Faculty of Exact Sciences, Tel Aviv University, Israel

Abstract. Three versions of significance measures or False Alarm Probabilities (FAPs) for periodogram peaks are presented and compared for sinusoidal and box-like signals, with specific application on large-scale surveys in mind.

Keywords. methods: data analysis, methods: statistical, stars: planetary systems

---

The detection of tiny periodic signals in noisy and irregularly sampled time series of large-scale surveys is a challenging task due to sparse time sampling which is neither regular nor fully irregular, oversampling in frequency space and the ensuing strong dependency among periodogram values at different frequencies. We compare three recent propositions for the computation of the FAP from the literature. The F^M method (Paltani 2004; Schwarzenberg-Czerny 2012) is based on an equivalent independent frequency set. Baluev (2008) draws on the extreme-value theory of stochastic processes. The GEV method (Süveges 2014) uses univariate extreme-value theory.

The F^M method can be applied to any periodogram type for which the marginal distribution is known; the Baluev method, only to those with some specific margins; the GEV method, to any periodogram. All assume uncorrelated errors in the time series. The Baluev method does not need the estimation of any parameters and its CPU requirements are negligible. The other two methods must estimate some parameters from simulations. The needs of the GEV method however can be reduced in two ways (Süveges 2014; Süveges et al. 2015) and so it requires less CPU than the F^M method.

Süveges et al. (2015) describe the performance of the three methods on sinusoids. For transit-like signals analysed with the box least squares method, we show that the GEV method approximates the true distribution of the periodogram peak better than the F^M method, similarly to the results on sinusoids. At a desired confidence level α, the fraction of false signal detections on noise (Type I error) by the GEV method is close to α, whereas the F^M method results in too many false positives. Consequently, the F^M method detects a larger fraction of shallow-transit signals than the GEV method. However, among the detections, the fraction of those with a *correctly recovered* period is smaller than for the GEV method. When time series with no periodicity (pure noise) are also present among the data, this fraction decreases further, more so for the F^M method than for the GEV method, due to the general permissiveness of the former.

In summary, both the Baluev and the GEV methods are better suited to the needs of large-scale surveys than the F^M method, due to their more favourable rate of correct frequency identifications among all detections and to their lower CPU needs.

References

Baluev, R. V. 2008, *MNRAS*, 385, 1279
Paltani, S., 2004, *A&A*, 420, 789
Schwarzenberg-Czerny, A. 2012, Proceedings of the 285th IAU Symposium, 81
Süveges, M. 2014, *MNRAS*, 440, 2099
Süveges, M., Guy, L., Eyer, L. *et al.* 2015, *MNRAS*, 450, 2052

Planet Frequency beyond the Snow Line from MOA-II Microlensing Survey

Daisuke Suzuki[1], David P. Bennett[1] and the MOA collaboration

[1]Department of Physics, University of Notre Dame, Notre Dame, IN 46556, USA
email: dsuzuki@nd.edu

Abstract. We present the first statistical analysis of the exoplanet frequency using planets found by a microlensing survey rather than follow-up observations. We present an analysis of 2007-2012 MOA (Microlensing Observations in Astrophysics) survey data to derive the planet frequency as a function of the planet/star mass ratio, q and separation, s, relative to the Einstein radius. Our sample includes 1472 microlensing events, including 22 planetary events and 1 ambiguous event with possible planetary and stellar binary solutions. The detection efficiency is calculated for each event and we employ a Bayesian analysis to deal with the ambiguous event. A broken power law model is used to fit the mass ratio function and we find a break and likely peak at $q \sim 1.0^{-4}$.

Keywords. gravitational lensing: micro - planetary systems

Detection Efficiency

The exoplanet detection efficiency is calculated for each event following the method of Rhie *et al.*(2000) using a survey data detection threshold of $\Delta\chi^2 \geqslant 100$ between single and binary lens models. Follow-up observations are excluded from these threshold calculations, but are used to help characterize the binary lens signals. The survey sensitivity (the sum of the detection efficiencies for all 1472 events) and the 22-23 planetary events are used to determine the exoplanet mass ratio function as a function of q and s.

Results

We fit a mass ratio function of the form $f = A(q/q_{\rm break})^n s^m$ for $q \geqslant q_{\rm break}$ and $f = A(q/q_{\rm break})^p s^m$ for $q < q_{\rm break}$, where $q_{\rm break}$ is the mass ratio for a break in the mass ratio function, and search for the best fit model with a Markov Chain Monte Carlo. We find that the mass ratio function rises steeply toward lower masses as $n \approx -1$, and then has a rather sharp break at $q_{\rm break} \sim 1.0^{-4}$. Below this, the mass ratio decrease or is flat. A mass ratio function that increases with separation, $m \approx 0.5$, is somewhat favored, but $m = 0$ is only disfavored by 1-σ. The mass ratio function changes only slightly when the previous microlens studies Gould *et al.*(2010) and Cassan *et al.*(2012) are added. Since the typical host star mass is $\sim 0.5 M - \odot$, this mass ratio function break corresponds to about a Neptune mass, suggesting that "failed Jupiter" cores are the most common planets beyond the snow line.

References

Gould, A., Dong, S., Gaudi, B. S., *et al.* 2010, *ApJ*, 720, 1073
Cassan, A., Kubas, D., Beaulieu, J.-P., *et al.* 2012, *Nature*, 481, 167
Rhie, S. H., Bennett, D. P., Becker, A. C., *et al.* 2000, *ApJ*, 533, 378

An Analytic Model Approach to the Frequency of Exoplanets

Wesley A. Traub

Jet Propulsion Laboratory, California Institute of Technology
4800 Oak Grove Dr., M/S 321-100, Pasadena, CA 91109, USA
email: wtraub@jpl.nasa.gov

Abstract. The underlying population of exoplanets around stars in the Kepler sample can be inferred by a simulation that includes binning the Kepler planets in radius and period, invoking an empirical noise model, assuming a model exoplanet distribution function, randomly assigning planets to each of the Kepler target stars, asking whether each planet's transit signal could be detected by Kepler, binning the resulting simulated detections, comparing the simulations with the observed data sample, and iterating on the model parameters until a satisfactory fit is obtained. The process is designed to simulate the Kepler observing procedure. The key assumption is that the distribution function is the product of separable functions of period and radius. Any additional suspected biases in the sample can be handled by adjusting the noise model or selective editing of the range of input planets. An advantage of this overall procedure is that it is a forward calculation designed to simulate the observed data, subject to a presumed underlying population distribution, minimizing the effect of bin-to-bin fluctuations. Another advantage is that the resulting distribution function can be extended to values of period and radius that go beyond the sample space, including, for example, application to estimating eta-sub-Earth, and also estimating the expected science yields of future direct-imaging exoplanet missions such as WFIRST-AFTA.

Keywords. methods: statistical

---

1. Overview

There are at least two different approaches that can be taken in order to derive the frequency (or occurrence rate) of exoplanets in the population, from a sample as observed by the Kepler mission. The approach taken by Mulder, Pascucci, & Apai (2015), for example, is the one that is taken by most authors. This requires starting with the probability of detecting a planet (given the transit geometry of the system, a model for the noise, and a model for the instrument bias), binning the observed planets in convenient bins of period and radius, and estimating the population of planets in each bin by inverting the sum of the probability of detection for each observed planet in each bin. An alternative approach Traub (2015) is to start with similarly calculated probabilities and binned observed planets, assume a parameterized functional form for the population, simulate the detection of randomly selected planets from the assumed population, compare the resulting population to the observed sample (e.g., as in Fig. 1), and adjust the parameters so as to achieve an optimized fit.

Each method has its merits. The latter one is worth exploring because it provides an analytical expression for the occurrence rate at every period and radius within the range of inputs, and by extrapolation it yields an occurrence rate for points outside that range. These estimates may provide insights into the formation and evolution processes of the system that cannot be obtained by the former method.

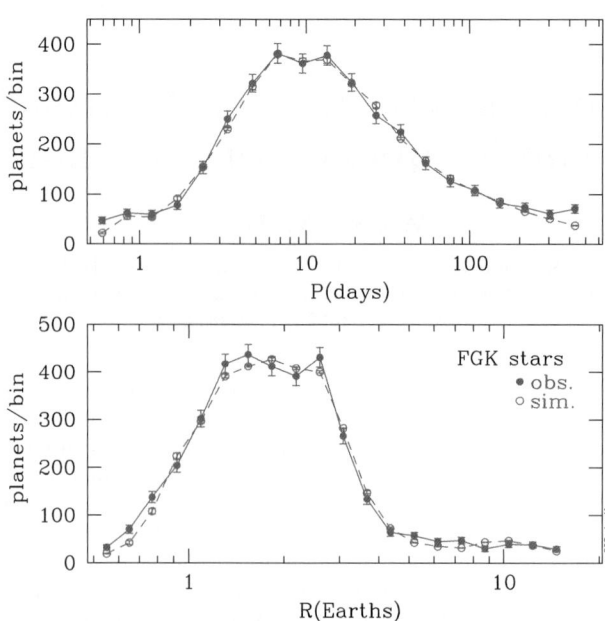

Figure 1. Example showing an observed sample of Kepler planets projected onto period and radius axes, along with a simulated sample derived from a model of the population.

References

Mulders, G. D., Pascucci, I., & Apai, D. 2015, *ApJ*, 798:112
Traub, W. A. 2015, *ApJ*, submitted

The Small Exoplanet Mass-Radius Relation: Quantifying the Astrophysical Scatter

Angie Wolfgang[1], Leslie A. Rogers[2] and Eric B. Ford[3]

[1] NSF Postdoctoral Fellow, Dept. of Astronomy & Astrophysics, Penn State University

[2] Sagan Fellow, Dept. of Astronomy, University of California, Berkeley

[3] Dept. of Astronomy & Astrophysics; Center for Astrostatistics, Penn State University

Abstract. The *Kepler Mission* has discovered thousands of planets with radii < 4 R$_\oplus$, paving the way for the first statistical studies of super-Earth dynamics, formation, and evolution. These calculations often require planetary masses, and yet the vast majority of *Kepler* planet candidates do not have theirs measured. A key concern is therefore how to map the measured radii to mass estimates in a size range that lacks Solar System analogs. While previous works have derived one-to-one relationships between radius and mass, a realistic mass-radius (M-R) relation should account for the range of compositions that we expect within the population. This compositional diversity creates astrophysical scatter in the relation, which we quantify here.

We have applied a hierarchical Bayesian model to quantify the intrinsic population dispersion in the small planet M-R relation, in the spirit of Kelly (2007). This method has several added benefits over other techniques, including: effortless inclusion of large mass uncertainties and upper limits; self-consistent incorporation of uncertainties on the independent variable — radii in this case — without the need for elaborate bootstrapping schemes; and the production of posterior distributions, allowing easy characterization of the uncertainties in both the population-wide and planet-specific parameters.

In Wolfgang, Rogers, & Ford (2015), we analyze how the constraints on this relation depend on the radius range of the sample, and on the method used to provide the mass measurements. Assuming that the M-R relation can be described as a power law with a dispersion that is constant and normally distributed, we find that $M/M_\oplus = 2.7(R/R_\oplus)^{1.3}$ is the best fit "average" relation for the sample of RV-measured transiting sub-Neptunes ($R_{pl} < 4$ R$_\oplus$), and that the astrophysical scatter around this "average" relation is $1.9 M_\oplus$.

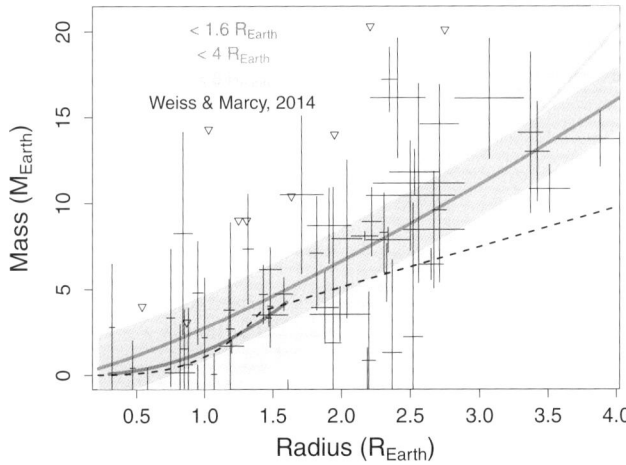

Figure 1. The best-fit M-R relations for data sets spanning different ranges of planetary radii. For each, the solid line denotes the "average" relation while the faded region denotes the standard deviation of the intrinsic, astrophysical scatter (note there is not yet sufficient observational evidence for scatter among the smallest planets). The dataset is overplotted as the thin black lines with triangles for the upper limits; a commonly used one-to-one relation is the dotted black line.

GATE (Gaia Transiting Exoplanets): Detecting Transiting Exoplanets with Gaia

Shay Zucker[1], Laurent Eyer[2], Simon Hodgkin[3] and Gisella Clementini[4]

[1] Dept. of Geosciences, Faculty of Exact Sciences, Tel Aviv University, Tel Aviv, Israel [2] Dept. of Astronomy, University of Geneva, Geneva, Switzerland [3] Institute of Astronomy, Cambridge University, Cambridge, UK [4] INAF – Osservatorio Astronomico di Bologna, Bologna, Italy

Gaia will have a revolutionary impact on most fields of astronomy. However, its scanning law is too sparse for traditional transit detection approaches (de Bruijne 2012). Practically, only stars brighter than 16th magnitude are relevant for follow-up of transiting exoplanets. For those stars, Gaia's precision is of the order of 1 mmag (Eyer *et al.* 2015). On average, Gaia will have sampled each target 70 times, but certain stars may be observed as many as 200 times (Voss *et al.* 2013). Hipparcos scanning law was similar, but its precision much worse. Nevertheless the transit of HD209458 could be seen, aposteriori, in Hipparcos' data (Söderhelm 1999). This inspired our *GATE* initiative.

For each star, GATE will use Gaia data and Bayesian inference to produce probability distributions of potential transits. These distributions will be used to infer Instantaneous Transit Probability (ITP) – the probability for a hypothetical transit to occur in each future time (Dzigan & Zucker 2011 and 2013). ITP will be calculated by Coordination Unit 7 (CU7) of Gaia DPAC (Data processing and Analysis Consortium). The times of highest ITP are the best times to observe a star, in order to determine if transits do occur. Best times will be disseminated to the community using the facilities of Gaia Science Alerts (GSA). GSA operates a website, a Twitter account, and other channels, to alert the astronomical community about variable sources to follow (Hodgkin *et al.* 2013).

GATE will use the DPAC/CU7 network of telescopes for Supplementary Observations. Participation of the global observer community, using the GSA alerts will improve significantly the detection yield of GATE. All observatories with the ability to observe transits of extrasolar giant planets are welcome to contribute their observing power.

GATE is essential for full exploitation of Gaia's potential to detect transiting exoplanets. Using GATE, we estimate that Gaia will be able to detect a few thousand transiting exoplanets. It is important to start observations as soon as alerts become possible, since the information contained in Gaia data will degrade over time. The DFU approach can be utilized to other low-cadence photometric surveys, such as Pan-STARRS, ZTF, NGTS, etc.

References

de Bruijne, J. 2012, *Ap&SS*, 341, 68
Dzigan, Y. & Zucker, S. 2011, *MNRAS*, 415, 2513
Dzigan, Y. & Zucker, S. 2013, *MNRAS*, 428, 3641
Eyer, L., *et al.* 2015, in Living Together Planets, Host Stars and Binaries, eds. S.M. Rucinski, G. Torres & M. Zejda, ASP Conf. Ser V. 496 (San Francisco: ASP), p. 121
Hodgkin, S., *et al.* 2013, *Phil. Trans. R. Soc. A*, 371, 20239
Söderhelm, S. 1999, *Inf. Bull. Var. Stars*, 4816, 1
Voss, H., *et al.* 2013, in Highlights of Spanish Astrophysics VII, eds. J.C. Guirado, L.M. Lara, V. Quilis & J. Gorgas, p. 738

FM 9:
Highlights in the Exploration of Small Worlds

Comet composition and Lab

Dominique Bockelée-Morvan

LESIA, Observatoire de Paris, LESIA/CNRS, UPMC, Université Paris-Diderot,
F-92195 Meudon, France
email: dominique.bockelee@obspm.fr

The XXIX IAU General Assembly took place during the golden year of the exploration of small solar system bodies. With the Rosetta ESA mission around comet 67P, NASA Dawn and New Horizons missions nearby dwarf planets Ceres and Pluto, respectively, and the NASA/Cassini mission in Saturn neighborhood, year 2015 marked an important step towards further understanding of small solar system bodies. On August 11-13, Focus meeting 9 "Highlights in the exploration of small worlds" gathered scientists of all over the world to present and discuss the spectacular results obtained from these missions, as well as recent achievements obtained from past missions, comprehensive spectroscopic surveys from space (e.g., Herschel, NEOWISE, Gaia), ground-based observations, and geochemical analyses. This meeting was also the opportunity to discuss the state of our understanding of the nature of the various populations of small bodies in the Solar System, including icy satellites, in a cosmo-chemistry perspective.

Investigating in detail the physical and chemical properties of asteroids, comets, trans-neptunian objects and dwarf planets is indeed of tremendous importance for understanding the formation of the Solar System, and overall the process of star and planet formation. These bodies are the remnants - either fragments or "survivors"- of the swarm of planetesimals from which the planets were formed. They are thus primitive leftover building blocks of the Solar System formation process that can offer clues to the chemical mixture from which the planets formed some 4.6 billion years ago.

It is now clear that there was a general mixing of the different populations of small bodies at the earlier stages of the Solar System, as a result of planetary migration. Indeed, geochemical evidence obtained from analyses of extraterrestrial samples and the recent discovery of active and icy asteroids, show that differences between primitive asteroids and comets, are much less sharp than previously thought. During this Focus Meeting, the interrelationships between the various populations of small bodies were investigated through a detailed comparison of their physical and chemical properties, which is an mandatory step towards understanding their formation conditions and their evolutionary paths.

The interpretation of the isotopic, molecular and mineralogical properties of primitive solar system material is complex, and can only be achieved through a multi-disciplinary approach. A full session of Focus Meeting 9 discussed on solar system formation in the light of proto-planetary disk models, experimental works and recent insights on the composition of comets and proto-planetary disks.

Dominique Bockelée-Morvan, Daniel Hestroffer, Paola Caselli
Chairs
Michael A'Hearn, Elisabetta Dotto,
Ian Franchi, Karen Meech, Hideyo Kawakita, Diane Wooden, Jorge M. Carvano, Javier Licandro, Gabriel Tobie, Richard Binzel, Mikael Granvik, Wing Ip, Steven Charnley
SOC

Chemical diversity in the comet population

Nicolas Biver and Dominique Bockelée-Morvan

LESIA, Observatoire de Paris, CNRS, PSL Research University, UPMC, Université
Paris-Diderot, 5 place Jules Janssen, F-92195 Meudon, France
email: nicolas.biver@obspm.fr

Abstract. For the last 3 decades, infrared and microwave techniques have enabled the detection of up to 27 parent molecules in the coma of comets. Several molecules have been detected in over 40 different comets. A large diversity of composition is seen in the sample, comprising comets of various dynamical origins. Abundances relative to water for the molecules can vary by a factor 3 to more than 10. The taxonomic study of a sample of comets in which the abundance of several molecules (e.g., HCN, CH_3OH, CO, CH_4, C_2H_6, H_2S, H_2CO, CH_3CN, CS,...) has been measured does not show any clear grouping. Except for fragments of a common parent comet, every observed comet shows a different composition. The absence of any clear correlation between the volatile content of the comets and their dynamical origin (Kuiper Belt versus Oort Cloud) is consistent with a common origin for these two populations. Their diversity in composition may also suggest that radial and temporal mixing in the early proto-planetary nebula may have played an important role.

Keywords. comets: general, infrared: solar system, radio lines: solar system

1. Introduction

Comets are the most pristine remnants of the formation of the Solar System. Investigating the composition of cometary nuclei ices provides clues to the physical conditions and chemical processes at play in the primitive solar nebula. Comets may also have played a role in the delivery of water and organic material to the early Earth (Hartogh *et al.* 2011). The first taxonomy studies concerning the composition of the comets has been done (and his pursuing) by A'Hearn *et al.*(1995) on the basis of visible spectroscopy of the comets, sampling OH, CN, C_2, C_3 and NH radicals. Here we will focus on observations at longer wavelengths which probe the vibrational and rotational lines of the parent molecules, i.e., directly escaping from the sublimation of nuclear ices.

2. Infrared and radio observation of comets

During the past 10-20 years, the infrared and radio techniques made significant steps forward. The high spectral resolution in the infrared ($\lambda/\delta\lambda > 20\,000$) is necessary to isolate the narrow cometary lines from telluric absorption whereas in the millimeter to submillimeter domain the very high spectral resolution ($\lambda/\delta\lambda > 10^6$) enables the resolution of the velocity structure of the line to derive the gas expansion velocity and outgassing pattern. The recent achievements in the infrared and radio have been the extension of the instantaneous frequency coverage with the multiple order echelle spectrographs (e.g. NIRSPEC at Keck II telescope or CRIRES at VLT), and wide-band receivers coupled to fast Fourier-transform spectrometers in the radio, covering several GHz of bandwidth. Combined with an increased sensitivity of the instruments, these enable the simultaneous observations of lines of several molecular species and derive more precise relative abundances in the atmosphere of the comets. Figure 1 shows the sample of comets and

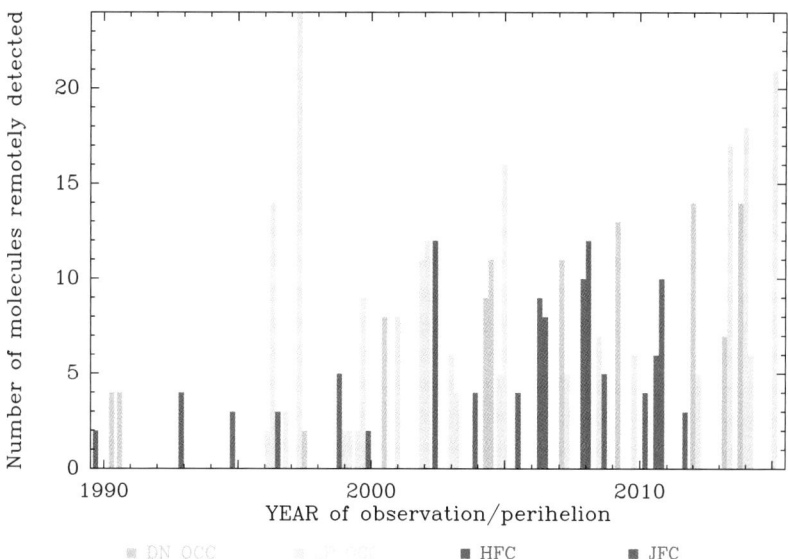

Figure 1. Number of molecules detected at infrared to radio wavelengths in each comet since 1990. Jupiter-family comets are in red, Halley-family comets in dark blue, long-period dynamically old and new comets are in light blue and green, respectively.

number of molecules observed in the IR to radio domains. We can notice that there is a deficiency in detections of a large number of molecules in Jupiter-family comets, due to their average lower activity level.

3. Molecules in cometary comae

Thanks to the increased sensitivity of the instruments, we now regularly observe a growing number of molecules in each comet, which will help to compare comets and try to infer some taxonomic grouping. In 1997, about 25 molecules were detected in the great comet Hale-Bopp. It has been since possible to observe all these molecules in at least one more comet and even two new cometary molecules were recently identified in comet C/2014 Q2 (Lovejoy) (Biver *et al.* 2015). Table 1 provides the list of the 27 molecules (other than H_2O) identified in comets and the range of abundances relative to water that has been measured.

4. Molecules with an unusual behavior

CO is one of the most volatile cometary molecule and can escape from the nucleus far from the Sun, beyond the distance at which water can sublime. So its abundance relative to water must be investigated with caution as comets initially rich in CO might become CO-depleted, as this is likely the case for Jupiter-family comets. In 2011–2012 comet C/2009 P1 (Garradd) actually showed a CO/H_2O ratio varying with time independently of the heliocentric distance (Feaga *et al.* 2014). CO_2, only observed from space, might also behave the same way and is also overabundant in comae far from the Sun (Ootsubo *et al.* 2012). HNC is unlikely a parent cometary molecule but likely produced in the coma as suggested by maps (Cordiner *et al.* 2014) and the strong dependence of its abundance with heliocentric distance (Lis *et al.* 2008). Therefore HNC cannot be considered for taxonomic studies.

Table 1. Range of abundances relative to water (in %) of molecules detected in comets.

Abundances relative to water (%), technique used: U = ultraviolet, I = infrared, R = radio (20–600 GHz)								
CHO molecules (~ 4%)			Nitrogenous molecules (~ 1%)			Sulfureted molecules (~ 1.5%)		
CH_3OH	0.6–6.2	I,R	HCN	$0.08-0.25^a$	R,I^a	H_2S	0.13–1.5	R
H_2CO	0.13–1.4	I,R	HNC	0.002–0.035	R	OCS	0.03–0.40	I,R
HCOOH	0.028–0.18	R	HNCO	0.009–0.08	R	H_2CS	0.009–0.09	R
$(CH_2OH)_2$	0.07–0.35	R	CH_3CN	0.008–0.036	R	CS	0.02–0.20	U, R
$HCOOCH_3$	0.07–0.08	R	HC_3N	0.002–0.068	R	SO	0.04–0.30	R
CH_3CHO	0.047–0.08	R	NH_2CHO	0.008–0.021	R	NS	0.006–0.012	R
CH_2OHCHO	0.016	R	NH_3	0.3–0.7	I,R	S_2	0.001-0.25	U
C_2H_5OH	0.12	R						
C-O molecules			hydrocarbons (~ 2%)					
CO	0.2–23	U,I,R	CH_4	0.12–1.5	I			
CO_2	2.5–30	U,I	C_2H_6	0.14–2.0	I			
			C_2H_2	0.04–0.5	I			

^{a}: HCN is also commonly observed in the infrared, but abundances tend to be 2–3 times higher. References to individual measurements and previous reviews can be found in the Reference section.

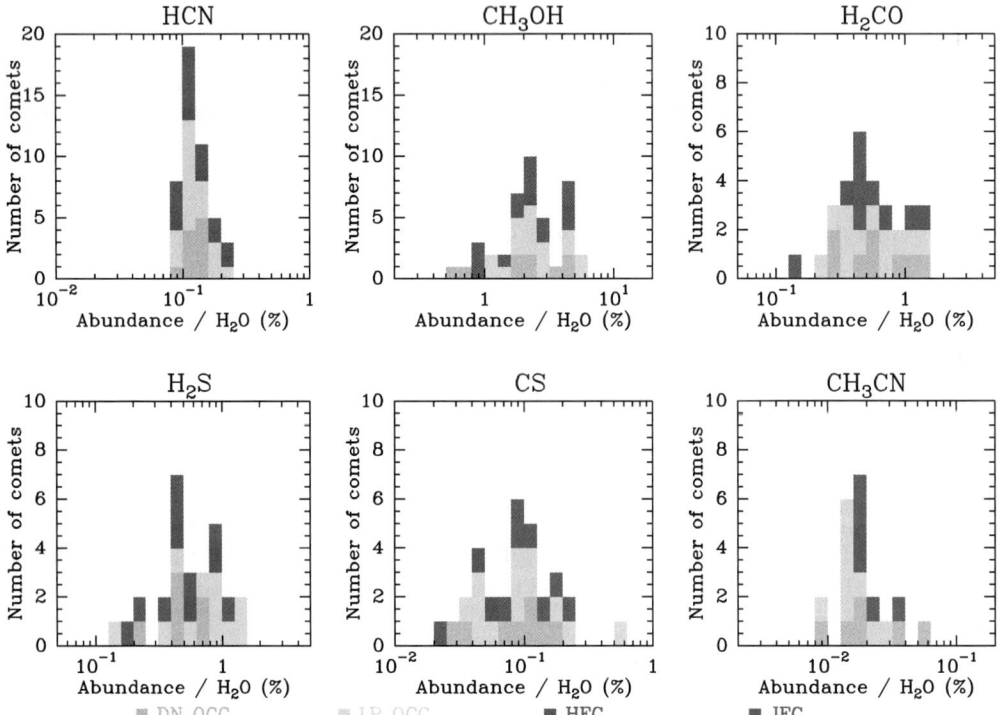

Figure 2. Histograms of the abundances relative to water (in %) of molecules observed in more than 20 comets. Jupiter family comets are in red, Halley family comets in dark blue, long period dynamically old and new are in light blue and green respectively.

5. Composition diversity

The ultimate objective of these study would be to discern grouping of comets according to their composition. Our initial attempt based on a dozen of comets and 6 molecules observed in the radio and infrared did not yield any statistically very significant subgroups. The only exception is the two fragments B and C of comet 73P observed in 2006 which revealed a similar composition (Dello Russo *et al.* 2007). This is on the other hand an indication that comets might be homogeneous in composition. Figures 2,3 show the

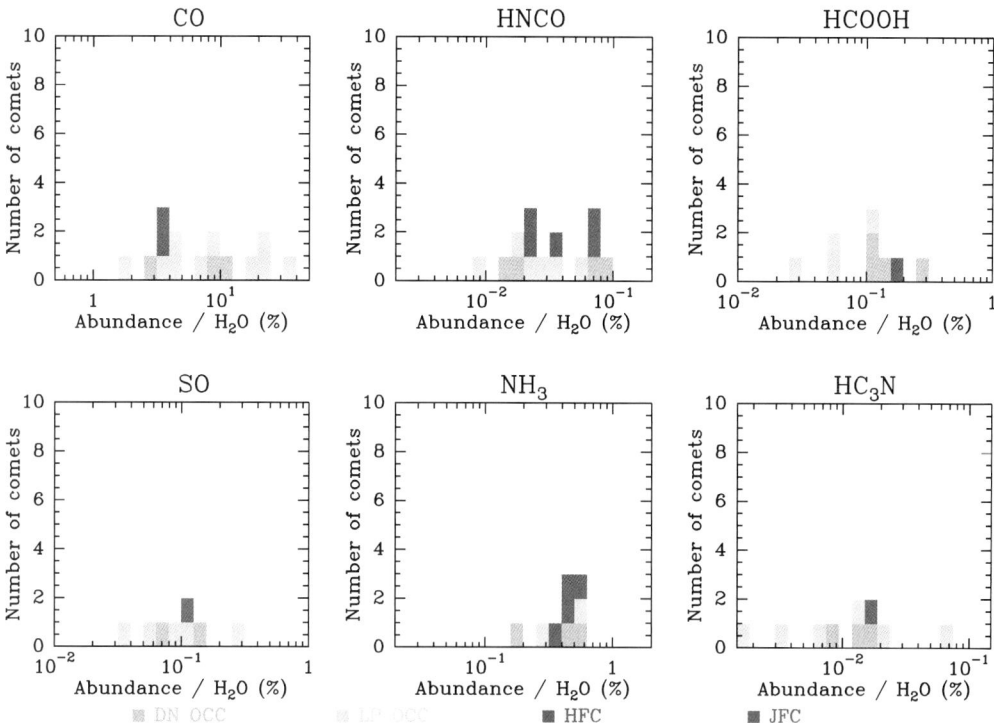

Figure 3. Same as Fig. 2: five other parent molecules and SO (being assumed to be produced from the photodissociation of SO_2) observed at radio wavelengths in more than 8 comets.

dispersion of the abundances for 12 molecules in the full sample of comets observed until 2015. The various colors correspond to different dynamical groups. When the sample gets larger (e.g. molecules of Fig. 2) the distribution of abundances seem to get closer to a Gaussian distribution with no grouping of comets according to their dynamical origin. There is a clear dispersion in abundances, which can be larger than one order of magnitude, but it looks more like a continuum is present in all dynamical class of comets. This is consistent with the dynamical calculations in the frame of the Nice model which suggest that both the Oort cloud and the scattered disk of the Kuiper belt were populated with comets formed in the same regions of the Solar System (Brasser & Morbidelli 2013).

References

A'Hearn, M. F., Millis, R. L., Schleicher, D. G., Osip, D. J., & Birch, P. V. 1995, *Icarus*, 118, 223–270
Agúndez, M., Biver, N., Santos-Sanz, P., Bockelée-Morvan, D., & Moreno R. 2014, *Astron. Astrophys.*, 564, L2
Biver, N., Bockelée-Morvan, D., Crovisier, J., et al. 1999, *Astron. J.*, 118, 1850
Biver, N., Bockelée-Morvan, D., Crovisier, J., et al. 2000, *Astron. J.*, 120, 1554
Biver, N., Bockelée-Morvan, D., Crovisier, J., et al. 2006, *Astron. Astrophys.*, 449, 1255
Biver, N., Bockelée-Morvan, D., Boissier, J., et al. 2009, *Icarus*, 187, 253–271
Biver, N., Bockelée-Morvan, D., Colom, P., et al. 2011, *Astron. Astrophys.*, 528, A142
Biver, N., Crovisier, J., Bockelée-Morvan, D., et al., 2012, *Astron. Astrophys.*, 539, A68
Biver, N., Bockelée-Morvan, D., Crovisier, J., et al. 2014, *Astron. Astrophys.*, 566, L5
Biver, N., Bockelée-Morvan, D., Moreno, R., et al. 2015, *Science Advances*, 23 october 2015

Bockelée-Morvan, D., Crovisier, J., Mumma, M. J., & Weaver, H. A. 2005, *Comets II*, ed. M. C. Festou, H. U. Keller, & H. A. Weaver (Tucson, AZ: Univ. Arizona Press), 391–423
Bockelé-Morvan, D., Hartogh, P., Crovisier, J. , *et al.* 2010, *Astron. Astrophys.*, 518, L49
Brasser, R. & Morbidelli, A. 2013, *Icarus*, 225, 40–49
Cochran, A. L., Levasseur-Regourd, A.-C., Cordiner, M., *et al.* 2015, *Space Science Reviews*,
Cordiner, M. A., Remijan, A. J., Boissier, J. *et al.* 2014, *Astrophy. J. Lett.*, 792, L2
Crovisier, J., Biver, N., Bockelée-Morvan, D., & Colom, P. 2009, *Planet. Space Sci.*, 57, 1162–1174
Crovisier, J., Biver, N., Bockelée-Morvan, D., *et al.* 2009, *Earth, Moon, and Planets*, 105, 267–272
Dello Russo, N., Mumma, M. J., DiSanti, M. A., *et al.* 2006, *Icarus*, 184, 255–276
Dello Russo, N., Vervack, R. J., Jr., Weaver, H. A., *et al.* 2007, *Nature*, 448, 172–175
Dello Russo, N., Vervack, R. J., Weaver, H. A., *et al.* 2008, *Astrophys. J.*, 680, 793–802
Dello Russo, N., Vervack, R. J., Weaver, H. A., *et al.* 2009, *Astrophys. J.*, 703, 187–197
Dello Russo, N., Vervack, R. J., Lisse, C. M., *et al.* 2011, *Astrophys. J. Lett.*, 734, L8
DiSanti, M. A., Villanueva, G. L., Milam, S. N. *et al.* 2009, *Icarus*, 203, 589–598
DiSanti, M. A., Bonev, B. P., Villanueva, G. L., & Mumma, M. J. 2013, *Astrophy. J.*, 763, A1
Feaga, Lori M.; A'Hearn, Michael F.; Farnham, Tony L.; *Astron. J.*,147, A24
Gibb, E. L., Bonev, B. P., Villanueva, G. L., *et al.* 2012, *Astrophys. J.*, 750, A102
Hartogh, P., Lis, D. C., Bockelée-Morvan, D., *et al.* 2011, *Nature*, 478, 218
Kawakita, H., Kobayashi, H., Dello Russo, N, *et al.* 2013, *Icarus*, 222, 723–733
Kawakita, H., Dello Russo, N., Vervack, R., Jr., *et al.* 2014, *Astrophys. J.*, 788, A110
Kobayashi, H., Bockelée-Morvan, D., Kawakita, H., *et al.* 2010, *Astron. Astrophys.*, 509, A80
Lis, D. C., Bockelée-Morvan, D., Boissier, J., *et al.* 2008, *Astrophy. J.*, 675, 931–936
Mumma, M. J., DiSanti, M. A., Dello Russo, N., *et al.* 2003, *Adv. Space Res.*, 31, 2563–2575
Ootsubo, T., Kawakita, H., Hamada, S. *et al.* 2012, *Astrophy. J.*, 752, A15
Paganini, L., Mumma, M. J., Bonev, B. P. *et al.* 2012, *Icarus*, 218, 644–653
Paganini, L., Mumma, M. J., Villanueva, G. L. *et al.* 2012, *Astrophy. J. Lett.*, 748, L13
Paganini, L., DiSanti, M. A., Mumma, M. J. *et al.* 2014, *Astron. J.*, 147, A15
Paganini, L., Mumma, M. J., Villanueva, G. L. *et al.* 2014, *Astrophy. J.*, 791, A122
Radeva, Y. L., Mumma, M. J., Bonev, B. P., *et al.* 2010, *Icarus*, 206, 764–777
Radeva, Y. L., Mumma, M. J., Villanueva, G. L., *et al.* 2013, *Icarus*, 223, 298–307
Villanueva, G. L., Mumma, M. J., DiSanti, M. A., *et al.* 2011, *Icarus*, 216, 227–240

Measuring the Distribution and Excitation of Cometary CH$_3$OH Using ALMA

M. A. Cordiner[1], S. B. Charnley[1], M. J. Mumma[1], D. Bockelée-Morvan[2], N. Biver[2], G. Villanueva[1], L. Paganini[1], S. N. Milam[1], A. J. Remijan[3], D. C. Lis[6], J. Crovisier[2], J. Boissier[4], Y.-J. Kuan[5] and I. M. Coulson[7]

[1] NASA Goddard Space Flight Center, 8800 Greenbelt Road, Greenbelt, MD 20771, USA.
email: martin.cordiner@nasa.gov
[2] LESIA, Observatoire de Paris, CNRS, UPMC, Université Paris-Diderot, Meudon, France.
[3] National Radio Astronomy Observatory, Charlottesville, VA 22903, USA.
[4] IRAM, 300 Rue de la Piscine, 38406 Saint Martin d'Heres, France.
[5] National Taiwan Normal University, Taipei 116, Taiwan, ROC.
[6] LERMA, Observatoire de Paris, PSL Research University, CNRS, Sorbonne Universités, UPMC Univ. Paris 06, F-75014, Paris, France.
[7] East Asian Observatory, Hilo 96720, USA.

Abstract. The Atacama Large Millimeter/submillimeter Array (ALMA) was used to obtain measurements of spatially and spectrally resolved CH$_3$OH emission from comet C/2012 K1 (PanSTARRS) on 28-29 June 2014. Detection of 12-14 emission lines of CH$_3$OH on each day permitted the derivation of spatially-resolved rotational temperature profiles (averaged along the line of sight), for the innermost 5000 km of the coma. On each day, the CH$_3$OH distribution was centrally peaked and approximately consistent with spherically symmetric, uniform outflow. The azimuthally-averaged CH$_3$OH rotational temperature (T_{rot}) as a function of sky-projected nucleocentric distance (ρ), fell by about 40 K between $\rho = 0$ and 2500 km on 28 June, whereas on 29 June, T_{rot} fell by about 50 K between $\rho = 0$ km and 1500 km. A remarkable (~ 50 K) rise in T_{rot} at $\rho = 1500$-2500 km on 29 June was not present on 28 June. The observed variations in CH$_3$OH rotational temperature are interpreted primarily as a result of variations in the coma kinetic temperature due to adiabatic cooling, and heating through Solar irradiation, but collisional and radiative non-LTE excitation processes also play a role.

Keywords. Comets: individual (C/2012 K1 (PanSTARRS), techniques: interferometric, submillimeter, molecular processes

---

1. Introduction

Comets are believed to have formed around 4.5 Gyr ago and contain ice, dust and debris left over from the formation of the Solar System. Their compositions can therefore reveal information on the physical and chemical properties of the protosolar disk and prior interstellar cloud. Much of our knowledge on cometary compositions comes from remote (ground-based) observations, for which relatively low angular resolution and spatial coverage limits the amount of information that can be obtained. In particular, there is a lack of understanding concerning the physical and chemical structure of the near-nucleus coma, at distances less than a few thousand kilometres from the nucleus.

The first cometary observations with ALMA were reported by Cordiner *et al.* (2014), who measured the distributions of HCN, HNC and H$_2$CO in the inner comae of comets C/2012 F6 (Lemmon) and C/2012 S1 (ISON). By virtue of its large abundance in comets and its complex energy level structure, methanol (CH$_3$OH), is the most readily-detectable molecule for probing the coma temperature at radio/sub-mm wavelengths. Here, we

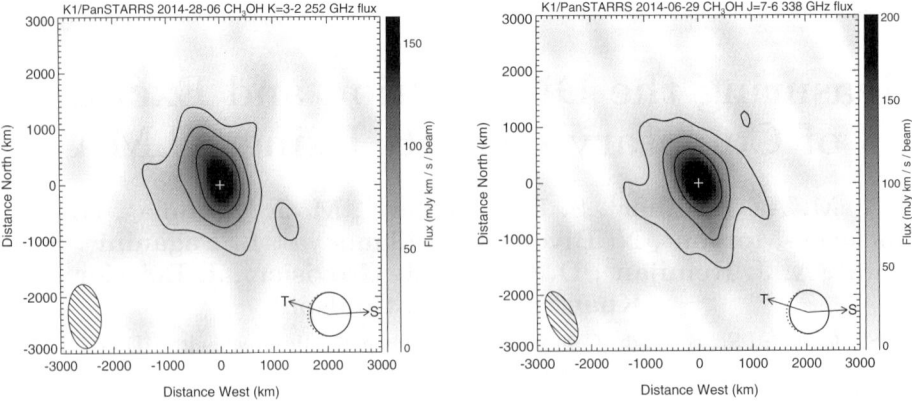

Figure 1. ALMA flux maps of CH_3OH in comet C/2012 K1 (PanSTARRS) observed 2014-06-28 at 252 GHz (left) and 2014-06-29 at 338 GHz (right). White crosses indicate the emission peaks, which are employed as the origin of the coordinate axes. Contours are plotted at 3σ intervals, where σ is the RMS noise in each map. The PSF FWHM are shown lower left. Direction of Sun (S) and orbital trail (T) are indicated lower right along with the illumination phase.

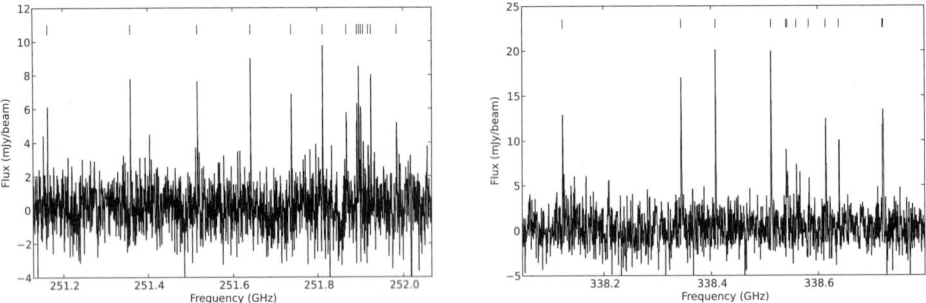

Figure 2. CH_3OH spectra observed on 2014-06-28 of the $K = 3-2$ band (left) and on 2014-06-29 of the $J = 7-6$ band (right). Ticks indicate detected CH_3OH lines. These spectra were averaged over a $0.1''$-thick annulus centered on the nucleus, with an inner radius of $0.2''$

briefly summarise some early results that exploit the unprecedented resolution and sensitivity of ALMA to provide new information on the distribution and temperature of CH_3OH in the inner coma of comet C/2012 K1 (PanSTARRS).

2. Observations

ALMA observations of C/2012 K1 (PanSTARRS) were obtained on 2014-06-28 19:07-20:05 and 2014-06-28 17:37-18:26 at a heliocentric distance $r_H = 1.43$ AU and geocentric distance $\Delta = 1.97$ AU. The CH_3OH $K = 3-2$ band near 251.9 GHz was observed on 28 June and the $J = 7-6$ band near 338.5 GHz was observed on 29 June. Atmospheric conditions were outstanding throughout (with zenith PWV < 0.4 mm). Thirty 12-m antennae, with baseline lengths 20-650 m, resulted in an angular resolution of $0.80'' \times 0.43''$ at 252 GHz and $0.71'' \times 0.33''$ at 338 GHz, with 488 kHz spectral resolution. The data were flagged, calibrated and imaged using standard CASA routines (see Cordiner et al. 2014). Twelve individual CH_3OH emission lines (originating from the $K = 3$ rotational level, with $\Delta K = -1$, $\Delta J = 0$) were detected on 28 June and fourteen lines (originating from the $J = 7$ level, with $\Delta J = -1$, $\Delta K = 0$) were detected on 29 June.

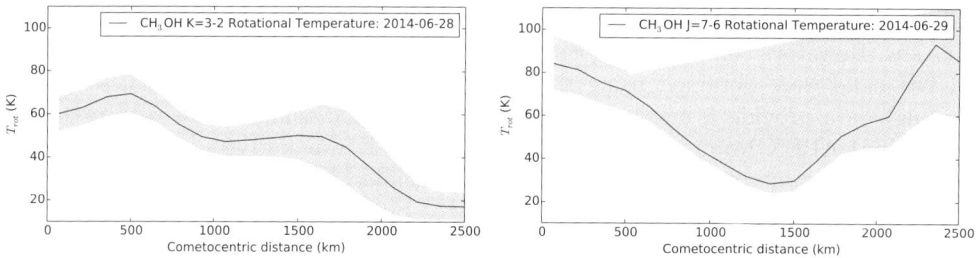

Figure 3. Azimuthally-averaged temperature profiles for the $K = 3 - 2$ (252 GHz) band (left) and the $J = 7 - 6$ (338 GHz) band (right). Statistical 1σ error envelopes are shown.

3. Results

Fig. 1 shows the observed CH_3OH flux maps integrated over the strongest lines detected in each band. To first order (given the noise), these ALMA maps are consistent with azimuthal symmetry about the peak, and can be modeled under the assumption of an isotropic outflow from the nucleus (Cordiner et al. 2014). To obtain the highest spectroscopic signal-to-noise ratio as a function of radius, the spectral data cubes were azimuthally averaged about the emission peak, allowing for a significant improvement in sensitivity (particularly towards larger radii). The data were binned into a series of $0.1''$-thick annuli centered on the emission peak, and the average flux was taken in each annulus, resulting in azimuthally-averaged spectra ($\bar{S}_\nu$) as a function of sky-projected radius ρ from the optocenter. The $\bar{S}_\nu$ for $\rho = 0.25''$ (≈ 360 km) are shown in Fig. 2.

From the observed emission line strengths, an excitation diagram technique was used to derive the CH_3OH rotational temperature (T_{rot}) as a function of ρ, following a similar method to that described by Disanti et al. (2006). Integrated line intensities (and their statistical errors) for each annulus were first converted to brightness temperature units (K km s^{-1}) using the Rayleigh-Jeans approximation. Then using the standard (optically thin) relationship between brightness temperature and column density (e.g. Cummins et al. 1986), the column density (N_i) corresponding to each line flux ($S_i dv$) was plotted as a function of upper-state energy (E_i), and the gradient (m) of this diagram was derived. Assuming LTE, the excitation temperature for which $m = 0$ represents T_{rot} (the mean rotational temperature along the line of sight), and the error on T_{rot} is calculated from $\sigma_m dT_{rot}/dm$. This technique has the advantage over the conventional rotational diagram method that blended lines can be included through use of the weighted mean energy ($<E_i>$) of the blended transitions. Further details of this analysis will be provided in a future article (Cordiner et al., in prep.).

4. Discussion and Conclusion

Fig. 3 shows the azimuthally-averaged temperature profiles as a function of sky-projected distance from the emission peak, for both sets of CH_3OH observations. Both datasets show a general trend for decreasing rotational temperatures as a function of distance between $\rho = 0$ and 1500 km, although the decrease was much slower on June 28th than June 29th. In fact, on June 28th (for the $K = 3 - 2$ band), a possible increase in T_{rot} occurred (from 60.2 ± 7.8 K to 69.4 ± 8.6 K) between $\rho = 0$ and 500 km. This was followed by a relatively slow decrease with radius to 17.6 ± 6.4 K at $\rho = 2500$ km. Conversely, on June 29th (for the $J = 7 - 6$ band), there was a rapid decrease in T_{rot} from 84.0 ± 2.4 K to 28.6 ± 4.2 K between $\rho = 0$ and 1400 km, followed by a significant increase to 90 ± 30 K at $\rho \approx 2400$ km.

The maximum radial extent of the CH$_3$OH flux detected by ALMA was approx. 2500 km. However, contemporaneous position-switched observations of the 338 GHz CH$_3$OH band were obtained using the APEX telescope on June 29th. These (single-pointing) data provide a measure of the mean rotational temperature within the 18″ (26,000 km) APEX beam. A value of $T_{rot} = 20 \pm 7$ K was derived, which shows that over the larger values of ρ probed by APEX, the temperature was significantly less than in the inner few thousand km probed by ALMA. Such variations in temperature with distance from the nucleus have been observed previously by long-slit IR spectroscopy (*e.g.* Bonev *et al.* 2013), and are predicted by theoretical studies due to adiabatic cooling of the outflowing gas (see Rodgers *et al.* 2005, and references therein).

Interpretation of CH$_3$OH rotational temperatures is non-trivial due to the low density of the coma, as demonstrated by Bockelée-Morvan *et al.* (1994). Radiative cooling combined with decreased collision rates as a function of radius results in departure from local thermodynamic equilibrium (LTE), such that with increasing ρ, T_{rot} falls increasingly below the coma kinetic temperature T_{kin}. Detailed excitation modelling (including radiative pumping, cooling and collisions between CH$_3$OH, H$_2$O and electrons), shows that $T_{rot}(J = 7-6)$ falls more rapidly as a function of ρ than $T_{rot}(K = 3-2)$, due to the larger radiative relaxation rates of the $J = 7$ levels. On both dates, however, non-LTE effects alone are insufficient to explain the observed temperature drops with increasing ρ, so these are attributed to adiabatic cooling of the coma. The increasing temperature with distance between 1500-2500 km on June 29th is difficult to explain, but a possible source of heating may be through the sublimation of icy grains in the coma, heated by Solar radiation (see Fougere *et al.* 2012).

As a caveat to our analysis, the ALMA observations are not sensitive to large-scale flux from the outer coma (on angular scales $\gtrsim 5''$ or $\rho \gtrsim 4000$ km), which could bias our derived $T_{rot}(\rho)$ curves. The magnitude of this effect will be quantified through detailed simulations in a future article, but the optimal method to deal with this problem will be through direct (line by line) modelling of the interferometric visibilities.

Our ALMA observations of CH$_3$OH emission at millimeter and submillimeter wavelengths have permitted the first 'instantaneous', spatially-resolved 2-D measurements of rotational temperatures in the inner few thousand kilometers (sky-projected distance) of a cometary coma. Large variations in the CH$_3$OH rotational temperature in C/2012 K1 (Panstarrs) over distances ~ 1000 km are likely due to variations in the coma kinetic temperature, the cause of which will be investigated in more detail in a future article. This study demonstrates that spatial temperature variations may need to be considered when deriving coma molecular abundances from spectral line data. Further high resolution observations and modelling are required in order to better understand the coma thermal physics and molecular excitation, and to assist in the determination of more accurate cometary compositions.

References

Bockelée-Morvan, D., Crovisier, J., Colom, P., & Despois, D. 1994, *A&A*, 287, 647
Bonev, B. P., Villanueva, G. L., Paganini, L. *et al.* 2013, *Icarus*, 222, 740
Cordiner, M. A., Remijan, A. J., Boissier, J. *et al.* 2014, *ApJ*, 792, L2
Cummins, S. E., Linke, R. A., & Thaddeus, P. 1986, *ApJS*, 60, 819
DiSanti, M. A., Bonev, B. P., Magee-Sauer, K. *et al.* 2006, *ApJ*, 650, 470
Rodgers, S. D., Charnley, S. B., Huebner, W. F., & Boice, D. C. 2005, Comets II, eds. M. C. Festou, H. U. Keller & W. A. Weaver, Univ. Arizona, Tucson, p505
Fougere, N., Combi, M. R., & Tenishev, V. 2012, *Icarus*, 221, 174

Active Asteroids: Main-Belt Comets and Disrupted Asteroids

Henry H. Hsieh[1,2]

[1] Academia Sinica Institute of Astronomy & Astrophysics
11F, Astronomy-Mathematics Building, National Taiwan University
No. 1, Sec. 4, Roosevelt Road, Taipei 10617, Taiwan
email: hhsieh@asiaa.sinica.edu.tw

[2] Planetary Science Institute
1700 East Fort Lowell Road, Suite 106
Tucson, Arizona 85719, United States of America
email: hhsieh@psi.edu

Abstract. The study of active asteroids has attracted a great deal of interest in recent years since the recognition of main-belt comets (which orbit in the main asteroid belt, but exhibit comet-like activity due to the sublimation of volatile ices) as a new class of comets in 2006, and the discovery of the first disrupted asteroids (which, unlike MBCs, exhibit comet-like activity due to a physical disruption such as an impact or rotational destabilization, not sublimation) in 2010. In this paper, I will briefly discuss key areas of interest in the study of active asteroids.

Keywords. comets: general; minor planets, asteroids; astrobiology; methods: n-body simulations; methods: numerical; techniques: photometric; techniques: spectroscopic; surveys

1. Background

The first object in the main asteroid belt observed to eject mass like a comet (133P/Elst-Pizarro) was discovered in 1996. Despite initial debate over whether sublimation or an impact event could best explain 133P's activity, impact scenarios were eventually ruled out by observations of the object's reactivation in 2002, leaving sublimation as the most plausible explanation (Hsieh *et al.* 2004). Main-belt comets (MBCs), which exhibit comet-like mass loss as the result of the sublimation of volatile ice even though they occupy orbits in the main asteroid belt, and of which 133P was the first known example, were recognized as a new class of comets ten years later in 2006 (Hsieh & Jewitt, 2006).

Perhaps the most significant shift in this field occurred in 2010 with the discoveries of the first "disrupted asteroids", P/2010 A2 and (596) Scheila, whose activity was determined to be due to impact disruption, where a collision by another asteroid produces a visible ejecta cloud, or rotational destabilization, where an asteroid rotates faster than the limit at which self-gravity and internal cohesion can prevent mass loss due to centrifugal forces (e.g., Jewitt *et al.* 2010, 2011; Snodgrass *et al.* 2010; Bodewits *et al.* 2011). As the term "main-belt comet" was initially chosen to refer to objects that are physically "cometary" (i.e., containing sublimating ice), the use of a new term, "active asteroids" (cf. Jewitt 2012), was adopted to refer to all observationally comet-like objects on asteroid-like orbits, regardless of the dust ejection mechanisms involved.

Dynamically, active asteroids have orbits with Tisserand parameters with respect to Jupiter of $T_J > 3$, where $T_J > 3$ for most main-belt asteroids, and $T_J < 3$ for classical comets (Kresák, 1972), and have semimajor axes smaller than that of Jupiter, i.e., $a \leqslant a_J$. While most active asteroids are found in the main asteroid belt, use of T_J as the formal dynamical criterion for identifying active asteroids means that a few near-Earth objects with $T_J > 3$ that exhibit comet-like activity are also included.

MBCs have attracted interest due to their potential use as tracers of the ice content of the inner solar system (cf. Hsieh 2014). They may also be useful for investigating hypotheses that objects from the main asteroid belt may have played a significant role in the primordial delivery of water to the terrestrial planets (cf. Morbidelli *et al.* 2000, 2012; Mottl *et al.* 2007). Meanwhile, rotational and impact disruptions of asteroids have been extensively studied using numerical models, laboratory experiments, and large statistical studies (e.g., Holsapple *et al.* 2002; Bottke *et al.* 2005; Walsh *et al.* 2012), but the discoveries of real disrupted asteroids has opened up new opportunities for observationally constraining these models, as well as gaining unique insights into the internal structure and physical properties of these objects (e.g., Hirabayashi *et al.* 2014).

In this proceedings paper, I highlight key areas of interest in this rapidly growing and diverse field of research. For more detailed discussions of active asteroids, the reader is referred to reviews by Jewitt (2012) and Jewitt *et al.* (2015b). For more details of MBCs in particular, the reader is referred to Bertini (2011) and Hsieh (2015), the related proceedings paper from IAU Symposium 318, which was also held as part of the XXIXth IAU General Assembly. Discussions of observational diagnostics for distinguishing MBCs from disrupted asteroids can be found in Hsieh *et al.* (2012a) and Jewitt *et al.* (2015b).

2. Main-Belt Comets

To date, sublimation-driven activity has actually only been inferred for all objects currently considered to be MBCs. Despite many attempts using ground-based 8-10 m telescopes and the space-based *Herschel* telescope, gas products from sublimation have not yet been successfully spectroscopically confirmed for a MBC. Most ground-based observations have searched for CN emission at 3889Å (Jewitt *et al.* 2009, 2014, 2015a; Hsieh *et al.* 2012b, 2012c, 2013; Licandro *et al.* 2011, 2013), finding 3σ upper limit production rates of $Q_{CN} \sim 10^{21} - 10^{23}$ mol s^{-1}, from which water production rates of $Q_{H_2O} < 10^{24} - 10^{26}$ were inferred, usually assuming JFC-like Q_{CN}/Q_{H_2O} ratios (e.g., A'Hearn *et al.* 1995). Hypervolatile species like CN may be more depleted than H_2O in main-belt objects compared to JFCs though (Prialnik & Rosenberg 2009). As such, the real upper limit water production rates implied by these CN limits could be much higher. *Herschel* observations directly targeting the 557 GHz $1_{10} - 1_{01}$ ground state rotational transition of H_2O did set limits of $Q_{H_2O} < 4 \times 10^{25}$ mol s^{-1} for 176P/LINEAR and $Q_{H_2O} < 8 \times 10^{25}$ mol s^{-1} for P/2012 T1 (de Val-Borro *et al.* 2012; O'Rourke *et al.* 2013). No dust emission was observed for 176P at the time though (Hsieh *et al.* 2014), and P/2012 T1's activity was well past its peak (Hsieh *et al.* 2013), so their water production rates may have both been higher during periods of stronger activity.

Detection of sublimation or even ice on a MBC observed to exhibit visible dust emission is challenging due to the faintness of both MBCs and their associated activity, and the difficulty of scheduling observations at times of peak activity when the chances of detecting sublimation are maximized. Water vapor has been detected from dwarf planet (1) Ceres (Küppers *et al.* 2014) and surface ice has been detected on large main-belt asteroids (24) Themis and (90) Antiope (Rivkin & Emery 2010; Campins *et al.* 2010; Hargrove *et al.* 2015), but these objects are much larger (diameters of $D > 100$ km) and brighter than any of the km-scale MBCs, and visible dust emission has never been observed for any of them. Meanwhile, the point of peak activity is impossible to immediately determine for newly discovered MBCs, meaning that deep spectroscopic observations of a new MBC can easily be conducted too early or too late. Spectroscopic observations could instead be scheduled for a known MBC during its next perihelion passage, but observability restrictions may apply (e.g., the object could be obscured by the Sun or far from the Earth when activity is expected to peak), the object could exhibit much weaker activity than before, and at least one orbit period must elapse following the first active episode, limiting short-term observing opportunities. As such, thus far, a successful verification of sublimation on a MBC observed to exhibit visible dust emission remains elusive.

3. Disrupted Asteroids

Perhaps the most unambiguous example of an impact-driven dust ejection event is that observed for Scheila in 2010, where numerical dust modeling of radiation pressure acting on a hollow ejecta cone and a down-range plume from an oblique impact by a decameter-sized impactor provided an excellent match to the observed multi-plumed dust structure around the asteroid (Ishiguro *et al.* 2011). Dust emission events for P/2010 A2 and P/2012 F5 (Gibbs) were both initially assumed to be caused by impacts (e.g., Jewitt *et al.* 2010; Snodgrass *et al.* 2010; Stevenson *et al.* 2012), but later analyses suggest that both events may have been caused by rotational destabilization instead (Agarwal *et al.* 2013; Drahus *et al.* 2015). Other objects suspected of being rotationally disrupted have also been identified (e.g. 311P/PANSTARRS, P/2013 R3, and (62412) 2000 SY_{178}; Jewitt *et al.* 2013, 2014; Sheppard & Trujillo 2015), although of these, only (62412) has had its rotation period determined and been confirmed to be a fast-rotator.

Disruption events provide opportunities to probe the physical properties of asteroids in various ways. Impact events like the one experienced by Scheila may alter the surface of the impacted asteroid, depending on the size of the impactor and other details of the impact event, where these changes can potentially then be used to infer properties of the impacted surface material by placing observational constraints of physical models of the impact, as well as providing direct compositional information about the excavated material (e.g., Bodewits *et al.* 2014). In principle, the rates at which impacts occur could also place constraints on the otherwise unobservable population of m- or 10-m-scale main-belt asteroids (cf. Hsieh 2009). By placing constraints on numerical models and laboratory experiments, impact and rotational disruption events can also be used to infer material properties such as the internal strength and porosity of the affected asteroid (e.g., Housen & Holsapple 2003; Holsapple 2009; Hirabayashi 2015). As rotational disruptions may potentially be as important in shaping the size distribution of main-belt asteroids as collisions (Jacobson *et al.* 2014), it will be crucial to understand as much as possible about the practical aspects of this process, where continued observational studies of rotationally disrupted asteroids will be very valuable in this regard.

4. Future Outlook

Despite the separate classifications of MBCs and disrupted asteroids, it should be emphasized that multiple mechanisms could contribute to any given dust emission event. For example, an impact could trigger sublimation by excavating subsurface ice, or dust could be ejected by both weak sublimation and rapid rotation. As such, as more objects are discovered, detailed studies to understand their unique characteristics will be essential before broader generalizations can be made. Otherwise, perhaps the largest obstacle in the study of active asteroids is the extremely small number of known examples. Surveys equipped with effective detection algorithms capable of finding these objects are thus vital to progress in this field (e.g., Waszczak *et al.* 2013; Hsieh *et al.* 2015).

References

A'Hearn, M. F., Millis, R. L., Schleicher, D. G., et al. 1995, *Icarus*, 118, 223-270
Agarwal, J., Jewitt, D., & Weaver, H. 2013, *Astrophys. J.*, 769, 46
Bertini, I. 2011, *Planet. Space Sci.*, 59, 365-377
Bodewits, D., Kelley, M. S., Li, J.-Y., et al. 2011, *Astrophys. J. Letters*, 733, L3
Bodewits, D., Vincent, J.-B., & Kelley, M. S. P. 2014, *Icarus*, 229, 190-195
Bottke, W. F., Jr., Durda, D. D., Nesvorný, D., et al. 2005, *Icarus*, 179, 63-94
Campins, H., Hargrove, K., Pinilla-Alonso, N., et al. 2010, *Nature*, 464, 1320-1321
Drahus, M., Waniak, W., Tendulkar, S., et al. 2015, *Astrophys. J. Letters*, 802, L8
Hargrove, K. D., Emery, J. P., Campins, H., et al. 2015, *Icarus*, 254, 150-156
Hirabayashi, M., Scheeres, D. J., Sánchez, D. P., et al. 2014, *Astrophys. J. Letters*, 789, L12
Hirabayashi, M. 2015, *Mon. Not. R. Astron. Soc.*, 454, 2249-2257
Holsapple, K. A. 2009, *Planet. Space Sci.*, 57, 127-141
Holsapple, K. A., et al. 2002, *Asteroids III*, (Univ. of Arizona Press: Tucson, AZ), 443-462
Housen, K. R. & Holsapple, K. A. 2003, *Icarus*, 163, 102-119
Hsieh, H. H. 2009, *Astron. Astrophys.*, 505, 1297-1310
Hsieh, H. H. 2014, *Proc. IAU Symp.* 293, 212-218
Hsieh, H. H. 2015, *Proc. IAU Symp.* 318, in press
Hsieh, H. H. & Jewitt, D. 2006, *Science*, 312, 561-563
Hsieh, H. H., Jewitt, D., & Fernández, Y. R. 2004, *Astron. J.*, 127, 2997-3017
Hsieh, H. H., Yang, B., & Haghighipour, N. 2012a, *Astrophys. J.*, 744, 9
Hsieh, H. H., Yang, B., Haghighipour, N., et al. 2012b, *Astrophys. J. Letters*, 748, L15
Hsieh, H. H., Yang, B., Haghighipour, N., et al. 2012c, *Astron. J.*, 143, 104
Hsieh, H. H., Kaluna, H. M., Novaković, B., et al. 2013, *Astrophys. J. Letters*, 771, L1
Hsieh, H. H., Denneau, L., Fitzsimmons, A., et al. 2014, *Astron. J.*, 147, 89
Hsieh, H. H., Denneau, L., Wainscoat, R. J., et al. 2015, *Icarus*, 248, 289-312
Ishiguro, M., Hanayama, H., Hasegawa, S., et al. 2011, *Astrophys. J. Letters*, 741, L24
Jacobson, S. A., Marzari, F., Rossi, A., et al. 2014, *Mon. Not. R. Astron. Soc.*, 439, L95-L99
Jewitt, D. 2012, *Astron. J.*, 143, 66
Jewitt, D., Yang, B., & Haghighipour, N. 2009, *Astron. J.*, 137, 4313-4321
Jewitt, D., Weaver, H., Agarwal, J., et al. 2010, *Nature*, 467, 817-819
Jewitt, D., Weaver, H., Mutchler, M., et al. 2011, *Astrophys. J. Letters*, 733, L4
Jewitt, D., Agarwal, J., Weaver, H., et al. 2013, *Astrophys. J. Letters*, 778, L21
Jewitt, D., Agarwal, J., Li, J., et al. 2014, *Astrophys. J. Letters*, 784, L8
Jewitt, D., Agarwal, J., Peixinho, N., et al. 2015a, *Astron. J.*, 149, 81
Jewitt, D., Hsieh, H. H., & Agarwal, J. 2015b, *Asteroids IV* (Univ. of Arizona Press: Tucson, AZ), in press (arXiv:1502.02361)
Kresák, L. 1972, *The Motion, Evolution of Orbits, and Origin of Comets*, 45, 503
Küppers, M., O'Rourke, L., Bockelée-Morvan, D., et al. 2014, *Nature*, 505, 525-527
Licandro, J., Campins, H., Tozzi, G. P., et al. 2011, *Astron. Astrophys.*, 532, A65
Licandro, J., Moreno, F., de León, J., et al. 2013, *Astron. Astrophys.*, 550, A17
Morbidelli, A., Chambers, J., & Lunine, J. I. 2000, *Meteoritics & Planet. Sci.*, 35, 1309-1320
Morbidelli, A., Lunine, J. I., O'Brien, D. P., Raymond, S. N., & Walsh, K. J. 2012, *Ann. Rev. Earth Planet. Sci.*, 40, 251-275
Mottl, M., Glazer, B., Kaiser, R., et al. 2007, *Chemie der Erde - Geochemistry*, 67, 253-282
O'Rourke, L., Snodgrass, C., de Val-Borro, M., et al. 2013, *Astrophys. J. Letters*, 774, L13
Prialnik, D. & Rosenberg, E. D. 2009, *Mon. Not. R. Astron. Soc.*, 399, L79-L83
Rivkin, A. S. & Emery, J. P. 2010, *Nature*, 464, 1322-1323
Sheppard, S. S. & Trujillo, C. 2015, *Astron. J.*, 149, 44
Snodgrass, C., Tubiana, C., Vincent, J.-B., et al. 2010, *Nature*, 467, 814-816
Stevenson, R., Kramer, E. A., Bauer, J. M., et al. 2012, *Astrophys. J.*, 759, 142
de Val-Borro, M., Rezac, L., Hartogh, P., et al. 2012, *Astron. Astrophys.*, 546, L4
Walsh, K. J., Richardson, D. C., & Michel, P. 2012, *Icarus*, 220, 514-529
Waszczak, A., Ofek, E. O., Aharonson, O., et al. 2013, *Mon. Not. R. Astron. Soc.*, 433, 3115-3132

Icy Dwarf Planets: Colored Popsicles in the Outer Solar System

Noemi Pinilla-Alonso[1,2]

[1] Florida Space Institute: 12354 Research Parkway, partnership 1 Building, Suite 214
Orlando, 32826-0650
email: npinilla@ucf.edu

[2] Department of Earth and Planetary Sciences, University of Tennessee.
306 EPS Building, 1412 Circle Dr Knoxville TN 37996-1410

Abstract. We update the list of candidates to be considered by the IAU as dwarf planets using the criterium suggested by Tancredi & Favre (2008). We add here the information collected in the last 10 years (mostly the sizes and albedos by the herschel hey program TNOs Are Cool). We compare the physical characteristics of these candidates with the physical characteristics of the rest of the TNOs. Our goal is to study if there are common physical properties among the candidates that enable the identification of a dwarf planet.

Keywords. solar system: general, icy bodies, Kuiper Belt

---

1. Introduction

In 1992 the discovery of 1992 QB_1 (Jewitt & Luu, 1992) was the trigger of a race to characterize the newly discovered trans-Neptunian belt. The existence of a swarm of "icy asteroids" similar to Pluto orbiting in the outer Solar System, that would be the origin of the short-period comets, had been largely hypothesized but its detection was also an elusive goal (Leonard, 1930, Edgeworth 1943, Kuiper 1951, Fernandez, 1980). This icy belt was for long considered by the planetary scientists as the icy promised land, the largest reservoir of primordial ices in the Solar System (Schmitt *et al.* 1998, Cruikshank 2005).

From 1992 to 2005 about 1000 trans-Neptunian objects and Centaurs were discovered and a lot of "first ever" science was published e.g. 1996 TO_{66}, first ever detection of the water ice bands in a TNO's spectrum (Brown *et al.* 1999); 1998 WW_{31}, first detection of a binary (Veillet *et al.* 2002); first estimation of size and albedo from thermal and visible observations, Varuna (Jewitt *et al.* 2001).

Almost 10 years ago, 2005 was the year of the announcement of another unique discovery: the existence of three large icy objects, (136108) Haumea, (136472) Makemake and (136199) Eris (Brown *et al.* 2005a, Santos-Sanz *et al.* 2005), only comparable in size with Pluto. One year after, the International Astronomical Union revisited the definition of a planet and introduced a new category of objects in the Solar System, the "dwarf planets" (objects large enough to be in hydrostatic equilibrium but not large enough to have cleaned their orbit of other minor bodies (see Resolution 5 and 5b of the XXVI General Assembly of the IAU for more detail on the definition). With only four icy objects at this moment (the three mentioned above plus Pluto) the exclusive club of the icy dwarf planets† is formed by the TNOs at the higher end of the size distribution and hence we may expect that they exhibit unique characteristics.

† Ceres, in the asteroid belt, is another Dwarf Planet. It is not included in this description as we are focusing on icy objects.

By virtue of their size and low surface temperatures, these bodies can probably retain most of their original inventory of ices (Schaller & Brown, 2007; Levy & Podolak, 2009). As a consequence, their visible and near-infrared spectra should show evidences of ices (e.g. nitrogen, methane). Accordingly, the early characterization of the surface composition of these bodies revealed how special they were compared to other TNOs (e.g. Brown *et al.* 2005b, Licandro *et al.* 2006a, Licandro *et al.* 2006b, Pinilla-Alonso *et al.* 2009, Brown *et al.* 2007). In addition, these objects are by far the best candidates to have at the moment or occasionally develop on their orbit around the Sun, a bonded atmosphere that exists in equilibrium with the ices on the surface. This atmosphere may or may no collapse on the periods when the object moves further from the Sun, at temperatures such that all the gases condensate onto the surface (for information regarding the detection of an atmosphere around Eris and Makemake read Sicardy *et al.* 2011; Ortiz *et al.* 2012; for information on Pluto's atmosphere read Lellouch *et al.* 2015 and references therein). Moreover, they show some of the highest geometric albedos in the visible, a characteristic that is highly influenced by the physical characteristics of their surface composition i.e. relative abundance of ices vs. silicates or carbonaceous residues, and size of the particles. Also the accrecional and radiogenic heating for these bodies was likely more than sufficient to have caused their internal differentiation (McKinnon *et al.* 2008).

It is clear that these four giants are peculiar objects what gives the icy dwarf planets an aura of exclusivity. But, are these the only objects in the TNb that, according the IAU definition, can be considered dwarf planets? And if not, which are their physical characteristics? In this paper we review the list of known TNOs and cross it with the most recent estimations of albedo and size to update the list of TNOs candidates to be dwarf planets (CDPs). We also study their main physical characteristics to search for common attributes that can help define other candidates, in the lack of more accurate size estimations.

2. How many icy dwarf planets are in the Solar System?

Tancredi & Favre (2008) review the geophysical criteria to separate a dwarf planet from a regular TNO. They adopt a set of criteria presented as a decision tree. The main idea is to use the data available at that moment to check which minor objects are large enough to overcome the material strength and be in hydrostatic equilibrium. The main parameters affecting the classification are the estimations of the size and shape, and the assumption of a density. Summarizing, their models find a minimum critical size of D = 450 km for a TNO to be considered a dwarf planet. After applying this criterium to the list of TNOs, they propose a list of 12 very probable and six possible TNOs (always depending on an improvement in the accuracy of the shape or size estimation). They also add that there my be several tens up to more than one hundred objects larger than 450 km not yet discovered or with no estimation of their size (Tancredi 2009).

In that respect, a recent work (Brown *et al.* 2015) shows the results of a 7-years survey in search for bright objects covering most of the Northern and Southern hemisphere. The survey probes to be 100% effective in detecting bright objects (V $\lesssim$ 19) beyond 25 AU, but no new discovery is made. According to this complete study, the remaining probability of finding one or two of this kind of objects in the galactic plane, where the background makes detections more difficult, stays bellow 35%.

3. Update of the list of candidates to be dwarf planets

With no new discoveries, the natural extension of this study is to take advantage of the new estimations of the size and shape of the TNOs and use them to apply the Tancredi & Favre (2008) criterium†. In the last years, the Herschel open time key program "TNOs are Cool!" has improved our knowledge of the physical and thermal properties of a large sample of TNOs (Mullet et al. 2010). When put together with the results of the Spitzer Space Telescope (Stansberry et al. 2008) the improvement in the study of the size of the TNOs is large, not only in the number of estimations (119 objects at this moment) but also in their precision. A compilation of the results of this key program can be accessed via internet at the TNOs Are Cool public database: http://public-tnosarecool.lesia.obspm.fr/Published-results.html. It includes 119 TNOS and centaurs. We use these results to determine which TNOs have a diameter above 450km. We use a conservative criterium so we will only consider candidates those whose diameter minus the error is above 450 km. The results are listed in table 1.

We find 25 CDPs (including the four already known). We discard three out of the 12 candidates (2002 TX_{300}, Huya and 1996 TL66) in Tancredi & Favre (2008), and other three (1999 TC_{36}, 1999DE_9 and 2001 QF_{298}) out of the six possible dwarf planets. One could not be compared because there are no new estimations.

3.1. Physical Characteristics and Conclusions

Here we compare the main physical characteristics (table 1; figure 1) of the CDPs with the sample of TNOs included in the TNOs Are Cool database.

Dynamical classification: The absence of cold classical objects and centaurs in the sample is normal considering that these populations contain small objects (Vilenus et al. 2014; Duffard et al. 2014). The absence of scattered objects could be affected by a bias in the "TNOs Are Cool sample", as that number of scattered objects observed is smaller than for the other dynamical groups.

Albedo: The values of the albedo in the visible range from 0.1 to 0.96, with a mean value of 0.21. If we remove from the sample the four known icy dwarf planets, then the maximum is 0.4 and the group of candidates, with a mean value of 0.12, cannot be distinguish from the rest of the sample of TNOs (figure 1), ranging from 0.1 to 0.35, with a mean value of 0.12.

Beaming factor: The values of the beaming factor are also very similar for both groups, with η_{median}=(1.08, 1.11) for the TNOs and the CDPs, respectively (figure 1). The incidence of large values ($\eta \gtrsim 2$), indicative of surfaces with high thermal inertia and low values ($\eta \lesssim 1$), surfaces dominated by roughness effects, is similar among the TNOs and the CDPs.

Binarity: the ratio of binaries detected among the CDPs is 40%. This is double than the ratio of binaries in the rest of the population. However this could be a bias due to the fact that the largest size of the CDPs could produce better data to detect multiple systems (most of them detected from the light curve) making it easier to detract a binary among the CDPs than among the TNOs.

Surface composition: The ratio of clear, tentative and no detections of water ice in the list of CDPs is (37, 21, 42) % respectively, very similar to the ratio in the sample of TNOs studied in Barucci et al. (2011; 36, 24, 40 %) respectively (figure 1). The peculiar surface composition of Eris, Makemake and Pluto has no comparison in the TNb. We can say the same if we consider Haumea and the family of the carbon depleted objects.

† we leave for future consideration the light-curves of the bodies and stick here to the size criterium

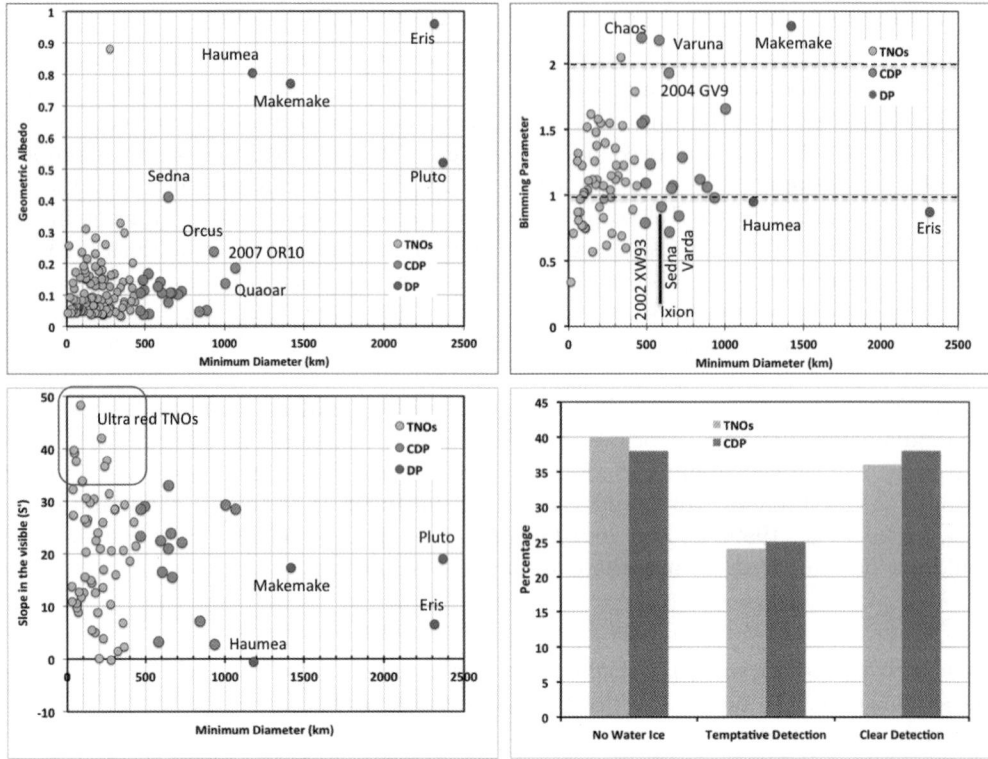

Figure 1. Physical parameters of TNOs, DP and CDPs. See caption of table 1 for references. The distribution of the visible albedo, bumming parameters, color in the visible and detection of water ice for CDP is similar to the distribution for the regular TNOs. None of these parameters could be use to classify a TNO as DP in the absence of a good estimation of size and/or shape. The peculiar characterstics of the DP have no paragon among the TNb.

Neither there is a difference in the slope in the visible of both samples (figure 1). The only possible difference is the absence of extremely red objects among the CDPs, as well as among the DPs, what suggests that the surface of the visible colors of these objects are not dominated by red-dark complex organics.

In conclusion, we have reviewed and compiled here the main physical parameters that can be used to characterize a TNO. We then compare the behavior of two samples, the CDPs (those with $D > 450$ km) and the regular TNOs (all from the list of targets of the TNOs Are Cool program, the ones with a good estimation of the size). We find that, with the exception of the largest ones, the ones classified at the present moment as dwarf planets, there is not a charactenistic, other than the size, that is common among the sample. Neither the albedo, or the surface composition, or the dynamical classification, or the thermal properties (beaming factor) help in classifying a regular TNO into the DP class. Only the binarity incidence shows some differences, but we cannot assure that this is not a bias due to the highest quality of the data used in the detection of these multiple systems among the brightest/largest objects.

The actual list of CDPs includes 25 icy objects (see Table 1). It can be refined on the future using the shape information that can be derived from the study of the light-curves.

Table 1. List of candidate to be dwarf planets and their main physical characteristics

number	name	$D^{(1)}$ - err [km]	$pv^{(1)+err}_{-err}$	Dyn. Class$^{(2)}$	$S'^{(3)}$ [%/1000Å]	Water Ice$^{(4)}$	multiple$^{(5)}$
	Pluto†	2370	0.44 - 0.59	Res	12.3	clear	yes
136199	Eris	2314	$0.96^{0.04}_{-0.04}$	Det	6.5	no	yes
136472	Makemake	1421	$0.77^{0.02}_{-0.02}$	Hot	17.3	no	
136108	Haumea	1181.7	$0.80^{0.06}_{-0.1}$	Hot	-0.61	clear	yes
225088	2007 OR$_{10}$	1070	$0.19^{0.08}_{-0.05}$	SDO	28.5	clear	
50000	Quaoar	1005.6	$0.14^{0.01}_{-0.01}$	Hot	29.2	clear	yes
90482	Orcus	933.6	$0.24^{0.01}_{0.1}$	Plut	2.8	clear	yes
307261	2002 MS$_4$	887	$0.05^{0.04}_{-0.02}$	Hot			
120347	Salacia	842	$0.05^{0.01}_{0.00}$	Hot	4.1		yes
55565	2002 AW$_{197}$	730	$0.11^{0.01}_{-0.01}$	Hot	22.1	no	
174567	Varda	708	$0.10^{0.02}_{-0.02}$	Hot		no	yes
55637	2002 UX$_{25}$	669	$0.11^{0.01}_{-0.01}$	Hot	15.5	no	yes
208996	2003 AZ$_{84}$	660.5	$0.11^{0.02}_{-0.02}$	Plut	23.9	clear	yes
90377	Sedna	648	$0.41^{0.39}_{-0.19}$	IOO	33.0	clear	
90568	2004 GV$_9$	646	$0.08^{0.01}_{-0.01}$	Hot	21.0	no	
145452	2005 RN$_{43}$	606	$0.11^{0.03}_{-0.02}$	Hot	16.5	no	
28978	Ixion	597	$0.14^{0.01}_{-0.01}$	Plut	22.5	temp	
20000	Varuna	582	$0.13^{0.04}_{-0.04}$	Hot	3.2	clear	
	2002 XV$_{93}$	526.2	$0.04^{0.02}_{-0.02}$	Plut			
229762	2007 UK$_{126}$	522	$0.17^{0.06}_{-0.04}$	Det		temp	yes
84522	2002 TC$_{302}$	496.1	$0.12^{0.05}_{-0.03}$	Res	29.0	temp	
78799	2002 XW$_{93}$	492	$0.04^{0.04}_{-0.03}$	Hot			
84922	2003 VS$_2$	488.6	$0.15^{0.06}_{-0.04}$	Plut		clear	yes
120348	2004 TY$_{364}$	472	$0.11^{0.02}_{-0.02}$	Plut	28.4	temp	
19520	Chaos	470	$0.05^{0.03}_{-0.02}$	Hot	23.3	no	

Notes:
† References for Pluto: size and water ice detection Stern *et al.* 2015; S', p_v Lorenzi *et al.* (2015);
^{1} For the exact reference of the Diameter, and albedo estimation use the *TNOs Are Cool public database* ^{2}: Hot: Hot classical; Plut: plutino; Det: detatched; Res: resonant; IOO: Inner Oort object ^{3} For the reference of the S' or the dynamical class we follow the *MBOSS-2 database* ^{4} We considered no water ice, clear and tentative detections, see Barucci *et al.* (2011) ^{5} For multiple systems use Grundy *et al.* webpage: http://www2.lowell.edu/users/grundy/tnbs/status.html where a detailed compilation of the refereed publications can be found.

4. Acknowledgements

NPA acknowledges funding support from the American Astronomical Society through the AAS International travel Grant program. She also acknowledges the SOC of the Focus Meeting 9: *Highlights in the Exploration of Small Worlds*, for the invitation to participate in this meeting.

References

Barucci, M. A., Alvarez-Candal, A. Merlin, F. et al. 2011, *Icarus*, 214, 297
Brown, R. H., Cruikshank, D. P., & Pendleton, Y. 1999, *ApJ*, 519, 101
Brown, M. E., Trujillo, C. A., & Rabinowitz, D. 2005, *IAU Circ.*, 8577, 1
Brown, M. E., Trujillo, C. A., & Rabinowitz, D. L. 2005b *ApJ*, 635, L97
Brown, M. E., Bannister, M. T., & Schmidt, B. P. 2015, *AJ*, 149, id. 69
Cruikshank, D. P. 2005, *Space Sci. Rev.*, 116, 421
Duffard, R., Pinilla-Alonso, N., Santos-Sanz, P. et al. 2014, *A&A*, 564, 17
Edgeworth, K. E. 1943, *Journal of the British Astronomical Association*, 53, 181
Fernandez, J. A. 1980, *MNRAS*, 192, 481
Jewitt, D. & Luu, J. 1993, *Nature*, 362, 730
Jewitt, D., Aussel, H., & Evans, A 2001, *Nature*, 411, 446
Kuiper, G. P. 1951, in *50th Anniversary of the Yerkes Observatory and Half a Century of Progress in Astrophysics*, ed. J. A. Hynek, 357
Lellouch, E., de Bergh, C., & Sicardy, B. 2015, *Icarus*, 246, 268
Leonard, F. C. 1930, *Leaflet of the Astronomical Society of the Pacific*, 1, 121
Levi, A. & Podolak, M 2009, *Icarus*, 202, 681
Licandro, J., Pinilla-Alonso, N., & Pedani, M. 2006, *A&A*, 445, L35
Licandro, J., Grundy, W. M., & Pinilla-Alonso, N 2006, *A&A*, 458, L5
Lorenzi, V., Pinilla-Alonso, N., Licandro, J 2015 *A&A* Accepted
McKinnon, W. B., Prialnik, D., Stern, S. A. & Coradini, A. 2008 *The Solar System Beyond Neptune*, University of Arizona Press, Tucson, 592, 213
Muller, M. Lellouch, E., et al. 2010, *A&A*, 518, L146
Ortiz, J. L., Sicardy, B. Braga-Ribas, F. et al. 2012 *Nature*, 491, 566
Pinilla-Alonso, N., Brunetto, R., Licandro, J. et al. 2009, *A&A*, 496, 547
Santos-Sanz, P., Ortiz, J. L., Aceituno, F. J., Brown, M. E., & Rabinowitz, D. 2005, *IAU Circ.*, 8577, 2
Schaller, E. & Brown, M. E. 2007, *ApJ*, 659, L61
Schmitt, B., De Bergh, C., & Festou, M. (eds) 1998, *Solar System Ices, Astrophysics and Space Science Library*, vol 227
Sicardy, B., Ortiz, J. L., Assafin, M. et al. 2011 *Nature*, 478, 493
Stansberry, J., Grundy, W., Brown, M., et al. 2008 *The Solar System Beyond Neptune*, University of Arizona Press, Tucson, 592, 161
Stern, S. A., Bagenal, F., & Ennico, K. 2015, *Science*, 350, id.aad1815
Tancredi, G. & Favre, S 2008, *Icarus*, 195, 851
Tancredi, G 2009 *IAU 2009: Icy Bodies in the Solar System*, 33
Veillet, C., Parker, J. W., Griffin, I., Marsden, B., Doressoundiram, A., Buie, M., Tholen, D. J., Connelley, M., & Holman, M. J. 2002, *Nature*, 416, 711
Vilenius, E., Kiss, C., Muller, T. et al. 2014, *A&A*, 564, 18

The Chelyabinsk event

Jiří Borovička

Astronomical Institute of the Czech Academy of Sciences, Fričova 298, CZ-25165 Ondřejov,
Czech Republic
email: jiri.borovicka@asu.cas.cz

Abstract. On February 15, 2013, 3:20 UT, an asteroid of the size of about 19 meters and mass of 12,000 metric tons entered the Earth's atmosphere unexpectedly near the border of Kazakhstan and Russia. It was the largest confirmed Earth impactor since the Tunguska event in 1908. The body moved approximately westwards with a speed of 19 km s^{-1}, on a trajectory inclined 18 degrees to the surface, creating a fireball of steadily increasing brightness. Eleven seconds after the first sightings, the fireball reached its maximum brightness. At that point, it was located less than 40 km south from Chelyabinsk, a Russian city of population more than one million, at an altitude of 30 km. For people directly underneath, the fireball was 30 times brighter than the Sun. The cosmic body disrupted into fragments; the largest of them was visible for another five seconds before it disappeared at an altitude of 12.5 km, when it was decelerated to 3 km s^{-1}. Fifty six second later, that ~ 600 kg fragment landed in Lake Chebarkul and created a 8 m wide hole in the ice. Small meteorites landed in an area 80 km long and several km wide and caused no damage. The meteorites were classified as LL ordinary chondrites and were interesting by the presence of two phases, light and dark. More material remained, however, in the atmosphere forming a dust trail up to 2 km wide and extending along the fireball trajectory from altitude 18 to 70 km. The dust then circled the Earth within few days and formed a ring around the northern hemisphere. In Chelyabinsk and its surroundings a very strong blast wave arrived 90 – 150 s after the fireball passage (depending on location). The wave was produced by the supersonic flight of the body and broke $\sim 10\%$ of windows in Chelyabinsk ($\sim 40\%$ of buildings were affected). More than 1600 people were injured, mostly from broken glass. The whole event was well documented by video cameras, seismic and infrasonic records, and satellite observations. The total energy was 500 kT TNT (2×10^{15} J).

Keywords. Meteors, Meteoroids, Asteroids

1. Introduction

It is now widely acknowledged that impacts of cosmic bodies (asteroids and comets) played important role in the history of Earth's life. The most significant impacts, of multikilometer bodies, occur only on geological timescales. The largest impact in modern history was the Tunguska event in Siberia on June 30, 1908 (Vasilyev 1998). The asteroid of a size of about 50 meters exploded 5 – 10 km above the surface and its radiation ignited the forest beneath. The blast wave arrived somewhat later, ceased the fire but flattened the forest on an area of 2150 km^2. The total energy of the event was estimated about 15 MT TNT (1 kT TNT = 4.184×10^{12} J). For comparison, the larges thermonuclear test ever conducted (in the USSR in 1961) had an energy of 50 MT TNT, while the Hiroshima bomb had only 15 kT TNT. More recently detected impacts, such as those near Marshall Islands in 1994 (McCord et al. 1995) and near Sulawesi, Indonesia, in 2009 (Silber et al. 2011) had an energy of the order of tens of kilotons. There was, nevertheless, one unconfirmed event of the energy of 1.5 MT TNT over Indian Ocean in 1963 (Silber et al. 2009).

On February 15, 2013, the citizens of the Russian city Chelyabinsk of more than one

Table 1. Energy estimates from various types of data.

Method	Energy (kt TNT)	Reference
Seismic	430	Brown et al. (2013)
Infrasound	600	"
US government sensors	530	"
Video-derived light curve	> 470	"
Infrasound	570	Popova et al. (2013)

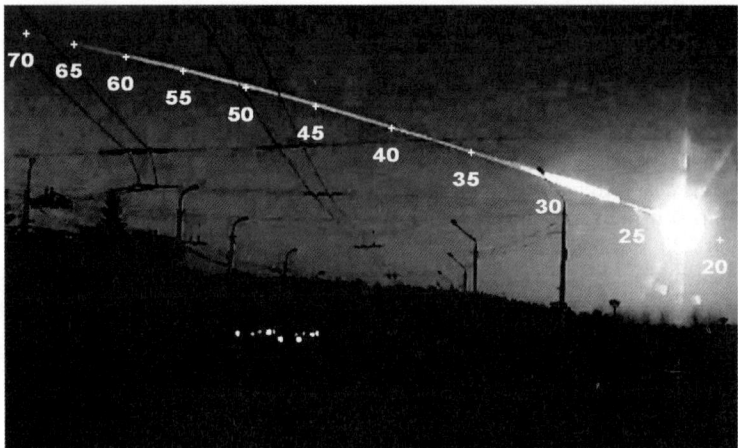

Figure 1. The bolide and fresh dust trail as seen from north. Frame from video taken by A. Ivanov in Kamensk-Uralskyi. The numbers are altitudes in km above ground.

million of inhabitants and the wide surroundings were surprised by bright bolide on the clear morning sky (Fig. 1). Although two small impactors had been discovered (by chance) in space the day before their impacts (Jenniskens et al. 2009; Chesley et al. 2015), the much larger Chelyabinsk impact came as a surprise. In fact, there was no chance to discover the impactor since it came from the direction close to the Sun (Borovička et al. 2013). The atmospheric entry was well documented and we can reconstruct in much more detail what happened than in previous cases. More than 400 casual video records of the bolide, from dashboard cameras in cars, security cameras, and traffic cameras, were posted on the Internet (Borovička et al. 2015). Additional hundreds of videos showed the bolide light, dust trail in the atmosphere, or the damage caused by the blast wave. The arrival of the blast wave and secondary sonic booms were recorded in the sound tracks of the videos. Further data came from seismic records and infrasonic records from around the world. The dust trail was imaged from the orbit by meteorological satellites (Proud 2013; Miller et al. 2013), see Fig. 2. The US Government sensors also recorded the event. Finally, the recovered meteorites were analyzed.

2. The results

The results of the analyses of various data have been already published in a number of papers, although some more detailed studies are still underway. The estimates of the total energy of the whole event obtained by various methods by different authors are summarized in Table 1. There is good agreement of 500 ± 100 kT TNT. The bolide trajectory was computed from calibrated videos by Borovička et al. (2013) and Popova et al. (2013) and are also in good agreement. Other computations found in the literature

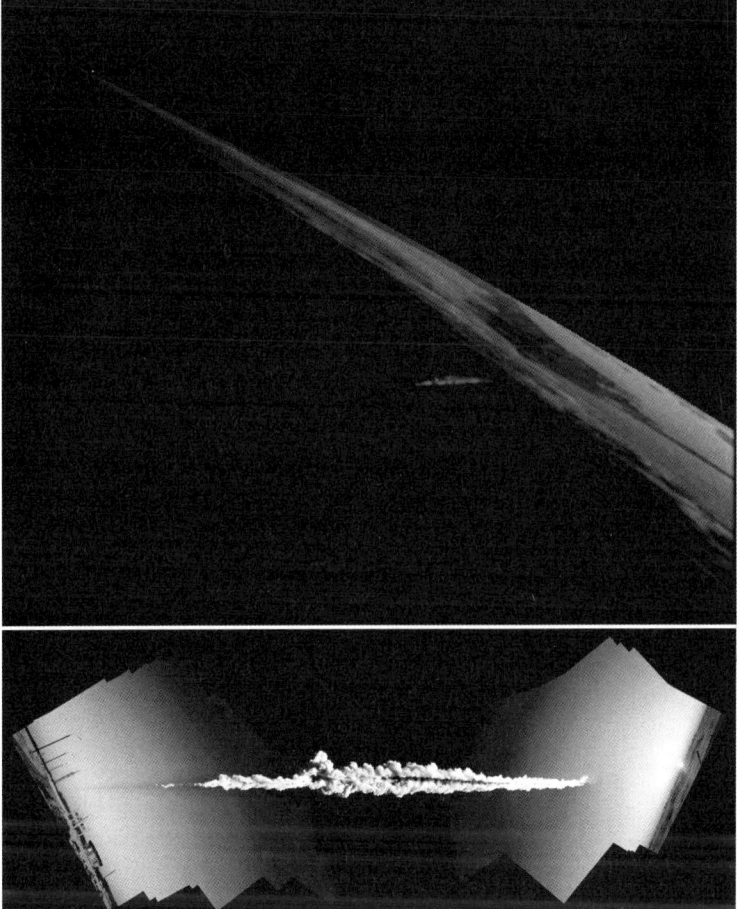

Figure 2. Dust trail from space and from ground. The upper image was taken by MSG2 satellite from geostationary orbit. Combination of visible and infrared channel. Courtesy Eumetsat and CHMI (Z. Charvát). The lower image was composed by L. Shrbený from video taken by A. Vazhenin in Borisovka, where the trail was seen directly overhead and was illuminated by the rising Sun (on the right).

are less reliable. Borovička *et al.* (2013) gave the observed height span 95.1 – 12.6 km, the slope of the trajectory (272 km long) relatively to the horizontal 18.5° at the beginning and 17° at the end (the slope changes primarily due to Earth's curvature, although the trajectory itself was also not straight), the initial velocity 19.03 ± 0.13 km s^{-1}, terminal velocity 3.2 km s^{-1}, and bolide duration 17 seconds.

Combining the known trajectory with the arrival times of the blast wave at various sites, it was proven that the blast wave causing damage was cylindrical not spherical. The wave originated at various heights between 25 – 35 km, not in a single point (Brown *et al.* 2013). It was therefore produced by the supersonic flight of the fragmenting asteroid. Secondary, weaker shocks after the main arrival were spherical waves from various fragmentation points. The region of damage extended perpendicularly from the part of the trajectory, where most energy was deposited (Popova *et al.* 2013). According to Popova *et al.* (2013), windows of 7,230 buildings were affected. Brown *et al.* (2013) examined more than 5000 windows in the city of Chelyabinsk and found that nearly 10% of them broke due to initial shock and 40% of buildings were affected. The window glass velocity

Figure 3. The hole in the ice of Lake Chebarkul caused by the impact of the largest fragment, the largest fragment displayed in Chelyabinsk museum, the collapsed roof and wall of the Chelyabinsk zinc plant, and windows in Chelyabinsk destroyed by the blast wave.

was measured to be 7 – 9 m s^{-1}. The roof one building collapsed (Fig. 3). The pressure was a few percent of atmospheric pressure. Popova et al. (2013) reported that 1,613 people asked for medical assistance at hospitals, 112 people were hospitalized, 2 in serious condition. There were, fortunately, no fatalities. Most injuries were from broken glass. Other inconveniences reported by the people were heat, sunburn, painful eyes, temporal deafness, and stress. No significant damage or injuries was caused by falling meteorites.

From the known energy and velocity, the mass of the impacting asteroid was found to be 12,000 kg. Assuming that the density of the meteorites (3300 kg m^{-3}) was valid for the whole body gives the asteroid equivalent diameter 19 ± 2 m. The asteroid severely fragmented in the atmosphere. The fragmentation was modeled by Borovička et al. (2013) using the observed light curve (total bolide brightness as a function of time), times of arrivals of secondary sonic booms, and deceleration toward the end of trajectory. Fresh dust trail images were also considered. It was found that intensive dust release (from near-surface) started at height about 70 km. The first fragmentation occurred at 45 km, where 1% of mass was lost. Large scale disruption with 95% mass loss occurred at heights 39 – 30 km. By 29 km the asteroid was fragmented into 10 – 20 boulders of sizes 1 – 3 m. These boulders then broke again at 26 – 22 km. Only one large (~ 0.7 m) fragment survived and landed in Lake Chebarkul (Fig. 3), from where it was lifted up 8 months later (Popova et al. 2013).

We can compare the dynamic pressure acting at the fragmentations ($p = \rho v^2$, where ρ is atmospheric density and v is velocity) with the typical tensile strength of meteorites, which is about 50 MPa (Popova et al. 2011). The first fragmentation occurred at 0.5 MPa, severe destruction at 1 – 5 MPa, and secondary fragmentation of boulders at 10 – 18 MPa. The maximum pressure encountered by the largest surviving fragment was 15 MPa. The bulk strength of few megapascals is obviously much lower than the strength

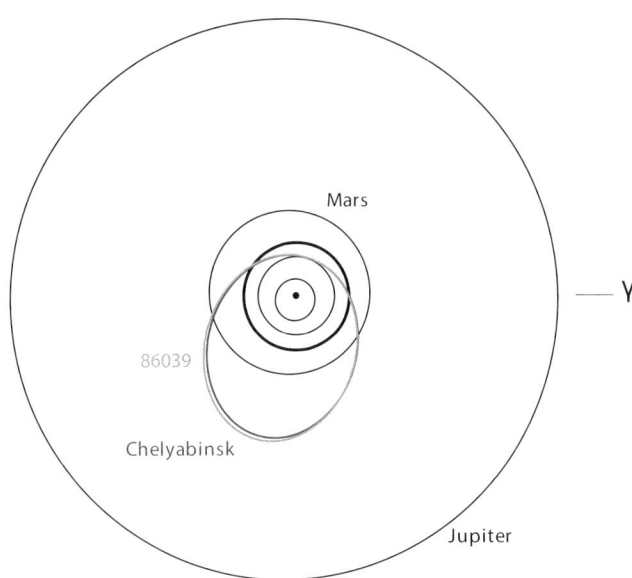

Figure 4. Orbits of Chelyabinsk and asteroid 86039 (1999 NC43).

of meteoritic material but is similar to other meteoroids (Popova et al. 2011). The low strength was probably caused by internal cracks from previous collisions in interplanetary space. The parts with strength larger than 10 MPa represented only few percent of the body.

At least 1923 meteorites recovered in a strewn field 80 km long and 7 km wide (Badyukov et al. 2014). Most of them very small (mass $\sim$ 1 g). More than 625 kg of meteoritic material was found in Lake Chebarkul (Popova et al. 2014). A 24.3 kg fragment was found near Travniki following the prediction of Borovička et al. (2013). Other cataloged meteorites have masses 0.04 g – 3.4 kg and total mass 94 kg. Total fallen mass was estimated to 4,000 – 10,000 kg. i.e. only 0.03 – 0.08% of the initial mass (Popova et al. 2014).

The meteorites were classified as LL type ordinary chondrite breccia (Kohout et al. 2014). LL is a common type of meteorite (represents 9% of all falls). However, two lithologies (light and dark) are present, together with impact melt. All three phases have identical composition. The dark lithology was produced by shock-darkening (Kohout et al. 2014; Reddy et al. 2014; Richter et al. 2015). Its reflectance spectrum can mimic carbonaceous material. The cosmic ray exposure age was measured to be 1.2 Myr, one of the lowest among LL chondrites (Popova et al. 2013; Povinec et al. 2015).

The dust trail left in the atmosphere was mostly formed by micron sized dust. It represents unablated residuals of tiny fragments. Total mass of the dust may be 25% of the initial mass (Popova et al. 2013). Within few days after the event, the dust circled the globe forming a optically thin dust ring around the northern hemisphere (Gorkavyi et al. 2013). The dust remained detectable in the atmosphere for three months (Rieger et al. 2014).

The pre-impact orbit of the asteroid was found to be very similar to that of asteroid 86039 (1999 NC43) with diameter of about 2 km (Borovička et al. 2013). Although there is only $\sim$ 1:10,000 chance that the proximity of Chelyabinsk orbit (Fig. 2) to an asteroid of this size is due purely to chance, spectral comparison did not confirm genetic relation

(Reddy *et al.* 2015). Detailed analysis of the reflectance spectrum of 1999 NC43 showed that it is of type L rather than LL.

The statistics of large bolides (Brown *et al.* 2013) and asteroid discoveries (Harris & D'Abramo 2015) now agree better than in the past and suggest that the impacts of Chelyabinsk size occur globally once per 40 ± 20 years on average.

3. Summary

The Chelyabinsk event was the first asteroid disaster in (at least modern) history. The damage was from the blast wave. If the body were stronger and penetrated deeper intact, the blast wave would be more damaging. In any case Chelyabinsk demonstrated that 20-m asteroids are dangerous and that asteroids of this size are more numerous than was thought several years ago. The new survey telescopes like ATLAS (Tonry 2011) and LSST (Jones *et al.* 2009) can provide advance warning if the impactor comes from the night side. Day side can be covered only from space.

Acknowledgements

This work was supported by project P209/11/1382 from the Czech Science Foundation (GAČR). The institutional project was RVO:67985815.

References

Badyukov, D. D., Dudorov, A. E. & Khaibrakhmanov, S. A. 2014. *Vestnik Chelyab. Gosudar. Univ.*, 1/2014, Fizika, Vyp. 19, p. 40 (in Russian)
Borovička, J., Spurný, P., Brown, P. *et al.* 2013. *Nature*, 503, 235
Borovička, J., Shrbený, L., Kalenda, P. *et al.* 2015. *Astron. Astrophys.*, in press, doi: 10.1051/0004-6361/201526680
Brown, P. G., Assink, J. D., Astiz, L. *et al.* 2013. *Nature*, 503, 238
Chesley, S. R., Farnocchia, D., Brown, P.,G., & Chodas, P.,W. 2015. In Aerospace Conference, 2015 IEEE , 8 pp., 7-14 March 2015, doi: 10.1109/AERO.2015.7119148
Gorkavyi, N., Rault, D. F., Newman, P. A. *et al.* 2013. *Geophys. Res. Lett.* 40, 4728
Harris, A. W. & D'Abramo, G. 2015. *Icarus*, 257, 302
Jenniskens, P., Shaddad, M. H., Numan, D. *et al.* 2009. *Nature*, 458, 485
Jones, R. L., Chesley, S. R., Connolly, A. J., *et al.* 2009, *Earth Moon and Planets*, 105, 101
Kohout, T., Gritsevich, M., Grokhovsky, V. I. *et al.* 2014. *Icarus*, 228, 78
McCord, T. B., Morris, J., Persing, D. *et al.* 1995. *J. Geophys. Res.* 100 (E2), 3245.
Miller, S. D., Straka III, W. C. , Scott Bachmeier, A. *et al.* 2013. *PNAS*, 110, 18092
Popova, O., Borovička, J., Hartmann, W. K. *et al.* 2011. *Meteorit. Plan. Sci.*, 46, 1525
Popova, O. P., Jenniskens, P., Emel'yanenko, V. *et al.* 2013. *Science*, 342, 1069
Popova, O. P., Jenniskens, P., & Glazachev, D. O. 2014. In: Geofiz. effekty padeniya Chelyab. Meteorita (Moscow: IDG RAS), *Dynam. Proc. Geospher.* 5, 59 (In Russian)
Povinec, P. P., Laubenstein, M., Jull, A. J. T. *et al.* 2015. *Meteorit. Plan. Sci.*, 50, 273
Proud, S. R. 2013. *Geophys. Res. Lett.* 40, 3351
Reddy, V., Sanchez, J. A., Bottke, W. F. *et al.* 2014. *Icarus*, 237, 116
Reddy, V., Vokrouhlický, D., Bottke, W. F. *et al.* 2015. *Icarus*, 252, 129
Richter, K., Abell, P., Agresti, D. *et al.* 2015. *Meteorit. Plan. Sci.*, 50, 1790
Rieger, L. A., Bourassa, A. E., & Degenstein, D. A. 2014. *Atmos. Meas. Tech.*, 7, 777
Silber, E. A., ReVelle, D. O., Brown, P. G. & Edwards, W. N. 2009. *J. Geophys. Res.*, 114, E08006.
Silber, E. A., Le Pichon, A. & Brown, P. G. 2011. *Geophys. Res. Lett.*, 38, L12201.
Tonry, J. L. 2011. *Publ. Astron. Soc. Pacific*, 123, 58
Vasilyev, N. V. 1998. *Plan. Space Sci.*, 46, 129.

The asteroid-comet continuum from laboratory and space analyses of comet samples and micrometeorites

Cécile Engrand[1], Jean Duprat[1], Noémie Bardin[1], Emmanuel Dartois[2], Hugues Leroux[3], Eric Quirico[4], Karim Benzerara[5], Laurent Remusat[5], Elena Dobrică[6], Lucie Delauche[1], John Bradley[7], Hope Ishii[7], Martin Hilchenbach[8] and the COSIMA team

[1]CSNSM, CNRS/IN2P3-Univ. Paris-Sud, Université Paris-Saclay, 91405 Orsay, France
email: cecile.engrand@csnsm.in2p3.fr

[2]IAS, CNRS/INSU-Univ. Paris-Sud, Université Paris-Saclay, 91405 Orsay, France

[3]UMET, CNRS/Univ. Lille1, 59655 Villeneuve d'Ascq, France

[4]IPAG, UMR 5274 CNRS/INSU-UJF-Grenoble 1, 38041 Grenoble, France

[5]IMPMC, MNHN, Case postale 115, 4 place Jussieu, 75252 Paris Cedex 05, France

[6]Dpt Earth Planetary Sciences MSC03 2040, Univ. New Mexico, Albuquerque NM 87131, USA

[7]Hawai'i Inst. Geophysics & Planetology, University of Hawai'i, Honolulu, HI 96822, USA

[8]Max-Planck-Institut für Sonnensystemforschung, 37077 Göttingen, Germany

Abstract. Comets are probably the best archives of the nascent solar system, 4.5 Gyr ago, and their compositions reveal crucial clues on the structure and dynamics of the early protoplanetary disk. Anhydrous minerals (olivine and pyroxene) have been identified in cometary dust for a few decades. Surprisingly, samples from comet Wild2 returned by the Stardust mission in 2006 also contain high temperature mineral assemblages like chondrules and refractory inclusions, which are typical components of primitive meteorites (carbonaceous chondrites - CCs). A few Stardust samples have also preserved some organic matter of comet Wild 2 that share some similarities with CCs. Interplanetary dust falling on Earth originate from comets and asteroids in proportions to be further constrained. These cosmic dust particles mostly show similarities with CCs, which in turn only represent a few percent of meteorites recovered on Earth. At least two (rare) families of cosmic dust particles have shown strong evidences for a cometary origin: the chondritic porous interplanetary dust particles (CP-IDPs) collected in the terrestrial stratosphere by NASA, and the ultracarbonaceous Antarctic Micrometeorites (UCAMMs) collected from polar snow and ice by French and Japanese teams. Analyses of dust particles from the Jupiter family comet 67P/Churyumov-Gerasimenko by the dust analyzers on Rosetta orbiter (COSIMA, GIADA, MIDAS) suggest a relationship to interplanetary dust/micrometeorites. A growing number of evidences highlights the existence of a continuum between asteroids and comets, already in the early history of the solar system.

Keywords. solar system: general, comets: general, minor planets, asteroids

1. Introduction

The structure and composition of the protoplanetary disk can be investigated by the study of the small bodies that have escaped planetary formation in the solar system: asteroids and comets. Samples from asteroid 25143 Itokawa returned by the Hayabusa mission and from comet 81P/Wild 2 returned by Stardust mission provide ground truth analyses for these small bodies (e.g. Nakamura *et al.* 2011; Brownlee 2014). Meteorites recovered on Earth can also give information on the formation and evolution of asteroids.

Cosmic dust particles originate from both comets and asteroids, in proportions that are still debated. They can be efficiently collected in the stratosphere (as interplanetary dust particles - IDPs) by NASA (Brownlee 1985), and as micrometeorites (MMs) from snow and ice in polar caps (e.g. Maurette et al. 1991; Duprat et al. 2007; Taylor et al. 2000; Yada & Kojima 2000). They can also be studied as microxenoliths trapped in meteorites (e.g. Bradley et al. 2014; Engrand et Maurette 1998; Briani et al. 2011). Analyses of comet 81P/Wild 2 (Stardust samples), meteorites and cosmic dust, as well as recent data obtained with the Rosetta mission on comet 67P/Churyumov-Gerasimenko suggest the presence of a continuum between asteroids and comets.

2. The asteroid-comet continuum seen from space and laboratory analyses

Carbonaceous chondrites (especially the CI type) have preserved the solar photosphere's composition for non volatile elements, and are taken as reference composition of the material initially present in the protosolar nebula (Lodders et al. 2010). They however have lost the most volatile elements, that are in turn preserved in comets. Carbonaceous chondrites are related to carbonaceous asteroids, but they represent only about 5% of the meteorites. A reexamination of the compositional classes in the main asteroid belt has shown that carbonaceous asteroids are actually present throughout the whole main belt, including the inner regions (DeMeo & Carry 2014). Meteorites thus give a biased view of the types and abundances of interplanetary matter, as more than 80% of the meteorites are classified as ordinary chondrites that are related to S-type asteroids. This bias is probably due to preferential destruction of carbonaceous chondrites during atmospheric entry or by terrestrial weathering at the Earth's surface.

Comets contain crystalline Mg-rich silicates, olivines and pyroxenes, initially identified by infrared spectroscopy (e.g. Wooden et al. 2000; Hanner & Zolensky 2010), and show a large variety of organic compounds (e.g. Bockelée-Morvan et al. 2004). Samples from comet 81P/Wild 2 (brought back by the Stardust mission in 2006) show significant similarities with carbonaceous chondrites (CCs): a chemical composition compatible (within a factor of 2) with that of CI chondrites, the presence of high temperature components: refractory inclusions (Ca-Al-rich inclusions or CAIs) and chondrule-like objects showing CC-like oxygen isotopic compositions, and remnant organics showing high level of complexity (Flynn et al. 2006; Zolensky et al. 2006; Nakamura et al. 2008; Nakashima et al. 2012; Ogliore et al. 2015; De Gregorio et al. 2011 and references therein). These results further outlined the need for radial transport from the inner regions of the protoplanetary disk where high temperature minerals are formed to the outer regions of the solar system where comets are stored (e.g. Bockelée-Morvan et al. 2002). They also showed for the first time a possible link between Jupiter family comet samples and carbonaceous chondrites - and therefore carbonaceous asteroids. Further Al-Mg dating of refractory components of 81P/Wild 2 samples (CAIs and chondrules) suggest a late formation of these components, constraining the formation age of Jupiter to be less than 3 Myr after the formation of CAIs, as radial transport from the inner to the outer regions of the protoplanetary disk would be inhibited in the presence of Jupiter (Nakashima et al. 2015 and references therein).

It has also been proposed for a long time that some meteorites (or meteorite classes) could come from comets. These candidates mostly belong to the CI or CR groups (e.g. Anders 1975; McSween & Weissman 1989; Campins & Swindle 1998; Gounelle et al. 2004; Gounelle et al. 2008; Nakashima et al. 2012). These assumptions are based on orbital considerations, abundance of organic compounds (up to 5wt%), similarity in oxygen isotopes, elevated D/H ratios or $^{15}N/^{14}N$ ratios, and abundance of glassy phases. Results from the Stardust mission cannot confirm (or not) the cometary origin of these meteorites.

Cosmic dust particles around 200 µm currently represent the dominant input of extraterrestrial matter on Earth, and show similarities only with carbonaceous chondrites (less than 1% of the micrometeorites in that size range are related to ordinary chondrites). These cosmic dust particles probably originate from the outer regions of the asteroid belt and/or from the Jupiter family comets (JFCs) (Nesvorný et al. 2010). The characteristics of Stardust samples generally match those of micrometeorites and IDPs (e.g. Dobrică et al. 2009), in particular in terms of oxygen isotopes (Engrand et al. 1999; McKeegan et al. 1987; Aléon et al. 2009). Two relatively rare classes of cosmic dust particles collected on Earth have a very probable cometary origin: the chondritic-porous (CP) IDPs–MMs (Bradley et al. 2014; Noguchi et al. 2015), and the ultracarbonaceous Antarctic micrometeorites (UCAMMs) that show similarities with the CHON particles detected in comet 1P/Halley (Duprat et al. 2010; Nakamura et al. 2005). These particles are characterized by a high carbon content (with up to 80 vol% in the case of UCAMMs) showing anomalous (D-rich) hydrogen isotopic compositions, an anhydrous mineralogy showing a high abundance of pyroxene minerals with regard to olivines (while olivine is the most abundant anhydrous minerals in meteorites, with the exception of the CR clan), and the widespread occurence of glassy phases coined GEMS (Glass with Embedded Metals and Sulfides) that have a debated presolar origin (Keller & Messenger 2013; Bradley 2013). These particles could be even more primitive than sample brought back by Stardust (e.g. Ishii et al. 2008). The organic matter of UCAMMs is enriched in nitrogen (with atomic N/C ratios up to 0.2) and could have been formed by irradiation at the surface of a $CH_4 - N_2$ rich icy body in the Oort cloud comet reservoir (Dartois et al. 2013).

One of the highlights of the Rosetta mission on comet 67P/Churyumov-Gerasimenko (67P/C-G) is the water D/H ratio of $(5.3\pm0.7)\times 10^{-4}$, that is about 3 times the terrestrial value (Altwegg et al. 2015), a value for a JFC that is even higher than that measured for the Oort Cloud comet Hale-Bopp (Bockelée-Morvan et al. 2015). This result confirms the fact that D/H (and $^{15}N/^{14}N$) ratios in comets are not characteristic of the comet reservoir. The D/H ratio measured in cometary HCN in Hale Bopp ($(2.3 \pm 0.4) \times 10^{-3}$), is compatible with that of organic matter in UCAMMs (Duprat et al. 2010). Analyses of dust particles with COSIMA on Rosetta show the association of rock forming elements with carbonaceous matter and highlights the enrichment of the dust particles in Na. They suggest similarities between the refractory (i.e. non-icy) component of 67P/C-G dust and cosmic dust collected on Earth (e.g. Schulz et al. 2015).

3. Implications

The analyses of samples from space missions (e.g. Stardust, Rosetta), of meteorites and cosmic dust highlight the existence of a continuum between primitive asteroids and comets. The presence of active asteroids main-belt comets also confirms the existence of this continuum. The presence of refractory minerals in comets invokes the need for radial transport from the inner to the outer regions of the protoplanetary disk, very early in the history of the solar system, and lasting at least a few Myr. Cosmic dust particles collected on Earth preferentially sample the carbonaceous asteroids and the cometary reservoir, thus being relevant samples to study this continuum. There is no clear evidence of large compositional differences between comets originating from the Kuiper Belt, and from the Oort cloud, supporting a common place for the formation of all comets. Comet samples are probably present in the terrestrial cosmic dust collections, in the form of chondritic porous IDPs collected in the stratosphere, and ultracarbonaceous Antarctic micrometeorites. The formation of N-rich organic matter of UCAMMs could result from irradiation of $CH_4 - N_2$-rich surface of icy bodies, in the comet formation regions.

References

Aléon, J. et al. 2009, *Geochim. Cosmochim. Acta* 73, 4558-4575
Altwegg, K., et al. 2015, *Science* 347
Anders, E. 1975, *Icarus* 24, 363-371
Bockelée-Morvan, D. et al. 2002, *A&A* 384, 1107-1118
Bockelée-Morvan, D. et al. 2004, *In Comets II* Univ. Arizona Press, pp. 391-423
Bockelée-Morvan, D., et al. 2015, *Space Sci. Rev.*, 1-37
Bradley, J. P. 2014, *In Treatise on Geochemistry (Second Edition)*, Elsevier, Oxford. pp. 287-308
Bradley, J. P. 2013, *Geochim. Cosmochim. Acta* 107, 336-340
Briani, G. et al. 2011, *Meteoritics Planet. Sci.* 46, 1863-1877
Brownlee, D.E. 1985 *Ann. Rev. Earth Planet. Sci.* 13, 147–173
Brownlee, D. E. 2014, *Annual Review of Earth and Planetary Sciences* 42, 179-205
Campins, H. & Swindle, T. D. 1998, *Meteoritics Planet. Sci.* 33, 1201-1211
DeMeo, F. E. & Carry, B. 2014, *Nature* 505, 629-634
Dartois, E. et al. 2013, *Icarus* 224, 243–252
De Gregorio, B. T. et al. 2011, *Meteoritics Planet. Sci.* 46, 1376-1396
Dobrică, E. et al. 2009, *Meteoritics Planet. Sci.* 44, 1643-1661
Duprat, J. et al. 2007, *Adv. Space Res.* 39, 605-611
Duprat, J. et al., 2010 *Science* 328, 742–745
Engrand, C. & Maurette, M. 1998, *Meteoritics Planet. Sci.* 33, 565-580
Engrand, C., McKeegan, K. D., & Leshin, L. A. 1999, *Geochim. Cosmochim. Acta* 63, 2623-2636
Flynn, G. J., et al. 2006, *Science* 314, 1731-1735
Gounelle, M., Spurny, P., & Bland, P. A. 2004, *Meteoritics Planet. Sci.* 39, #5174
Gounelle, M. et al. 2008, *In The Solar System Beyond Neptune*, Arizona Univ. Press. 525-541
Hanner, M. S. & Zolensky, M. E. 2010, *In Astromineralogy*, Springer-Verlag. 203–226.
Ishii, H.A. et al., 2008 *Science* 319, 447–450.
Keller, L. P., Thomas, K. L., & McKay, D. S. 1992, *Geochim. Cosmochim. Acta* 56, 1409-1412
Keller, L. P. & Messenger, S. 2013, *Geochim. Cosmochim. Acta* 107, 341-344
Lodders, K. 2010, *In Principles and Perspectives in Cosmochemistry*(eds. A. Goswami and B. E. Reddy), Springer Berlin Heidelberg. pp. 379-417
Maurette, M. et al. 1991, *Nature* 351, 44-47
McKeegan, K. D. 1987, *Science* 237, 1468-1471
McSween, H. Y., Jr. & Weissman, P. R. 1989, *Geochim. Cosmochim. Acta* 53, 3263-3271
Nakamura, T. et al. 2011, *Science*, 333, 113
Nakamura, T. et al. 2008, *Science* 321, 1664-1667
Nakamura, T. et al. 2005, *Meteoritics Planet. Sci.* 40 Suppl., #5046
Nakashima, D. et al. 2012, *Earth Planet. Sci. Lett.* 357-358, 355-365
Nakashima, D. et al. 2015, *Earth Planet. Sci. Lett.* 410, 54-61
Nesvorný, D. et al. 2010, *Astrophys. J.* 713, 816-836
Noguchi, T. et al. 2015, *Earth Planet. Sci. Lett.* 410, 1-11
Ogliore, R. C. et al. 2015, *Geochim. Cosmochim. Acta* 166, 74-91
Schulz, R. et al. 2015, *Nature* 518, 216-218
Taylor, S., Lever, J. H., & Harvey, R. P. 2000, *Meteoritics Planet. Sci.* 35, 651-666
Yada, T. & Kojima, H. 2000, *Antarctic Met. Res.* 13, 9-18
Wooden, D. H., Butner, H. M., Harker, D. E., & Woodward, C. E. 2000, *Icarus* 143, 126-137
Zolensky, M. E. et al. 2006, *Science* 314, 1735-1739

Carbonaceous Material in Extra-terrestrial Matter

Zita Martins

Dept of Earth Science and Engineering, Imperial College London, South Kensington Campus, London SW7 2AZ, UK
email: z.martins@imperial.ac.uk

Abstract. Comets, asteroids, meteorites, micrometeorites, interplanetary dust particles (IDPs), and ultra-carbonaceous Antarctic micrometeorites (UCAMMs) may contain carbonaceous material, which was exogenously delivered to the early Earth. Carbonaceous chondrites have an enormous variety of extra-terrestrial compounds, including all the key compounds important in terrestrial biochemistry. Comets contain several carbon-rich species and, in addition, the hypervelocity impact-shock of a comet can produce several α-amino acids. The analysis of the carbonaceous content of extra-terrestrial matter provides a window into the resources delivered to the early Earth, which may have been used by the first living organisms.

Keywords. meteors, meteoroids

1. Introduction

The inner solar system was bombarded from around 4.5 to 3.8 billion years ago (Schidlowski 1988; Schopf 1993; Chyba and Sagan 1992). These impacts may be observed by the impact craters on planetary bodies such as the Moon. Comets, asteroids and their fragments (i.e. meteorite, micrometeorites and interplanetary dust particles (IDPs)) are known to contain carbonaceous material, which was exogenously delivered to the early Earth during that period of time (Anders 1989; Chyba *et al.* 1990; Chyba and Sagan 1992). Interplanetary dust particles (IDPs) can contain up to 50% of organic carbon by mass, but typically have 10% organic carbon by mass (Anders 1989; Schramm *et al.* 1989). Micrometeorites and IDPs are known to contain organic molecules, such as polycyclic aromatic hydrocarbons (PAHs) and alkylated derivatives, aliphatic hydrocarbons, ketones and amino acids (Clemett *et al.* 1993; Matrajt *et al.* 2004, 2005). Furthermore, some aromatic compounds present in IDPs are associated with D/H anomalies (Wopenka 1988), while the long aliphatic chains with little ramifications present in IDPs may have originated at the edges of the protoplanetary disk (Matrajt *et al.* 2013). However, it is not certain whether all these molecules, in particular the amino acids are indeed indigenous or terrestrial contamination (Clemett *et al.* 1993; Brinton *et al.* 1998; Flynn 2003; Matrajt *et al.* 2004). Ultra-carbonaceous Antarctic micrometeorites (UCAMMs) contain up to 90% of carbonaceous material (Duprat *et al.* 2010; Dartois *et al.* 2013). UCAMMs contain unusually high nitrogen- and deuterium-rich organic matter, with bulk atomic N/C ratios of 0.05 and 0.12 (locally exceeding 0.15). They were proposed to be formed in very low temperature regions of the Solar System, such as the surface of small objects beyond the trans-neptunian region (Dartois *et al.* 2013). Carbonaceous meteorites contain up to 5wt% of organic carbon (Alexander *et al.* 2013), which is either locked in an insoluble kerogen-like polymer, or in a rich organic inventory of soluble organic compounds (Cronin and Chang 1993; Cody and Alexander 2005; Martins and Sephton 2009). There are several differences in abundance and distribution of the soluble organic content between different carbonaceous meteorites. This may result from different physical and chemical processes such as thermal metamorphism (Burton et al. 2011; Chan et al. 2012), or aqueous alteration on the meteorite parent body (Glavin et al. 2006; Martins et al. 2015). Comets also have several extra-terrestrial organic molecules (Crovisier and Bockelée-Morvan 1999; Ehrenfreund et al. 2002), including the simplest amino acid glycine (Elsila et al. 2009). Most recently the results from the Rosetta mission of the surface of comet 67P/Churyumov-Gerasimenko were published. Carbon-rich species, including alcohols, carbonyls, amines, nitriles, amides, isocyanates, and the radiation-induced polymer polyoxymethylene were detected (Goesmann et al. 2015; Wright et al. 2015). In addition, the impact-shock of comets produces complex organic molecules (Martins et al. 2013).

2. Synthesis of amino acids via the impact-shock of comets

Goldman et al. (2010) performed ab initio molecular dynamics simulations that showed that shock waves passed into ice mixtures representative of comets, could theoretically yield amino acids. Amino acid precursors (such as ammonia, methanol and carbonyl compounds) have been observed in comets, e.g. Halley, Hyakutake, Tempel 1, Giacobini-Zinner, Hartley 2 and Hale-Bopp (Festou et al. 2005; DiSanti et al. 2013; Crovisier and Bockelée-Morvan 1999; Ehrenfreund et al. 2002; Ehrenfreund and Charnley 2000; Mumma et al. 2003; Bockelée-Morvan et al. 2000). Therefore, it has been experimentally tested in the laboratory whether the hypervelocity impact-shock of a typical comet ice mixture would synthesize amino acids (Martins et al. 2013). Results show that the impacts of comets onto rocky surfaces, and the impacts of meteorites onto icy surfaces (such some of the Jovian and Saturnian satellites) produce several α-amino acids. Several lines of evidence prove that the detected amino acids were indeed synthesized by impact-shock (Martins et al. 2013):

• A racemic mixture of alanine was detected, with a D/L ratio of 0.99, and a racemic mixture of norvaline was also detected with a D/L ratio of 0.97;
• Structural diversity and the decrease of amino acid abundances with increasing number of carbon atoms were observed;
• The non-protein amino acids α-aminoisobutyric acid (α-AIB) and isovaline were obtained;
• The synthesis is in agreement with ab initio molecular dynamics simulations modelling (Goldman et al. 2010);
• The control ice samples and the extraction procedural blanks were completely free from organic compounds.

The synthetic pathway to create amino acids via impact-shock is not completely clear. Suggested synthetic pathway includes the Strecker-cyanohydrin synthesis using α-amino acid precursors (carbonyl compounds, ammonia, and hydrogen cyanide). In this case, the carbonyl precursors (aldehydes and ketones) would be synthesised by the oxidation of methanol (present in the simulated ice mixture), which would generate the α-amino acid precursor carbonyl compounds. Support for this is indicated by an ice mixture containing only ammonia and carbon dioxide and lacking methanol. Even with an impact velocity of 7.12 km/s, no detectable quantities of amino acids were produced (Martins et al. 2013). To note, however that the impact shock occurs on timescales of nanoseconds to milliseconds, which should not be consistent with the Strecker-cyanohydrin synthesis. Alternatively, a high shock pressure (i.e. at 7 km/s the peak shock pressure experienced by the ice mixture is 50 GPa (Martins et al. 2013; Goldman et al. 2010)) results in the formation of ions and radicals, which will then be involved in the post-shock reaction to form amino acids.

In summary, the exogenous delivery of carbonaceous material in our Solar System expands the inventory of locations where life could have potentially originated. In fact, the detection by future space missions of an amino acid racemic ratio on the surface of icy Jovian and Saturnian satellites may be the result of impact shock events as well as the contribution from meteorites.

References

Alexander, C. M. O.'D., Howard, K. T., Bowden, R., & Fogel, M. L. 2013, *GCA*, 123, 244-260
. Anders, E. 1989, *Nature*, 342, 255-257.
Bockelée-Morvan, D., Lis, D. C., Wink, J. E., Despois, D., Crovisier, J., Bachiller, R., Benford, D. J., Biver, N., Colom, P., Davies, J. K., Gérard, E., Germain, B., Houde, M., Mehringer, D., Moreno, R., Paubert, G., Phillips, T. G., & Rauer, H. 2000, *A&A*, 353, 1101-1114.
Brinton, K. L. F., Engrand, C., Glavin, D. P., Bada, J. L., & Maurette, M. 1998. *OLEB*, 28, 413-424.
Burton, A. S., Glavin, D. P., Callahan, M. P., Dworkin, J. P., Jenniskens, P., & Shaddad, M. H. 2011, *Meteoritics & Planetary Science*, 46, 1703-1712.
Chan, H.-S., Martins, Z., & Sephton, M. A. 2012. *Meteoritics and Planetary Science*, 47, 1502-1516.
Chyba, C. F. & Sagan, C. 1992, *Nature*, 355, 125-132.
Chyba, C. F., Thomas, P. J., Brookshaw, L., & Sagan, C. 1990, *Science*, 249, 366-373.
Clemett, S. J., Maechling, C. R., Zare, R. N., Swan, P. D., & Walker, R. M. 1993. *Science*, 262, 721-725.
Cody, G. D. & Alexander, C. M. O.'D. 2005, *Geochimica et Cosmochimica Acta*, 69, 1085-1097.
Cronin, J. R. & Chang, S. 1993, In The Chemistry of Life's Origin, edited by Greenberg J. M., Mendoza-Gomez C. X. and Pirronello V. Dordrecht: Kluwer. pp. 209-258.
Crovisier, J. & Bockelee-Morvan, D. 1999, *Space Science Reviews*, 90, 19-32.
Dartois, E., Engrand, C., Brunetto, R., Duprat, J., Pino, T., Quirico, E., Remusat, L., Bardin, N., Briani, G., Mostefaoui, S., Morinaud, G., Crane, B., Szwec, N., Delauche, L., Jamme, F., Sandt, Ch., & Dumas, P. 2013, *Icarus*, 224, 243-252.
DiSanti M. A., Bonev, B. P., Villanueva, G. L., & Mumma M. J. 2013, *The Astrophysical Journal*, 763, 1.
Duprat, J., Dobrica, E., Engrand, C., Aléon, J., Marrocchi, Y., Mostefaoui, S., Meibom, A., Leroux, H., Rouzaud, J.-N., Gounelle, M., & Robert, F. 2010, *Science*, 328, 742-745.
Ehrenfreund, P., Irvine, W., Becker, L., Blank, J., Brucato, J. R., Colangeli, L., Derenne, S., Despois, D., Dutrey, A., Fraaije, H., Lazcano, A., Owen, T., & Robert, F., *International Space Science Institute ISSI-Team 2002*, Reports on Progress in Physics, 65, 1427-1487.
Ehrenfreund, P. & Charnley, S. B. 2000, *Annual Review of Astronomy and Astrophysics*, 38, 427-483.
Elsila, J. E., Glavin, D. P., & Dworkin, J. P. 2009, *Meteoritics and Planetary Science*, 44, 1323.
Festou M., Uwe-Keller, H. & Weaver, H. A. 2005, Comets-I. I. University of Arizona Press.
Flynn, G. J., Keller, L. P., Feser, M., Wirick, S., & Jacobsen, C. 2003, GCA, 67, 4791-4806.
Glavin, D. P., Dworkin, J. P., Aubrey, A., Botta, O., Doty, J. H., Martins, Z., & Bada, J. L. 2006, *Meteoritics & Planetary Science*, 41, 889-902.
Goesmann, F. *et al.* 2015, *Science*, 349, aab0689-1- aab0689-3.
Goldman, N., Reed, E. J., Fried, L. E.; William Kuo, I.-F., & Maiti, A. 2010, *Nature Chemistry*, 2, 949-954.
Matrajt, G., Pizzarello, S., Taylor, S., & Brownlee, D. 2004, *Meteoritics & Planetary Science*, 39, 1849-1858.
Matrajt, G., Muñoz Caro, G. M., Dartois, E., D'Hendecourt, L., Deboffle, D., & Borg, J. 2005, *A&A*, 433, 979-995.
Matrajt, G., Flynn, G., Brownlee, D., Joswiak, D., & Bajt, S. 2013, *ApJ*, 765, 145, 18 pp. Martins,

Z. M. & Sephton, A. 2009, In Hughes, A.B. (Ed.), Amino acids, peptides and proteins in organic chemistry. Wiley-VCH Verlag GmbH & Co. KGaA, Weinheim, pp. 3-42.

Martins, Z., Price, M. C., Goldman, N., Sephton, M. A., & Burchell, M. J. 2013, *Nature Geoscience*, 6, 1045-1049.

Martins, Z., Modica, P., Zanda, B., & Le Sergeant D'Hendecourt, L. 2015, *Meteoritics & Planetary Science*, 50, 926-943.

Mumma, M. J., DiSanti, M. A., Dello Russo, N., Magee-Sauer, K., Gibb, E., & Novak, R. 2003, *AdSR*, 31, 2563-2575.

Schidlowski, M. 1988, *Nature*, 333, 313-318.

Schopf, J. W. 1993, *Science*, 260, 640-646.

Schramm, L. S., Brownlee, D. E., & Wheelock, M. M. 1989. *Meteoritics*, 24, 99-112.

Wopenka, B. 1988, *EPSL*, 88, 221-231.

Wright, I. P., Sheridan, S., Barber, S. J., Morgan, G. H., Andrews, D. J., & Morse, A. D. 2015, *Science*, 349, aab0673-1-aab0673-3.

Water-Rock Differentiation of Icy Bodies by Darcy law, Stokes law, and Two-Phase Flow

Wladimir Neumann, Doris Breuer and Tilman Spohn

Deutsches Zentrum für Luft- und Raumfahrt (DLR), Institut für Planetenforschung, Planetenphysik, Rutherfordstr. 2, 12489 Berlin, Germany, email: wladimir.neumann@dlr.de

Abstract. The early Solar system produced a variety of bodies with different properties. Among the small bodies, objects that contain notable amounts of water ice are of particular interest. Water-rock separation on such worlds is probable and has been confirmed in some cases. We couple accretion and water-rock separation in a numerical model. The model is applicable to Ceres, icy satellites, and Kuiper belt objects, and is suited to assess the thermal metamorphism of the interior and the present-day internal structures. The relative amount of ice determines the differentiation regime according to porous flow or Stokes flow. Porous flow considers differentiation in a rock matrix with a small degree of ice melting and is typically modelled either with the Darcy law or two-phase flow. We find that for small icy bodies two-phase flow differs from the Darcy law. Velocities derived from two-phase flow are at least one order of magnitude smaller than Darcy velocities. The latter do not account for the matrix resistance against the deformation and overestimate the separation velocity. In the Stokes regime that should be used for large ice fractions, differentiation is at least four orders of magnitude faster than porous flow with the parameters used here.

Keywords. Dwarf planets, Asteroids, Planetesimals, Differentiation, Accretion, Ceres

1. Introduction

Objects with an ice-silicate composition (comets, asteroids, and dwarf planets) can be found in the outer asteroid belt (e.g., the dwarf planet Ceres) and populate the outer range of the Solar system. We present a numerical model that calculates the evolution of small icy bodies and combines accretion and differentiation into a rocky core and a water mantle (and further possible metal-silicate separation of the rocky core). To simulate differentiation, we use a multiphase model for which phase changes and separation by matrix compaction are taken into account in spherical symmetry, considering simultaneously ice and rock in solid and liquid states. The model provides a numerical solution of a moving boundary problem for the temperature $T(r,t)$ as a function of radius r and time t on a moving ("accreting") domain of a planetesimal's interior. In particular, we compare the theoretical predictions based on Darcy and Stokes law with the differentiation velocities calculated with the numerical model. The basic structure of differential equations is given below and is built upon the thermal evolution and accretion code from Šrámek et al. (2012) (the system of equations we use here, however, is applied to an icy body). The model is designed to address the evolution of Ceres, but it is applicable to any small icy object.

2. Model

The model calculates the thermal evolution and differentiation in 1D for an object that consists of ice and dust and has the size and the density of Ceres. The separation of water and rock is modelled within the formalism of the two-phase flow theory. Thereby, the modelled processes are coupled with the continuous growth of the asteroid due to the accretion of dust.

Frame of reference and accretion: All differential equations involved are transformed from the formulations on (r, t) with the radial coordinate r and time t to those on a moving frame (η, t): $\eta := r \cdot R(t)^{-1}$, where $R(t)$ is the radius evolution. Here, we only present models of instantaneous formation with $R(t) = R_0 = R_f = 480$ km (with the initial radius R_0 and the final radius R_f).

Energy balance: Energy balance equation is solved with parameters that depend on temperature, porosity and composition (with bulk density ρ, space coordinate η, bulk heat capacity c_p, temperature T, current radius of the planetesimal R, bulk thermal conductivity k, heat sources Q, latent heat L, melting rate Γ, water w and dust d):

$$\rho(\eta)c_p(\eta,T)\left(\frac{\partial T}{\partial t} - \eta\frac{\dot{R}}{R}\frac{\partial T}{\partial \eta}\right) = \frac{1}{\eta^2}\frac{\partial}{\partial \eta}\left(\frac{k(\eta)}{R^2}\eta^2\frac{\partial T}{\partial \eta}\right) + Q(\eta,t) - L_w\Gamma_w - L_d\Gamma_d.$$

Water-rock and metal-silicate differentiation: Water percolation / particle settling is modelled by solving advection equations for the matrix and the liquid phases within the two-phase flow formalism (with solid fraction ϕ_s^i or liquid fraction ϕ_l^i of component i, index i equal to d or w, average separation velocity $v(\eta)$, and density ρ_i of component i):

$$\frac{\partial \phi_s^i}{\partial t} - \eta\frac{\dot{R}}{R}\frac{\partial \phi_s^i}{\partial \eta} + \frac{1}{R\eta^2}\frac{\partial}{\partial \eta}\left(\eta^2\phi_s^i v\right) = -\frac{\Gamma_i}{\rho_i}, \quad \frac{\partial \phi_l^i}{\partial t} - \eta\frac{\dot{R}}{R}\frac{\partial \phi_l^i}{\partial \eta} + \frac{1}{R\eta^2}\frac{\partial}{\partial \eta}\left(\eta^2\phi_l^i v\right) = -\frac{\Gamma_i}{\rho_i}.$$

After the separation of water, the solidus of metal can be reached. Then, the same equations can be used for the components Fe (metal) and Si (silicates) with the respective densities and melting rates.

Differentiation velocity: The water percolation velocity is computed according to the two-phase flow theory (with viscosities of the liquid μ_l and of the solid μ_s, permeability K_ϕ, fraction of the liquid $\phi_l = \phi_l^w + \phi_l^d$, fraction of the solid $\phi_s = \phi_s^w + \phi_s^d$, critical liquid fraction ϕ_c, and gravity g):

$$\frac{\mu_l}{K_\phi}v = \frac{1}{R^2}\frac{4}{3}\mu_s\frac{\partial}{\partial \eta}\left(\frac{\phi_s}{\eta^2}\frac{\partial}{\partial \eta}\eta^2 v\right) + \frac{1}{R^2}\frac{\partial}{\partial \eta}\left(\frac{\mu_s\phi_s}{\phi_l - \phi_c}\frac{1}{\eta^2}\frac{\partial}{\partial \eta}\eta^2 v\right) + \frac{4\mu_s}{R^2}\frac{\partial \phi_l}{\partial \eta} - (\phi_s^d + \phi_l^d)(\rho_d - \rho_w)g.$$

For the metal-silicate separation the velocity is computed analogously using suitable parameters. The permeability is computed from the reference permeability K_0 at the reference liquid fraction ϕ_0 (equal to initial volume fraction of ice), with $K_0 = b^2/\tau$, solid grain size b, coefficient $\tau = 72\pi$, and exponent $n = 2$:

$$K_\phi = K_0\left(\phi_l^w - \phi_c\right)^n / \left(\phi_0 - \phi_c\right)^n.$$

Material properties: A mixture of ice and dust is considered. The initial mass fraction of dust is $0.865 - 0.737$ (ice mass fractions of $0.135 - 0.263$), corresponding to the rock density variation of $2540 - 3450$ kg m^{-3} (from a low-density serpentine-like to a high-density Vesta-like material). The associated ice volume fraction is $\phi_0 = 0.284 - 0.552$. The ice density $\rho_w = 1000$ kg m^{-3} is fixed, the volume-weighted average density of the asteroid is $\rho = 2100$ kg m^{-3} for the above variation. The thermal conductivity k is volume-weighted arithmetic mean of the heat conductivities of ice ($k_d = 0.4685 + 488.12T^{-1}$) and rock ($k_w = 2.2$ W m^{-1}K^{-1}). The specific heat capacity is a mass-weighted arithmetic mean of temperature dependant heat capacities of ice ($c_{p,w} = \min(4200, 185 + 7.037T)$ J K^{-1}kg^{-1}) and rock ($c_{p,d} = 800 + 0.25T - 1.5 \times 10^7 T^{-2}$ J K^{-1}kg^{-1}).

Heat sources: Both short- (^{26}Al, ^{60}Fe, ^{53}Mn) and long-lived (^{40}K, ^{232}Th, ^{235}U, ^{238}U) radiogenic nuclides are included in the rock fraction, while the ice fraction contains no heat sources.

Initial conditions: The initial and surface temperature is 180 K. The surface temperature remains constant, the central heat flux is zero.

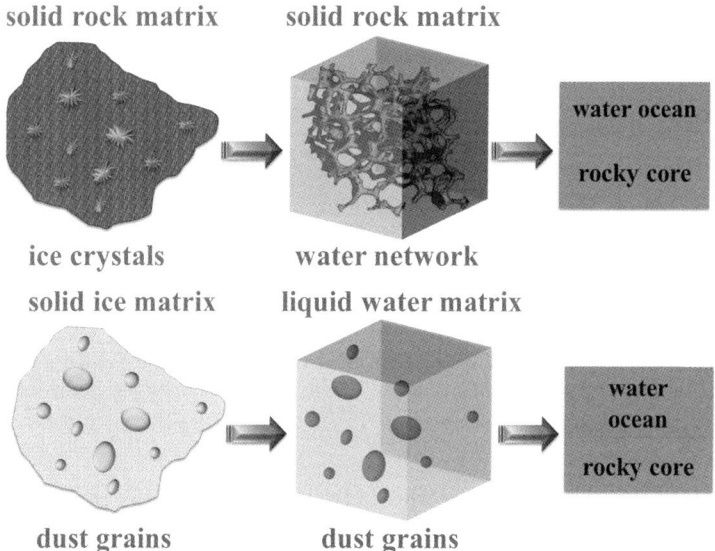

Figure 1. Overview of the differentiation regimes. Top panel: The composition is dominated by the dust. Melting of ice creates a water network. Percolation of water results in the formation of a water ocean and a rocky core. Bottom panel: The composition is dominated by ice. Upon melting of the ice matrix a water ocean forms and a rocky core grows by sinking dust particles.

3. Results

In the following, we first describe two differentiation regimes and compare their differentiation velocities. To this end, we consider two extreme ice mass fractions of 0.135 and 0.2629.

Ice-rock separation regimes: Assuming accretion of ice and dust, the rheology of the matrix is dominated by one of the two components, depending on their relative proportions. Two differentiation regimes arise (see Fig. 2): (a) Upon melting of ice in a rocky matrix water ascends via Darcy flow (Fig. 2, top); (b) Upon melting of an icy matrix the dust grains settle via Stokes flow (Fig. 2, bottom). For (a) water will percolate if the matrix deforms sufficiently to squeeze the water out of the matrix. Because temperatures of up to 700 K are needed for this, water will first remain immobile until the matrix deforms and then percolate, or until vapour forms and gets mobilised by the matrix deformation. On its way to the surface water and ice will condense or crystallise in cooler layers.

Separation velocity: Estimates of the particle settling velocity and water percolation velocity can be obtained from Stokes and Darcy law, respectively. Free parameters are: Grain size b, gravity $g < 0.25$ m s^{-2} (Ceres' surface gravity), dust density ($2500 < \rho_d < 3600$ kg m^{-3}, see McCord & Sotin (2005)), permeability parameters n and τ, and liquid volume fraction ϕ_l^w. Fixed parameters are water density $\rho_w = 1000$ kg m^{-3} and water viscosity $\eta_l = 0.002$ Pa s. As an upper bound, a gravity $g = 0.25$ m s^{-2} is used. Darcy and Stokes velocities v_D and v_S follow from vspace-0.2cm

$$v_D = K_\phi (\rho_d - \rho_w) g / \phi_l^w \mu_l, \quad v_S = 2(\rho_d - \rho_w) g b^2 / 9 \mu_l.$$

In Fig. 3, Darcy and Stokes velocities are given as function of ϕ_l^w and b. Solid lines show velocity in m s^{-1} and dashed lines in m year^{-1}, blue lines correspond to the dust density of $\rho_d = 2500$ kg m^{-3} (serpentine) and red ones to $\rho_d = 3600$ kg m^{-3} (Vesta-like). The Darcy velocity varies between 10^{-12} and 10^{-10} m s^{-1} with $\phi_l^w = 0.01 - 0.5$ for a fixed grain size of $b = 10^{-6}$ m (and between 10^{-6} and 10^{-4} m s^{-1} for $\phi_l^w = 0.01 - 0.5$ and

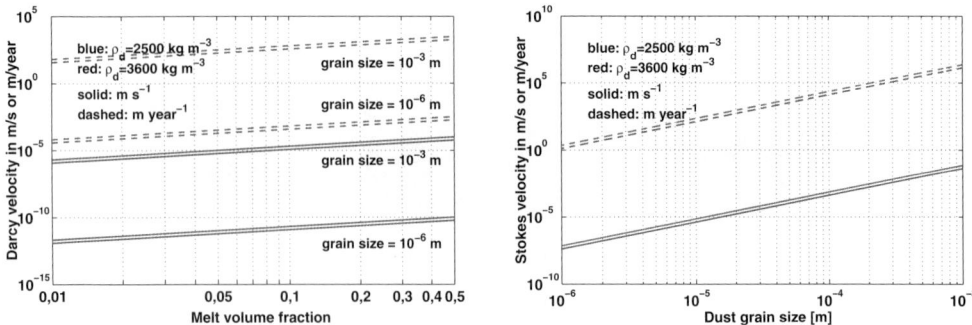

Figure 2. Darcy velocity (left panel) and Stokes velocity (right panel) calculated for $\rho_d = 2500$ kg m^{-3} (blue lines) and $\rho_d = 3600$ kg m^{-3} (red lines) and varying values of b and ϕ_l^w.

$b = 10^{-3}$ m). The Stokes velocity varies between 10^{-8} and 10^{-2} m s^{-1} for grain sizes between 10^{-6} and 10^{-3} m. In general, v_S is two orders of magnitude higher than v_D for an equal grain size. Thus, if the water fraction is low (for instance $\phi_0 = 0.284$) liquid water percolates comparatively slow in the Darcy regime. When liquid water is produced rapidly in large amounts before it can percolate upwards (probable for $\phi_0 = 0.552$), a muddy water-dust ocean will form. In this case grains settle very fast at the bottom of the ocean on the time scale shown in Fig. 3, right panel. For a serpentine-like matrix, Ceres' differentiation velocity is can be estimated by the ice volume fraction ($\phi_l^w = \phi_0 = 0.284$) and is $\approx 3 \times 10^{-5}$ m s^{-1} (≈ 1 km a^{-1}) for $b = 10^{-3}$ (Fig. 3, left panel) - differentiation is completed within ≈ 500 years. For a high density dust fraction, complete melting of ice leads to $\phi_l^w = \phi_0 = 0.552$. At such liquid fraction Stokes regime occurs. For $b = 10^{-3}$, the Stokes velocity is 0.072 m s^{-1} or $\approx 2 \times 10^6$ m a^{-1}. This extreme velocity implies a complete differentiation of Ceres within less than one year. During the differentiation, the fractions of dust and water and the grain size vary with radius and with time. For this reason, both regimes with the velocity ranges showed in Fig. 3 can occur locally.

Two-phase flow: In addition, we calculate water-rock differentiation of Ceres self-consistently as described in section 2. As an example, we consider instantaneous formation at $t_0 = 1$ Ma relative to the formation of the calcium-aluminium-rich inclusions (CAIs). The viscosity of a serpentine-like matrix is temperature-dependent but not well know and represents an uncertainty in our model. We use $\mu_s = 10^{18}$ as a lower bound. In addition, the gravity is depth-dependent and the liquid fraction is computed self-consistently during the differentiation. It can be shown that the separation velocity is not equal to the Darcy velocity, since the resistance of the matrix against the deformation is a counter-acting factor. The velocities calculated for both matrix densities are shown in Fig. 3. Values obtained for the grain size $b = 10^{-3}$ are at most $\approx 2.5 \times 10^{-6}$ and are, thus, smaller than the Darcy velocity of $\approx 4 \times 10^{-6}$ for $\phi_l^w = 0.02$. The timing of the core and ocean formation is similar for both dust composition and varies around $10-20$ thousand years. The Darcy law overestimates the differentiation velocity for $\mu_s = 10^{18}$ since it does not account for the resistance of the matrix against the deformation. Because we use a rather small value for the viscosity of the serpentine matrix, even a slower core and ocean formation is probable. Nevertheless, the phases separate very fast compared to the time scale of the decay of radiogenic nuclides, leading to a complete separation of water from the dust. Only a very thin undifferentiated layer remains at the surface. The water layer is ≈ 50 km thick for a serpentine-like dust composition and ≈ 112 km for a Vesta-like dust composition.

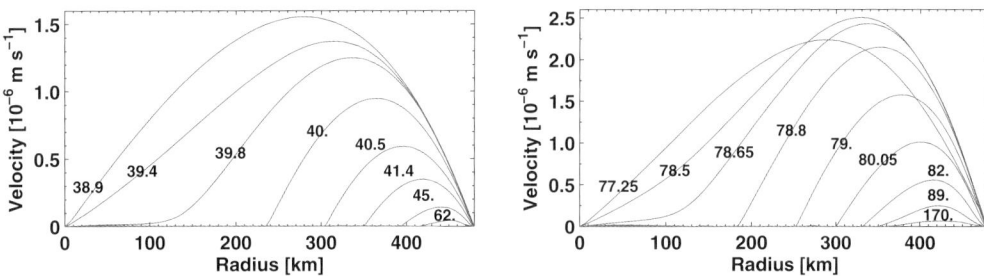

Figure 3. Separation velocity during the differentiation of Ceres. Left panel: Dust density $\rho_d = 2500$ kg m^{-3} (serpentine), initial ice volume fraction $\phi_0 = 0.284$. Right panel: Dust density $\rho_d = 3600$ kg m^{-3} (Vesta-like), initial ice volume fraction $\phi_0 = 0.552$. Further parameters: Grain size $b = 0.001$ m, water viscosity $\mu_l = 0.002$ Pa s, matrix viscosity $\mu_s = 10^{18}$ Pa s.

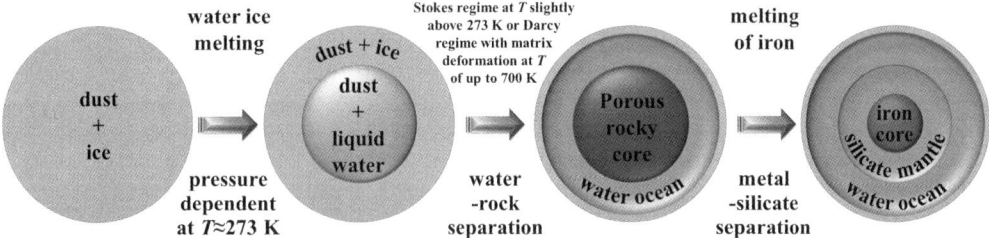

Figure 4. The evolution of Ceres' structure. Early radiogenic heating leads to the melting of ice. Water-rock separation results in a rocky core and water ocean below a solid crust. Further heating leads to the differentiation of a metallic core and a silicate mantle for an early accretion.

Structure evolution of Ceres: The water-rock differentiation occurs early in the evolution of Ceres and should begin during the accretion. An early formation within ≈ 1 Ma after the CAIs results in a rapid temperature increase and an early differentiation. For a later formation time, for instance at 3 Ma after CAIs, the temperature increases slowly, but water melting is still expected within 1 Ma after the formation. In both cases, a water ocean forms fast within ≈ 0.01 Ma after the onset of water ice melting. Subsequent heating of the iron-silicate core depends on the formation time. For an early formation ($1-2$ Ma after CAIs) solidus and even liquidus of metal and silicates is reached within the subsequent 1 Ma and melting is extensive. With a higher concentration of radionuclides in the core after the separation of water, the temperature increases above 2000 K. A complete differentiation into a metal core and a silicate mantle is inevitable, but in such a case the overlying water ocean would boil and evaporate, leading to a dry Vesta-like asteroid. If, however, Ceres forms later (for instance, 3 Ma after CAIs), no metal melt is produced. The ocean contains no heat sources, it is heated by the core from below and isolated by the solid ice-dust crust above. The temperature of the ocean is higher than the melting point of water ice but remains globally below 670 K (and locally below the evaporation point of water at higher pressures).

4. Discussion and conclusions

We modelled the evolution of temperature and the water-rock differentiation for a Ceres-like body using two extreme fractions of ice. We compared theoretical predictions of the separation velocity (Darcy and Stokes law) and, thus, of the timing of the differentiation, with the velocities obtained in two-phase flow calculations. The assumed relative amount of ice (i.e., the density of the silicats, see McCord & Sotin (2005)) determines the predominant differentiation regime and timing according to Darcy or Stokes law. In the

Stokes regime, the differentiation proceeds at least three orders of magnitude faster than in the Darcy regime. Both regimes can occur simultaneously in a single body at some intermediate differentiation stages, e.g., Darcy at a depth with a small liquid fraction and Stokes in a water ocean. Using two-phase flow the differentiation is at least one order of magnitude slower than in the Darcy regime. While the relative amount of water determines the differentiation regime (Stokes law should be used for amounts of liquid above a critical fraction for which the rheology of a mixture is dominated by water), two-phase flow calculations should be used instead of Darcy law for small liquid fractions. For the parameters adopted, an early forming Ceres differentiates a water ocean within ≈ 0.01 Ma. Because we used a lower bound on the matrix viscosity, a slower differentiation is expected for a higher μ_s. The temperature evolution indicates metal-silicate separation for an early accretion within the first 3 Ma after CAIs. However, a plethora of additional processes is expected to influence the thermal evolution further. Evolution of the dust porosity, hydration of silicates, convection in the water ocean, for instance, will change the temperature profile in the interior and in the rocky core influencing potential metal differentiation. The interplay of these factors must certainly has been a very complex process that requires further detailed studies.

References

Šrámek, O., Milelli, L., Ricard, Y., & Labrosse, S. 2012, *Icarus*, 217, 339-354
McCord, T. B. & Sotin, C. 2005, *JGR*, 110, E05009

Ice-gas interactions during planet formation

Karin I. Öberg

Harvard-Smithsonian Center for Astrophysics
60 Garden St, Cambridge, MA 02138
email: koberg@cfa.harvard.edu

Abstract. Planets form in disks around young stars. In these disks, condensation fronts or snowlines of water, CO_2, CO and other abundant molecules regulate the outcome of planet formation. Snowline locations determine how the elemental and molecular compositions of the gaseous and solid building blocks of planets evolve with distance from the central star. Snowlines may also locally increase the planet formation efficiency. Observations of snowlines have only become possible in the past couple of years. This proceeding reviews these observations as well as the theory on the physical and chemical processes in disks that affect snowline locations.

Keywords. astrochemistry, astrobiology, molecular processes

1. Introduction

Planets form in disks around young stars. These disks are composed of dust and gas, where both the dust and gas composition varies with disk radius and height. These variations come in two flavors. First, a combination of molecules inherited from the protostellar stage and ongoing chemistry in disks determines the composition of volatiles in different disk regions and at different times. Second, a balance between freeze-out and sublimation, where the sublimation rate is set by a combination of volatile binding energies on (icy) dust grains, the thermal structure and the flux of high-energy radiation, determines how specific volatiles are divided between gas and dust/ice phases (Öberg *et al.* 2011b). This balance is generally achieved at short time scales compared to chemical and dynamical timescales, but exceptions exist as discussed in §3.

In disks the main reservoirs of volatiles after hydrogen are expected to be water ice, CO_2 and CO ice and gas. Apart from gas-phase CO, the most abundant volatiles are effectively hidden from view. Ices are generally not possible to observe in disks, and CO_2 has no strong rotational transitions because of its lack of a permanent dipole moment. The volatile composition in the preceding protostellar stage is better understood because of surveys of ice absorption spectra toward low and high-mass protostars where the protostar is used as the background IR source toward which ice absorption is measured. Based on such surveys, it is well established that water is the most important ice constituent, followed by CO and CO_2 and trace amounts of CH_3OH, CH_4 and NH_3 (Gibb *et al.* 2004, Öberg *et al.* 2011a, Boogert *et al.* 2015). N_2 is probably also an important ice constituent in the coldest lines of sight, but cannot be observed. Figure 1 illustrates the median ice composition as well as the range of compositions observed in different lines of sight. It is clear that for a disk that does not experience a major chemical evolution, H_2O, CO_2 and CO will be the most important snowlines for carbon and oxygen containing species.

This proceeding summarizes recent successful attempts at observing snowlines of H_2O and CO in protoplanetary disks using ALMA and infrared observations (§2). §3 reviews and discusses the processes that regulate snowline locations in disks.

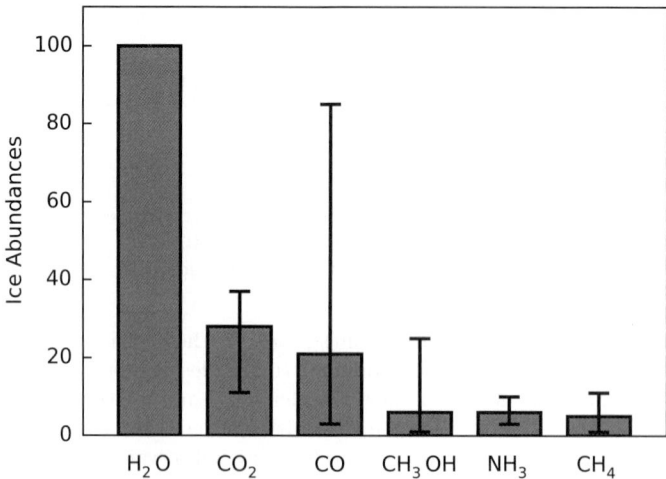

Figure 1. The median ice composition with respect to water toward low-mass protostars (blue bars). The black 'error bars' represent the minimum and maximum abundances observed with respect to water ice in low-mass protostellar lines of sight. Values are taken from Boogert et al. (2015).

2. Observations of snowlines in disks

Three disk snowlines have been observed during the past few years, the water snowline in the disk around TW Hya (Zhang et al. 2013), and the CO snowlines in the disks around TW Hya (Qi et al. 2013) and HD 163296 (Qi et al. 2011, Mathews et al. 2013, Qi et al. 2015). The water snowline toward TW Hya was extracted from infrared and far-infrared observations of water vapor emission lines. Based on detailed modeling of the lines, Zhang et al. (2013) found that water vapor in the disk is present in a narrow ring with the inner rim close to the 4 AU dust hole in the disk. The outer 4.2 AU rim was interpreted as the water snowline.

The CO snowline has been resolved using millimeter interferometry, by direct CO observations and through imaging of other trace species whose chemistry is closely linked to CO freeze-out. Early attempts to use ^{13}CO observations to constrain the CO snowline location seems to have overestimated the CO snowline radius by 50-60 AU, highlighting the potential traps of this method (Qi et al. 2011). To avoid such traps, Qi et al. (2013) used N_2H^+ to determine the CO snowline location in the TW Hya disk. N_2H^+ is expected to be anti-correlated with CO gas because CO gas inhibits formation and speeds up destruction of N_2H^+. The inner rim of N_2H^+ emission should therefore mark the CO midplane snowline (Fig. 2). Another potential tracer of the CO snowline, DCO^+ (Mathews et al. 2013) has since been found to have a more complicated relationship with CO freeze-out. This conclusion is based on comparison between DCO^+, N_2H^+ and $C^{18}O$ emission in the HD 163296 disk (Qi et al. 2015) as well as the presence of a pair of concentric DCO^+ rings in the disk of IM Lup (Öberg et al. 2015).

Based on the N_2H^+ observations toward TW Hya and HD 163296, the CO snowline does not appear to occur at the same disk temperature in the two disks. Toward TW Hya, the derived CO snowline location corresponds to a disk midplane temperature of 17 K, while in HD 163296 the CO snowline location corresponds to 25 K (Qi et al. 2013, Qi et al. 2015). These observations imply that there is not a single temperature that defines the CO snowline location in all disks, which should be the case if disk dynamics

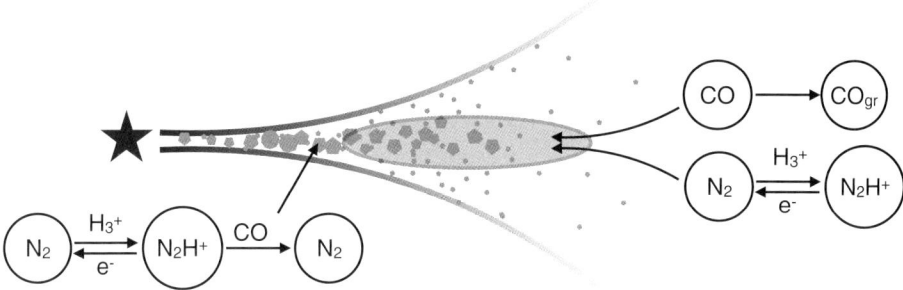

Figure 2. Abundant N_2H^+ is only expected in disk regions where CO is depleted due to CO freeze-out, i.e. in the disk midplane exterior to the CO snowline (blue region). Interior to the CO snowline N_2H^+ formation is inhibited by CO reactions with H_3^+ while N_2H^+ is rapidly destroyed by CO reactions with N_2H^+. In the absence of CO, N_2H^+ is mainly destroyed by much slower electron recombination reactions.

is always slow compared to sublimation *and* if the ice structure and therefore ice binding energy is the same in all disks.

3. Theory of snowline locations

Figure 3 shows the C/O ratio in gas and dust across a disk due to water, CO_2 and CO snowlines, assuming a typical disk density and temperature profile, a static disk, a lack of chemical evolution, and standard assumptions on the binding energies of these three molecules (Öberg *et al.* 2011b). Disks are dynamic systems however, and they do chemically evolve and may present different volatile binding energies dependent on the local ice environment. Major snowlines may therefore occur at different disk midplane temperatures in different disks.

Fundamentally a snowline location is set by the balance of desorption (sublimation) and adsorption of a volatile. In a static disk, the temperature at which the adsorption overtakes desorption can be calculated if the density and binding energy are known. The dependence on density is weak (logarithmic), however (Hollenbach *et al.* 2009) and will at most change the snowline temperature by a few degrees for 'reasonable' disk midplane density profiles. The dependence on binding energy is stronger (linear), and the CO binding energy is known to depend sensitively on the local ice environment. In particular CO interactions with porous water ice can be almost twice as strong as the interaction between CO molecules in a pure CO ice. If CO is sufficiently abundant to form a pure CO ice on grains, the CO snowline should thus occur far out in the disk at locations with dust temperatures of 20 K or less (cf. TW Hya). In disks where CO is mixed with water ice, CO will, by contrast, begin to freeze out at 25–30 K.

Ices can also desorb non-thermally. UV ice photodesorption is efficient (Fayolle *et al.* 2011). The presence of strong UV fields may push CO snowlines outward, corresponding to a lower freeze-out temperature than would be expected from a balance between adsorption and thermal desorption alone (Fig. 3). Finally, there are several dynamical processes that can affect snowline locations. Accretion flows and drift of grains can move snowlines inward by ∼50% compared to the expected locations in static disks (Piso *et al.* ApJ subm.), if grains have grown to cm size or larger. For these larger grains drift timescale are shorter than sublimation time scales. Considering the number of processes that can affect snowline locations, it is perhaps not strange that CO snowlines have been detected

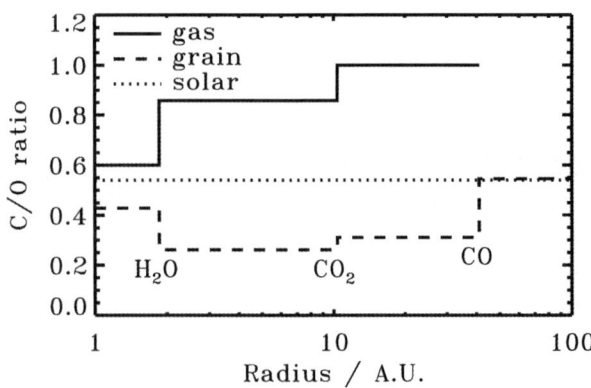
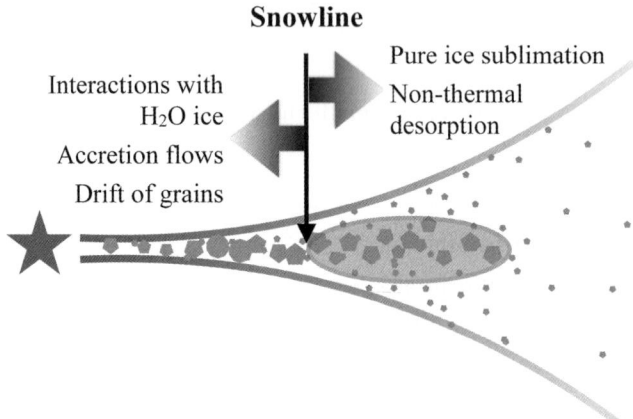

Figure 3. *Top*: The C/O gas and dust ratio in a disk assuming a static disk and standard prescriptions for the balance between desorption and adsorption for the most abundant C and O volatiles. *Bottom*: The qualitative effects of some disk processes on snowline locations in disks.

at locations that correspond to different CO freeze-out temperatures. With more data it should be possible to tell whether chemical and dynamical effects regulate (CO) snowline locations and thus how snowlines evolve during planet formation.

References

Collings, M. P., Dever, J. W., & Fraser, H. J. 2003, *ApJ*, 583, 1058
Hollenbach, D. J., Kaufman, M. J., Bergin, E. A., & Melnick, G. J. 2009, *ApJ*, 690, 1497
Mathews, G. S, Klaassen, P. D, Juhasz, A, *et al.* 2013, *A&A*, 557, 132
Öberg K. I., Boogert A. C. A., Pontoppidan K. M. *et al.* 2011 *ApJ*, 740, 109
Öberg, K. I., Murray-Clay, R., Bergin, E. A. 2011 *ApJL*, 743, L16
Öberg, K. I., Furuya, K. and Loomis, R. 2015 *ApJ*, 810, 112
Qi, C., Öberg, K. I., Wilner, D. J. *et al.* 2013, *Science*, 341, 630
Qi, C., Öberg, K. I., Andrews, S. M. *et al.* 2015, *eprint arXiv*, 1510.00968
Zhang, K., Pontoppidan, K. M., Salyk, C., & Blake, G. A. 2013, *ApJ*, 766, 82
Fayolle, E. C., Bertin, M., & Romanzin, C. 2011, *ApJL*, 739, L36

^{15}N fractionation in star-forming regions and Solar System objects

E. S. Wirström[1], G. Adande[2], S. N. Milam[2], S. B. Charnley[2] and M. A. Cordiner[2,3]

[1] Department of Earth and Space Sciences, Chalmers University of Technology, Onsala Space Observatory, SE-439 92 Onsala, Sweden. email: eva.wirstrom@chalmers.se

[2] Astrochemistry Laboratory, Mailstop 691, NASA Goddard Space Flight Center, 8800 Greenbelt Road, Greenbelt, MD 20770, USA

[3] Department of Physics, Catholic University of America, 620 Michigan Avenue NE, Washington, DC 20064, USA

Abstract. We briefly review what is currently known of ^{14}N/^{15}N ratios in interstellar molecules. We summarize the fractionation ratios measured in HCN, HNC, CN, N_2 and NH_3, and compare these to theoretical predictions and to the isotopic inventory of cometary volatiles.

Keywords. astrochemistry, molecular data, solar system: formation, ISM: abundances, ISM: molecules, radio lines: ISM, astrobiology

1. Introduction

Cometary ices are believed to be the least modified in the Solar System and as such may have the closest connection to the pristine interstellar material which comprised the presolar molecular cloud. Molecular isotopic ratios offer the prospect of understanding the provenance and processing of the volatile material found in comets, meteorites and interplanetary dust particles (Mumma & Charnley 2011; Marty 2012; Bockelée-Morvan et al. 2015). For example, the enhanced deuterium fractionation in primitive Solar System matter has long been considered a marker for low-temperature ($\sim$10 K) fractionation chemistry and recent theoretical work indicates that even the relatively modest D/H ratios measured in comets (a few times 10^{-4}) could not have been generated in the Sun's protoplanetary disk (Cleeves et al. 2014).

Anomalies (i.e. ^{15}N enrichments) in nitrogen isotope ratios are found in both the soluble and insoluble organic matter in meteorites and IDPs (Marty 2012). Isotopic measurements indicate that ^{15}N and D enrichments are present in the bulk material and can be even higher in small localized regions (Aléon 2010). Observations of HCN, CN and NH_3 in cometary comae also indicate enhanced ^{15}N fractionation (Jehin et al. 2009; Rousselot et al. 2014), with ^{14}N/^{15}N ratios lower than both the Solar value (440) and the nominal value of the local ISM (300, Adande & Ziurys 2012).

In this article we briefly summarize our theoretical and observational understanding of ^{14}N/^{15}N ratios in interstellar molecules and their possible connection to cometary molecules.

2. Interstellar Nitrogen Fractionation: Theory and Observations

As with deuterium fractionation, it was posited that the ^{15}N enrichments in Solar System matter could have their origin in low-temperature ion-molecule reactions (e.g. Bernatowicz 1997). Candidate ion-neutral isotope exchange reactions were identified and evaluated although only modest enhancements were found in related model calculations of molecular cloud chemistry (Terzieva & Herbst 2000). The most important of these are those that can lead to fractionation in N_2, and consequently in ammonia, and in the nitriles HCN, HNC and CN:

$$^{15}N + {}^{14}N_2H^+ \leftrightharpoons {}^{14}N + {}^{14}N^{15}NH^+ \qquad (2.1a)$$

$$\leftrightharpoons {}^{14}N + {}^{15}N^{14}NH^+ \qquad (2.1b)$$

$$^{15}N + HC^{14}NH^+ \leftrightharpoons {}^{14}N + HC^{15}NH^+ \qquad (2.1c)$$

These molecular ion-neutral atom processes and other ion-molecule reactions proposed by Terzieva & Herbst (2000) have formed the basis of most subsequent theoretical studies of fractionation in molecular clouds. These studies have shown that enhanced gas-phase and solid-phase $^{14}N/^{15}N$ ratios can be attained depending on the initial atomic/molecular N ratio and, when all the nitrogen is initially molecular, on the level of CO depletion (Charnley & Rodgers 2002; Rodgers & Charnley 2008). More recent calculations show that the spin state of molecular hydrogen can play an important role in the combined evolution of molecular D/H and $^{14}N/^{15}N$ ratios (Wirström et al. 2012).

These time-dependent models show that ^{15}N fractionation in the nitriles occurs on a much shorter timescale than that which occurs in N_2 and NH_3. Overall, they predict $^{14}N/^{15}N$ ratios in interstellar molecules consistent with the meteoritic values, and with the nitrile ratios measured in comets (see below).

Until recently, both the theoretical models and the proposed ISM-comet connection were largely unconstrained because of lack of $^{14}N/^{15}N$ ratio measurements in interstellar clouds (cf. Ikeda et al. 2002). However, motivated by the meteoritic and cometary measurements, in the last few years, molecular $^{14}N/^{15}N$ ratios have now been determined for HCN, HNC, CN, N_2 and NH_3 in a variety of interstellar and protostellar environments. Table 1 summarizes these observations and also lists the corresponding range of cometary $^{14}N/^{15}N$ ratios. For each molecule, we can compare the observed interstellar $^{14}N/^{15}N$ ratios with the theoretical predictions and each of these with the range of cometary values.

HCN and HNC show a range of $^{14}N/^{15}N$ ratios in (pre-stellar) dark clouds that are consistent both with the cometary values and the predictions of published models; in protostars, these molecules appear to be less-enriched in ^{15}N. The CN ratios tend to be larger which is difficult to understand if all three nitriles are produced mainly from dissociative recombination of protonated HCN.

Across the sample, $^{14}NH_3/^{15}NH_3$ ratios do not exhibit enhancements comparable to that in the nitriles and in fact show evidence for depletion in ^{15}N, the most extreme cases being in L1544 and L1689N. These trends are consistent with calculations of ammonia fractionation (Wirström et al. 2012) but none of the interstellar $^{14}NH_3/^{15}NH_3$ ratios come within a factor of two of the highly-enriched cometary values.

The molecular nitrogen $^{14}N/^{15}N$ ratios, as measured via N_2H^+, $^{15}N^{14}NH^+$ and $^{14}N^{15}NH^+$, exhibit an extremely wide range of values across all types of sources. Theoretical models of dense cores actually predict that N_2 should be as enriched as the nitriles but never produce high $^{14}N/^{15}N$ ratios (i.e. ^{15}N depleted). Molecular nitrogen has only been measured in one comet (Rubin et al. 2015) and the prospects for detection of the associated $^{14}N/^{15}N$ ratio are slim.

However, the most serious issue for studies of nitrogen fractionation is that recent quantum-chemical calculations and chemical modeling by Roueff et al. (2015) indicate that ^{15}N atomic exchange reactions do not fractionationate N_2H^+ and $HCNH^+$ as efficiently as previously assumed (see Eq's 2.1). Have previous theoretical models been

Table 1. Interstellar Nitrogen Isotope Ratios

Source	Type	NH_3	N_2H^+ [§]	HCN	HNC	CN	Ref.
L1544	dark core	>700	1000±200	69-154	>27	500±75	4,1,3,3,9
			1000±200	140-360			1,2
L1498	dark core	619±100	...	>75	>90	500±75	3,3,3,9
				>813			5
L1521E	dark core	...	...	151±16	...	...	5
L1521F	dark core	539±118	...	>51	24-31	...	3,3,3
L1262-core	dark core	356±107	>297	...	...	...	3,3
			175±79				3
L183	dark core	530^{+570}_{-180}	...	140-250	...	...	4,2
NGC 1333-DCO$^+$	dark core	360^{+260}_{-110}	...	...	...	...	4
NGC 1333-4A	Class 0 protostar	344±173	...	...	...	...	6
		>270					4
B1	Class 0 protostar	300^{+55}_{-40}	>600	165^{+30}_{-25}	75^{+25}_{-15}	240	10,10,10,10,9
		334±50	400^{+100}_{-65}				6,10
		470^{+170}_{-100}					4
L1689N	Class 0 protostar	810^{+600}_{-250}	...	...	...	...	4
Cha-MMS1	Class 0 protostar	...	729^{+212}_{-135}	...	135	...	16,7
IRAS 16293A	Class 0 protostar	...	...	163±20	242±32	...	13
R Cr A IRS7B	Class 0 protostar	...	...	287±36	259±34	...	13
OMC-3 MMS6	Class 0 protostar	...	...	366±86	460±65	...	13
L1262-YSO	Class I protostar	453±247	>410	...	...	...	3,3
			>410				3
Several	Massive starless cores	...	65-1100	...	...	330-400	15,15
			180-1445[#]				15
Several	Massive protostars	...	190-1000	...	...	190-450	15,15
			180-1300				15
Several	UC HII regions	...	320-900	...	...	230-430	15,15
			350-700				15
Comets	JFC & Oort Cloud	127[‡]	...	139±26	...	135-170[†]	11,12,8

References: (1) Bizzocchi et al. (2013); (2) Hily-Blant et al. (2013a); (3) Milam & Charnley (2012), Adande et al. (2015); (4) Gerin et al. (2009); (5) Ikeda et al. (2002); (6) Lis et al. (2010); (7) Tennekes et al. (2006); (8) Hutsemékers et al. (2008); (9) Hily-Blant et al. (2013b); (10) Daniel et al. (2013); (11) Rousselot et al. (2014); (12) Bockelée-Morvan et al. (2008); (13) Wampfler et al. (2014); (15) Fontani et al. (2015); (16) Cordiner et al., private communication.

[§] In each N_2H^+ entry the uppermost value is for the $^{15}N^{14}NH^+$ isotopologue. [#] Larger value is a lower limit.
[‡] 'Average' based on optical observations of NH_3 daughter molecule NH_2 in an ensemble of comets. [†] This range can be taken as a surrogate for the HCN ratio, however in comets there may be additional sources of CN (see Mumma & Charnley 2011). Only 2 measurements have been made for HCN itself, in OC comets Hale-Bopp and 17P/Holmes.

able to predict and reproduce the meteoritic, cometary and interstellar $^{14}N/^{15}N$ ratios of nitriles and amines merely by chance?

3. Summary

The study of nitrogen isotopic fractionation in primitive matter, in comets and in astronomical environments has been the focus of much recent activity. For nitriles, $^{14}N/^{15}N$ ratios measured in interstellar and protostellar sources are comparable with cometary values, as are recent measurements of $HC^{14}N/HC^{15}N$ in disks (Öberg et al., these proceedings). Interstellar ammonia does not appear to be as enriched as the nitriles and the interstellar $^{14}NH_3/^{15}NH_3$ ratios are significantly higher than those found in cometary ammonia. Molecular nitrogen exhibits the largest range of values with enrichments similar to the nitriles, but also very marked ^{15}N depletions. The theoretical perspective is rather puzzling and clearly something fundamental is missing from our understanding of interstellar nitrogen isotope fractionation. This is most likely not connected to isotope-selective photodissociation (Heays et al. 2014), but some proposed neutral-neutral processes remain viable (Rodgers & Charnley 2008; Roueff et al. 2015).

More measurements of the complete suite of important molecules - HCN, HNC, CN, N_2H^+ and NH_3 - in cold clouds, regions of low-mass and massive star formation, and in comets are necessary to confirm and explore these trends further.

ESW acknowledges generous support from the Swedish National Space Board. This work was supported by NASA's Origins of Solar Systems Program.

References

Adande, G. R. et al. 2015, in preparation
Adande, G.R. & Ziurys, L. M. *Astrophys. J.* 744, 194
Aléon, J. 2010, *Astrophys. J.* 722, 1342
Bernatowicz, T. J. 1997, in *From Stardust to Planetesimals*, eds. Y. Pendleton, A.G.G.M. Tielens, 122, 227. (Provo: Utah. Astron. Soc. Pac.)
Bizzocchi, L. et al. 2013, *Astron. Astrophys.* 555, A109
Bockelée-Morvan, D, et al. 2008. *Astrophys. J.* 679, L49
Bockelée-Morvan, D. et al. 2015, *Space Sci. Rev.*, Published online, DOI 10.1007/s11214-015-0156-9
Charnley, S. B. & Rodgers, S. D. 2002. *Astrophys. J.* 569, L133
Cleeves, L. I. et al. 2014, *Science* 345, 1590
Daniel, F. et al. 2013, *Astron. Astrophys.* 560, A3
Fontani, F. et al. 2015, *Astrophys. J.* 808, L46
Gerin, M. et al. 2009, *Astron. Astrophys.* 498, L9
Heays, A. N. et al. 2014. *Astron. Astrophys.* 562, 61
Hily-Blant, P. et al. 2013a, *Icarus*, 223, 582
Hily-Blant, P. et al. 2013b, *Astron. Astrophys.* 557, A65
Hutsemékers, D. et al. 2008. *Astron. Astrophys.* 490, L31
Ikeda, M., et al. 2002. *Astrophys. J.* 575, 250
Jehin, E. et al. 2009. *Earth, Moon, and Planets* 105, 167
Lis, D. C., et al. 2010, *Astrophys. J.* 710, L49
Marty, B. 2012, *EPSL*, 313-314, 56
Milam, S. N. & Charnley, S. B. 2012, *LPSC Meeting XLIII*, #2618
Mumma, M. J. & Charnley, S. B. 2011, *Annu. Rev. Astron. Astrophys.* 49, 471
Rodgers, S. D. & Charnley, S. B. 2008, *Astrophys. J.* 689, 1448
Roueff, E. et al. 2015, *Astron. Astrophys.* 576, A99
Rousselot, P. et al. 2014. *Astron. Astrophys.* 780, L17
Rubin, M. et al. 2015, *Science* 348, 232
Tennekes, P. P. et al. 2006, *Astron. Astrophys.* 456, 1037
Terzieva, R. & Herbst, E. 2000. *Mon. Not. R. Astron. Soc.* 317, 563
Wampfler, S. et al. 2014, *Astron. Astrophys.* 572, A24
Wirström, E. S. et al. 2012, *Astrophys. J.* 757, L11
Öberg, K. I. et al. 2015, these proceedings

FM 11:
Global Coordination of Ground and Space Astrophysics and Heliophysics

PART III

Global Coordination: What are the Next Steps?

David Spergel[1] and Robert Williams[2]

[1] Princeton University Observatory, Peyton Hall, Ivy Ln, Princeton NJ 08544-1001, New Jersey (NJ) United States
email: dns@astro.princeton.edu

[2] Space Telescope Science Institute, 3700 San Martin Drive,Baltimore MD 21218-2410,Maryland (MD),United States
email: wms@stsci.edu

From the transit expeditions of 1761 to JWST, ALMA, and the SKA, international projects have played an important role in driving astronomy and heliophysics. Over the past two decades, the increasing complexity and cost of new facilities, the constrained amount of funding available from individual sources, and the rapidly increasing volume of data produced by newer facilities have made international collaboration on large ground- and space-based facilities essential to moving the fields forward. As international cooperation becomes commonplace, data-sharing policies have become ever more important. All IAU members have a stake in the policy decisions made by nations and various scientific consortiums concerning data access and international collaborations. This focus meeting provided a forum to discuss how to improve coordination of global strategic planning in astronomy, astrophysics, and heliophysics in order to maximize the scientific return from research facilities.

How do we coordinate these international planning efforts? How do we balance national prioritizations with the increasing multi-national structures of our projects? How can and should we share the data produced both by these international collaborations and by other projects? How should we provide access to these facilities? Furthermore, the huge volume of data produced by current and future observation systems necessitates modes of research that have not heretofore figured prominently in astronomical and heliophysical research enterprises. The Advanced Technology Solar Telescope (ATST) will collect 3.65 petabytes in its first year of science operations while the Large Synoptic Survey Telescope (LSST) alone will produce 30 TB of data per night.

The potential benefit of enhanced international coordination is high. Much can be learned in astrophysics by adopting a broad-scoped approach, in which ground and space-based facilities look at the same target with different wavelengths, timescales and technologies. Such an approach requires more resources than a single nation could maintain. Heliophysics has the added issue of coordinating truly global ground-based systems and space missions in various regions of the Sun-Earth system. In this context, Earth is an additional spacecraft embedded in its own space plasma environment. For the first time in history, we are capable of looking at a complicated coupled space system in its entirety, from the sun through the heliosphere, magnetosphere, ionosphere, and atmosphere down to the biosphere, in which we try to survive the present climate change. To study and understand the system around us is the ultimate benchmark to be able to understand other star-planet systems.

As the scope and scale of international collaboration on space science projects increases, challenges concerning managing data access and reciprocity between stakeholders have so far been dealt with on an ad hoc basis, project by project. The meeting consisted of a series of panels that bring together a diverse set of stakeholders representing the astronomical and heliophysics disciplines from around the world. Each of the sessions had significant audience participation and there were many fruitful exchanges between the panelists and between the panelists and the audience.

The FM11 focus meeting discussed the process that led successful international collaborations on projects such as the Cassini probe, ALMA, and Astro-H. Many national organizations and the European community have created processes of long-term prioritization, often based on synthesizing community views. The role of decadal reviews, now being undertaken by a number of countries, was emphasized, and agreed upon as the necessary basis for international coordination because of the national priorities that they establish. This said, Stuart Wyithe of Australia described the different philosophy that Australia's decade review undertook, in not prioritizing but rather setting out their strategy for achieving the goals of advancing astrophysics in the country.

Separate sessions were held on ground-based and space astronomy, with various speakers emphasizing the different situations that apply to ground vs. space telescopes, such as the crucial role of the space agencies in determining which space missions are approved. They concluded that global strategies for international coordination of ground projects should be handled separately and differently than global collaboration on space priorities and missions. An important aspect of international collaboration that was emphasized repeatedly by speakers from developing countries was that their participation as partners in 'big science' projects is the best way for small countries to enter the world of 'big time' astronomy. For developing countries international collaboration is essential for their meaningful introduction to design, fabrication, and operation of world class facilities.

Individual sessions were also devoted to the concepts of Open Skies and Open Data, with general agreement that science is advanced more when access to facilities is permitted on the basis of peer review without limitations based on other criteria such as cost compensation when necessary (largely for facilities constructed and operated from private funds). An obvious result from the discussions of all the above topics was that sharing of information in an open forum is very important for progress to be made on every aspect of global cooperation. The IAU should accept this as an important part of its mission in the future, in addition to being proactive in taking concrete steps to foster international collaboration.

The goal of the final session of FM11, devoted to 'Next Steps', was to conduct a panel discussion among five astronomers (N. Kaifu, R. Kraan-Korteweg, M. Colless, R. Bonnet, & P. Ho) experienced in international collaborations and strategizing and have them give their assessment of ideas presented during the meeting. The panel was then charged with engaging the audience of interested participants in a discussion of specific actions that the IAU might take to facilitate cooperation and collaboration among nations, including the possible development of a common roadmap for future projects that would require international partners.

The panelists led off the discussion by proposing specific activities the IAU should undertake to enhance collaborations. These can be summarized as:

(*a*) sponsoring meetings, which could be full symposia or sessions of symposia devoted to research themes, where information on large projects being planned or under development is given,

(*b*) having the Division B Working Group on Future Large Scale Facilities taken on increased responsibility in helping develop international roadmaps in astronomy and infrastructure models for large collaborations, and to coordinate planning and information exchange between groups interested in committing to collaborations on large forefront facilities,

(c) creating a host website that serves as an information gathering and dissemination site for projects and groups seeking partnerships, and

(d) working with other organizations such as COSPAR to formulate an international roadmap, e.g., as has been done in Europe with Astronet, taking the priorities of the different national decadal reviews into account in a way that can serve as a focus for future mission and project planning.

The discussion between the panelists and the audience confirmed that IAU adoption of the above suggestions would provide significant support for projects of importance that, because of complexity and cost almost certainly require international collaboration, would move astronomy forward. It should be noted, however, that a few audience participants felt that the current system of establishing collaborations has worked well in the past and is not in need of improvement. The large majority of meeting attendees disagreed with this thinking, insisting that however successful astronomers may have been in forming collaborations in the past, the urgency of moving ahead with forefront facilities that are becoming prohibitively expensive requires serious efforts by the community.

An assessment by the FM11 SOC of the options available to the IAU to identify common priorities and facilitate international collaborations leads us to recommend that Division B and the Executive Committee agree to authorize the following course of actions:

• Given the importance of the WG on Future Large-Scale Facilities it may be appropriate to shift the WG from Div. B to the EC. This has been discussed with WG Chair Roger Davies and Div. B President Pietro Ubertini.

• The FL-SF WG is advised to create a mechanism, e.g., a website, that provides easy access to current information on large-scale facilities and includes announcements, offers of collaboration, etc., to be sent out to the community.

• Periodic symposia that are devoted to technological development and new facilities should be conducted by the IAU, much as SPIE holds meetings dedicated to instrumentation. The EC should also encourage sessions at selected symposia that are dedicated to future facilities.

• The IAU should consider a collaboration with COSPAR to define a process that could produce a roadmap identifying important scientific questions and mission concepts that would address them.

David Spergel
Chair

—— Matthew Colless, Sarah Gibson, Lynne Hillenbrand,Cristina Mandrini, Hermann Opgenoorth, Xue Suijian,Yasushi Suto,Jean-Pierre Swings, Saku Tsuneta, Oskar von der Lühe, Patricia Whitelock,Robert Williams, Mei Zhang

SOC

FM 12:
Bridging Laboratory Astrophysics and Astronomy

FM12: A Focus Meeting on Bridging Laboratory Astrophysics and Astronomy

Farid Salama[1], Lyudmila Mashonkina[2] and Steve Federman[3]

[1] NASA Ames Research Center, Moffett Field, CA, United States
email: farid.salama@nasa.gov
[2] Institute of Astronomy RAS, Russia
email: lima@inasan.ru
[3] University of Toledo, United States
email: steven.federman@utoledo.edu

Laboratory astrophysics is the Rosetta stone that enables astronomers to understand and interpret the cosmos. The IAU Commission 14, the predecessor of the new IAU Laboratory Astrophysics Commission C. B5 and the AAS Laboratory Astrophysics Division (LAD) decided to coordinate their efforts this summer to hold a joint meeting at the IAU General Assembly.

This joint effort was in the form of a Focus Meeting that helped bridge Laboratory Astrophysics and Astrochemistry with Astronomy by bringing together expert providers and users of laboratory and astronomical data. This multidisciplinary meeting brought together astronomers with theoretical and experimental chemists and physicists to discuss the state-of-the-art research in their respective disciplines and how their combined expertise address important open questions in astronomy and astrophysics. Attendees were encouraged to bring and to discuss their data needs to improve the interpretations of astronomical phenomena.

The Focus Meeting 12, Bridging Laboratory Astrophysics and Astronomy, discussed the strong interplay between astronomy, astrophysics, astrochemistry and planetary science with theoretical and experimental studies of the chemical and physical processes that drive our Universe.

The meeting was divided into seven topical sessions that discussed atomic and molecular data, dust and ices, plasma, nuclear, and particle physics and their application to astronomy and astrophysics/astrochemistry.

Twenty-five invited talks were presented, distributed between eleven in-depth reviews of the field, twelve presentations that focused on recent, hot topics and two summary talks that concluded the meeting. In addition, close to 70 research contributions were presented at the meeting, distributed between 12 oral presentations and 57 posters. A rich discussion resulted from this wide variety of topics as illustrated in the program of the sessions detailed below (Figs. 1, 2).

All presentations were very well received and the meeting ended on a general discussion of the future directions for laboratory astrophysics. Given the current major development of next-generation facilities and projects, Focus Meeting 12 stressed how laboratory studies can best address the needs of astronomy and stimulate new observations and discussed open questions to be solved in the next decade. All the presentations that were made at the meeting are included in the new Proceedings series of the IAU, entitled "Focus on Astronomy.

Wed like to thank all the people who participated and contributed to make this meeting a real success, highlighting the growing importance and recognition of Laboratory

Astrophysics in today's astronomy. We are particularly indebted to the members of the Scientific Organizing Committee who generously contributed with their time and their expertise to the making of this first Laboratory Astrophysics Meeting at an IAU General Assembly.

The editors, Farid Salama, Lyudmila Mashonkina and Steve Federman

Scientific Organizing Committee (SOC) of FM12:

Martin Asplund (Australian National University, Australia)
Beatriz Barbuy (University of Sao Paulo, Brazil)
Paul Drake (University of Michigan, United States)
Steven Federman (University of Toledo, United States), co-Chair
Karlheinz Langanke (GSI Helmholtzzentrum fr Schwerionenforschung Gmb, Germany)
Harold Linnartz (Leiden Observatory, the Netherlands)
Xiaowei Liu (Peking University, Kavli Institute for Astronomy and Astrophysics, China)
Lyudmila Mashonkina (Institute of Astronomy RAS, Russia), co-Chair
Tom Millar (Queen's University, United Kingdom)
Evelyne Roueff (Observatoire de Paris, France)
Farid Salama (NASA-Ames Research Center, United States), co-Chair
Daniel Savin (Columbia University, United States)

References

IAU Atomic and Molecular Data Commission 14, `http://oldwww.inasan.ru/iau14/`
IAU Laboratory Astrophysics Commission C.B5, `http://www.iau.org/science/scientific_bodies/commissions/B5/`
AAS Laboratory Astrophysics Division (LAD), `http://lad.aas.org/`
IAU General Assembly, `http://astronomy2015.org/`
Focus Meeting 12, `http://astronomy2015.org/focus_meeting_12`

FM 12: Bridging Laboratory Astrophysics and Astronomy

Monday, 3 August 2015
FM12.1: Atoms, 10:30 am - 12:30 pm (Session Chairs: Lyudmila Mashonkina and Beatriz Barbuy)
10:30 am: Farid Salama SOC opening remarks FM12.1.00
10:35 am: Christopher Sneden (Invited Review) Atomic Data for Stellar Nucleosynthesis FM12.1.01
11:05 am: Jelle Kaastra (Invited Review) Atomic processes in optically thin plasmas FM12.1.02
11:35 am: Karin Lind (Invited Topical) Modeling cool star spectra with inadequate input physics FM12.1.03
11:50 am: Michael Murphy (Invited Topical) Quasar searches for variations in fundamental constants: the need for laboratory spectroscopy FM12.1.04
12:05 pm: Natalie Hell (Contributed) K-shell transitions in L-shell ions with the EBIT calorimeter spectrometer FM12.1.05
12:20 pm: Norbert Przybilla (Contributed) Quantitative spectroscopy of hot stars: accurate atomic data applied on a large scale as driver of recent breakthroughs FM12.1.06

FM12.2: Molecules I, 2:00 pm - 3:30 pm (Session-Chair: Steve Federman)
2:00 pm: Ewine van Dishoeck (Invited Review) The molecular universe: from astronomy to laboratory astrophysics and back FM12.2.01
2:30 pm: Svetlana Berdyugina (Invited Review) Molecules in Magnetic Fields FM12.2.02
3:00 pm: Leen Decin (Contributed) Evolved stars as complex chemical laboratories – the quest for gaseous chemistry FM12.2.03
3:15 pm: Annemieke Petrignani (Contributed) Anharmonicity and infrared bands of Polycyclic Aromatic Hydrocarbon (PAH) molecules FM12.2.04

Tuesday, 4 August 2015
FM12.3: Molecules II and Dust and Ices I, 10:30 am - 12:30 pm (Session Chair: Harold Linnartz)
10:30 am: Ralf I Kaiser (Invited Topical) Probing the Formation of Complex Organic Molecules in Interstellar Ices FM12.3.01
10:45 am: James Lyons (Contributed) CO isotopologue ratios in the solar photosphere FM12.3.02
11:00 am: Karin Oberg (Invited Review) Laboratory constraints on ice formation, restructuring and desorption FM12.3.03
11:30 am: Anthony Jones (Invited Review) Interstellar dust: interfacing laboratory, theoretical and observational studies FM12.3.04
12:00 pm: Pascale Ehrenfreund (Contributed) Organics in Space: Results from SPace Exposure Platforms and Nanosatellites FM12.3.05
12:15 pm: Gianfranco Vidali (Contributed) Nitrogen chemistry on dust grains: the formation of hydroxylamine, precursor to glycine FM12.3.06

FM12.4: Dust and Ices II and Planetary I, 2:00 pm - 3:30 pm (Session Chairs: Gianfranco Vidali and Farid Salama)
2:00 pm: Adwin Boogert (Invited Topical) Telescope Observations of Interstellar and Circumstellar Ices: Successes of and Need for Laboratory Simulations FM12.4.01
2:15 pm: Takashi Onaka (Invited Topical) AKARI nearHinfrared spectroscopy of interstellar ices FM12.4.02
2:30 pm: Dominique Bockelee-Morvan (Invited Review) Comets and Laboratory Astrophysics FM12.4.03

Figure 1. Schedule of the Focus Meeting on Bridging Laboratory Astrophysics and Astronomy.

FM 12: Bridging Laboratory Astrophysics and Astronomy

Tuesday, 4 August 2015
FM12.4: Dust and Ices II and Planetary I, 2:00 pm - 3:30 pm (Session Chairs: Gianfranco Vidali and Farid Salama)
3:00 pm: Athena Coustenis (Invited Topical) Laboratory and theoretical work in the service of planetary atmospheric research FM12.4.04
3:15 pm: Robert Nelson (Contributed) Laboratory Simulations of Planetary Surfaces: Understanding Regolith Physical Properties from Astronomical Photometric Observations FM12.4.05

Wednesday, 5 August 2015
FM12.5: Planetary II and Plasma I, 10:30 am - 12:30 pm (Session Chairs: Helen Fraser and Daniel Savin)
10:30 am: Peter Jenniskens (Invited Topical) Meteorites FM12.5.01
10:45 am: Ella Sciamma-O'Brien (Contributed) The THS: Simulating Titan's atmospheric chemistry at low temperature FM12.5.02
11:00 am: William Fox (Invited Review) Magnetic field generation, Weibel-mediated collisionless shocks, and magnetic reconnection in colliding laserHproduced plasmas FM12.5.03
11:30 am: Frederico Fiuza (Invited Topical) Generation of collisionless shock in laser-produced plasmas FM12.5.04
11:45 am: James Drake (Invited Review) The emerging understanding of magnetic reconnection through laboratory experiments, theory and modeling and satellite measurements FM12.5.05
12:15 pm: Michael Hahn (Contributed) Influence of Multiple Ionization on Charge State Distributions FM12.5.06

FM12.6: Plasma II and Nuclei and Particles I, 2:00 pm - 3:30 pm (Chairs: Lyudmila Mashonkina, Xiaowei Liu)
2:00 pm: Jan Egedal (Invited Topical) The Wisconsin Plasma Astrophysics Laboratory (WiPAL): A New Experimental User Facility FM12.6.01
2:15 pm: Maëlle Le Pennec (Contributed) Testing stellar opacities with laser facilities FM12.6.02
2:30 pm: Weiping Liu (Invited Review) Progress of Jinping Underground laboratory for Nuclear Astrophysics experiment JUNA FM12.6.03
3:00 pm: Elisabete de Gouveia Dal Pino (Invited Topical) Cherenkov Telescope Array: Unveiling the Gamma Ray Universe and its Cosmic Particle Accelerators FM12.6.04
3:15 pm: Yong-Zhon Qian (Invited Topical) Neutrinos, Nuclei, and Nucleosynthesis: Implications for Chemcial Evolution of the Early Galaxy FM12.6.05

FM12.7: Nuclei and Particles II and Summary, 4:00 pm - 6:00 pm (Session Chairs: Steve Federman and Farid Salama)
4:00 pm: Kei Kotake (Invited Review) Multi-D Core-Collapse Supernova Models and the Multi-messenger Observables FM12.7.01
4:30 pm: Rubén López-Coto (Contributed) MACHETE: A transit Imaging Atmospheric Cherenkov Telescope to survey half of the VHE gamma ray sky FM12.7.02
4:45 pm: Marie-Lise Dubernet (Invited Topical) Atomic and Molecular Databases, VAMDC (Virtual Atomic and Molecular Data Centre) FM12.7.03
5:00 pm: Beatriz Barbuy (Invited Summary) Summary & outstanding questions FM12.7.04
5:25 pm: Alexander Tielens (Invited Summary) Summary & outstanding questions FM12.7.05
5:50 pm: Farid Salama SOC closing remarks FM12.7.06

Figure 2. Schedule of the Focus Meeting on Bridging Laboratory Astrophysics and Astronomy.

Atomic Data for Stellar Nucleosynthesis

Christopher Sneden[1], James E. Lawler[2], Elizabeth A. Den Hartog[2] and Michael E. Wood[2]

[1] Department of Astronomy and McDonald Observatory, The University of Texas, Austin, TX 78712
email: chris@verdi.as.utexas.edu

[2] Department of Physics, University of Wisconsin-Madison, 1150 University Ave., Madison, WI 53706
email: jelawler@wisc.edu, eadenhar@wisc.edu, michael.wood@nist.gov

Abstract. Stellar chemical composition analyses can only yield reliable abundances if the atomic transition parameters are accurately determined. During the last couple of decades a renewed emphasis on laboratory spectroscopy has produced large sets of useful atomic transition probabilities for species of interest to stellar spectroscopists. In many cases the transition data are of such high quality that they play little part in the abundance uncertainties. We summarize the current state of atomic parameters, highlighting the areas of satisfactory progress and noting places, where further laboratory progress will be welcome.

Keywords. atomic data, stars: abundances, stars: Population II

1. Introduction

Detailed understanding of the Milky Way's chemical evolution leads to insights on the lives and deaths of element-giving stars, on the return of newly-minted elements to the interstellar medium, on stellar populations, and ultimately on the complex elemental and even isotopic composition of our solar system. Galactic chemical evolution depends on a variety of inputs, but is fundamentally limited by the accuracies of observed abundances of individual stars. Stellar abundance accuracies depend on many factors, such as the quality of observed spectra, the reliability of stellar model atmospheres, the completeness of physical descriptions of atomic and molecular species and energy level populations, and the understanding of radiative transfer in stellar atmospheres. Again there is a fundamental limitation: stellar abundances can be no better than the accuracies of line data (transition probabilities, hyperfine and isotopic substructures) of spectroscopic features that are detectable in stellar spectra.

Here we briefly describe recent laboratory advances in atomic transition data. We concentrate on contributions by the Wisconsin atomic physics group on lines from species of most interest to spectroscopists working on cool stars. Important work has been done by several other groups, in particular those in Lund, Mons, and London, and we comment briefly on their results.

2. Neutron-Capture Elements

More than three decades ago came the discovery of low metallicity ([Fe/H] < -2)† stars with abnormal enhancements of the heavy neutron-capture (n-capture; atomic number

† We adopt the standard spectroscopic notation: for elements A and B, $[A/B] \equiv \log_{10}(N_A/N_B)_\star - \log_{10}(N_A/N_B)_\odot$. We use the definition $\log \epsilon(A) \equiv \log_{10}(N_A/N_H) + 12.0$, and equate metallicity with the stellar [Fe/H] value.

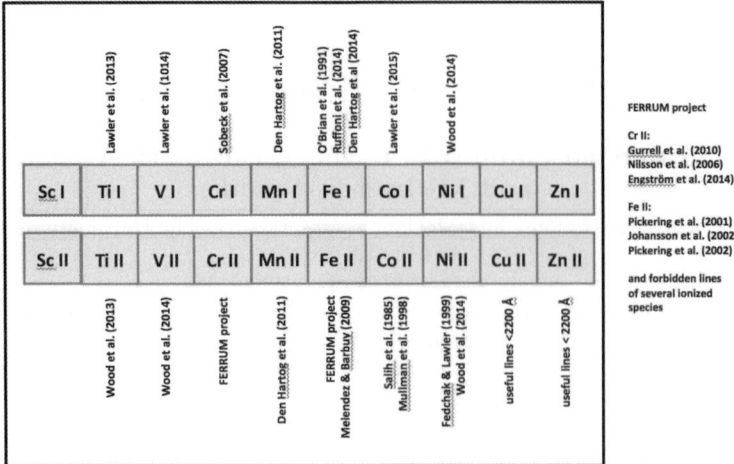

Figure 1. Visual display of neutral and ionized Fe-group species, with labels citing mostly recent extensive laboratory studies. Lack of a reference for a species indicates no recent lab publication, but does not necessarily indicate lack of good older data.

$Z > 30$) elements (e.g., Griffin et al. 1982). Now many metal-poor stars are known that exhibit enrichments by rapid (the r-process) and slow (the s-process) n-capture synthesis events. A summary of the general properties of r- and s-process-rich stars was given in Sneden et al. (2008). For some stars, particularly those with r-process enrichments, there are impressive numbers of published n-capture elemental abundances. Just to name two examples: CS 31082-001 with 37 elements (Siqueira Mello et al. 2013) and HD 221170 (Ivans et al. 2006) with 35 elements. These stars' n-capture abundances show excellent agreement with the scaled solar-system r-process-only element distribution. This conclusion is made possible by improvements in observed stellar spectra, and by significant upgrades in the quality and quantity of transition data, especially for the rare-earth elements (summarized in Sneden et al. 2009) and for lighter n-capture elements (e.g., Malcheva et al. 2006, Biémont et al. 2011). The excellent agreement between the n-capture abundances of r-rich metal-poor stars and the solar-system values suggests that the major observational work in this area may have now been accomplished, and further advances mostly may be with theoretical r-process predictions. Observers might want to concentrate more on detailed abundance investigations of so-called "r-process truncation" stars (Boyd et al. 2012 and references therein) and the s-process-rich stars (Placco et al. 2015 and references therein).

3. Fe-Group Elements

In contrast to the n-capture elements, abundances of Fe-group ($Z = 21-30$, Sc–Zn) elements have not reached a level of accuracy to really constrain stellar nucleosynthesis models. This is disappointing because theoretical models can predict the Fe-group elemental abundances in some detail (e.g., Kobayashi et al. 2011 and references therein).

The Fe-group observed abundance reliability has been limited by (1) lack of transitions of both neutral and ionized species of several elements in the usually studied optical spectral region ($\lambda > 4000$ Å); (2) concerns about large departures from local thermodynamic equilibrium in lines of especially the neutral species; (3) and most importantly, less-than-desired quality in laboratory gf values and hyperfine substructures. Fortunately, several groups are making much progress in lab spectroscopy of these elements. We depict the

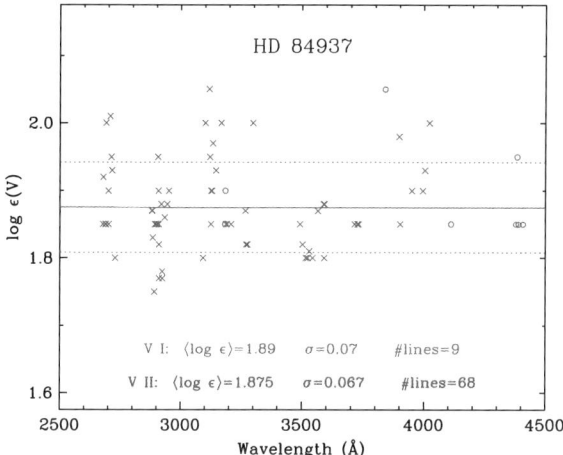

Figure 2. Vanadium abundances in the warm main-sequence very metal-poor star HD 84937, from V I and V II transitions. This figure is taken from Lawler *et al.* (2014). The abundance statistics are written in the figure legend.

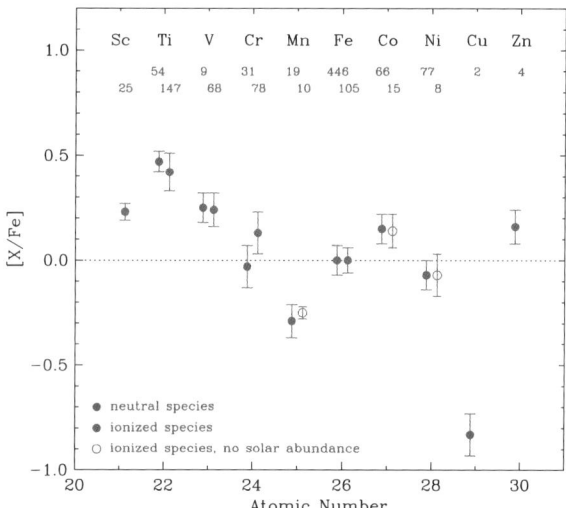

Figure 3. Abundance ratios of Fe-group elements with respect to Fe in the warm main-sequence very metal-poor star HD 84937. The symbol meanings are given in the figure legend. The number of lines that have contributed to each species abundance is written beneath its element name.

current state of published Fe-group transition papers in Fig. 1. By inspection one can easily see that most Fe-group species have received renewed lab scrutiny in the last decade.

As an example of application of new lab data to the stellar spectra, Fig. 2, taken from Lawler *et al.* (2014), shows abundances from individual V I and V II lines in the spectrum of the very metal-poor ([Fe/H] $\simeq -2.3$) warm ($T_{\rm eff} \simeq 6250$K) main sequence star HD 84937. Several conclusions can be drawn here. (1) Mean abundances from the two species are in agreement, and the line-line abundance scatters are relatively small. (2) Only a few V I lines are detectable; V is mostly ionized, and the metallicity of HD 84937 is low. Both effects conspire to weaken V I transitions. (3) A rich spectrum of V II can be seen, but only by obtaining spectra in the near-UV (3000–4000 Å) and the vacuum-UV (2300–3000 Å).

We have recently completed a comprehensive Fe-group abundance study of HD 84937 (Sneden et al. 2015). A summary of our results is shown in Fig. 3. It is clear that neutral and ionized species yield the same elemental abundances for seven of the Fe-group elements; Saha ionization balance holds in this stellar atmosphere. More interesting astrophysically, there are pronounced departures from the solar chemical composition. This is especially seen in the (unexpected) overabundances of the three lightest elements Sc, Ti, and V. This is not predicted in standard massive-star nucleosynthesis models. Further discussion of these abundance anomalies are given in Sneden et al. A general (but detailed) Fe-group abundance survey in other low metallicity stars should be undertaken.

This work has been supported in part by NASA grant NNX10AN93G (J.E.L.), by NSF grant AST-1211055 (J.E.L.) and NSF grant AST-1211585 (C.S.).

References

Biémont, É., Blagoev, K., Engström, L., et al. 2011, MNRAS, 414, 3350
Boyd, R. N., Famiano, M. A., Meyer, B. S., et al. 2012, ApJL, 744, L14
Den Hartog, E. A., Ruffoni, M. P., Lawler, J. E., et al. 2014, ApJS, 215, 23
Den Hartog, E. A., Lawler, J. E., Sobeck, J. S., Sneden, C., & Cowan, J. J. 2011, ApJS, 194, 35
Engström, L., Lundberg, H., Nilsson, H., Hartman, H., & Bäckström, E. 2014, A&A, 570, A34
Fedchak, J. A. & Lawler, J. E. 1999, ApJ, 523, 734
Griffin, R., Gustafsson, B., Vieira, T., & Griffin, R. 1982, MNRAS, 198, 637
Gurell, J., Nilsson, H., Engström, L., et al. 2010, A&A, 511, A68
Ivans, I. I., Simmerer, J., Sneden, C., et al. 2006, ApJ, 645, 613
Johansson, S., Derkatch, A., Donnelly, M. P., et al. 2002, Phys. Scr. T, 100, 71
Kobayashi, C., Karakas, A. I., & Umeda, H. 2011, MNRAS, 414, 3231
Lawler, J. E., Guzman, A., Wood, M. P., Sneden, C., & Cowan, J. J. 2013, ApJS, 205, 11
Lawler, J. E., Sneden, C., & Cowan, J. J. 2015, ApJS, 220, 13
Lawler, J. E., Wood, M. P., Den Hartog, E. A., et al. 2014, ApJS, 215, 20
Malcheva, G., Blagoev, K., Mayo, R., et al. 2006, MNRAS, 367, 754
Meléndez, J. & Barbuy, B. 2009, A&A, 497, 611
Mullman, K. L., Cooper, J. C., & Lawler, J. E. 1998, ApJ, 495, 503
Nilsson, H., Ljung, G., Lundberg, H., & Nielsen, K. E. 2006, A&A, 445, 1165
O'Brian, T. R., Wickliffe, M. E., Lawler, J. E., Whaling, W., & Brault, J. W. 1991, Journal of the Optical Society of America B Optical Physics, 8, 1185
Pickering, J. C., Johansson, S., & Smith, P. L. 2001, A&A, 377, 361
Pickering, J. C., Donnelly, M. P., Nilsson, H., Hibbert, A., & Johansson, S. 2002, A&A, 396, 715
Placco, V. M., Beers, T. C., Ivans, I. I., et al. 2015, ApJ, 812, 109
Ruffoni, M. P., Den Hartog, E. A., Lawler, J. E., et al. 2014, MNRAS, 441, 3127
Siqueira Mello, C., Spite, M., Barbuy, B., et al. 2013, A&A, 550, A122
Sneden, C., Cowan, J. J., & Gallino, R. 2008, ARA&A, 46, 241
Sneden, C., Cowan, J. J., Pignatari, M., Kobayashi, C., Lawler, J. E., Den Hartog, E. A., & Wood, M. P. 2015, ApJS, submitted
Sneden, C., Lawler, J. E., Cowan, J. J., Ivans, I. I., & Den Hartog, E. A. 2009, ApJS, 182, 80
Sobeck, J. S., Lawler, J. E., & Sneden, C. 2007, ApJ, 667, 1267
Wood, M. P., Lawler, J. E., Den Hartog, E. A., Sneden, C., & Cowan, J. J. 2014a, ApJS, 214, 18
Wood, M. P., Lawler, J. E., Sneden, C., & Cowan, J. J. 2013, ApJS, 208, 27
Wood, M. P., Lawler, J. E., Sneden, C., & Cowan, J. J. 2014b, ApJS, 211, 20

Atomic processes in optically thin plasmas

Jelle S. Kaastra[1,2,3], Liyi Gu[1], Junjie Mao[1,2], Missagh Mehdipour[1], Ton Raassen[1] and Igone Urdampilleta[1,2]

[1]SRON Netherlands Institute for Space Research, Sorbonnelaan 2, 3584 CA Utrecht, The Netherlands
email: j.kaastra@sron.nl
[2]Leiden Observatory, Leiden University, PO Box 9513, 2300 RA Leiden, the Netherlands
[3]Department of Physics and Astronomy, Universiteit Utrecht, P.O. Box 80000, 3508 TA Utrecht, the Netherlands

Abstract. The Universe contains a broad range of plasmas with quite different properties depending on distinct physical processes. In this contribution we give an overview of recent developments in modeling such plasmas with a focus on X-ray emission and absorption. Despite the fact that such plasmas have been investigated already for decades, and that overall there is a good understanding of the basic processes, there are still areas, where improvements have to be made that are important for the analysis of astrophysical plasmas. We present recent work on the update of atomic parameters in the codes that describe the emission from collisional plasmas, where older approximations are being replaced now by more accurate data. Further we discuss the development of models for photo-ionised plasmas in the context of outflows around supermassive black holes and models for charge transfer that are needed for analyzing the data from the upcoming ASTRO-H satellite.

Keywords. Atomic data, atomic processes, radiation mechanisms: general, X-rays: general

1. Introduction

X-ray emitting plasmas are found everywhere in the Universe, from the small scales of the Solar system up to the large-scale cosmic web filaments. These plasmas show a broad range of environments and physical conditions, for instance collisionally ionised, photo-ionised or transiently ionised. High-resolution X-ray spectroscopy is the key tool to understand these sources. While a lot of progress has been made in the last 15 years using the grating spectrometers on the XMM-Newton and Chandra observatories for point sources, high-resolution X-ray spectroscopy for extended sources has been limited to only the most compact sources on the sky, typically smaller than a few arcmin and using the XMM-Newton RGS detector.

In 2016 Japan will launch ASTRO-H with the SXS (soft X-ray spectrometer) calorimeter, the first detector able to obtain high-resolution spectra of extended sources and with unsurpassed spectral sensitivity in the Fe-K band around 6 keV. The older models that are being used to model astrophysical sources are not always up to date with respect to these future capabilities, and therefore efforts are being made to improve the atomic data and models that are being used for X-ray spectroscopy.

There is a strong need to model these different sources with the same, consistent set of atomic parameters. In this contribution we give an outline of recent work in this field done by the authors at SRON.

2. Strategy

The first X-ray spectroscopic models at SRON were developed by Rolf Mewe. With his background as atomic physicist he was hired in 1970 by SRON to model the Solar X-ray spectra obtained by people from this group. His first paper dealt with He-like line ratio's (Mewe 1972a) followed one month later by his first paper with a list of 260 lines in the 1–60 Å range (Mewe 1972b). Both papers and follow-up work showed a basic effective strategy to model such spectra, by parameterising the relevant physical processes using simple formulae that are accurate enough relative to the original calculations, and interpolate along isoelectronic sequences, where data are lacking.

The work by Mewe and co-workers gradually evolved and culminated in two papers about continuum emission (Mewe, et al. 1986) and line emission (Mewe, et al. 1985), the latter paper with 2167 lines between 1 Å and 300 Å. Some minor corrections, major improvements in the Fe-L complex and addition of lines up to 2000 Å (now totalling 5408 lines) were done later in the *mekal* model that is implemented in e.g., the XSPEC package.

Starting in 1992, the spectral code was incorporated in a broader scope spectral analysis package, named SPEX (Kaastra, et al. 1996), that allows the user to do spectral data analysis (fitting), plotting, and diagnostic output of physical quantities. In late 2015, version 3 of this package is expected to be released.

Similar developments took place in the USA led by John Raymond, evolving eventually in the APEC model, and in Italy by Landini and Monsignori-Fossi, evolving eventually in the CHIANTI code.

Driven by the requirements of new instrumentation, or by the extension of the present models to more extreme environments, updates of the atomic data are needed.

For example, in collisional ionisation equilibrium plasmas, for most lines collisional excitation by free electrons is the dominant line emission process, and other processes such as radiative recombination contribute only a small fraction of the total line power. For this reason, Mewe had approximated the radiative recombination contribution to the line power as a function of temperature T by a local power-law, proportional to $T^{-0.7}$. This is a reasonable approximation for temperatures close to the temperature, where the relevant ion has its peak concentration. However, radiative recombination rates scale approximately as $T^{-0.5}$ and $T^{-1.5}$ at very low and high temperatures, respectively. Thus, for instance for photo-ionised plasmas, where the temperature is relatively low compared to the typical ionisation potentials, the recombination rates are overestimated using this approach, and a similar situation occurs at high energies for ionising plasmas such as found for example in supernova remnants. Thus, more accurate approximations are needed.

Despite all this evolution of the models, the basic strategy remains unchanged. The code must allow options for fast calculation and yet be accurate enough. This means that the number of mathematical operations and data storage for the basic cross sections or plasma rates need to be minimised. Therefore we follow the same strategy as Mewe used long ago: we use simple, accurate and fast approximations, but more accurate and complete than before.

In addition, since the spectral resolution in the X-ray band in the near future will still be limited as compared for instance to the optical band, we restrict our models to all elements with nuclear charge $Z < 30$. Higher elements have too low abundances to give detectable X-ray lines.

Finally, for a full model, we need updates for many processes, such as the collisional and photo-ionisation cross sections, radiative transition probabilities, auto-ionisation rates, recombination rates, line energies, etc.

Since this is a process, where it easy to make mistakes, and sometimes small changes in specific parameters can have a large effect on parts of the spectrum, a careful comparison of our results with both the older calculations and other plasma models will be made.

3. Examples

3.1. *Collisional ionisation rates*

In the past, several compilations of direct collisional ionisation rates have been made (here collisions with free electrons are assumed). One of the most recent compilations was the one given by Dere (2007). He has given a thorough review of cross sections and has calculated total ionisation rates for each ion in a plasma, listing it at typically 20 spline points on a scaled temperature grid.

Since the review by Dere newer data have become available. But more importantly, since Dere lists total rates, the individual contributions of the atomic subshells are not available, and these are needed for instance in order to calculate the fluorescence contribution to spectral lines after inner-shell ionisation. We have therefore started a project of updating the database of Dere, but obtaining parameterised forms of the cross-sections for individual subshells. These parameterised forms are chosen in such a way that they can be integrated analytically over a thermal Maxwell-distribution or other simple forms of electron distributions. This work will be reported by Urdampilleta *et al.* (in prep.).

3.2. *Radiative recombination*

For radiative recombination, extensive material exists that gives the total recombination rates, added over all possible higher shells, into which the recombining electron is captured. Also in this case we need rates to individual levels in order to be able to calculate the corresponding emission line spectrum. In addition, not only the capture rates to individual levels are of interest, but also the associated cooling rates, the kinetic energy of the captured electron averaged over the recombination rate. This quantity is of importance for photo-ionised plasmas.

We have started collecting data from the literature or made our own calculations, and parameterise the plasma rates by a relatively simple seven-parameter equation that gives an accuracy of $< 5\,\%$ for most recombinations. This work will be presented by Mao *et al.* (in prep.).

3.3. *Photo-ionised plasmas*

In recent work on photo-ionised plasmas, it appeared to us that in some cases there are apparently large differences between the calculated photo-ionisation equilibria using different spectral models like *Cloudy*, *XSTAR*, and our own *pion* model in the SPEX package. Photo-ionised plasmas are the most challenging environments, because of the large number of physical processes that play a role. We have started comparing the individual processes that are used in these models in detail. For example, it appears that in some work the radiative recombination approximation of Seaton (1959) for hydrogen is used, which gives an accurate approximation up to kT of about 300 eV, but overestimates the recombination rate significantly at higher energies. As a consequence, the cooling due to radiative recombination is overestimated by more than an order of magnitude for temperatures of a few keV or higher. Other important differences concern the treatment of Compton scattering in the heating or cooling balance. A systematic comparison will be presented elsewhere (Mehdipour et al., in prep.).

3.4. Helium-like line ratios in active galactic nuclei

The so-called R-ratio between the forbidden and intercombination line in helium-like ions is known to be a good density indicator. At low densities, R is high, and then it quickly drops above a critical density that depends on the ion that is considered. However, in some AGN R-ratios have been found that are even higher than the low-density limit, and therefore cannot be explained. But because the intercombination line of the He-like triplet is close in energy to energy of the $1s^2 2s$ to $1s2s(^1S)2p$ ^2P doublet in Li-like ions, and for sufficiently large Doppler broadening and column density of the Li-like ion, the Li-like ions can absorb a significant fraction of the intercombination line flux, thereby artificially increasing the apparent R-ratio (Mehdipour, et al. 2015). Thus, depending on the precise physical conditions, care must be taken in interpreting R-ratios.

3.5. Charge transfer modelling

Charge transfer is the process, where an ion collides with another ion, neutral atom or molecule, and where the ion either obtains an electron from the other particle, or looses an electron. Charge transfer recombination of ions with neutral hydrogen atoms is an important process in many situations, in particular at the interface of cold and hot material, like comets, the solar wind, supernova remnant shocks, etc. We have started a project (Gu et al., in prep.), where we collect all available cross section data for this process from the literature and databases, derive from them scaling laws for the distribution over n and ℓ, dependent of the velocity difference between both particles, and apply these scaling laws to the other elements and ions, for which no direct data exists. Radiative cascade calculations are taken into account up to $n = 16$, $\ell = 15$. Preliminary tests of this cx model in SPEX show that it gives an excellent fit to the X-ray spectrum of comet C2000 WM1.

4. Conclusions

Astrophysical sources are sometimes found in remarkable areas of parameter space. New X-ray missions like ASTRO-H (launch 2016) demand more details and accuracy. We showed examples of the work in progress at SRON: updates of the atomic parameters in our X-ray spectral models to account for this. SPEX (www.sron.nl/spex) Version 3 will contain these updates (release late 2015).

References

Dere, K. P. 2007, A&A, 466, 771
Kaastra, J. S., Mewe, R., & Nieuwenhuijzen, H. 1996, in: *11th Colloquium on UV and X-ray Spectroscopy of Astrophysical and Laboratory Plasmas*, p. 411
Mehdipour, M., Kaastra, J. S., & Raassen, A. J. J. 2015, A&A, 579, A87
Mewe, R. 1972a, *Solar Physics*, 22, 114
Mewe, R. 1972b, *Solar Physics*, 22, 459
Mewe, R., Gronenschild, E. H. B.. M., & van den Oord, G. H. J.. 1985, A&AS, 62, 197
Mewe, R., Lemen, J. R., & van den Oord, G. H. J.. 1986, A&AS, 65, 511
Seaton, M. J. 1959, *MNRAS*, 119, 81

K-shell transitions in L-shell ions with the EBIT calorimeter spectrometer

Natalie Hell[1,2], Greg V. Brown[2], Jörn Wilms[1], Peter Beiersdorfer[2], Richard L. Kelley[3], Caroline A. Kilbourne[3] and F. Scott Porter[3]

[1] Dr. Remeis-Sternwarte & ECAP, Friedrich-Alexander Universität Erlangen-Nürnberg, Sternwartstr. 7, 96049 Bamberg, Germany
email natalie.hell@sternwarte.uni-erlangen.de

[2] Lawrence Livermore National Laboratory, 7000 East Ave., Livermore CA 94550, USA

[3] NASA-GSFC, 8800 Greenbelt Road, Greenbelt, MD 20771, USA

Abstract. With the large improvement in effective area of Astro-H's micro-calorimeter soft X-ray spectrometer (SXS) over grating spectrometers, high-resolution X-ray spectroscopy with good signal to noise will become more commonly available, also for faint and extended sources. This will result in a range of spectral lines being resolved for the first time in celestial sources, especially in the Fe region. However, a large number of X-ray line energies in the atomic databases are known to a lesser accuracy than that expected for Astro-H/SXS, or have no known uncertainty at all. To benchmark the available calculations, we have therefore started to measure reference energies of K-shell transition in L-shell ions for astrophysically relevant elements in the range $11 \leqslant Z \leqslant 28$ (Na to Ni), using the Lawrence Livermore National Laboratory's EBIT-I electron beam ion trap coupled with the NASA/GSFC EBIT calorimeter spectrometer (ECS). The ECS has a resolution of $\sim 5\,\mathrm{eV}$, i.e., similar to Astro-H/SXS and Chandra/HETG. A comparison to crystal spectra of lower charge states of sulfur with $\sim 0.6\,\mathrm{eV}$ resolution shows that the analysis of spectra taken at ECS resolution allows to determine the transition energies of the strongest components.

Keywords. atomic data; X-rays: general; methods: laboratory

1. Introduction

The microcalorimeter soft X-ray spectrometer (SXS) onboard the upcoming Astro-H mission (Takahashi *et al.* 2010) will bring a large improvement in effective area over the currently available grating spectrometers on the XMM-Newton and Chandra X-ray observatories. This will make high-resolution spectroscopy more commonly available and extend its application to faint and extended sources. Consequently, a plethora of new spectral features is expected to be observed with the new instrument. For example, in a variety of brighter sources – including X-ray binaries, AGN and solar flares – K-shell transitions of L-shell ions of Si and S have already been observed with grating spectrometers. With the larger effective area, similar features should become apparent also for other elements and additional sources. Detailed analysis of the Fe K lines in CCD spectra of supernova remnants reveals a complex ionization structure (Yamaguchi *et al.* 2014) and some SNRs show signatures of the trace elements Cr and Mn (Hughes *et al.* 2014). Astro-H SXS will not only be able to resolve the charge balance for these Fe group elements; it will also detect and resolve weak K-shell lines from lower odd-Z trace elements that are more abundant than Cr and Mn, but generally too close in energy to stronger lines for CCD resolution (Hughes *et al.* 2014). To take full advantage of the diagnostic capabilities of these lines, accurate atomic reference data are crucial, starting with transition energies. However, inner-shell transitions are not commonly available in

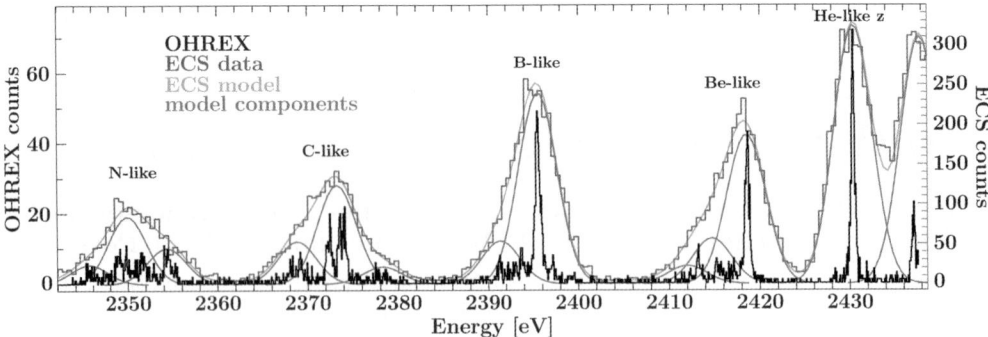

Figure 1. Crystal spectra of sulfur in comparison with ECS spectra and its fit model.

current databases, and although some calculations do exist, their accuracy is unknown. To benchmark the theoretical reference energies, we are measuring K-shell transitions in L-shell ions of astrophysically relevant ions, described here on the example of Si and S.

2. Measurements

We use the Lawrence Livermore electron beam ion trap EBIT-I (Marrs et al. 1988), where the electron beam ionizes and excites injected neutral material. The ions are trapped by the potential well created through a set of three drift tubes. Employing short cycles of dumping and refilling the trap together with a large supply of neutrals from the gas injector, allows us to create a low charge balance. We record the photons emitted from the trapped ions with the EBIT calorimeter spectrometer (ECS; Porter et al. 2008). The ECS is similar to the calorimeter on-board Astro-H and has a resolution of about 5 eV. At this resolution, the principle lines of individual L-shell ions are clearly separated (Fig. 1). The line profiles and the near constant energy resolution allow us to further disentangle the strongest transitions in these line blends. The energy scale is calibrated using well-known reference lines like the Rydberg series of He- and H-like ions. The accuracy of the measured strong lines in Si and S is $\lesssim 0.5$ eV, corresponding to a Doppler shift of $90\,\mathrm{km\,s^{-1}}$, i.e., slightly better than the calibration of Chandra HETG. For the S spectra, a comparison of the model components of the fit to the ECS data with a higher resolution (~ 0.6 eV FWHM) spherical crystal spectrometer, dubbed OHREX, shows that the strong transitions are correctly reproduced by our analysis of the ECS data (Fig. 1). The ECS measurements are available for a few of the astrophysically relevant elements between $Z = 11$ and $Z = 28$, including Si and S (Hell et al. 2013), and we are continuing the effort to fill in the gaps.

This work was supported by LLNL under Contract DE-AC52-07NA27344, by NASA grants to LLNL and NASA/GSFC, and by the European Space Agency under ESA contract No. 4000114313/15/NL/CB.

References

Hell, N., Miškovičová, I., Brown, G. V., et al. 2013, *Phys. Scr.*, T156, 014008
Hughes, J. P., Safi-Harb, S., Bamba, A., et al. 2014, arxiv:1412.1169v1
Marrs, R. E., Levine, M. A., Knapp, D. A., & Henderson, J. R. 1988, *Phys. Rev. Lett.*, 60, 1715
Porter, F. S., Beiersdorfer, P., Brown, G. V., et al. 2008, *J. Low Temp. Phys.*, 151, 1061
Takahashi, T., Mitsuda, K., Kelley, R., et al. 2010, in: M. Arnaud, S. S. Murray, T. Takahashi (eds.), *Space Telescopes and Instrumentation 2010: UV to Gamma Ray*, Proc. SPIE 7732
Yamaguchi, H., Eriksen, K. A., Badenes, C., et al., 2014, *ApJ* 780, 136

Quantitative spectroscopy of hot stars: accurate atomic data applied on a large scale as driver of recent breakthroughs

N. Przybilla[1], V. Schaffenroth[1], M. F. Nieva[1] and K. Butler[2]

[1]Institut für Astro- und Teilchenphysik, Universität Innsbruck,
Technikerstr. 25/8, A-6020 Innsbruck, Austria
email: norbert.przybilla@uibk.ac.at, veronika.schaffenroth@uibk.ac.at,
maria-fernanda.nieva@uibk.ac.at

[2]Universitäts-Sternwarte München, Scheinerstr. 1, D-81679 München, Germany
butler@usm.lmu.de

Abstract. OB-type stars present hotbeds for non-LTE physics because of their strong radiation fields that drive the atmospheric plasma out of local thermodynamic equilibrium. We report on recent breakthroughs in the quantitative analysis of the optical and UV-spectra of OB-type stars that were facilitated by application of accurate and precise atomic data on a large scale. An astrophysicist's dream has come true, by bringing observed and model spectra into close match over wide parts of the observed wavelength ranges. This allows tight observational constraints to be derived from OB-type stars for a wide range of applications in astrophysics. However, despite the progress made, many details of the modelling may be improved further. We discuss atomic data needs in terms of laboratory measurements and also *ab-initio* calculations. Particular emphasis is given to quantitative spectroscopy in the near-IR, which will be the focus in the era of the upcoming extremely large telescopes.

Keywords. atomic data, radiative transfer, stars: abundances, stars: atmospheres, stars: early-type, stars: fundamental parameters

---

OB-type stars are important targets for addressing a wide range of astrophysical questions as their high luminosity allows high-resolution spectra to be obtained even if they are situated at large distances. The atmosphere of an OB star is exposed to an intense radiation field that drives the plasma out of local thermodynamic equilibrium (non-LTE). This leads to a complication in the spectral analysis as the usually adopted Saha-Boltzmann formulae for the description of excitation and ionization are no longer applicable. Instead, a detailed treatment of the individual atomic processes (de-)populating the energy levels and the radiation field is required, i.e. a simultaneous solution of the radiative transfer and statistical equilibrium equations. Among other factors, the extent to which all relevant processes are included and the accuracy and precision of the atomic data employed in the modelling – the model atoms – determine the quality and realism of the synthetic spectra to be compared with observation (see e.g. Przybilla 2010).

We have adopted a hybrid non-LTE approach for the modelling of the atmospheres of OB-type stars (and their evolved progeny) in the mass range $\sim$7–18 $M_\odot$, as described in detail by Przybilla *et al.* (2006a) and Nieva & Przybilla (2007, 2008). The modelling combines hydrostatic, plane-parallel, and line-blanketed LTE model atmospheres with non-LTE line formation, which allows highly detailed model atoms to be used. The model atoms and line-formation input rely substantially on high precision and accurate data from laboratory measurements and *ab-initio* computations. The latter comprise applications of the multi-configuration Hartree-Fock method (e.g. Froese Fischer & Tachiev

2004) and the R-matrix method in the close-coupling approximation as employed within the Opacity and IRON Projects (Seaton 1987, Hummer et al. 1993), own work (e.g. Przybilla & Butler 2004) and numerous diverse studies in the literature.

The models facilitate high-quality observations of the OB stars under consideration to be closely matched over wide parts of the optical wavelength region once the atmospheric parameters are tightly constrained. Examples of the high quality of the fits are given by Nieva & Przybilla (2012). As a consequence, elemental abundances in individual OB stars can be constrained to a precision of $\sim$0.05-0.10 dex and an accuracy of $\sim$0.07-0.12 dex, coming close to corresponding values in photospheric abundance studies for the Sun. First results on an extension of the modelling to the UV, based on high-quality HST/STIS spectra (Schaffenroth 2015), and the near-IR, based on high-resolution VLT/CRIRES data (Nieva et al. 2011) for a sub-set of objects are highly promising. The latter wavelength regime will be of particular interest in the era of extremely large telescopes.

Data requirements. Despite all the progress made, quantitative spectroscopy of hot stars still requires large amounts of data for non-LTE modelling and line-formation computations to be provided. A summary of the most desired data sorted by processes follows.

Radiative bound-bound and bound-free. Extended laboratory measurements of oscillator strengths for iron-group elements at UV wavelengths are required to benchmark large-scale *ab-initio* computations that are needed for improved non-LTE modelling of UV spectra. *Ab-initio* data on photoionization cross-sections are missing for most of the relevant ions (ionization stages II-V) of the iron-group elements other than iron.

Collisional bound-bound and bound-free. *Ab-initio* data for electron-impact excitation for most of the lighter elements (carbon, nitrogen and oxygen, α-process elements, but also including hydrogen and helium) are missing for transitions to and among energy levels with principal quantum number $n \gtrsim 5$-7. Even basic data are missing for most ions of the iron-group elements that could be used to replace simple approximations. Few measurements and *ab-initio* computations on electron-impact ionization are available. Reliable computations are challenging, and they would be required in particular for ionization of excited levels, where collisions are effective because of favourable threshold energies.

Line broadening. Reliable Stark broadening data are missing for many stronger transitions in the spectra of hot stars. These comprise in particular (resonance) lines of CNO and α-elements in the UV, (diffuse) He I lines involving upper levels with $n \geqslant 6$ in the optical, and hydrogen Brackett and Pfund lines and many of the helium lines in the near-IR. The lack of data on helium lines in particular limits comprehensive spectrum synthesis for helium-rich objects, see e.g., Przybilla et al. (2005, 2006b).

References

Froese Fischer, C. & Tachiev, G. 2004, *At. Data Nucl. Data Tables*, 87, 1
Hummer D. G., Berrington K. A., Eissner W., et al. 1993, *A&A*, 279, 298
Nieva, M. F. & Przybilla, N. 2007, *A&A*, 467, 295
Nieva, M. F. & Przybilla, N. 2008, *A&A*, 481, 199
Nieva, M. F. & Przybilla, N. 2012, *A&A*, 539, A143
Nieva, M. F., Przybilla, N., Seifahrt, A., et al. 2011, *Bull. Soc. R. Sci. Liège*, 80, 175
Przybilla, N. 2010, *EAS Publ. Ser.*, 43, 115
Przybilla, N. & Butler, K. 2004, *ApJ*, 609, 1181
Przybilla, N., Butler, K., Heber, U., & Jeffery, C. S. 2005, *A&A*, 443, L25
Przybilla, N., Butler, K., Becker, S. R., & Kudritzki, R. P. 2006a, *A&A*, 445, 1099
Przybilla, N., Nieva, M. F., & Edelmann, H. 2006b, *Baltic Astronomy*, 15, 107
Schaffenroth, V. 2015, PhD Thesis (University Erlangen-Nuremberg)
Seaton, M. J. 1987, *J. Phys. B*, 20, 6363

The molecular universe: from observations to laboratory and back

Ewine F. van Dishoeck[1,2]

[1] Leiden Observatory, Leiden University, the Netherlands
email: ewine@strw.leidenuniv.nl

[2] Max Planck Institute for Extraterrestrial Physics, Garching, Germany

Abstract. This brief overview stresses the importance of molecular processes in modern astrophysics and provides examples where the availability of new laboratory or theoretical data proved crucial in the analysis. This includes basic data such as spectroscopy and collisional rate coefficients, but also an improved understanding of reactions and photoprocesses in the gaseous and solid state. In spite of many lingering uncertainties, the future of molecular astrophysics is bright with new facilities such as ALMA, JWST and ELTs on the horizon. Together, they will allow increased understanding of the journey of gas and solids from clouds to stars and planets, and back to the interstellar medium.

Keywords. molecular data, molecular processes, astrochemistry, ISM: molecules, techniques: spectrocopic

1. Introduction

Molecules are found in a wide range of astronomical environments, from our Solar system to distant starburst galaxies at the highest redshifts. Thanks to the opening up of the infrared and (sub)millimeter wavelength regime, culminating with *Herschel* and ALMA, more than 180 different species (not counting isotopologs) have now been found throughout the various stages of stellar birth and death. This includes diffuse and dense interstellar clouds, protostars and disks, the envelopes of evolved stars and planetary nebulae, and exo-planetary atmospheres. Molecules and solid-state features are now also routinely detected in the interstellar medium of external galaxies, near and far. Thus, we can now truly speak of a 'Molecular Universe' (Tielens 2011).

From a chemical perspective, interstellar space provides a unique laboratory to study basic processes under very different conditions than those normally found in a laboratory on Earth. For astronomers, molecules are unique probes of the many environments where they are found, providing information on density, temperature, dynamics, ionization fractions and magnetic fields. Molecules also play an important role in the cooling of clouds, allowing them to collapse, including the formation of the very first stars and galaxies in the universe. Finally, the molecular composition is sensitive to the history of the material, and ultimately provides critical information on our origins.

This paper briefly summarizes a number of recent observational highlights and provides examples of cases where the availability of new laboratory data proved crucial in the analysis. This includes (i) laboratory spectroscopy; (ii) collisional rate coefficients; (iii) gas-phase chemistry; (iv) photodissociation; (v) solid-state chemistry. There has been no shortage of much more detailed reviews of astrochemistry and the importance of laboratory experiments and theory over the last few years, see Tielens (2013), van Dishoeck (2014), van Dishoeck *et al.* (2013), Herbst (2014), Caselli & Ceccarelli (2012),

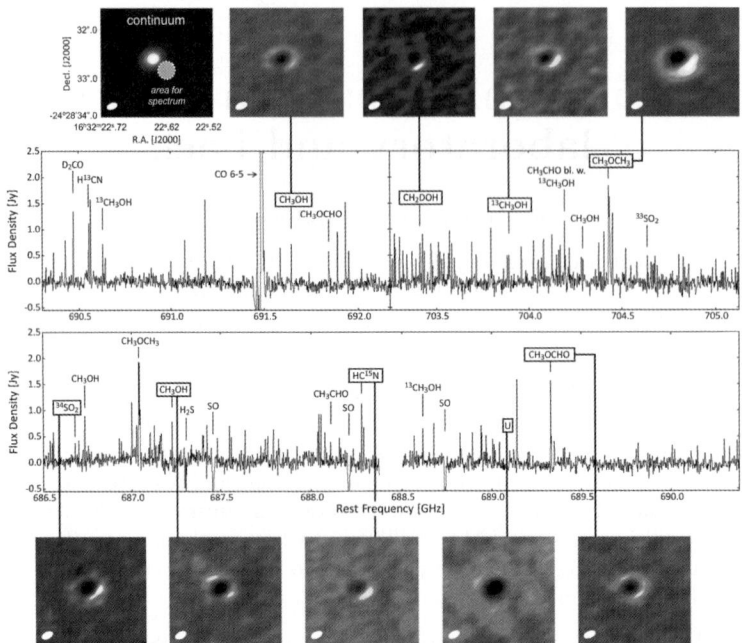

Figure 1. ALMA spectrum at 690 GHz of the low-mass protostar IRAS16293-2422 B taken over a region of $0.3'' \times 0.3''$). Images of the continuum and selected lines are shown as well (Baryshev et al. 2015). A significant fraction of these lines have not yet been identified.

Bergin (2013), Savin et al. (2012), Herbst & van Dishoeck (2009) and the December 2013 special issue of Chemical Reviews for further information.

2. Gas-phase processes

Spectroscopy. Herschel was particularly well suited to observe light molecules such as hydrides, whose primary transitions occur at far-infrared wavelengths (THz frequencies). New molecules include H_2O^+ which is surprisingly widespread in the diffuse ISM, HCl^+, H_2Cl^+ and $^{36}ArH^+$ (Gerin et al. 2016). These identifications were possible thanks to dedicated laboratory work. In addition, the THz range is well suited to observe high excitation lines of heavy molecules with upper energy levels up to 1000 K. *Herschel* line surveys of well known sources such as Orion-KL and SgrB2(N) beautifully reveal the chemical composition of hot gas but they also show that 5-12% of the channels are unidentified (Crockett et al. 2014, Neill et al. 2014). Most of the lines are likely to be due to isotopologs and/or vibrationally excited states of known complex organic molecules.

Line surveys with ALMA are only just starting and are more suited to detect new complex molecules because of the higher angular resolution and sensitivity, combined with the fact that the strongest lines of large molecules occur at lower frequencies (Fig. 1). One early highlight is de detection of i-propyl cyanide toward SgrB2(N), the first branched molecule (Belloche et al. 2014). Such side chains are characteristics of amino acids, which have not yet been detected in interstellar space. This detection was made toward a position in SgrB2(N) with particularly narrow lines, minimizing line confusion. Similarly, the low-mass source IRAS 16293-2422B has very narrow lines which allows identification of prebiotic molecules toward solar-mass protostars for the first time (Jørgensen et al. 2012, Baryshev et al. 2015, Fig. 1). The ALMA spectrum of the carbon-rich AGB star

IRC+10216 shows many U-lines, in contrast with the *Herschel* spectrum (Cernicharo *et al.* 2013). All these surveys point to the continued need for laboratory spectroscopy at (sub)millimeter wavelengths, both of known molecules and of new increasingly more complex organic molecules.

At optical and near-IR wavelengths there has been great progress in measuring the spectra of long carbon chains and aromatic molecules that could give rise to the diffuse interstellar bands (Steglich *et al.* 2010, 2013, Contreras *et al.* 2013, Zhao *et al.* 2014) (see also Cox & Cami 2014). Silicon-containing chains have been a recent focus (Steglich *et al.* 2015), triggered by the detection of SiCSi (Cernicharo *et al.* 2015). The gas phase spectrum of C_{60}^+ has finally been measured and shown to be indeed responsible for two DIB features at near-IR wavelengths (Campbell *et al.* 2015), as proposed originally by Foing & Ehrenfreund (1997).

Collisional rate coefficients. Much progress has been made in this area thanks to coordinated efforts between various groups. State-to-state collisional rate coefficients with H_2 for pure rotational transitions have been computed in the last decade for H_2O, H_2CO, HCN, HNC, CN, CS, SO, SO_2, HF, HCl, OH^+, NH_2D, CH_3CN, and CH_3OH, among others. For ions and molecules with large dipole moments such as HF, CH^+ and ArH^+, collisions with electrons are also important (Hamilton *et al.* 2016). Another new development is the calculation of rate coefficients for vibration-rotation transitions seen at infrared wavelengths (Pontoppidan *et al.* 2010, Brown *et al.* 2013). New data have been provided for H_2O and CO with H_2 and H (Faure *et al.* 2008, Song et al. 2015). These calculations of the potential energy surfaces and subsequent dynamics are very computationally and labor intensive, sometimes taking a decade of dedicated work. The data can be accessed through the BASECOL database (Dubernet *et al.* 2013) and the LAMDA database (Schöier *et al.* 2005, van der Tak 2011).

Chemical reactions. The main databases for gas-phase chemical reactions, UMIST (McElroy *et al.* 2013) and KIDA (Wakelam *et al.* 2015), continue to be updated with new results from the chemical physics literature. Particular attention has been paid to neutral-neutral reactions with C and N atoms leading to carbon chain molecules (Daranlot *et al.* 2012, Loison *et al.* 2014). There are also new merged beam experiments on atom-molecular ion reactions at a range of temperatures (O'Connor *et al.* 2015).

Photodissociation. An update of the wavelength dependence of the photodissociation and photoionization cross sections for many astrophysically important molecules has been made (Heays *et al.*, submitted), making use of databases such as the MPI-Mainz UV/VIS spectral atlas of gaseous molecules. Rates are provided both for the interstellar radiation field and for the cosmic ray induced field (important inside dark cores and disks). Photodissociation rates by the solar radiation field (important for comets) have been updated by Huebner & Mukherjee (2015). The photodissociation of CO and its isotopologs has been further studied and self- and mutual shielding factors quantified (Visser *et al.* 2009, and refs cited). The photodissociation of N_2 and $^{14}N^{15}N$ are finally well understood thanks to decades of experimental and theoretical work (Heays *et al.* 2014, and refs cited). Figure 2 illustrates the diversity in cross sections and their wavelength dependence for various interstellar molecules. Taking this wavelength dependence into account is particularly important for the chemistry in protoplanetary disks around different types of stars.

The photostability of PAHs is much larger than that of small molecules (see review by van Dishoeck & Visser 2015) but is important in protoplanetary disks where UV fields

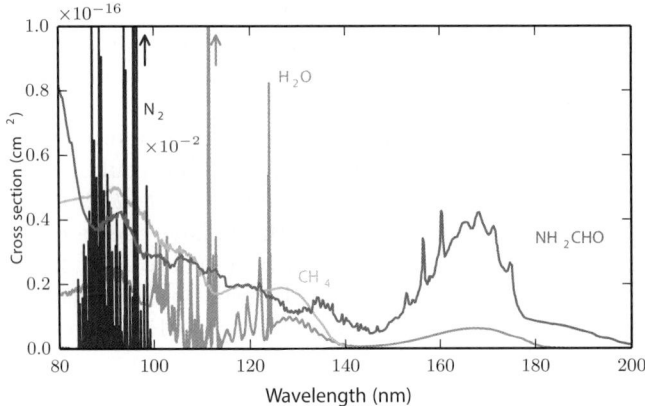

Figure 2. Examples of photodissociation cross sections of interstellar molecules. Note that some cross sections are highly structured (e.g., N_2) whereas others have broad absorption features over a large range of wavelengths. Figure by Alan Heays.

can be enhanced by six orders of magnitude compared with the standard interstellar radiation field. There are new experiments to measure and quantify the loss of hydrogen and C_2 fragments by photodestruction in the laboratory (Zhen et al. 2015).

3. Ice processes

Freeze out and desorption. At low grain temperatures of ~ 10 K, the probability for molecules to freeze out on dust grains is unity on every collision. The main question is how to get molecules off the grains at temperatures well below their thermal desorption rates. One option is photodesorption, a process that has been studied both in the laboratory (Öberg et al. 2009a, Fayolle et al. 2011, Fillion et al. 2014) and through molecular dynamics modeling (Andersson & van Dishoeck 2015, Arasa et al. 2015). It is imporant to note that the processes of photodesorption and photodissociation in the ice are linked for most molecules (exceptions are CO and N_2 below 11 eV), but that this is usually not treated correctly in astrochemical models. As for gas-phase processes, the dissociation and/or desorption efficiencies also have a strong wavelength dependence.

The process of chemical desorption, in which part of the energy liberated by formation of the chemical bond is used to desorb the molecule, is potentially important for explaining the observed gas-phase abundances in cold clouds. New experiments to quantify this process have just been reported (Minissale et al. 2015). Alternative ways to get molecules back into the gas include impulsive spot heating by cosmic rays (Ivlev et al. 2015).

Formation of complex organic molecules. Much (but not all) of the chemical complexity in star-forming regions is thought to be due to reactions occuring on top of, and within, ices that coat interstellar dust grains. One of the recent observational surprises is the fact that complex organic molecules such as $HCOOCH_3$ are detected not just in hot cores surrounding protostars (Herbst & van Dishoeck 2009) but also on larger scales in the low temperature parts of envelopes (Öberg et al. 2010) and even in the coldest pre-stellar cores (Bacmann et al. 2012, Vastel et al. 2014). Ultra-high vacuum solid-state laboratory experiments have demonstrated that the formation of complex molecules can indeed occur at temperatures as low as 10 K without any energy input (Hidaka et al. 2004, Fuchs et al. 2009, Fedoseev et al. 2015, Fresneau et al. 2015, Chuang et al. 2016). UV irridiation of ices breaks bonds, produces radicals which become mobile upon heating and this leads to further complexity, with experiments on mixed ices dating back to the 1980s (d'Hendecourt et al. 1986, Öberg et al. 2009b).

The challenge for the next decade is to translate these experiments into binding energies and reaction barriers that can be used in gas-grain models. The overall efficiency of surface reactions depends on the probability that species stick to the grains, their mobility on the surface, and the probability that molecule formation occurs. Diffusion of at least one of the two reactants is thought to control the rate of the reaction. This requires a prescription to hop from site to site, in which the crucial parameter is the energy barrier $E_{\rm hop}$. Usually $E_{\rm hop} = c^{st} \times E_{\rm bind}$ is assumed with c^{st} varying between 0.3 and 0.7. For rough surfaces, the hopping barriers may actually vary from site to site. Also, the importance of tunneling at the lowest temperatures is still debated and the formulation for the competition between diffusion and reaction is not clear. Recent Monte Carlo simulations have found values of 0.31 and 0.39 for CO and CO_2 on water ice, respectively (Karssemeijer & Cuppen 2014).

4. Outlook

In summary, there is steady progress in spectroscopy and in pinning down the key processes that determine the excitation, formation and evolution of molecules in the Universe. Compared with atomic data studies, molecular physics lags behind by several decades and needs to address a wider variety of species and processes. Concerted efforts within and between various countries, such as being discussed by the US AAS Laboratory Astrophysics Division, the ACS Astrochemistry Division and the European task force on Laboratory Astrophysics, are key to make the next step forward. With ALMA still going strong for the next 25+ years, JWST being launched late 2018, and Extremely Large Telescopes becoming operational in the mid-2020s, there is a growing need for accurate molecular data. The IAU, and in particular Commission B5 on Laboratory Astrophysics and Commission H2 on Astrochemistry, can play a role in bringing astronomers, chemists and molecular physicists together to make all these experiments and calculations happen!

References

Andersson S. & van Dishoeck E. F. 2008, *Astron. Astrophys.*, 491, 907
Arasa C. et al. 2015, *Astron. Astrophys.*, 575, A121
Bacmann A. et al. 2012, *Astron. Astrophys.*, 541, L12
Baryshev A. M. et al. 2015, *Astron. Astrophys.*, 577, A129
Belloche A. et al. 2014, *Science*, 345, 1584
Bergin E. A. et al. 2013, *Nature*, 493, 644
Brown J. M. et al. 2013, *Astrophys. J.*, 770, 94
Campbell E. K. et al. 2015, *Nature*, 523, 322
Caselli P. & Ceccarelli C. 2012, *Astron. Astrophys. Review*, 20, 56
Cernicharo J. et al. 2013, *Astrophys. J. Lett.*, 778, L25
Cernicharo J. et al. 2015, *Astrophys. J. Lett.*, 806, L3
Chuang K.-J. et al. (2016) *Mon. Not. R. Astron. Soc.*, 455, 1702
Contreras C. S. & Salama F. 2013, *Astrophys. J., Suppl. Ser.*, 208, 6
Cox N. L. J. & Cami J. 2014, eds., *Diffuse Interstellar Bands*, IAU Symposium 297, Cambridge Univ. Press
Crockett N. R. et al 2014, *Astrophys. J.*, 787, 112
Daranlot J. et al. 2012, *Proc. Natl. Acad. Sci., U.S.A.*, 109, 10233
D'Hendecourt L. B. et al. 1986, *Astron. Astrophys.*, 158, 119
Dubernet M.-L. et al. 2013, *Astron. Astrophys.*, 553, A50

Faure A. & Josselin E. 2008, *Astron. Astrophys.*, 492, 257
Fayolle E. C. et al. 2011, *Astrophys. J. Lett.*, 739, L36
Fedoseev G. et al. 2015, *Mon. Not. R. Astron. Soc.*, 448, 1288
Fillion J.-H. et al. 2014, *Faraday Discussions*, 168, 533
Foing B. H. & Ehrenfreund P. 1997, *Astron. Astrophys.*, 317, L59
Fresneau A. et al. 2015, *Mon. Not. R. Astron. Soc.*, 451, 1649
Fuchs G. W. et al. 2009, *Astron. Astrophys.*, 505, 629
Gerin M. et al. 2016, *Annu. Rev. Astron. Astrophys.*, 54, in press
Hamilton J. R. et al. 2016, *Mon. Not. R. Astron. Soc.*, 455, 3281
Heays A. N. et al. 2014, *Astron. Astrophys.*, 562, A61
Herbst E. 2014, *Phys. Chem. Chem. Phys.*, 16, 3344
Herbst E. & van Dishoeck E. F. 2009, *Annu. Rev. Astron. Astrophys.*, 47, 427
Hidaka H. et al. 2004, *Astrophys. J.*, 614, 1124
Huebner W. F. & Mukherjee J. 2015, *Plan. Space Sci.*, 106, 11
Ivlev A. V. et al. 2015, *Astrophys. J.*, 805, 59
Jørgensen J. K. et al. 2012, *Astrophys. J. Lett.*, 757, L4
Karssemeijer L. J. & Cuppen H. M. 2014, *Astron. Astrophys.*, 569, A107
Loison J.-C. et al. 2014, *Mon. Not. R. Astron. Soc.*, 437, 930
McElroy D. et al. 2013, *Astron. Astrophys.*, 550, A36
Minissale M. et al. 2015, *Astron. Astrophys.*, in press
Neill J. L. et al. 2014, *Astrophys. J.*, 789, 8
Öberg K. I. et al. 2009a, *Astron. Astrophys.*, 504, 891
Öberg K. I. et al. 2009b, *Astrophys. J.*, 693, 1209
Öberg K. I. et al. 2010, *Astrophys. J.*, 716, 825
O'Connor A. P. et al. 2015, *Astrophys. J., Suppl. Ser.*, 219, 6
Pontoppidan K. M. et al. 2010, *Astrophys. J.*, 720, 887
Savin D. W. et al. 2012, *Rep. Prog. Phys.*, 75, 3
Schöier F. L. et al. 2005, *Astron. Astrophys.*, 432, 369
Song L. et al. 2015, *Astrophys. J.*, 813, 96
Steglich M. & Maier J. P. 2015, *Astrophys. J.*, 801, 119
Steglich M. et al. 2010, *Astrophys. J. Lett.*, 712, L16
Steglich M. et al. 2013, *Astrophys. J., Suppl. Ser.*, 208, 26
Tielens A. G. G. M. 2011, in: *The Molecular Universe*, IAU Symposium 280, ed. J. Cernicharo and R. Bachiller), Cambridge Univ. Press, p. 3
Tielens A. G. G. M. 2013, *Rev. Mod. Phys.*, 85, 1021
van der Tak F. 2011, in *The Molecular Universe*, IAU Symposium 280, ed. J. Cernicharo and R. Bachiller), Cambridge Univ. Press, p. 449
van Dishoeck E. F. 2014, *Faraday Discussions*, 168, 9
van Dishoeck E. F. & Visser R. 2015, in: *Laboratory Astrochemistry*, ed. S. Schlemmer *et al.*, Wiley, p. 229
van Dishoeck E. F. et al. 2013, *Chemical Reviews*, 113, 9043
Vastel C. et al. 2014, *Astrophys. J. Lett.*, 795, L2
Visser R. et al. 2009, *Astron. Astrophys.*, 495, 881
Wakelam V. et al. 2015, *Astrophys. J., Suppl. Ser.*, 217, 20
Zhao D. et al. 2014, *Astrophys. J. Lett.*, 791, L28
Zhen J. et al. 2015, *Astrophys. J. Lett.*, 804, L7

Untangling the Formation Mechanisms of Biorelevant Molecules in the ISM with Photoionization Reflectron Time-of-Flight Mass Spectrometry

Marko Förstel[1,2] and Ralf I. Kaiser[1,2]

[1] W. M. Keck Research Laboratory in Astrochemistry, University of Hawaii at Manoa, Hawaii, HI, 96822, USA

[2] Department of Chemistry, University of Hawaii at Manoa, Honolulu, Hawaii, HI, 96822, USA
email: `ralfk@hawaii.edu`

Abstract. Exploiting reflectron time of flight mass spectrometry coupled with single photon ionization of the subliming molecules (PI-ReTOF-MS) during the temperature programmed desorption (TPD) and combining these data with on line and in situ infrared spectroscopy (FTIR), a versatile experimental approach has been established to elucidate the formation pathways of complex organic molecules in interstellar analog ices upon interaction with ionizing radiation at astrophysically relevant temperatures as low as 5 K.

Keywords. laboratory astrophysics, isomer specific detection, tunable photoionization

---

An understanding of the formation of complex organic molecules (COM) in the interstellar medium is of core value to the astronomy community upon multiple levels (Whittet et al. (2011), Walsh et al. (2014), Vasyunina et al. (2014), Bacmann et al. (2012)). Our experimental program advances the knowledge of the most fundamental processes leading to the synthesis of COMs, among them biorelevant molecules such as amino acids, dipeptides, sugars and vitamins, on ice-coated interstellar grains (ICIGs) in cold molecular clouds and in star forming regions (Charnley, Ehrenfreund & Kuan (2001), Aikawa et al. (2008)). Our experimetal approach also allows us to differentiate between structural isomers of specific molecule and to measure their branching ratios and rate constants of formation. Together with this information, these isomers can serve as a molecular clock and tracers in defining the evolutionary stage of cold molecular clouds and star forming regions (Garrod, Weaver & Herbst (2008)). Finally, since COMs have also been identified in the Murchison meteorite (Botta, Bada & Ehrenfreund (2002)), our project might bring us closer to understanding the extent to which key classes of COMs might have been synthesized exogenously in the ISM, (partly) incorporated into parent bodies of, e.g., Murchison, and then delivered to early Earth.

The aforementioned goals are achieved by systematically replicating the conditions of ICIGs as present in cold molecular clouds and toward high- and low-mass star forming regions in a next-generation ultra-high vacuum surface scattering machine. Interstellar ice analog samples at astrophysically relevant temperatures as low as 5 K are exposed to ionizing radiation in the form of energetic electrons. We follow a radically different approach than traditional infrared spectroscopy (FTIR) and electron impact mass spectrometry by interfacing complementary detection schemes to a single machine. Individual COMs subliming into the gas phase upon warm-up of the samples are identified via a reflectron time-of-flight mass spectrometer (ReTOF-MS) coupled with soft photoionization by simultaneously monitoring the decay signal of relevant functional groups in the condensed phase via FTIR (Maity, Kaiser & Jones (2015), Jones, Kaiser & Strazzulla (2014)). The energy of the ionizing photon can be controlled within a resolution of at least 0.01 eV allowing to photoionize subliming molecules according to their ionization energies. The ideally fragment-free isomer-specific detection of individual isomers represents the key advantage of soft single photon ionization compared to traditional mass spectrometry with electron impact ionization. Since the subliming molecules and structural isomers can also be separated via fractionated sublimation, the sublimation sequence along with the correlation of the ionization energy with the m/z ratio of the product represents a versatile and powerful approach to uniquely identify the complex organic molecules formed.

The exposure of simple astrophysical model ices containing carbon monoxide (CO), methanol (CH_3OH), methane (CH_4), and ammonia (NH_3) to energetic electrons leads to the formation of several key molecules as important representatives of biorelevant sugars (glycolaldehyde; $HCOCH_2OH$), polyalcohols [glycerol; $HOCH_2CH(OH)CH_2OH$], and amides (formamide; $HCONH_2$; urea; H_2NCONH_2) (Kaiser, Maity & Jones (2015), Abplanalp et al. (2015), Förstel et al. (2015)). Besides the generic identification of these structural isomers, we were also able to derive, in some instances, reaction mechanisms leading to COMs by fitting the system of coupled differential equations of the newly formed molecules. Most importantly, these COMS can be formed at 5 K within the ices via non-equilibrium processing involving, for instance, suprathermal hydrogen atoms and the excited triplet state of, for instance, carbon monoxide (CO) and/or acetylene (C_2H_2). To a certain degree, the low temperature matrix also stores reactive radicals such as formyl (HCO) and hydroxymethyl (CH_2OH). Upon annealing of the ices, these radicals can diffuse and might recombine. Based on our results, these thermal routes represent less prominent pathways leading to complex organic molecules within the processed ices.

Acknowledgement

The authors thank the W. M. Keck Foundation for financing the experimental setup. MF acknowledges Postdoctoral Funding from the DFG (FO 941/1). RIK thanks the US National Science Foundation (AST-1505502) for support.

References

Abplanalp, M. J., Borsuk, A., Jones, B. M., & Kaiser, R. I. 2015, *Astrophys. J.*, accepted
Aikawa, Y., Wakelam, V., Garrod, R. T., & Herbst, E. 2008, *ApJ*, 674, 13
Bacmann, A., Taquet, V., Faure, A., Kahane, C., & Ceccarelli, C. 2012, *A&A*, 541, L12
Botta, O., Bada, J. L., & Ehrenfreund, P. 2002, *ACM*, 500, 925
Charnley, S. B., Ehrenfreund, P., & Kuan, Y. J. 2001, *Spectrochim. Acta A: Molec. Biomolec. Spectrosc.*, 57, 4
Förstel, M., Maksyutenko, P., Jones, B. M., Sun, B.-J., Chang, A. H. H., & Kaiser, R. I. 2015, *manuscript subm.*
Garrod, R. T., Weaver, S. L. W., & Herbst, E. 2008, *Astrophys. J.*, 682, 1
Jones, B. M., Kaiser, R. I., & Strazzulla, G. 2014, *Astrophys. J.*, 781, 2
Kaiser, R. I., Maity, S., & Jones, B. M. 2015, *Angewandte Chemie Internatl. Ed.*, 54, 1
Maity, S., Kaiser, R. I., & Jones, B. M. 2015, *Phys. Chem. Chem. Phys*, 17, 5
Vasyunina, T., Vasyunin, A. I., Herbst, E., Linz, H., Voronkov, M., Britton, T., Zinchenko, I., & Schuller, F. 2014, *Astrophys. J.*, 780, 1
Walsh, C., Millar, T. J., Nomura, H., Herbst, E., Weaver, S. W., Aikawa, Y., Laas, J. C., & Vasyunin, A. I. 2014, *Astron. Astrophys.*, 563, A33
Whittet, D. C. B., Cook, A. M., Herbst, E., Chiar, J. E., & Shenoy, S. S. 2011, *Astrophys J.*, 742, 1

CO isotopologue ratios in the solar photosphere

James R. Lyons[1], Ehsan Gharib-Nezhad[2] and Thomas R. Ayres[3]

[1] School of Earth and Space Exploration, Arizona State University,
PO Box 871404, Tempe, AZ 85287-1404 USA
email: jimlyons@asu.edu

[2] School of Molecular Sciences, Arizona State University,
PO Box 871604, Tempe, AZ 85287-1604 USA
email: egharibn@asu.edu

[3] Center for Astrophysics and Space Astronomy, University of Colorado,
389 UCB, Boulder, CO 80309-0389 USA
email: thomas.ayres@colorado.edu

Abstract. We re-evaluate the CO dipole moment function in order to obtain more accurate isotope ratios for the solar photosphere using previous infrared observations. We used a new set of dipole moments from HITEMP which were accurately determined by both semi-empirical and ab initio methods. Preliminary values of isotope ratios using the new dipole moments are in better agreement with the inferred photosphere values from Genesis, showing that the solar photosphere is isotopically similar to primitive inclusions in meteorites.

Keywords. solar photosphere, solar system formation, solar abundances

1. Introduction

Determination of the oxygen isotope ratios in the bulk Sun is essential for understanding the formation environment of the solar system (Clayton 2003). The oxygen isotope composition of the photosphere is most readily determined from the infrared absorption lines of the isotopologues of CO. The solar CO fundamental ($\Delta v=1$) and first-overtone ($\Delta v=2$) bands were previously recorded by the shuttle-borne ATMOS Fourier transform spectrometer (FTS) (Abrams *et al.* 1996), and with the National Solar Observatory's FTS on the McMath-Pierce telescope at Kitt Peak. Analysis of the rovibrational bands from these photospheric spectra by 1D simulation models yielded a wide range of oxygen isotope ratios (Ayres *et al.* 2006), (Scott *et al.* 2006) none of which were consistent with the solar wind isotope ratios measured by the Genesis spacecraft (McKeegan *et al.* 2011). More recently, a CO5BOLD 3D convection model (Freytag *et al.* 2012) was employed to calculate ratios with lower uncertainties, $^{16}O/^{17}O = 2738\pm118$ and $^{16}O/^{18}O = 511\pm10$, which fall between terrestrial values and those reported by Genesis (Ayres *et al.* 2013). In that analysis a discrepancy in published CO dipole moment functions yielded a range of isotopic ratios spanning $\sim 30‰$ in $\delta^{18}O$. Here we re-evaluate the CO dipole moment function in order to obtain more accurate isotope ratios for the photosphere.

2. CO spectroscopy

In order to determine isotopologue abundances from the observations, the f-values (oscillator strengths) are needed for the rovibrational transitions of the ground electronic state of different isotopic C^xO for x=16, 17, 18. The two most commonly used oscillator strength scales are Hure & Roueff (1996) and Goorvitch (1994)(HR96 and G94, respectively). For a given rovibrational transition the f-value is proportional to the square of rovibrational dipole moment. According to Ayres *et al.* (2013), the derived $^{16}O/^{18}O$ ratios were 528 ± 11 for HR96 and 496 ± 7 for G94, respectively. The difference introduced by the two sets of dipole moments was too high to make a meaningful comparison to the Genesis values. Here, we have used a new set of dipole moments from Li *et al.* (2015)(LG15) which were accurately determined by both semi-empirical and ab initio methods. Using the spectroscopically determined potential energy function (Coxon & Hajigeorgiou 2004) of the electronic ground state of CO in the LEVEL 8.0 code, we employed the dipole moment function of LG15 to calculate the rovibrational dipole moments of $^{12}C^{16}O$, $^{12}C^{17}O$, and $^{12}C^{18}O$ isotopologues. Comparison of the f-value ratios of G94, HR96, and LG15 for $\Delta v=1$ for both $^{12}C^{16}O$, and $^{12}C^{18}O$ reveals a several percent difference between f-values of G94 and HR96 (Gharib-Nezhad *et al.* 2015). This difference produced a systematic offset in the 3D convection model results of Ayres *et al.* (2013). Our revised set of f-values is much closer to HR96 than G94, and thus the results of 3D simulations will be much be closer to the HR96 results of Ayres *et al.* (2013).

3. Implications

Using our revised f-values the new $\delta^{18}O$ values are within about 5‰ of the results obtained in Ayres *et al.* (2013) using the HR96 f-values. 3D radiative transfer calculations using the more accurate set of CO f-values are in progress. These new results are significant because the photospheric $\delta^{18}O$ using HR96 is the same, within uncertainties, as the Genesis inferred photospheric $\delta^{18}O$ value. To infer photospheric isotope ratios, the Genesis solar wind isotope ratios must be corrected for fractionation during acceleration of coronal ions to form the solar wind. The primary fractionation process is believed to be inefficient Coulomb drag, in which protons collide repeatedly with heavier ions, such as OVI and OVII, imparting a mass-dependent preference for escape of the lightest O ions (Bodmer & Bochsler 1998). McKeegan *et al.* (2011) estimated the mass-dependent fractionation of O ions to be ≈ 40‰ in $\delta^{18}O$. Thus, the observed Genesis solar wind $\delta^{18}O$ value of -102‰ corresponds to a photosphere $\delta^{18}O \sim -60$‰. Using our revised f-value, the photosphere $\delta^{18}O \sim -50$‰, with an uncertainty of about $10 - 15$‰. Direct observation of the solar photosphere is now consistent in $\delta^{18}O$ with the inferred values from Genesis, and suggests that inefficient Coulomb drag is the primary source of fractionation for heavy isotopes during formation of the solar wind.

References

Abrams, M. C., Goldman, A., Gunson, M. R., *et al.* 1996, *Applied Optics*, 35, 2747
Ayres, T. R., Plymate, C., & Keller, C. U. 2006, *ApJS*, 165, 618
Ayres, T. R., Lyons, J. R., Ludwig, H.-G., Caffau, E., & Wedemeyer-Böhm, S. 2013, *ApJ*, 765, 46
Bodmer, R. & Bochsler P. 1998, *A&A*, 337, 921
Clayton, R. N. 2003, *Space Sci. Rev.*, 106, 19
Coxon, J. A. & Hajigeorgiou, P. G. 2004, *JCP*, 121, 2992
Freytag, B., Steffen, M., & Ludwig, H.-G., *et al.* 2012, *JCoPh*, 231, 919
Gharib-Nezhad, E., Lyons, J. R., & Ayres, T. R. 2015, *abstract, 46th Lunar Planet. Sci. Conf.*
Goorvitch, D. 1994, *ApJS*, 95, 535
Hure, J. M. & Roueff, E. 1996, *A&AS*, 117, 561
Li, G., Gordon, I. E., & Rothman, L. S., *et al.* 2015, *ApJS*, 216, 15
McKeegan, K. D., Kalio, A. P. A., Heber, V. S., *et al.* 2011, *Science*, 332, 1528
Scott, P. C., Asplund, M., Grevesse, N., & Sauval, A. J. 2006, *A&A*, 456, 675

Laboratory constraints on ice formation, restructuring and desorption

Karin I. Öberg

Harvard-Smithsonian Center for Astrophysics
60 Garden St, Cambridge, MA 02138
email: koberg@cfa.harvard.edu

Abstract. Ices form on the surfaces of interstellar and circumstellar dust grains though freeze-out of molecules and atoms from the gas-phase followed by chemical reactions. The composition, chemistry, structure and desorption properties of these ices regulate two important aspects of planet formation: the locations of major condensation fronts in protoplanetary disks (i.e. snow lines) and the formation efficiencies of complex organic molecules in astrophysical environments. The latter regulates the availability of prebiotic material on nascent planets. With ALMA it is possible to directly observe both (CO) snowlines and complex organics in protoplanetary disks. The interpretation of these observations requires a detailed understanding of the fundamental ice processes that regulate the build-up, evolution and desorption of icy grain mantles. This proceeding reviews how experiments on thermal CO and N_2 ice desorption, UV photodesorption of CO ice, and CO diffusion in H_2O ice have been used to guide and interpret astrochemical observations of snowlines and complex molecules.

Keywords. astrochemistry, astrobiology, molecular processes, methods: laboratory, ISM: molecules

1. Introduction

In the cold and dark phases of star and planet formation the surfaces of interstellar dust grains become coated with icy mantles. These ices are built up in molecular clouds through a combination of direct freeze-out of gas-phase molecules (e.g. CO → CO_{gr}) and freeze-out followed by atom addition reactions (e.g. $O_{gr} + H_{gr} \rightarrow OH_{gr} \rightarrow H_2O_{gr}$). The resulting ice mantle is typically dominated by H_2O followed by CO and CO_2 (20–30% with respect to H_2O) and smaller amounts of CH_3OH, CH_4 and NH_3 (Öberg et al. 2011a, Boogert et al. 2015). When a cloud core collapses to form a star, some of the icy grains become incorporated into the circumstellar disk that is the formation site of planets. The volatile composition of forming planets depend intimately on the ice and gas composition in the disk (Öberg et al. 2011b), which is set by the desorption energies of the main ice constituents (together with initial conditions and disk chemistry). Major condensation fronts, of e.g. H_2O, CO_2, CO and N_2, may locally enhance planet formation.

Icy grain mantles are also important from a prebiotic perspective. Ice chemistry is the proposed main cause of the chemical complexity observed toward protostars (Garrod et al. 2008, Herbst & van Dishoeck 2009). The complex organics that become incorporated into protoplanetary disks and further into planets may seed the origins of life. The abundance and composition of this organic material depend on the efficiency at which simple ice mantles are converted into different kinds of complex organic ices.

Our ability to model both the organic content and the bulk volatile composition of nascent planets then depends fundamentally on our understanding of a small number of ice processes: accretion from the gas-phase, diffusion of atoms, radicals and molecules on top of and inside of ice, reactions when different species encounter one another in the

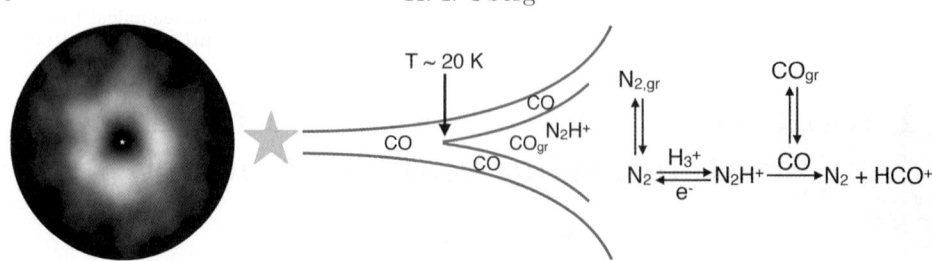

Figure 1. N_2H^+ observation toward the disk around TW Hya (left, after Qi et al. 2013), and cartoon and reaction diagram summarizing the relation between CO freeze-out (CO → CO_{gr}) and the appearance of N_2H^+ in disks. In essence the inner rim of the green N_2H^+ ring traces the onset of CO freeze-out in the disk midplane, i.e. the CO snowline.

ice, and thermal and non-thermal desorption. Quantifying these processes, including their energy barrier heights, requires experiments. This proceeding presents three observations related to disk snowlines and organic chemistry – a N_2H^+ ring in the disk around young star TW Hya, a double DCO^+ ring in the disk around young star IM Lup and the presence of complex organics in cold cloud cores – and three ice experiments that aid in interpreting these observations. The focus is on CO, both because it is an excellent model system, and because of its importance for ice organic chemistry and the volatile structure in disks during planet formation.

2. CO and N_2 ice desorption and observations of CO snowlines

Snowlines influence several aspects of planet formation, including the efficiency of the initial grain coagulation step and the bulk compositions of forming planets. Observational constraints on snowline locations in disks are therefore important. Based on laboratory experiments on CO desorption energy barriers (Collings et al. 2003), CO snowlines are expected at disk midplane temperatures of ~20 K, which corresponds to disk radii of ~30 AU in disks around young Solar type stars. This is readily resolvable by modern interferometers like the Atacama Large Millimeter and submillimeter Array (ALMA). Direct imaging of snowlines is challenging, however, due to the presence of vertical temperature gradients in disks, which maintains large quantities of CO in the gas-phase in the disk atmosphere at all disk radii, which can hide loss of CO emission from the CO freeze-out zone in the disk midplane (Fig. 1).

A potential solution is to identify a tracer that robustly anti-correlates with CO gas and therefore traces CO freeze-out regions. N_2H^+ is such a tracer as long as N_2 freeze-out occurs at (slightly) lower temperatures compared to CO. N_2H^+ forms through gas-phase reactions between N_2 and H_3^+ and is rapidly destroyed by proton transfer to CO, if there is any CO gas present. If CO is frozen out and N_2 is not, N_2H^+ can become quite abundant and N_2H^+ emission should trace the CO freeze-out zone, i.e. the inner radii of N_2H^+ emission should coincide with the CO snowline. Laboratory experiments have shown that there is a small difference in desorption energy barriers for CO and N_2 in pure ices (Öberg et al. 2005, Bisschop et al. 2006). More recent experiments have demonstrated that this difference persists for water-rich ices (Fayolle et al. in prep.). The utility of N_2H^+ as a CO snowline tracer was recently demonstrated observationally by the presence of a N_2H^+ ring in the disk around the young star TW Hya (Fig. 1), with an inner rim at 30 AU, the expected radius of the CO snowline in this disk (Qi et al. 2013). Snowline locations are thus accessible to observations through interferometric chemical imaging.

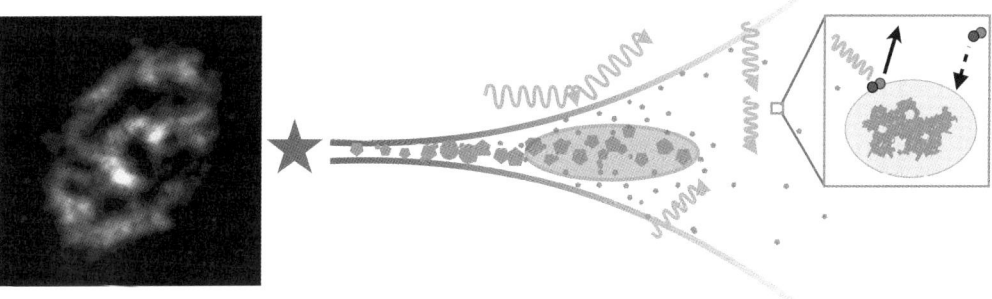

Figure 2. Observation of DCO$^+$ in the disk around young star IM Lup (left) and a cartoon illustrating the complete freeze-out zone of CO in a disk (light blue oval), limited by thermal desorption of CO inwards, and non-thermal desorption in the outer disk. DCO$^+$ is expected to be the most abundant just interior to the inner CO snowline and in the non-thermal desorption region, where the two ingredients of efficient DCO$^+$ formation – low temperatures and CO gas – are both present.

3. CO ice photodesorption and a DCO$^+$ double ring in a disk

In addition to the thermal desorption described in the previous section, ices can desorb non-thermally through interactions with UV, electrons, and cosmic rays, and through the release of chemical energy. In the dense, inner regions of protoplanetary disks, the disk mid planes are efficiently shielded from all external radiation. The division of molecules between ice and gas is therefore mainly set by gas adsorption onto grains and thermal desorption of ice, regulating e.g. the location of the CO snowline (Fig. 1). In the absence of non-thermal desorption, almost no CO is expected in the midplane exterior to the CO snowline. Further out yet in the disk some the gas can be partially repopulated by CO through a combination of low adsorption rate (caused by the low densities found in the outer regions of disks) and the activation of non-thermal desorption pathways, such as UV photodesorption of ice. The latter can occur if 1) the dust column becomes to small to efficiently shield the disk midplane from external radiation, and 2) UV photodesorption is an efficient process.

UV photodesorption of ices have been explored extensively through laboratory experiments (Öberg *et al.* 2009a, Munoz Caro *et al.* 2010, Fayolle *et al.* 2011), and the efficiencies are high, at least 10^{-3} per incident UV photon. These high desorption efficiencies imply that in the outer tenuous disk, UV photodesorption could partially repopulate the disk midplane with CO gas. The effects of ice photodesorption (or a related non-thermal desorption process) are manifest in the disk of IM Lup. The disk presents a set of concentric DCO$^+$ rings (Öberg *et al.* 2015), where the outer disk can be attributed to DCO$^+$ formation following desorption of (some) CO back into the gas-phase at large disk radii. Based on our understanding of isotopic fractionation chemistry and CO desorption we can interpret the two concentric DCO$^+$ rings as marking 4 distinct disk regions: an inner disk region where CO is in the gas-phase, but where it is too warm for efficient deuterium chemistry (inner hole), a region close to the CO snowline where CO is still in the gas phase and temperatures are sufficiently low for DCO$^+$ to be enhanced (the inner DCO$^+$ ring), a CO freeze-out region where both thermal and non-thermal desorption are very slow and all CO based chemistry is turned off, and an outer disk region where CO photodesorption becomes efficient and DCO$^+$ therefore reappears (the second ring).

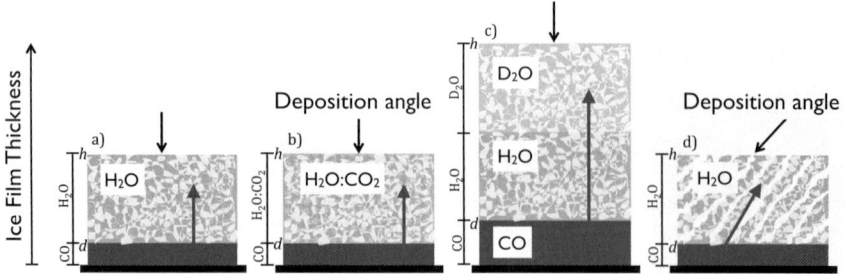

Figure 3. Cartoon illustrating CO diffusion experiments by Lauck *et al.* (2015), where CO diffusion into water ice was measured spectroscopically employing initially layered CO and H_2O ices of different ice thicknesses, porosities and isotopic compositions.

4. CO ice diffusion and a cold complex organic ice chemistry in space

A puzzle in astrochemistry is the presence of complex organic molecules in cold molecular cloud cores (Öberg *et al.* 2010, Bacmann *et al.* 2012, Cernicharo *et al.* 2012). In these regions the commonly invoked interstellar pathway for forming complex organics – photodissociation of CH_3OH and other ice molecules into radicals followed by radical diffusion and reactions – seem to fail. With the currently used prescriptions for diffusion in ices, the icy grain mantles in these cores are simply too cold for radical diffusion (e.g. Garrod *et al.* 2008). There are very few measurements of diffusion of radicals and molecules in ices, however, and it is therefore unclear whether existing assumptions on diffusion barriers are realistic. A number of recent studies have attempted to measure CO diffusion in H_2O ice to test these assumptions (e.g. Karssemeijer *et al.* 2014, Lauck *et al.* 2015). Our experimental approach is summarized in Fig. 3. The studies find consistently lower diffusion barriers than expected for CO in H_2O ice; we find a diffusion to desorption energy ratio that is ∼0.15, which can be compared to commonly assumed ratios of 0.3–0.7 (Lauck *et al.* 2015). This suggests that molecules and radicals can be mobile in ices at lower temperatures than previously assumed, perhaps explaining the existence of complex organics molecules in cold interstellar environments.

References

Bacmann, A, Taquet, V, Faure, A, *et al.* 2012, *A&A*, 541, 12
Bisschop, S. E. and Fraser, H. J., Öberg, K. I. *et al.* 2006, *A&A*, 449, 1297
Cernicharo, J, Marcelino, N, & Roueff, E, 2012, *A&A*, 759, L43
Collings, M. P., Dever, J. W., & Fraser, H. J. 2003, *ApJ*, 583, 1058
Fayolle, E. C., Bertin, M., & Romanzin, C. 2011, *ApJL*, 739, L36
Garrod R. T., Weaver S. L. W., & Herbst E. 2008, *ApJ*, 682, 283
Herbst, E., & van Dishoeck, E. F. 2009, *ARA&A*, 47, 427
Karssemeijer, L. J, Ioppolo, S, & van Hemert, M. C, 2014, *A&A*, 781, 16
Lauck, T., Karssemeijer, L., Shulenberger, K. *et al.* 2015, *ApJ*, 801, 118
Munoz Caro, G. M, Jimenez-Escobar, A, & Martin-Gago, J. A, 2010, *A&A*, 522, 108
Öberg, K. I., van Broekhuizen, F., Fraser, H. J. *et al.* 2005 *ApJL*, 621, L33
Öberg, K. I., van Dishoeck, E. F., & Linnartz, H. 2009 *A&A*, 496, 281
Öberg K. I., Boogert A. C. A., Pontoppidan K. M. *et al.* 2011 *ApJ*, 740, 109
Öberg, K. I., Murray-Clay, R., & Bergin, E. A. 2011 *ApJL*, 743, L16
Öberg, K. I., Furuya, K., & Loomis, R. 2015 *ApJ*, 810, 112
Qi, C., Öberg, K. I., Wilner, D. J. *et al.* 2013, *Science*, 341, 630

Interstellar dust modelling: Interfacing laboratory, theoretical and observational studies (The THEMIS model)

Anthony P. Jones

Institut dAstrophysique Spatiale, UMR8617, CNRS/Université Paris Sud, Université Paris-Saclay, Université Paris Sud, Orsay F-91405, France
email: Anthony.Jones@ias.u-psud.fr

Abstract. The construction of viable and physically-realistic interstellar dust models is only possible if the constraints imposed by laboratory data on interstellar dust analogue materials are respected and used within a meaningful theoretical framework. These "physical" dust models can then be directly compared to observations without the need for any tuning to fit the observations. Such models will generally fail to achieve the excellent fits to observations that "empirical" models are able to achieve. However, the physically-realistic approach will necessarily lead to a deeper insight and a fuller understanding of the nature and evolution of interstellar dust. The THEMIS modelling approach, based on (hydrogenated) amorphous carbons and amorphous silicates with metallic Fe and/or FeS nano-inclusions appears to be a promising move in this direction.

Keywords. dust, extinction; ISM: general;

---

1. Introduction

In order for a dust model to be viable it must be consistent with as wide a possible range of dust observables (*e.g.*, pre-solar grain compositions, elemental abundances/depletions, extinction, absorption, scattering, emission, infrared spectra, polarisation, x-ray absorption and scattering, ...) and the variations of those observables across the interstellar medium (ISM). However, each of these observables is necessarily selective because none of them are un-biased. For example, the analysed pre-solar grains (*e.g.*, Anders & Zinner 1993) and the analysis of the STARDUST interstellar grains, (Westphal, Stroud, Bechtel, *et al.* 2014) represent incomplete and/or selective samplings. Additionally, interstellar dust observations have not yet fully-sampled all dust in all environments.

As a minimum, but insufficient, requirement the optical properties of all dust components must be consistent with the Kramers-Kronig relations, *i.e.*, the real (n) and imaginary (k) parts of the complex index of refraction ($m = n + ik$) must be self-consistent over as wide as possible a range of wavelengths. However, just because this condition is fulfilled does not necessarily imply that the material is physically-realisable, *i.e.*, that it can actually exist. The only valid test of the physical-reasonableness of an interstellar dust analogue material is that it, or a very closely-related material, can be synthesised, characterised and analysed in the laboratory. Hence, interstellar dust models must, wherever possible, be guided by laboratory measurements of physically-realised materials.

2. More complex but more realistic dust modelling (THEMIS)

It is clear that the dust optical properties depend upon the material composition and structure. Some materials can show a wide range of structures for a given chemical composition and all optical properties depend upon the particle size below some limiting dimension. Amorphous carbons, a-C, and hydrogenated amorphous carbons, a-C:H, are a prime example of this. These materials, under the collective term a-C(:H) exhibit wide-ranging properties, from insulators to conductors (but are mostly semiconductors), from hydrogen-rich to hydrogen-poor and from highly absorbing to highly scattering. This impressive gamut of properties from the large family of a-C(:H) materials is rather impressive for a material consisting of only carbon and hydrogen atoms. The suite of a-C(:H) materials is now receiving some attention within the astrophysical community.

The structures, compositions and size-dependent optical properties for a-C(:H) materials were recently derived for a wide range of H/C ratios and particles sizes over a broad wavelength range (soft x-ray to cm, the optEC$_{(s)}$ and optEC$_{(s)}$(a) datasets, Jones 2012a,b,c). These data were built ground up from theoretical considerations and strongly-constrained by laboratory data. The application of these data within the framework of a diffuse ISM dust model (Jones et al. 2013; Köhler, Jones & Ysard 2014) appears to be consistent with most of the dust observables (e.g., see Fig. 1). This model comprises log-normal size distributions of large ($a \sim 150$ nm) core/mantle grains of amorphous silicate and a-C:H cores with a-C mantles, and a-C nano-particles with a power-law size distribution biased towards the smallest grains. The model satisfies the physically-reasonable requirement in the sense that it uses only optical properties that are either directly measured in the laboratory or that are firmly-anchored to laboratory measurements and that are consistent with detailed theoretical considerations (Jones 2012a,b,c). The Jones et al. (2013)/Köhler, Jones & Ysard (2014) dust model uses the laboratory-constrained a-C(:H) and olivine-type (a-Sil$_{ol}$) and pyroxene-type (a-Sil$_{px}$) amorphous silicate data as is without any tuning to match specific astronomical observations. The result is a model that matches reasonably well most of the interstellar dust observables without the need to "tweak" the input data. However, the model has evolved, much as dust does in the ISM, to encompass the changes in the dust optical and physical properties as it interacts with and responds to its immediate environment. We now refer to our standard dust model, all of the associated dust evolution studies and future developments and extensions of this work under the umbrella acronym THEMIS (The Heterogeneous dust Evolution Model at the IaS). THEMIS is a key part of the European FP7 DustPedia project (DustPedia.com). The following summarises key elements of our modelling work:

Dust evolution. Dust is clearly not the same everywhere and significant differences are observed within and between given phases of the ISM. The THEMIS dust modelling approach has the in-built capacity to account for dust differences in a physically-meaningful way through variations in the dust optical properties, size distributions and structure as a function of the local ISM conditions, principally the interstellar radiation field and the gas density (e.g., Jones 2012a,b,c, Jones et al. 2013).

Dust in the diffuse ISM. The Jones et al. (2013)/Köhler, Jones & Ysard (2014) dust model was developed to match the dust properties in the diffuse ISM and comparisons with recent Planck observations of dust in these regions appear to indicate that the model is consistent with the observed variations in the dust properties (Ysard et al. 2015a; Fanciullo et al. 2015). The principal processes encompassed by our model of large grain evolution in the ISM are variations in the mantling materials (i.e., depth and composition) and in the relative masses of the amorphous silicate and carbonaceous dust components (Ysard et al. 2015a).

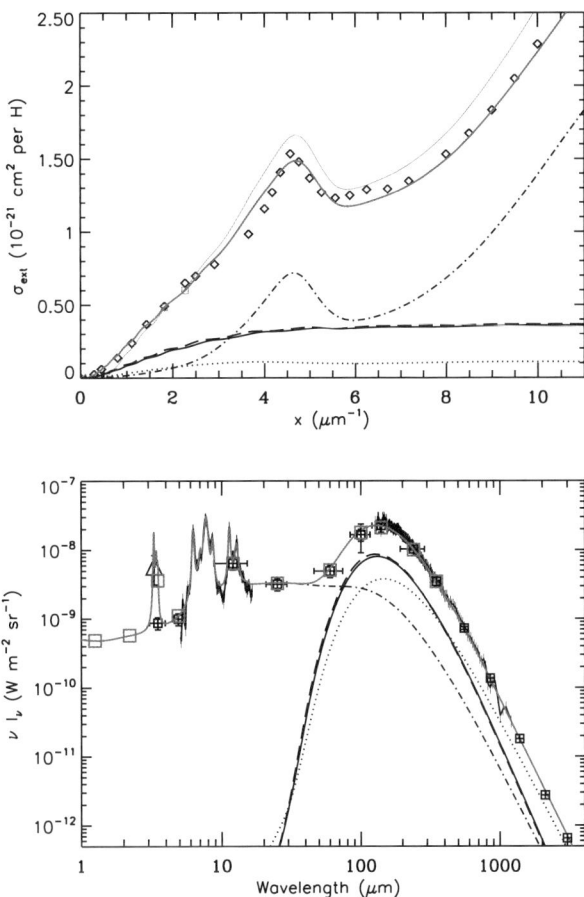

Figure 1. The Jones et al. (2013)/Köhler, Jones & Ysard (2014) diffuse ISM dust model visible-UV extinction cross-sections (top) and the dust IR-mm SED (for $N_H = 1 \times 10^{20}$ cm^{-2}, bottom). The red lines show the totals in each case, comprising contributions from a-C coated a-C:H (dotted), a-Sil$_{ol}$ (long-dashed) and a-Sil$_{px}$ (triple dot-dashed) grains, and small a-C grains (dot-dashed). The blue line in the upper plot shows the B and V band-normalised cross-section data.

Dust in low energy, high density regions. The model has also been extended to follow the evolution of the dust properties in the transition between diffuse and dense interstellar media (Jones et al. 2014; Köhler, Ysard & Jones 2015), including a self-consistent explanation of both cloud-shine and core-shine (Jones et al. 2015; Ysard et al. 2015b). In these environments grain growth through a-C(:H) mantle formation and grain-grain coagulation are the fundamental actors.

Dust in high energy, low density regions. The a-C(:H) properties also seem to provide viable routes to interstellar/circumstellar fullerene formation in planetary nebulæ (Micelotta et al. 2012) and to the formation of molecular hydrogen in moderately excited photo-dissociation regions (PDRs, Jones & Habart 2015). Further, dust evolution in energetic environments such as supernova-driven shock waves and a hot coronal gas has also been explored with the new dust model (Bocchio et al. 2012; Bocchio et al. 2013; Bocchio, Jones & Slavin 2014). This work is now being extended to a consideration of dust evolution in HII regions, galactic haloes and the intergalactic medium (IGM). Erosive processes involving high energy photons, electrons and ions are the major dust modifiers and destructors in the energetic regions of the ISM and the IGM.

The THEMIS approach provides a detailed framework within which the solutions to several interesting interstellar conundrums such as volatile silicon in PDRs, sulphur and nitrogen depletions, the origin of the blue and red photoluminescence and the diffuse interstellar bands (DIBs) may yet be found (Jones 2013; Jones 2014).

3. Conclusions

Interstellar dust is clearly not the same everywhere, it evolves in response to its immediate environment. For example in the transition from diffuse to denser regions of the ISM the dust will accrete a-C(:H) and ice mantles and coagulate to form aggregate grains with increasing density. Empirically-based models, which lack the necessary grounding in laboratory data, can never capture or fully uncover the physical origins of dust evolution even though they may provide excellent fits to the observational data. On the contrary, interstellar dust models strictly-anchored to laboratory measurements on physically-realisable and physically-realised dust analogue materials can provide a much deeper physical insight, even when the fit to the observational data is less that 'perfect'. The aim of such modelling should indeed not be a perfect fit to the observational data but the best-possible re-construction of those data with an astute use of measurement-constrained dust properties.

Acknowledgements: I would like to thank my collaborators, Melanie Köhler, Nathalie Ysard, Marco Bocchio, Aurélie Rémy-Ruyer, Elisabetta Micelotta, Lapo Fanciullo, Laurent Verstraete, Vincent Guillet, Emilie Habart, Alain Abergel, Emmanuel Dartois, Lisseth Gavilan and Marie Godard for innumerable hours of discussion about interstellar dust in all its guises.

References

Anders, E. & Zinner, E. 1993, *Meteoritics*, 28, 490
Bocchio, M., Micelotta, E. R., Gautier, A.-L., & Jones, A. P. 2012, *A&A*, 545, A124
Bocchio, M., Jones, A. P., Verstraete, L., et al. 2013, *A&A*, 556, A6
Bocchio, M., Jones, A. P., & Slavin, J. D. 2014, *A&A*, 570, A32
Fanciullo, L., Guillet, V., Aniano, G., et al.2015, *A&A*, 580, A136
Jones, A. P. 2012a, *A&A*, 540, A1
Jones, A. P. 2012b, *A&A*, 540, A2; 545, C2
Jones, A. P. 2012c, *A&A*, 542, A98; 545, C3
Jones, A. P. 2013, *A&A*, 555, A39
Jones, A. P., Fanciullo, L., Köhler, M., et al. 2013, *A&A*, 558, A62
Jones, A. P. 2014, *Planet. Space Sci.*, 100, 26
Jones, A. P., Ysard, N., Köhler, M., et al. 2014, *RSC Faraday Discuss.*, 168, 313
Jones, A. P. & Habart, E. 2015, *A&A*, 581, A92
Jones, A. P., Köhler, M., Ysard, N., et al. 2015, *A&A*, submitted
Köhler, M., Guillet, V., & Jones, A. 2011,
Köhler, M., Stepnik, B., Jones, A. P., et al. 2012, *A&A*, 548, A61 *A&A*, 528, A96
Köhler, M., Jones, A., & Ysard, N. 2014, *A&A*, 565, L9
Köhler, M., Ysard, N., & Jones, A. 2015, *A&A*, 579, A15
Micelotta, E. R., Jones, A. P., Cami, J., et al. 2012, *ApJ*, 761, 35
Westphal, A. J., Stroud, R., Bechtel, H. A., et al. 2014, *Science*, 345, 6198
Ysard, N., Köhler, M., Jones, A., et al. 2015a, *A&A*, 577, A110
Ysard, N., Köhler, M., Jones, A., et al. 2015b, *A&A*, submitted

Telescope Observations of Interstellar and Circumstellar Ices: Successes of and Need for Laboratory Simulations

A. C. A. Boogert

Universities Space Research Association, Stratospheric Observatory for Infrared Astronomy,
NASA Ames Research Center, MS 232-11, Moffett Field, CA 94035, USA
email: acaboogert@alumni.caltech.edu

Abstract. Ices play a key role in the formation of simple and complex molecules in dense molecular clouds and in the envelopes and protoplanetary disks surrounding young stars. Some fraction of the interstellar ices may become building blocks of comets, and thus be delivered to the early Earth. Laboratory simulations have proven to be crucial in the derivation of ice abundances, in quantifying reaction rates on cold grain surfaces, in determining the thermal and energetic processing history of the ices, and in understanding the interaction between the ices and the underlying refractory grain surfaces. In this invited topical paper I will review possible ways forward in improving our knowledge of the composition of the ices, as many signatures in the interstellar spectra are still poorly identified. I will also emphasize the observed importance of thermal processing of the ices (crystallization, segregation), which likely affects the chemistry after the initial dominance of grain surface reactions. Continued laboratory work is warranted in view of the upcoming observational data from, for example, the *James Webb Space Telescope* (JWST), which is ideally suited for ices studies. For an exhaustive review on this topic I refer to Boogert, Gerakines & Whittet (2015).

Keywords. astrochemistry, stars: formation, ISM: clouds, (ISM:) dust, extinction, ISM: abundances, ISM: molecules, molecular processes, methods: laboratory, infrared: ISM, telescopes

---

1. Introduction: Ice Formation

Observations of background stars have shown that H_2O and CO_2 ices form at visual extinctions $A_V \sim 1.5$ magnitude into dense clouds and cores, where they are shielded from the effects of photodesorption by the interstellar radiation field. Cold grain surface chemistry of O, H, and CO accreted from the gas is the dominant formation process. The intermediate product OH in H_2O formation also reacts with CO to form CO_2, resulting in an intimate mixture of CO_2 and H_2O ices. Deeper into the cloud ($A_V \sim 3$ magnitude), the gas becomes molecular (H_2/H increases) and a more volatile ($T < 10$ K) pure CO ice layer is formed. Observations have shown that at even greater depths ($A_V \sim 9$ magnitude), at high densities (10^5 cm^{-3}), CO ice hydrogenation leads to CH_3OH. The resulting mixture of CH_3OH with CO can be spectroscopically traced by a long wavelength wing of the 4.67 μm CO ice band. Penteado *et al.* (2015), expanding on work by Cuppen *et al.*(2011), show that the strength and shape of this wing agree with the strength and shape of the 3.53 and 9.7 μm bands of CH_3OH, as well as the ^{13}CO ice band toward the massive YSO AFGL 7009S. A gradient in mixing ratios of CO:CH_3OH=1:1-1:9 is possible, reflecting varying CH_3OH formation efficiencies over time.

2. Further Constraints on the Composition of Interstellar and Circumstellar Ices

The easily identifiable, strong modes of several simple ice species (CO, H_2O, CO_2, CH_3OH) are well studied. Absorption profile variations in samples of background stars and Young Stellar Objects (YSOs) are related to variations of the composition and the thermal history of the ices. These ice bands indeed have diagnostic value as to the nature and physical history of the line of sight observed. A number of absorption features are still not reliably identified, however. This is primarily because the infrared wavelength range traces vibrational modes of functional groups, e.g., the C-O stretch and N-H bending modes, which occur at similar wavelengths for different molecular species. Further observational constraints are possible, however, as the spectral resolving power of *Spitzer Space Telescope* observations in the 5-8 μm wavelength range is very low ($R = \lambda/\Delta\lambda \sim 60-100$). In particular for the distinct 7.25 and 7.41 μm ice features, the much improved spectral resolution of JWST ($R \sim 3000$) at two orders of magnitude better sensitivity will allow much more accurate measurements of the absorption profiles for much larger samples of sightlines (these features were so far reported in just a handful of sightlines). For the same reasons, further observational constraints can be expected for the elusive 6.85 μm ice band, the 7.58 μm "SO_2" band, and signatures of PAH species embedded in the ices.

A different approach to further constrain the composition of the ices is by measuring the ice lattice and torsional modes. These are more unique identifiers of specific molecules than the mid-infrared vibrational modes, as laboratory experiments have shown (e.g., Ioppolo et al. (2014)). The required telescope instrumentation, low-resolution spectrometers with wide instantaneous band-width and accurate broad-band spectral response calibration, are not available, however. This is a niche for future instruments on the *Stratospheric Observatory for Infrared Astronomy* (SOFIA).

3. Processing of Interstellar and Circumstellar Ices

In dense clouds and circumstellar environments, grain surface chemistry sets the bulk ice composition. Subsequently, thermal and energetic processing are expected to modify the ices. In particular, laboratory simulations have shown that energetic radiation (ultraviolet photons) and particles (cosmic rays) break molecular bonds and the radicals form more complex species. This has been hard to prove observationally however. Thermal processing on the other hand, is easily observed by the spectroscopic effects of crystallization, segregation, and sublimation. Considering the prominence of such heated ices in the envelopes and disks of YSOs, one must consider the efficiency of purely thermal reactions. For example, laboratory mixtures of H_2CO with NH_3 show an efficient formation of polyoxymethylene (POM), a polymer of the CH_2O group (Schutte et al. (1993)). The same mixture yields aminomethanol (NH_2CH_2OH), and the energy barriers and rates of this and other purely thermal reactions were measured by Theulé et al. (2013).

References

Boogert, A. C. A., Gerakines, P. A., & Whittet, D. C. B. 2015, *ARAA*, 53, 541
Cuppen, H. M., Penteado, E. M., Isokoski, K., van der Marel, N., & Linnartz, H. 2011, *MNRAS*, 417, 2809
Ioppolo, S., McGuire, B. A., Allodi, M. A., & Blake, G. A. 2014, *Faraday Discuss.*, 168, 461
Penteado, E. M., Boogert, A. C. A., Pontoppidan, K. M., et al. 2015, *MNRAS*, 454, 531
Schutte, W. A., Allamandola, L. J., & Sandford, S. A. 1993, *Icarus*, 104, 118
Theulé, P., Duvernay, F., Danger, G., et al. 2013, *Adv. Sp. Res.*, 52, 1567

AKARI NIR spectroscopy of interstellar ices

Takashi Onaka[1], Tamami I. Mori[1], Itsuki Sakon[1], Fumihiko Usui[1], Ronin Wu[1] and Takashi Shimonishi[2]

[1] Department of Astronomy, Graduate School of Science, The University of Tokyo
7-3-1 Hongo, Bunkyo-ku, Tokyo 113-0033, Japan
email: onaka@astron.s.u-tokyo.ac.jp

[2] Frontier Research Institute for Interdisciplinary Sciences, Tohoku University
6-3 Aramakiazaaoba, Aoba-ku, Sendai, Miyagi, 980-8578, Japan

Abstract. The Infrared Camera (IRC) onboard *AKARI* has a near-infrared (2–5 μm) spectroscopic capability with high sensitivity that allows us to study the major ice components in various objects. In particular, H_2O and CO_2 ice absorption features have been detected towards nearby galaxies, including several young stellar objects (YSOs) in the Large Magellanic Cloud (LMC), as well as a number of HII region-PDR complexes for the first time by IRC spectroscopy. While observations in the LMC show a high ratio (~ 0.34) of the CO_2 to H_2O ice column densities, the ratios in Galactic HII-region-PDR complexes are in the range of 0.1–0.2, being compatible with those found in Galactic massive YSOs in previous studies. The good correlation supports concurrent formation of the two ice species on the grain surface and the higher ratio in the low-metallicity LMC suggests possible environmental effects in the formation process.

Keywords. (ISM:) dust, extinction, ISM: lines and bands, infrared: ISM, astrochemistry

---

Ices of various species play important roles in interstellar chemistry (e.g., van Dishoeck 2014). Characteristic bands of ice species reside in the near- to mid-infrared (2–20 μm) and thus infrared observations are the most efficient means for the study of ices in the interstellar medium (ISM). *ISO* and *Spitzer* studied ice features towards deeply embedded young stellar objects (YSOs) and stars behind molecular clouds, making significant progress in our understanding of interstellar ices (e.g., Gibb *et al.* 2004; Pontoppidan *et al.* 2008). They indicate good correlations between the column densities of H_2O and CO_2 ices, suggesting concurrent formation of the two ice species on the grain surface (Ippolo *et al.* 2011; Oba *et al.* 2012).

The Infrared Camera (IRC) onboard *AKARI* offered high-sensitivity spectroscopy in 2.5–5 μm (Onaka *et al.* 2007), where major ice species have strong absorption bands, such as H_2O at 3.05 μm, CO_2 at 4.27 μm, XCN at 4.62 μm, and CO at 4.67 μm. The IRC allows us to study ice absorption features in faint objects, including nearby galaxies and relatively evolved diffuse objects. A recent study of ices in nearby galaxies with the IRC suggests no clear correlation between H_2O and CO_2 ices, which may be attributed to the relatively large beam ($\sim 5''$) that contains various components on the line-of-sight (Yamagishi et at. 2015). Studies of ice absorption towards YSOs in the Large Magellanic Cloud (LMC) with the IRC, on the other hand, show by a factor of 2 larger column densities of CO_2 ice relative to H_2O ice than towards Galactic massive YSOs, which must be related to the low-metallicity condition of the LMC and give important implication on the ice formation process (Shimonishi *et al* 2008, 2010). A study of Galactic molecular clouds with the IRC further suggests multi-stage ice formation (Noble *et al.* 2013).

In this report, we present latest results of NIR spectroscopy of ice species in Galactic HII-photodissociation region (PDR) complexes made with the IRC. Those targets are extended and relatively evolved compared to YSOs in previous studies and thus may

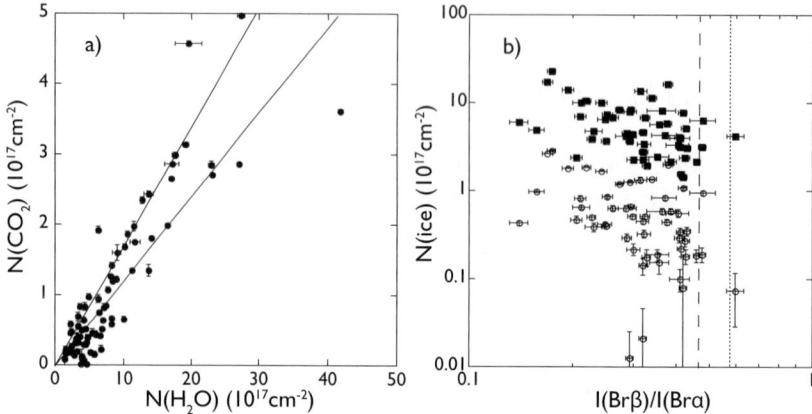

Figure 1. (a) Correlation of the column densities of CO_2 and H_2O ices for HII-PDR complexes observed with the IRC. The solid lines of the slopes of 0.17 and 0.12 are indicated as reference. The slope of 0.17 is suggested for Galactic massive YSOs in previous studies. (b) Correlations of the ice column densities and the ratio of $Br\beta$ to $Br\alpha$ line intensities. The filled squares show H_2O ice, while the open circles indicate CO_2 ice. The dotted line shows the ratio for Case B with the electron density of 10^4 cm^{-3} and the temperature of 10^4 K. The dashed line suggests a possible threshold for the presence of the two ice species, which corresponds to $A_v \sim 5$.

indicate evidence for interstellar processing of ice species. Previous studies of the emission bands related to polycyclic aromatic hydrocarbons (PAHs) indicate that IRC spectra of those objects indeed show ice absorption features (Mori et al.2014; Onaka et al. 2014). Figure 1a shows the correlation of the column densities between CO_2 and H_2O ices for the present samples, which is in agreement with previous results for Galactic massive YSOs except for several data located below the correlation line(s), suggesting possible interstellar processing. Figure 1b plots the ice column densities against the HI line intensity ratio, which suggests a threshold for the presence of these ices. The suggested threshold is $A_v \sim 5 \pm 1$, being compatible with those found in the Taurus cloud (Whittet et al. 2001, 2007). The new IRC results support concurrent formation of CO_2 and H_2O ices and suggest common physical conditions present for ice formation in the ISM.

This work is based on observations of *AKARI*, a JAXA project in participation of ESA. It is supported in part by a Grant-in-Aid from the JSPS (23244021).

References

Gibb, E. L., Whittet, D. C. B., Boogert, A. C. A., & Tielens, A. G. G. M. 2004, *ApJS*, 151, 35
Ioppolo, S., van Boheemen, Y., Cuppen, H. M., et al. 2011, *MNRAS*, 413, 228
Mori, T. I., Onaka, T., Sakon, I., et al. 2014, *ApJ*, 744, 68
Noble, J., Fraser, H., Aikawa, Y., Pontoppidan, K. M., & Sakon, I. 2013, *ApJ*, 775, 85
Oba, Y., Watanabe, N., Hama, T., et al. 2012, *ApJ*, 749, 67
Onaka, T., Matsuhara, H., Wada, T., et al. 2007, *PASJ*, 59, S401
Onaka, T., Mori, T. I., Sakon, I., et al. 2014, *ApJ*, 780, 114
Pontoppidan, K. M., Boogert, A. C. A., Fraser, H., et al. 2008, *ApJ*, 678, 1005
Shimonishi, T., Onaka, T., Kato, D., et al. 2008, *ApJL*, 686, L99
Shimonishi, T., Onaka, T., Kato, D., et al. 2010, *A&A*, 514, A12
van Dishoeck, E. F. 2014, *Faraday Discuss*, 168, 9
Whittet, D. C. B., Gerakines, P. A., Hough, J. H., & Shenoy, S. S. 2001, *ApJ*, 547, 872
Whittet, D. C. B., Shenoy, S. S., Bergin, E. A., et al. 2007, *ApJ*, 655, 332
Yamagishi, M., Kaneda, H., Ishihara, D., et al. 2015, *ApJ*, 807, 29

Comet composition and Lab

Dominique Bockelée-Morvan and Nicolas Biver

LESIA, Observatoire de Paris, LESIA/CNRS, UPMC, Université Paris-Diderot,
F-92195 Meudon, France
email: dominique.bockelee@obspm.fr

Abstract. Comet composition and properties provide information on chemical and physical processes that occurred in the early Solar system, 4.6 Gyr ago. The study of comets and of star-forming regions both help for a better understanding of the formation of planetary systems. A review of our present knowledge of cometary composition is presented. We also discuss laboratory studies that would be helpful for data analysis.

Keywords. comets: general, solar system: formation

1. Introduction

The solar system formed about 4.6 billion years ago from an infall of matter inside a molecular cloud. A wealth of processes happened, involving chemistry, and dynamics at all scales before the solar system reached its present state with its cortege of planets and small bodies. Comets formed in the outer skirts of the solar nebula and trapped volatile species formed either in the presolar cloud or in situ. Hence, the study of the composition of comets provides insights on the formation of the solar system, and proto-planetary disks, in general.

2. Comet composition from spectroscopy

The composition of cometary ices is best studied from observations of cometary atmospheres. At the present date, 28 molecules (not including isotopologues) have been identified by spectroscopy in the UV to the millimeter and submillimeter domains (Fig. 1). For a recent overview, see Cochran *et al.* (2015). The most recent discoveries are glycolaldehyde (CH_2OHCHO) and ethanol (C_2H_5OH), identified in comet C/2014 Q2 (Lovejoy) from lines in the millimeter range (Biver *et al.* 2015). Whereas most molecules are observed at millimeter wavelengths, the infrared domain gives access to molecules with no dipole moments as CO_2, CH_4, etc.

It is striking that the molecular complexity observed in comets resembles that observed in star-forming regions. A comparison between the abundances of complex organics observed in comets to values measured around high-mass and low mass proto-stars shows strong similarities, in line with their synthesis through grain-surface reactions in the presolar cloud or in the solar nebula (Biver *et al.* 2015).

3. Laboratory requests

Further progress in the characterization of cometary volatiles through infrared and mm-wave spectroscopy is expected in the near future, e.g., with ALMA and 30-m class optical telescopes. The following laboratory data and/or theoretical studies would be helpful for the analysis of cometary spectra:

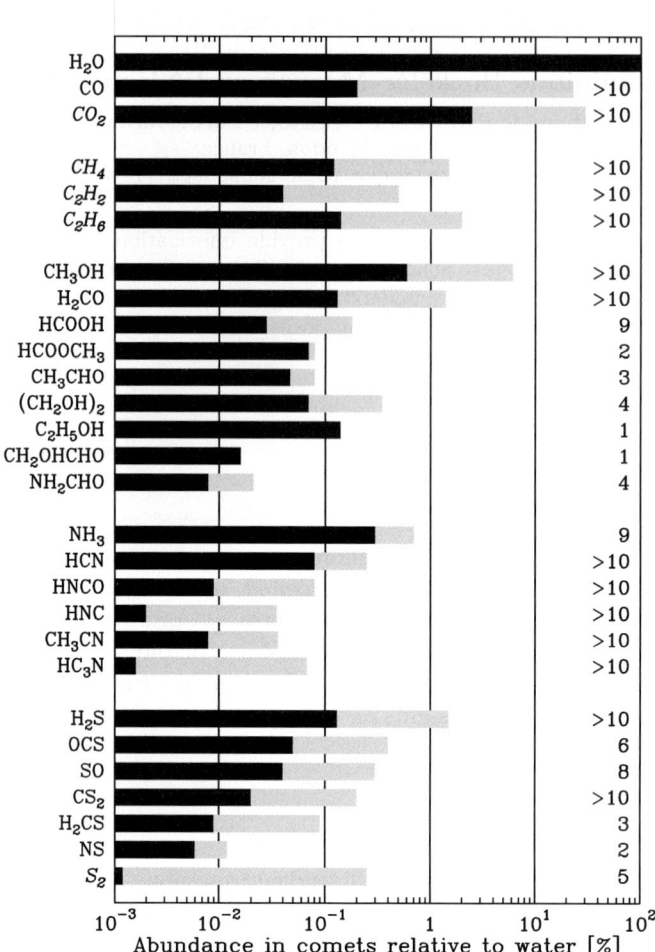

Figure 1. Molecules detected in comets and their abundances relative to water. Bars in blue (grey) show the range of measured abundances in comets, indicating composition diversity between comets. The number of comets in which abundance measurements are available is indicated in the right. This figure is an updated version of that published in Bockelée-Morvan et al. (2004)

- **Rotational lines**: lines frequencies and strengths for complex species. Molecules present in cometary atmospheres are similar to those found in molecular hot-cores and hot-corinos. Therefore, cometary science will benefit from spectral characterizations performed for assigning unidentified lines in ISM spectra. Specific requests may follow identifications made from the Rosina mass spectrometer onboard Rosetta (Balsiger et al. 2007).
- **Rotational lines**: In cometary atmospheres, the collisional partner is H_2O. Collisional cross-sections involving water are not available for most molecules, whereas they are an important parameter for determining accurate column densities from observed line intensities. Collisional cross-sections in the temperature range 20–120 K are needed.
- **Rotational lines**: Besides collisional excitation, rotational levels in the ground vibrational state are pumped through IR excitation of the vibrational bands by solar

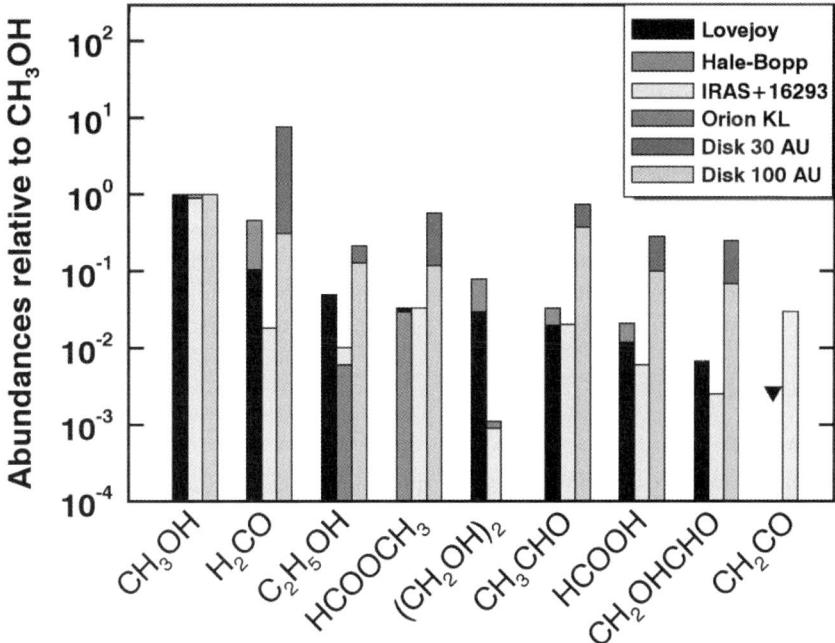

Figure 2. Abundances of complex organics in comets, molecular hot-core (Orion KL) and hot-corinos (IRAS+162293). Results from chemical modelling of protoplanetary disks (Walsh *et al.* (2014)) are also plotted. Figure from Biver *et al.* (2015).

Number and strength of molecular detections in Comet Lee on UT Aug. 21.6

Molecule	Number of single species emissions[a]					Number of multi-species emissions[a]				
	Total	Number of emissions by strength[b]				Total	Number of emissions by strength[b]			
		vs	s	m	w		vs	s	m	w
All	222	16	26	66	114	87	9	22	33	23
Identified	170	15	24	54	77	75	8	18	31	18
Unidentified[c]	52	1	2	12	37	12	1	4	2	5
Securely detected species										
H_2O	20	6	4	6	4	10	1	5	2	2
OH (1–0)	20	0	8	5	7	11	2	3	5	1
OH (2–1)	11	0	0	4	7	4	0	1	1	2
HCN	11	0	1	9	1	2	1	0	1	0
C_2H_2	4	0	0	0	4	3	1	0	1	1
NH_2	10	0	1	3	6	7	1	1	2	3
CH_4	4	2	1	0	1	2	0	1	1	0
C_2H_6	15	3	3	3	6	46	5	9	22	10
CH_3OH	67	4	6	23	34	63	6	15	24	18
H_2CO	6	0	0	1	5	11	0	4	3	4
Species noted but not securely detected[d]										
NH_3	0	0	0	0	0	1	0	0	0	1
CH_2	1	0	0	0	1	5	0	2	3	0
CH_3	1	0	0	0	1	0	0	0	0	0
CH	0	0	0	0	0	5	1	0	2	2

Figure 3. Infrared spectral survey of comet C/1999 H1 (Lee) covering the 3479–2702 cm^{-1} domain (extracted from Dello Russo *et al.* 2006). Strength: vs = very strong, s = strong, m = medium; w = weak

radiation. This fluorescent excitation cannot be computed for several detected molecules (e.g., HNCO, HCOOH, and complex organics) because information is missing, implying calculations in the LTE approximation.

• **Vibrational bands**: line-by-line ro-vibrational strengths and frequencies of many simple and complex CHO-bearing molecules are not available in the literature. This is the case for CH_3OH, the main CHO molecules in cometary atmospheres, which presents many strong ro-vibrational lines in the 3.3–3.5 μm domain (e.g., Villanueva *et al.* 2012,

Disanti et al. 2013). In addition, cometary spectra observed with high spectral resolution present many unidentified lines in this spectral region (e.g., Dello Russo et al. 2006, Dello Russo et al. 2013). In the spectral survey of comet C/1999 H1 (Lee) covering the 3479–2702 cm^{-1} domain, 101 out of 545 lines remain partially or completely unidentified, with the strongest lines in the 3300–3470 cm^{-1} range (Fig. 2, Dello Russo et al. 2006).

• **Molecular photodissociation rates**: Under the influence of solar radiation, molecules photodissociate in cometary atmospheres. Photodissociation rates are important parameters for the determination of molecular production rates. Compilations of determined values can be found in Huebner et al. (1992) and Crovisier (1994). Values are missing for several detected molecules. Photodissociation channels and quantum yields are also important quantities to investigate production mechanisms of detected molecules (Crovisier 1994).

4. Conclusion

In this paper, we focussed on laboratory data that would be helpful for analysing gas-phase signatures observed in the infrared and microwave domains. The optical domain, where many lines remain also unidentified in cometary spectra, would also benefit from laboratory investigations (Cochran & Cochran 2002). There are many other domains where laboratory experiments support cometary science, as illustrated in Gudipati et al. (2015). One can anticipate that the wealth of data acquired by the instruments onboard the Rosetta spacecraft will initiate many laboratory studies.

References

Balsiger, H., Altwegg, K., Bochsler, P., et al. 2007, *Space Science Reviews*, 128, 745
Biver, N., Bockelée-Morvan, D., Moreno, R., et al. 2015, *Science Advances*, in press
Bockelée-Morvan, D., Crovisier, J., Mumma, M. J., & Weaver, H. A. 2004, Comets II, 391
Cochran, A. L. & Cochran, W. D. 2002, *Icarus*, 157, 297
Cochran, A. L., Levasseur-Regourd, A.-C., Cordiner, M., et al. 2015, *Space Science Reviews*, 68
Crovisier, J. 1994, *JGR*, 99, 3777
Dello Russo, N., Mumma, M. J., DiSanti, M. A., et al. 2006, *Icarus*, 184, 255
Dello Russo, N., Vervack, R. J., Weaver, H. A., et al. 2013, *Icarus*, 222, 707
DiSanti, M. A., Bonev, B. P., Villanueva, G. L., & Mumma, M. J. 2013, *ApJ*, 763, 1
Huebner, W. F., Keady, J. J., & Lyon, S. P. 1992, *Astrophys. Space Sci.*, 195, 1-294, 1992.
Gudipati, M. S., Abou Mrad, N., Blum, J., et al. 2015, *Space Science Reviews*, 79
Villanueva, G. L., DiSanti, M. A., Mumma, M. J., & Xu, L.-H. 2012, *ApJ*, 747, 37
Walsh, C., Millar, T. J., Nomura, H., et al. 2014, *A&A*, 563, A33

Laboratory and theoretical work applied to planetary atmospheres

Athena Coustenis

Laboratoire d'Etudes Spatiales et d'Instrumentation en Astrophysique (LESIA),
Observatoire de Paris, PSL-Research Univ., CNRS, Univ. Paris 06, Sorbonne Univ., Univ.
Paris-Diderot, Sorbonne Paris-Cité, 5, place Jules Janssen, 92195 Meudon Cedex, France
email: athena.coustenis@obspm.fr

Abstract. We look at applications of recent work in theoretical and experimental spectroscopy for the analysis of IR data concerning giant planets, Titan and possibly exoplanets.

Keywords. planets and satellites: formation, methods: data analysis, techniques: spectroscopic

1. Introduction

The study of planetary atmospheres has various aims which include the :
- Cartography of the chemical composition therein with high precision
- Detection of new components with low abundance levels (including isotopes)
- Determination of the origin and evolution of the planetary bodies

For this, there is strong need for spectroscopic measurements at the right p, T conditions to be able to perform the spectroscopic analysis of data provided by different instruments on board of space missions or from telescopic ground-based observations. In the case of dense planetary atmospheres in the outer solar system (gas giants and Titan) what is needed is:
- Long atmospheric paths (at scale heights of a few tens of km)
- Temperatures difficult to achieve in a laboratory (for Titan : 70-94 K in troposphere; less than 200 K in stratosphere).

Spectroscopic data available in the infrared range for outer solar system objects include acquisitions by Voyager/IRIS, ground-based telescopes and ISO, Galileo, and more recently Cassini-Huygens (CIRS, VIMS, DISR, etc). The latter space mission has been exploring the Saturnian system since 2004 and will continue until 2017, providing a large amount of data on all the objects and in particular the planet and its main satellite, Titan. Hereafter we focus at the infrared spectrum of Titan and the giant planets and look at how theoretical and experimental spectroscopic data can help determine their composition for the objectives hereabove.

We study the physical properties in the atmosphere of the giant planets and Titan with radiative transfer calculations using a line-by-line code solving for : opacity sources; chemical abundances (gas and solid); haze/aerosols; clouds; and the temperature structure (Coustenis *et al.*, 2010; 2015). Spectroscopic data come from various sources but essentially from GEISA and HITRAN.

2. Titan and the giant planets

Titan is a satellite essentially composed of N_2, CH_4, and H_2, with an atmospheric pressure on the surface of 1,5 bar with a temperature of 94 K. In the thermal IR, Cassini Titan data are essentially derived from CIRS : 7μm-1mm (resolution up to 0.5 cm^{-1}).

In the near-IR the spectro-imager VIMS covers the 0,35-5,2 μm region (resolution 7-16 nm). Also, from the Huygens probe with DISR we get the images and spectra in the 0.48-1.7 μm region (resolution: 5-17 nm). For the current chemical composition of Titan see Table 2 in Coustenis, A. (2014) and references therein.

Several breakthroughs have been made in the determination of the chemical composition of Titan in recent years thanks to new spectroscopic data made available to the community. Examples are propane (for which multiple bands were identified in CIRS spectra, Nixon et al., 2012), ethane (Coustenis et al., 2010), isotopes and methane.

Isotopes. For isotopes we have the detections of C_2HD and of several isotopes of ^{13}C in HC_3N (Cyanoacetylene formed by substitution of -CN (from HCN) into C_2H_2 and C_2H_4), with a strong ν_2 band at 663.4 cm^{-1} due to bending of CH and $H^{13}CCCN = 658.7$ cm^{-1} ; $HC^{13}CCN = 663.1$ cm^{-1} ; $HCC^{13}CN = 663.1$ cm^{-1}. Also isotopes of CO_2, with essentially the ν_5 band at 667 cm^{-1} but also $^{13}CO_2$ at 648.5 cm^{-1} (6-σ detection) and probably the $C^{18}O^{16}O$ emission at 662.5 cm^{-1}.

Methane. Titan's near-IR spectrum contains information on the lower atmosphere and the surface of the satellite. It is equivalent to a 20 km cell with 75 mbar of CH_4 in 1.5 bar of N_2 and at T= 85 K. One needs to obtain the CH_4 absorption spectrum in Titan conditions in the several transparency windows at 0.83, 0.94, 1.075, 1.28, 1.6, 2.0, 2.6-2.8, and 4.9 μm. In recent efforts, with the goal to determine the transmission in these transparency windows at 80 K (in some cases, the different band models that existed before showed large disagreements in fitting the Cassini data) a combination of astronomer and experimental teams have produced new data to help in the analysis of the observations. On the basis of new experimental developments, as a first step the room-temperature (RT) and liquid-nitrogen temperature (LNT) spectra of CH_4 have been recorded in the full 1.2-1.6 μm region. They will be used on Titan to:

• re-measure the CH_3D/CH_4 ratio from much higher resolution ground-based spectra
• re-measure the CO abundance and vertical profile in Titan's atmosphere
• look for other minor atmospheric constituents
• study variations of the CH_4 mixing ratio over Titan with ground-based observations
• measure the surface albedo and its variations over the surface using VIMS data and ground-based data
• re-analyze DISR data (e.g. to study albedo variations near the Huygens site)

Giant planets. Other applications concern Uranus and Neptune (clouds levels, gaseous abundances, spatial variations). Simulations of Uranus spectra at 1.6 μm using the new datalist WKMC from Grenoble. The new linelists were also applied to Uranus in the 1.55 μm window (Bailey et al., 2012) on data from IRIS2 on the Anglo-Australian telescope at a resolving power of about 2400.

The new linelists for methane also apply to exoplanets.

We clearly need more adequate spectroscopic data for the analysis of the large sum of data that we retrieve from observations of the bodies in the outer solar system.

References

Bailey, A., et al. 2012, *Icarus*, 213, 218
Campargue, A., et al. 2012, *Icarus*, 219, 110
Coustenis, A., et al. 2015, *Icarus*, 207, 461
Coustenis, A. 2014, in:T. Spohn, D. Breuer, & T. V. Johnson (eds.), "Titan", in *Encyclopedia of the Solar System, Third Edition* (Elsevier), p. 831
Coustenis, A., et al. 2015, *Icarus*, in press
Nixon, A., et al. 2012, *ApJ*, 749, article id. 159

The THS experiment: Simulating Titan's atmospheric chemistry at low temperature (200 K)

Ella Sciamma-O'Brien[1,2] Kathleen T. Upton[3], Jack L. Beauchamp[3] and Farid Salama[1]

[1] NASA Ames Research Center, Moffett Field, CA
email: ella.m.sciammaobrien@nasa.gov

[2] Bay Area Environmental Research Institute, Petaluma, CA

[3] Noyes Laboratory of Chemical Physics and the Beckman Institute - Caltech, Pasadena, CA

Abstract. In the Titan Haze Simulation (THS) experiment, Titan's atmospheric chemistry is simulated by plasma discharge in the stream of a supersonic expansion, i.e. at low Titan-like temperature (150 K). Here, we present complementary gas and solid phase analyses of four N_2-CH_4-based gas mixtures that demonstrate the unique capability of the THS to monitor the chemical growth evolution in order to better understand Titan's chemistry and the origin of aerosol formation.

Keywords. astrochemistry, molecular processes, plasmas, methods: laboratory, planets and satellites: Titan.

1. Introduction

Titan, Saturn's largest moon, is the only solid body in the outer solar system with a dense atmosphere. In Titan's atmosphere, composed mainly of nitrogen (N_2 at 95-98%) and methane (CH_4 at 2-5%), a complex chemistry occurs at temperatures lower than 200 K, and leads to the production of heavy organic molecules and subsequently solid aerosols that form the orange haze surrounding Titan. Because the reactive carbon and nitrogen species present in Titan's aerosols could meet the functionality requirements for precursors to prebiotics, the study of Titan's aerosol has become a topic of extensive research in the fields of astrobiology and astrochemistry. Experiments have been developed in several laboratories in the past 30 years in order to understand the production processes and composition of these atmospheric aerosols.

2. The Titan Haze Simulation (THS) experiment

In the study presented here, we used the Titan Haze Simulation (THS) experiment, an experimental setup developed at the NASA Ames COSmIC simulation facility, to study Titan's atmospheric chemistry at low temperature (Ricketts *et al.* (2011)). In the THS, the chemistry is simulated by plasma in the stream of a pulsed supersonic jet-cooled expansion. With this unique design, the gas is cooled to Titan-like temperature (~ 150 K) before inducing the chemistry by pulsed plasma (Biennier *et al.* 2006), and remains at low temperature in the plasma discharge (~ 200 K). The pulsed nature of the plasma allows for a truncated chemistry that can be used to study the early stages of aerosol production, which has not been readily accomplished so far using other laboratory production methods. Different N_2-CH_4-based gas mixtures can be injected in the plasma. Due to the short residence time of the gas in the pulsed plasma discharge, only the first

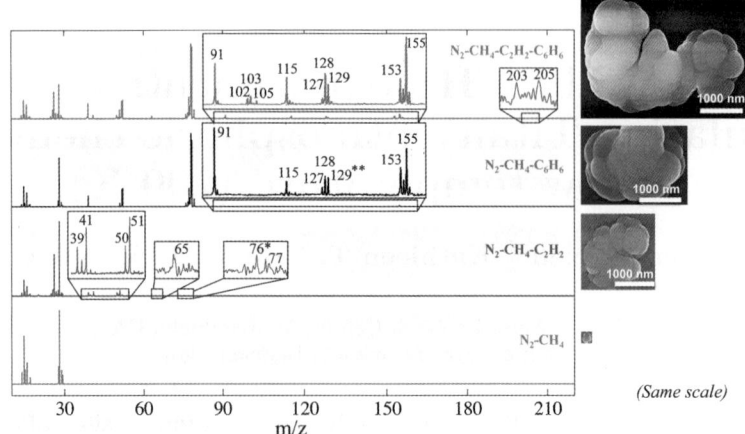

Figure 1. (Left): Mass spectra of four gas mixtures showing the observed chemical growth evolution when adding heavier precursors to the initial N_2-CH_4 gas mixture: N_2-CH_4 (red), N_2-CH_4-C_2H_2 (blue), N_2-CH_4-C_6H_6 (green), N_2-CH_4-C_2H_2-C_6H_6 (pink). (Right): SEM images of grains and aggregates produced in the same gas mixtures, demonstrating the consistency between gas and solid phase in COSmIC/THS.

steps of the chemistry have time to occur in a N_2-CH_4 discharge. However, by adding heavier precursors present as trace elements on Titan to the initial N_2-CH_4 mixture, we can observe a chemical growth evolution and study the intermediate steps of Titan's atmospheric chemistry. Both the gas- and solid phase products resulting from the plasma-induced chemistry can be monitored and analyzed using a combination of complementary in situ and ex situ diagnostics.

3. Results and Discussion

A recent Time-Of-Flight mass spectrometry study of the gas phase (Sciamma-O'Brien et al. 2014) has demonstrated that the COSmIC/THS is a unique tool to probe the first and intermediate steps of Titan's atmospheric chemistry at Titan-like temperature. In particular, the mass spectra obtained in a N_2-CH_4-C_2H_2-C_6H_6 mixture are relevant for comparison to the Cassini Plasma Spectrometer - Ion Beam Spectrometer (CAPS-IBS) instrument. The results of a complementary study of the solid phase are consistent with the chemical growth evolution observed in the gas phase, as shown in Fig. 1. Grains and aggregates form in the gas phase and can be jet deposited on various substrates for ex situ analysis. Scanning Electron Microscopy (SEM) images have shown that aggregates produced in more complex mixtures like N_2-CH_4-C_2H_2-C_6H_6 are much larger (up to 5 μm in diameter) than those produced in simpler mixtures like N_2-CH_4 (0.1-0.5 μm). Direct Analysis in Real Time mass spectrometry (DART-MS) combined with Collision Induced Dissociation (CID) of the solid phase have detected the presence of aminoacetonitrile, a precursor of glycine, in the COSmIC/THS aerosols. X-ray Absorption Near Edge Structure (XANES) measurements also show the presence of imine and nitrile functional groups, showing evidence of nitrogen chemistry. These complementary studies show the high potential of THS to better understand Titan's chemistry and the origin of aerosol formation.

References

Biennier, L., Benidar, A., & Salama, F. 2006, *Chem. Phys.*, 326, 445
Ricketts, C. L., Contreras, S. C. Walker, R., & Salama, F. 2011, *Int. J. Mass Spec.*, 300, 26
Sciamma-O'Brien, E., Ricketts, C. L., & Salama, F. 2014, *Icarus*, 243, 325

Magnetic field generation, Weibel-mediated collisionless shocks, and magnetic reconnection in colliding laser-produced plasmas

W. Fox[1,†], A. Bhattacharjee[1,2] and G. Fiksel[3,4]

[1] Princeton Plasma Physics Laboratory
P.O. Box 451, Princeton, NJ 08543 USA

[2] Princeton University, Dept. of Astrophysical Sciences
Princeton, NJ 08543 USA

[3] University of Rochester, Laboratory for Laser Energetics
Rochester, NY 14623, USA

[4] University of Michigan, Dept. of Nuclear Engineering and Radiological Science
Ann Arbor, MI 48109, USA

[†] email: wfox@pppl.gov

Abstract. Colliding plasmas are ubiquitous in astrophysical environments and allow conversion of kinetic energy into heat and, most importantly, the acceleration of particles to extremely high energies to form the cosmic ray spectrum. In collisionless astrophysical plasmas, kinetic plasma processes govern the interaction and particle acceleration processes, including shock formation, self-generation of magnetic fields by kinetic plasma instabilities, and magnetic field compression and reconnection. How each of these contribute to the observed spectra of cosmic rays is not fully understood, in particular both shock acceleration processes and magnetic reconnection have been proposed. We will review recent results of laboratory astrophysics experiments conducted at high-power, inertial-fusion-class laser facilities, which have uncovered significant results relevant to these processes. Recent experiments have now observed the long-sought Weibel instability between two interpenetrating high temperature plasma plumes, which has been proposed to generate the magnetic field necessary for shock formation in unmagnetized regimes. Secondly, magnetic reconnection has been studied in systems of colliding plasmas using either self-generated magnetic fields or externally applied magnetic fields, and show extremely fast reconnection rates, indicating fast destruction of magnetic energy and further possibilities to accelerate particles. Finally, we highlight kinetic plasma simulations, which have proven to be essential tools in the design and interpretation of these experiments.

Keywords. plasmas, magnetic fields, shock waves, acceleration of particles

1. Introduction

The collision of high velocity flowing plasmas allows the conversion of kinetic energy into other forms, including heat, magnetic fields, and perhaps most interestingly, the acceleration of particles to form high energy, super-thermal particle populations. In astrophysical contexts, such systems appear to be the sites, where particles are accelerated to extremely high energies to form the cosmic ray spectrum (Ackermann *et al.* 2013). Since astrophysical plasmas are typically collisionless due to the high energy of the plasma, kinetic plasma processes govern the interaction and particle acceleration processes, including shock formation, self-generation of magnetic fields by kinetic plasma instabilities, and magnetic field compression and reconnection. How each of these contribute to the

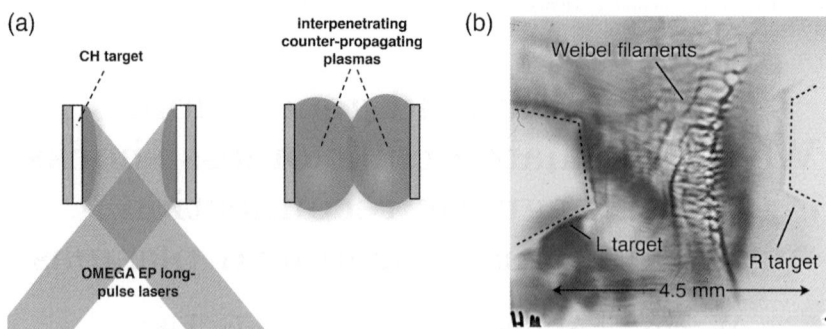

Figure 1. Diagram and proton radiography data from unmagnetized experiments showing development of filaments from ion-driven Weibel instability (Fox et al. 2013). The diagnostic uses a beam of protons to probe electromagnetic fields in the experiment volume; here, the filamentary magnetic fields of the Weibel instability lead to a characteristic filamentary proton intensity pattern on the detector film stack.

observed spectra of cosmic rays is not understood, and in particular two paradigms, shock acceleration (Bell 1978) and magnetic reconnection, have been proposed (Drake et al. 2010).

A recent generation of laboratory high-energy-density physics facilities, both laser and pulsed-power, has opened significant physics opportunities for experimentally modeling these processes. These experimental platforms are very interesting for laboratory astrophysics due to the high energy densities, which allows them to reach very low dissipation regimes, and to have good separation of scales from macroscopic to kinetic scales. This paper will highlight some recent results from these laboratory experiments toward understanding astrophysical magnetic field generation and particle acceleration mechanisms.

2. Collisionless shocks and ion-Weibel instability

The Weibel instability (Weibel 1959, Davidson et al. 1972) is among the few processes that generates magnetic field *de novo* in astrophysical and laboratory plasmas. This instability has received significant attention lately as a mechanism to mediate astrophysical collisionless shocks (e.g. Kato & Takabe 2008, Spitkovsky 2008). Coulomb collisions alone are typically too weak to sustain shocks in high-temperature astrophysical plasmas, since the ion mean-free-path is typically significantly larger than the observed shock width. Instead, collective electromagnetic fields are required. In cases, where ambient fields are weak or non-existent, the *Weibel instability* has been proposed to generate the fields required to sustain the shock. The interpenetration of plasmas in an initially unmagnetized shock front generates local counterstreaming beam conditions, which drive the Weibel instability. Weibel instability has been proposed to mediate shocks in systems ranging from gamma ray bursts to supernova remnants. Once a shock forms, particles are accelerated by repeated reflection off the plasma flows converging into the shock via the diffusive shock acceleration mechanism (Bell 1978).

Recent experiments made the first laboratory identification of the ion-driven Weibel instability between colliding unmagnetized plumes through experiments conducted on OMEGA EP (Fox et al. 2013) and OMEGA (Huntington et al. 2015), at the University of Rochester Laboratory for Laser Energetics, see Fig. 1. Large-scale particle-in-cell simulations provided *ab initio*, first principles simulations, which included the plasma ablation from the targets, collision of the two plumes, and in identifying the resulting Weibel instability (Fox et al. 2013). These results open up significant new questions and

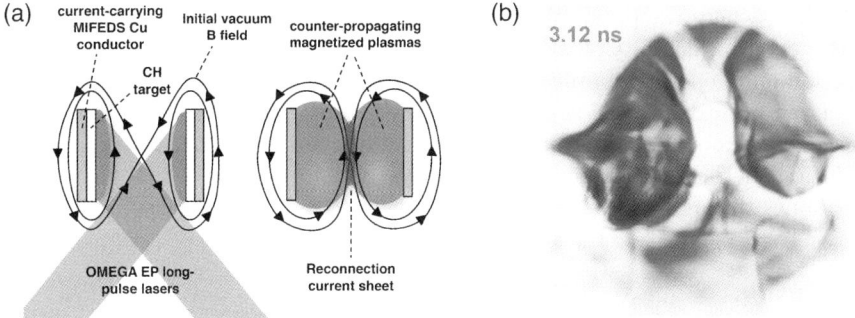

Figure 2. Diagram and sample proton radiography from magnetized experiment showing collision of magnetized plasmas and current sheet formation (Fiksel *et al.* 2014). The white bands reflect regions of highly-compressed magnetic field on the edge of the each plume. The magnetic topology is such that the collision of the plumes drives magnetic reconnection. The cellular pattern in the collision region possibly indicates the breakup of the current sheet into a few islands.

opportunities to study the astrophysical ion-Weibel instability. Future work should attempt to benchmark the dispersion relation in detail, and model the growth and saturation of the instability in 2-d and in 3-d under realistic scenarios. Experiments with embedded magnetic fields can study the suppression of Weibel instability by finite fields and thereby compare the shock mechanisms based on Weibel instability versus compression of a pre-existing upstream field.

3. Magnetic Reconnection

Throughout the Universe, magnetic reconnection is a ubiquitous process allowing the change of magnetic topology and the explosive release of stored magnetic energy (e.g. Yamada *et al.* 2010). Vivid examples of magnetic reconnection are solar and stellar flares, substorms in the Earth's magnetotail, which power the aurora, and sawtooth crashes and relaxation processes in tokamaks and reversed-field pinches. Key questions in magnetic reconnection research include the rate of reconnection, i.e. how fast can the magnetic energy be converted, the structure of reconnection current sheets, in particular a question of the importance of current sheet breakup into multiple magnetic islands, and finally the energization of particles by reconnection. In large reconnection systems, where reconnection occurs via multiple islands, a Fermi-type acceleration process again plays a role, where the particles interact with multiple reconnection outflows in a turbulent reconnecting current sheet (e.g. Drake *et al.* 2006).

In recent years a new set of magnetic reconnection experiments has been conducted in HED laser-produced plasmas (e.g. Nilson *et al.* 2006, Li *et al.* 2007). These laser-driven reconnection experiments collided plasma plumes and observed the reconnection of self-generated (e.g. Biermann battery) magnetic fields. Laser-driven experiments hold the possibility to observe reconnection dynamics in the large-system-size, turbulent regime; the large laser energy provides access to large energized volumes (when measured in fundamental plasma units such as the ion skin depth) and high-temperature, low dissipation regimes.

New experiments (Fiksel *et al.* 2014) have now demonstrated controlled magnetization of ablated plasma plumes using a choreography of externally applied magnetic fields, a "background" plasma, and "driver" plasma plumes. The results are shown in Fig. 2. Distinct white bands on the proton radiography observations indicate regions of strong magnetic field, where the pre-applied magnetic field is compressed from its initial $\sim$10 T at the targets to $\sim$25 T. Collision of oppositely magnetized plumes drives magnetic reconnection and generates bubble-like structures, which possibly reflect the breakup of the current sheet into multiple-island structures by a tearing instability. The experiments were successfully modeled by large-scale particle-in-cell simulation, which simulated the full evolution of the experiments starting from the formation of the two magnetized plasma plumes through to the collision and driven reconnection (Fiksel *et al.* 2014).

4. Conclusions

We have developed new techniques allowing laboratory experiments on the kinetic behavior of colliding magnetized plasmas, including magnetic field generation, compression, and reconnection. These offer significant opportunities for studying the mechanisms behind astrophysical collisionless shock formation and particle acceleration. Future work can study how both the reconnection and Weibel instability can heat and energize particles, both in the linear Weibel regime of the present experiments and eventually full-fledged shocks. In this case, comparison with magnetized experiments can allow comparison of shock-heating mechanisms (diffusive shock acceleration) with magnetic reconnection, on a common experimental platform (Deng *et al.* 2015). Large-scale, fully-kinetic particle simulation will continue to play a key role in designing and interpreting these experiments.

References

Ackermann, M., Ajello, M., Allafort, A., *et al.* 2013, *Science*, 339, 807
Bell, A. R. 1978, *MNRAS*, 182, 147
Davidson, R. C., Hammer, D. A., Haber, I., & Wagner, C. E. 1972, *Phys. Fluids*, 15, 317
Deng, W., Fox, W., & Bhattacharjee, A. 2015, *to be submitted*
Drake, J. F., Opher, M., Swisdak, M., *et al.* 2010, *ApJ*, 709, 963
Drake, J. F., Swisdak, M., Che, H., & Shay, M. A. 2006, *Nature*, 443, 553
Fiksel, G., Fox, W., Bhattacharjee, A., *et al.* 2014, *Phys. Rev. Lett.*, 113, 105003
Fox, W., Fiksel, G., Bhattacharjee, A., *et al.* 2013, *Phys. Rev. Lett.*, 111, 225002
Huntington, C. M., Fiuza, F., Ross, J. S., *et al.* 2015, *Nature Physics*, 11, 173
Kato, T. N. & Takabe, H. 2008, *ApJ*, 681, L93
Li, C. K., Séguin, F. H., Frenje, J. A., *et al.* 2007, *Phys. Rev. Lett.*, 99, 055001
Nilson, P. M., Willingale, L., Kaluza, M. C., *et al.* 2006, *Phys. Rev. Lett.*, 97, 255001
Spitkovsky, A. 2008, *ApJ*, 673, L39
Weibel, E. S. 1959, *Phys. Rev. Lett.*, 2, 83
Yamada, M., Kulsrud, R., & Ji, H. 2010, *Rev. Mod. Phys.*, 82, 603

Underground Nuclear Astrophysics in China

Liu WeiPing[1] for JUNA collaboration

[1] China Institute of Atomic Energy, P. O. Box 275(1), Beijing 102413, China

Abstract. Underground Nuclear Astrophysics in China (JUNA) will take the advantage of the ultra-low background in Jinping underground lab. High current accelerator with an ECR source and detectors will be set up. We plan to study directly a number of nuclear reactions important to hydrostatic stellar evolution at their relevant stellar energies, such as ^{25}Mg(p,γ)^{26}Al, ^{19}F(p,α)^{16}O, ^{13}C(α,n)^{16}O and ^{12}C(α,γ)^{16}O.

Keywords. direct measurement, underground laboratory, Gamow window, JUNA

1. Underground physics

Direct measurement of the cross sections for the key nuclear reactions crucial to hydrostatic stellar evolution within Gamow window is important for solving key scientific questions in nuclear astrophysics (Iliadis 2007). The direct measurement of astrophysical reaction rates on stable nuclei require high-intensity beams and extremely low background, which represents the major challenge at the frontiers of nuclear astrophysics. The largest challenge is the small cross section amid with large natural background. With the ultra-low background in deep underground environment, direct measurement of these key reactions in underground lab becomes a frontier in the field of experimental nuclear astrophysics. The first underground based low-energy accelerator facility, LUNA (Formicola *et al.* 2003; Costantini *et al.* 2009) at Gran Sasso underground laboratory, has successfully demonstrated the feasibility of meeting these challenges.

China JinPing underground Laboratory (CJPL) was established from a constructing hydro-power plant in the Jinping mountain, Sichuan, China (Chen 2010; Cheng *et al.* 2011). The facility is located near the middle of traffic tunnel. The facility is shielded by 2400 m of mainly marble overburden, with radioactively quiet rock. Its ultra-low cosmic ray background, which is about 2 orders of magnitude lower than that in Gran Sasso, makes it into an ideal environment for low background experiment. CJPL phase I (CJPL-I) now houses CDEX (Zhao *et al.* 2013) and PandaX dark matter experiments. CJPL phase II (Normile 2014) (CJPL-II) is expected to be available by the beginning of 2016 for much larger scale underground experiments (120,000 m^3 volume). JUNA will be one of its major research programs in CJPL-II.

2. Nuclear reactions

2.1. $^{12}C(\alpha,\gamma)^{16}O$ reaction

The ^{12}C(α,γ)^{16}O reaction is quoted as the holy grail in nuclear astrophysics (Rolfs & Rodney 1988). The uncertainty of this reaction affects not only the nucleosynthesis of elements up to iron, but also the evolution of the massive stars and their final fate (black hole, neutron star). The cross section of this reaction has to be known within an uncertainty less than 10% at helium burning temperatures (T$_9$=0.2), corresponding to a Gamow window around E$_{c.m.}$=300 keV. It is extremely difficult to determine the reaction cross section (about 10^{-17} barn) at this energy (Buchmann 2005). Current technology can only achieve 10^{-14} barn cross section level. A direct measurement at $E_{c.m.}$=600 keV near the Gamow window will be done in JUNA with high intensity ion beam of the experimental platform to provide better constraints for extrapolating models (Liu 2014).

For total cross section measurement at $E_{c.m.}$=600 keV, with the results of angular distribution measurement at $E_{c.m.}$=600 keV, we will optimize the experiment condition, including: 1) optimizing the beam transmission on the basis of the beam-optics calculation, adjusting the setup of shields to suppress the background coming from the beam, 2) confirming the origin of ^{13}C and improving the implantation condition of ^{12}C implantation target to reduce the disturbance of ^{13}C. The BGO detection array placed around the target chamber can significantly increase the detection efficiency (absolute efficiency 75% at $E_\gamma = 6$ MeV) of γ-rays. With the improvement above, an accurate total cross section will be obtained.

For total cross section test measurement at $E_{c.m.}$=380 keV, we will use ^{4}He^{2+} beam with an intensity of 2.5 cmA and an energy of 507 keV ($E_{c.m.}$=380 keV) and the high-efficiency BGO detection array. A direct measurement of the total cross section of ^{12}C(α,γ)^{16}O in the energy region of near Gamow window will be tested.

2.2. $^{13}C(\alpha,n)^{16}O$ reaction

The ^{13}C(α,n)^{16}O reaction is the key neutron source reaction for the stellar s-process nucleosynthesis. Due to the existence of sub-threshold resonances, there is a rather large uncertainty (30%) in this important reaction rate which limits our understanding to the nucleosynthesis of heavy elements. We plan to study directly this important reaction for the first time at energies down to $E_{c.m.} \sim 0.2$ MeV, within its relevant stellar energy range (Tang 2014).

We are designing a fast neutron detector consisting of 24 ^{3}He proportional counters and a liquid scintillator. The scintillator has a cylindrical shape with a length of 0.4 m and a diameter of 0.4 m. The 24 ^{3}He counters are distributed in the two circles with radii of 0.1 m and 0.15 m, respectively.

The energies of neutrons from the ^{13}C(α,n)^{16}O reaction are in the range of 2 to 3 MeV. The produced neutrons are firstly slowed down by the liquid scintillator. After their thermalization, some neutrons enter ^{3}He counters and are detected. With the coincidence between the fast signal from fast neutron slowing down inside the liquid scintillator and the delayed signal from the thermalized neutrons captured by the ^{3}He counters, we can effectively suppress the backgrounds in liquid scintillator and ^{3}He detectors. The detection efficiency after coincidence is estimated to be 20% for neutrons from the ^{13}C(α,n)^{16}O reaction.

2.3. $^{25}Mg(p,\gamma)^{26}Al$ reaction

The ^{25}Mg$(p,\gamma)^{26}$Al reaction is the main way to produce ^{26}Al in the Galaxy and its cross section is dominated by the capture process of the isolated resonances in ^{26}Al. The temperature range of astrophysical interest is T = 0.02-2 GK, so the levels between 50 keV and 310 keV are more important in the study of Galactic ^{26}Al. Many experiments have been performed to study the ^{25}Mg$(p,\gamma)^{26}$Al reaction since 1970 , but the experiment on the surface of earth can only reach to 190 keV energy level due to the small cross section and large background effects of the cosmic rays. In 2012, the laboratory of underground nuclear astrophysics (LUNA) in Italy successfully measured the resonance strength at 92 keV with the help of high shielding conditions in the underground laboratory (Strieder et al. 2012; Straniero et al. 2013). However, the ^{25}Mg$(p,\gamma)^{26}$Al cross section of 58 keV resonant capture is inaccessible for direct measurement in the shielding conditions of LUNA experiments. The underground laboratory of Jinping in China is covered with the

Table 1. Basic parameters of four reactions planned.

reaction	beam	intensity (emA)	c.m. energy (keV)	cross section	target thickness	efficiency %	CTS (/day)	BKD (/day)
$^{12}C(\alpha,\gamma)^{16}O$	$^4He^{2+}$	2.5	380	10^{-13} mb	10^{18} atoms/cm^2	75	0.7	0.7
$^{13}C(\alpha,n)^{16}O$	$^4He^{1+}$	10	200	10^{-12} mb	10^{21} atoms/cm^2	20	7	1
$^{25}Mg(p,\gamma)^{26}Al$	$^1H^{1+}$	10	58	$\omega\gamma$ 2.1×10^{-13} eV	0.6 μg/cm^2	38	1.4	0.7
$^{19}F(p,\alpha_\gamma)^{16}O$	$^1H^{1+}$	0.1	100	7.2×10^{-9} mb	4 μg/cm^2	75	27	0.7

Table 2. Comparison of the goal for four reaction with current status.

reaction	physics	current energy limit (keV)	precision (%)	ref.	JUNA energy limit (keV)	precision (%)
$^{12}C(\alpha,\gamma)^{16}O$	Massive star	890	60	Hammer et al. (2005)	380	test
$^{13}C(\alpha,n)^{16}O$	Heavy ion synthesis	279	60	Drotleff et al. (1993)	200	20
$^{25}Mg(p,\gamma)^{26}Al$	Galaxy ^{26}Al source	92	20	Strieder et al. (2012)	58	15
$^{19}F(p,\alpha_\gamma)^{16}O$	F overabundance	189	80	Lombardo et al. (2015)	100	10

marble rock of 2400 meters. Benefiting from the ultra low background and the high beam intensity, we will be able to measure the 58 keV resonance strength of $^{25}Mg(p,\gamma)^{26}Al$ with the new designed 4π BGO γ detectors array.

2.4. $^{19}F(p,\alpha)^{16}O$ reaction

The $^{19}F(p,\alpha)^{16}O$ reaction is considered to be an important reaction in the CNO cycle. Currently, the experimental cross sections of this reaction at Gamow energies are still incomplete, and the precision of its thermonuclear reaction rate does not yet satisfy the model requirement. The proposed experiment is targeting on direct cross section measurement of the key $^{19}F(p,\alpha)^{16}O$ reaction right down to the Gamow energies (70–350 keV in the center-of-mass frame) with a precision better than 10% (He 2014).

A 'lamp'-type Micron silicon array will be constructed for the charged particle measurement, which can cover about 4π solid angle. This universal detection array will set the base for studying the charged-particle-induced reactions at JUNA. It can not only measure the total (p,α_0) cross section but also the angular distribution. The experimental angular distribution is very useful for revealing nuclear structure of the low-energy resonances. In this experiment, a thin target of about 4 μg/cm^2 CaF$_2$ will be utilized, which is evaporated on a thin metal backings.

The counting rates are deduced from the CJPL-II environment and detector γ and neutron measurement data. The results are summarized in Table 1. We also compared our expected precision of data with current experimental results. The results are summarized in Table 2.

3. Accelerator, shielding system and summary

We adopted a design of 2.45 GHz ECR which is developed to CI-ADS project. This ion source is expected to deliver 12 emA proton, 6 emA He$^+$ and 2.5 emA He^{2+}. The maximum beam energy out of ion source is 50 keV/q with emittance less than 0.2 π·mm·mrad.

For the low energy and high intensity beam, the space-charge effect must be controlled during transmission in order to increase the transport efficiency. The high transport efficiency could not only ensure enough beam intensity on target, but also reduce the background brought by the beam itself. We plan to adopt segmental voltage for the

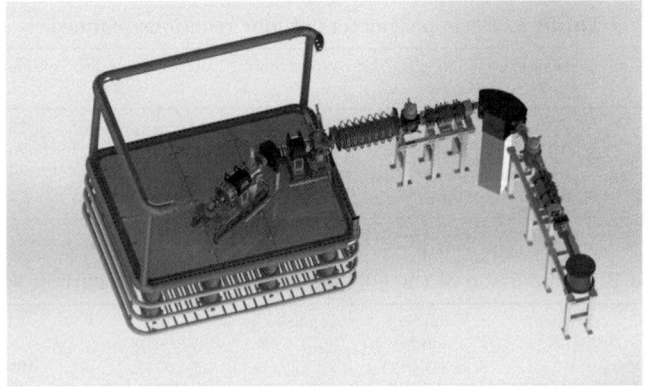

Figure 1. Design of the low energy and high current accelerator system.

accelerating tube and design an acceleration and deceleration structure for the accelerating tube electrode to reduce the space-charge effect.

The effect to background ratio of the nuclear reaction measurement will be significantly enhanced with the ultra-low background of CJPL and high current beam. But at the same time the high current beam will bring new background, which must be shielded. We plan to construct two shielding systems around the target chamber and the detectors, aiming at shielding γ-ray and neutron, respectively.

In summary, a new underground nuclear astrophysics experiment JUNA is planned for the expanded space CJPL-II. JUNA has the potential to take a favorable position among underground nuclear astrophysics labs. The accelerator system and detector array will be installed in 2017, experiment will be started in 2018 and the first batch of experimental results will be delivered in 2019.

References

Rolfs, C. & Rodney, W. S. 1988, Cauldrons in the Cosmos *The University Chicago Press*
Iliadis, C. 2007, Nuclear Physics of Stars, *Wiley-VCH Verlag GmbH*
Formicola, A., Imbriani, G., Junker, M, *et al.* 2003, *Nuclear Instruments & Methods in Physics Research*, 507(3):609-616.
Costantini, H., Formicola, A., Imbriani, G., *et al.* 2009, *Reports on Progress in Physics*, 972(8):086301.
Chen, H. S. 2010, *Science(in Chinese)*, 62: 4
Cheng, J. P., Wu, S. Y., Yue, Q., *et al.* 2011, *Physics (in Chinese)*, 03(03):149-154.
Zhao, W., Yue, Q., Kang, K. J., *et al.* 2013, *Physical Review D*, 88(5):1201-1205.
Normile, D. 2014, *Science*, 346(6213):1041-1041.
Buchmann, L. 2005, *Nuclear Physics A*, 758:335c-362c
Liu, W. P., The $^{12}C(\alpha,\gamma)^{16}O$ reaction proporsal, 2014, unpublished.
Tang, X. D., The $^{13}C(\alpha,n)^{16}O$ reaction proporsal, 2014, unpublished.
Strieder, F., Limata, B., Formicola, A., *et al.* 2012, *Physics Letters B*, 707:60-65.
Straniero, O., Imbriani, G., Strieder, F., *et al.* 2013, *Astrophysical Journal*, 763(2).
Li, Z. H., Su, J., Li, Y. J., *et al.* 2015, *Science China Physics, Mechanics & Astronomy*, 58.
He, J. J. 2014, The $^{19}F(p,\gamma)^{16}O$ reaction proporsal, unpublished.
Hammer, J. W., Fey, M., Kunz, R., *et al.* 2005, *Nuclear Physics A*, 758:363-366.
Drotleff, H. W., Denker, A., Knee, H., *et al.* 1993, *Astrophysical Journal*, 414:735-739.
Lombardo, I., Dell'Aquila, D., Leva, A. D., *et al.* 2015, *Physics Letters B*, 2015:178-182.

Cherenkov Telescope Array: Unveiling the Gamma Ray Universe and its Cosmic Particle Accelerators

Elisabete M. de Gouveia Dal Pino
(on behalf of the CTA Collaboration)[1]

[1]IAG, Universidade de São Paulo,
Rua do Matão 1226, São Paulo, SP, Brazil
email: dalpino@iag.usp.br

Abstract. Gamma-ray astronomy has a huge potential in astrophysics, particle physics and cosmology. The Cherenkov Telescope Array (CTA) is an international initiative to build the next-generation ground-based gamma-ray observatory which will have a factor of 5-10 improvement in sensitivity in the 100 GeV - 10 TeV range and an extension to energies well below 100 GeV and above 100 TeV. CTA is planned to consist of two arrays (one in the North and another in the South Hemisphere) and will provide the deepest insight ever reached into the non-thermal high-energy Universe and its particle accelerators.

Keywords. instrumentation: miscellaneous, gamma-rays, acceleration of particles

---

1. Overview

Ground-based gamma-ray astronomy is a young field with enormous scientific potential. The possibility of astrophysical measurements at tera-electronvolt (TeV) energies was demonstrated in 1989 with the detection of a signal from the Crab (Pulsar Wind) Nebula above 1 TeV with the Whipple Imaging Atmospheric Cherenkov Telescope (IACT). Since then, the instrumentation for astronomy with IACTs have evolved to the extent that a flourishing new scientific discipline has been established, with the detection of 150 sources and a major impact in astrophysics and more widely in physics. The current major arrays of IACTs (H.E.S.S., MAGIC, and VERITAS) have demonstrated the huge potential at these energies as well as the maturity of the detection technique. Many astrophysical source classes have been established, but there are indications that the known sources represent the tip of the iceberg both in terms of individual objects and source classes. The Cherenkov Telescope Array (CTA) will transform our understanding of the high-energy universe (see Actis *et al.*(2011) and Acharya *et al.*(2013) for reviews).

CTA will explore our Universe in depth in Very High Energy (VHE, E > 10 GeV) gamma-rays and investigate cosmic non-thermal processes, in close cooperation with observatories operating at other wavelength ranges of the electromagnetic spectrum, and those using other messengers such as cosmic rays, neutrinos and gravitational waves.

Besides anticipated high-energy astrophysics results, CTA will have a large discovery potential in key areas of astrophysics, cosmology and fundamental physics research. These include the study of the origin of cosmic rays and their impact on the constituents of the Universe, the investigation of the nature and variety of astrophysical accelerators including supernovae, neutron stars, merging stars, black holes, relativistic jets and explosions, and the inquiry into the ultimate nature of matter and physics beyond the Standard Model, searching for dark matter and effects of quantum gravity.

The major questions that CTA will address can be grouped into three broad themes:

(1) Understanding the Origin and Role of Relativistic Cosmic Particles:

What are the sites of high-energy particle acceleration in the universe? What are the mechanisms for cosmic particle acceleration? What role do accelerated particles play in feedback on star formation and galaxy evolution?

(2) Probing Extreme Environments:

What physical processes are at work close to neutron stars and black holes? What are the characteristics of relativistic jets, winds and explosions? How intense are the radiation and the magnetic fields in cosmic voids, and how do these evolve over cosmic time?

(3) Exploring Frontiers in Physics:

What is the nature of Dark Matter? How is it distributed? Are there quantum gravitational effects on photon propagation? Do axion-like particles exist?

CTA is planned to consist of two arrays, one in the South and another in the North, aiming to: (a) increase the sensitivity level of current instruments by a factor ~ 10 at 1 TeV; (b) boost significantly the detection area and hence the photon rate, providing access to the shortest timescale (transient) phenomena; (c) improve substantially the angular resolution and field of view and hence the ability to image extended sources, (d) provide energy coverage for photons from 20 GeV to at least 300 TeV, allowing the reach of high-redshifts and extreme accelerators; (e) enhance dramatically the surveying and monitoring capabilities, and the flexibility of operation, allowing for simultaneous observations of objects in multiple fields; and (f) provide access to the whole sky with sites in the two Hemispheres.

CTA will be operated as a proposal-driven open observatory, with a Science Data Centre providing access to data, analysis tools and user training. The main site will be in the southern hemisphere (in Chile), given the wealth of sources in the central region of our Galaxy and the richness of their morphological features. The northern site (in the Canary Islands) will be complementary and primarily devoted to the study of Active Galactic Nuclei (AGN), galaxy and star formation cosmological evolution, the extragalactic background light absorption (EBL), and the intergalactic magnetic fields.

The very wide energy range covered by CTA South will require the use of around 100 telescopes with at least three different sizes: large size telescopes (LSTs) with diameters of ~ 23 m, middle size telescopes (MSTs) with diameters of ~ 12 m, and small size telescopes (SSTs) with diameters ~ 4 to 6 m. In the North, a focus on extragalactic science is envisaged and hence (due to gamma-gamma absorption on Mpc scales) only limited sensitivity is required. The northern instrument can therefore be implemented with only LSTs and MSTs.

There are several strong motivations for the wide CTA energy range: the lowest energies provide access to the whole universe (avoiding significant gamma-gamma absorption on the EBL), while the highest energies are needed to study the extreme accelerators which are present in our Galaxy. Besides, a wide energy range maximises the chances of serendipitous detection of new source classes with unknown spectral characteristics, for example in the search for Dark Matter with an unknown WIMP mass.

Full sky coverage ensures that extremely rare but critically important events (for example a Galactic supernova explosion, bright gravitational wave transient, or very nearby gamma-ray burst) will be accessible to CTA.

CTA represents a genuinely world-wide effort, involving institutions of 32 countries in 5 continents. Currently, the collaboration is in the prototype development phase. A Mini-Array essentially constituted of 9 dual-mirror small size telescopes (named ASTRI), is under development and its installation in the South site - as a precursor of the CTA - is intended in the beginning of 2017. The full array should be deployed around 2020 (see http://portal.cta-observatory.org/Pages/Home.aspx to access most of the special reviews on the CTA and its vast science case).

Acknowledgements

We gratefully acknowledge support from the agencies and organizations listed under Funding Agencies at this website: http://www.cta-observatory.org/.

References

Actis, M., Agnetta, G., Aharonian, F., *et al.* 2011, *Experimental Astronomy*, 32, 193
Acharya, B. S., Actis, M., Aghajani, T., *et al.* 2013, *Astroparticle Physics*, 43, 3

Recent Status of Multi-Dimensional Core-Collapse Supernova Models

Kei Kotake[1], Ko Nakamura[2] and Tomoya Takiwaki[3]

[1]Department of Applied Physics, Fukuoka University, Jonan, Nanakuma, Fukuoka 814-0180, Japan

[2]Faculty of Science and Engineering, Waseda University, Ohkubo 3-4-1, Shinjuku, Tokyo 169-8555

[3]Astrophysical Big Bang Laboratory, RIKEN, Saitama, 351-0198, Japan

Abstract. We report a recent status of multi-dimensional neutrino-radiation hydrodynamics simulations for clarifying the explosion mechanism of core-collapse supernovae (CCSNe). In this contribution, we present two results, one from two-dimensional (2D) simulations using multiple progenitor models and another from three-dimensional (3D) rotational core-collapse simulation using a single progenitor. From the first ever systematic 2D simulations, it is shown that the compactness parameter ξ that characterizes the structure of the progenitors is a key to diagnose the explodability of neutrino-driven explosions. In the 3D rotating model, we find a new type of rotation-assisted explosion, which makes the explosion energy bigger than that in the non-rotating model. The unique feature has not been captured in previous 2D self-consistent rotational models because the growth of *non-axisymmetric* instabilities is the key to foster the explosion by enhancing the energy transport from the proto-neutron star to the gain region.

Keywords. Hydrodynamics—Neutrinos—Supernovae: general

1. Introduction

Core-collapse supernova (CCSN) mechanism is essentially an initial value problem. For low-mass progenitors with O-Ne-Mg core, the neutrino mechanism works successfully to explode in one-dimensional (1D) simulations because of the tenuous envelope (Kitaura et al.(2006)). For more massive progenitors with iron core, multi-dimensional (multi-D) effects such as neutrino-driven convection and the standing-accretion-shock-instability (SASI) have been suggested to help the onset of the neutrino-driven explosion. Recently this has been confirmed by a number of self-consistent two-(2D) and three-dimensional (3D) simulations (see collective references in Mezzacappa *et al.*(2015), Foglizzo *et al.*(2015), Janka(2012), Kotake *et al.*(2012)). Up to now, the number of these state-of-the-art models amounts to ~ 40 covering the zero-age main sequence (ZAMS) mass from 8.1 $M_\odot$ to 27 $M_\odot$ (Hanke *et al.* (2013)).

In this contribution, we first summarize our recent results (Nakamura *et al.*(2015)), in which we have computed first ever systematic 2D neutrino radiation hydrodynamics simulations using the whole progenitors (i.e., 378 models in total including 101 solar-metallicity models, 247 ultra metal-poor models, and 30 zero-metal models) in Woosley *et al.* (2002). By following a long-term evolution over 1.0 s after bounce, we have shown that most of the computed models exhibit neutrino-driven revival of the stalled bounce shock at ~ 200 - 800 ms postbounce, leading to the possibility of explosions.

This success, however, is now raising new questions. With the exception in Bruenn *et al.* (2013), the explosion energies obtained in these 2D models are typically smaller by one order of magnitude in explaining the canonical supernova kinetic energy of 10^{51} erg

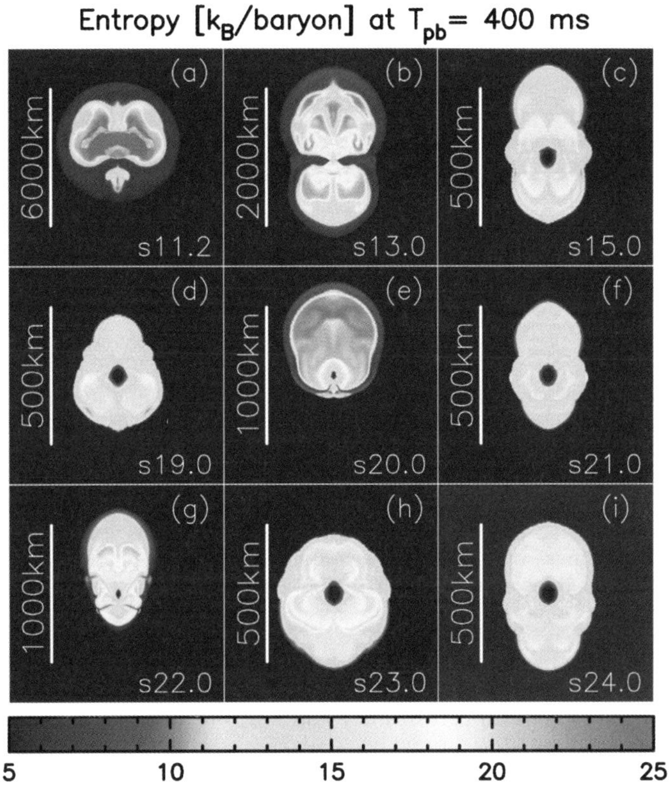

Figure 1. 2D Blast morphologies for multiple progenitor models with different masses from 11.2 (top left panel) to 24 $M_\odot$ (bottom left panel). Shown are entropy distributions in unit of k_B per baryon at $t_{pb} = 400$ ms after bounce. Note that each model presents a different scale as shown in the panel.

(Tanaka et al. (2009)). Most researchers are now seeking for some ingredients to make these underpowered explosions more energetic. In the latter half of this contribution, we report our most up-to-date results based on 3D rotational core-collapse simulations and discuss effects of stellar rotation on fostering the onset of neutrino-driven explosions (Takiwaki, Kotake, Suwa (2015)).

2. 2D Systematic Simulations Using Multiple Progenitors

In all of the computed models, the bounce shock stalls in a spherically symmetric manner and only after that, we observe a clear diversity of the multi-D hydrodynamics evolution in the postbounce (pb) phase. Figure 1 shows a snapshot of entropy distribution for selected nine solar-metallicity models at $t_{pb} = 400$ ms. For some less massive progenitors (e.g., s11.2 (panel (a))) the shock is reaching close to the outer boundary of the computational domain with developing pronounced unipolar and dipolar shock deformations. At this time, the shock of the most massive progenitor (s75.0) is reaching an average radius of $\langle r \rangle \sim 1000$ km, whereas the shock of s24.0 (panel (i)) still wobbles around at $\langle r \rangle \sim 200$ km. This demonstrates that the progenitor mass is not a good criterion to diagnose the possibility of explosion.

This is more clearly visualized in the left panel of Figure 2, showing time evolution of average shock radii for six models in the mass range between 19.0 $M_\odot$ and 24.0 $M_\odot$. The

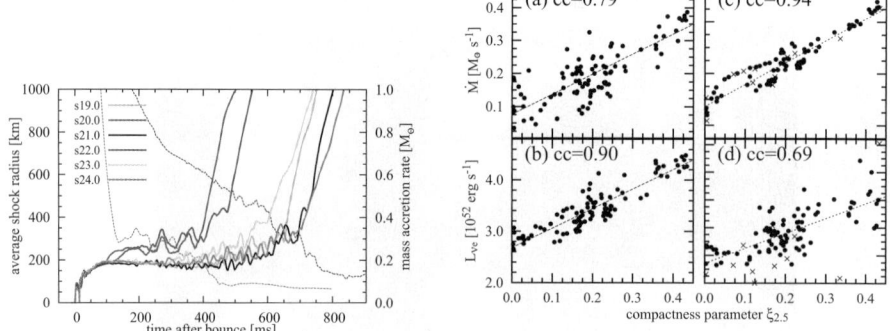

Figure 2. Left panel shows average shock radii (thick solid lines) and mass-accretion rate of the collapsing stellar core at 500 km (thin dashed lines) for some selected models (solar metallicity). Right panel summarizes explodability diagnostics versus the compactness parameter $\xi_{2.5}$ for the 101 solar-metallicity models (see Nakamura et al.(2015) for details about the rest of the progenitor series). Panel (a) is the mass accretion rate $\dot{M}$, panel (b) electron neutrino luminosity $L_{\nu e}$ estimated both at time of shock revival t_{400}, panel (c) is the mass of proto-neutron star (PNS) M_{PNS}, and (d) mass of nickel M_{Ni} in outgoing unbound material at final time of our simulations t_{fin}. Dashed lines present linear fitting with correlation coefficient denoted in each panel.

shock revival is shown to occur earlier for s20.0 (red line) and s22.0 (blue line) compared to the lighter progenitors s19.0 (green line) and s21.0 (black line). The progenitor compactness parameter ξ† is smaller for s20.0 and s22.0 in the chosen mass range. The smaller compactness is translated into smaller mass accretion rate onto the stalled bounce shock. For model s20.0 the relatively earlier shock revival coincides with the sharp decline of the accretion rate (dashed red line). After that, the accretion rate gradually decreases to $\sim 0.1\ M_\odot\ \mathrm{s}^{-1}$ till $t_{400} = 420$ ms at this time the revived shock has expanded to an average radius of $\langle r \rangle = 400$ km. Here t_{400} is a useful measure to qualify the vigour of the shock revival. On the other hand, model s21.0 has high compactness, which leads to the high accretion rate (black dashed line). It takes ~ 500 ms for the sloshing shock of model s21.0 (black solid line) to gradually turn into a pronounced expansion later on and 700 ms to arrive at $\langle r \rangle = 400$ km ($t_{400} = 700$ ms).

To summarize the explodability of the multiple progenitor models, we show in the right panel of Figure 2 various diagnostic properties as a function of the compactness parameter ($\xi_{2.5}$). Pushing the boundaries of expectations in previous 1D studies (e.g., Ugliano et al. (2012)), our results confirm that the compactness parameter ξ that characterizes the structure of the progenitors is also a key in 2D to diagnose the properties of neutrino-driven explosions. These results can be interpreted as follows: the core of high-ξ models is surrounded by high-density Si/O layers and the mass accretion rate therefore remains high long after the stalled shock has formed. This makes the PNS mass of the high-ξ models heavier (panel (c)). Due to the high accretion rate (panel (a)), the accretion neutrino luminosities become higher for models with high ξ (panel (b)). As a result, a stronger shock revival is obtained due to the more intense neutrino heating, which makes the amount of the synthesized nickel bigger (panel (d)). Our 2D simulations are terminated before the diagnostic (explosion) energies are saturated. However, a simple estimation regarding the mass accretion rate and the time derivative of the diagnostic

---

† Following O'Connor & Ott(2011), we define it as the ratio of mass M and the enclosed radius $R(M)$, $\xi_{2.5} \equiv \frac{M/M_\odot}{R(M=2.5\ M_\odot)/1000\mathrm{km}}$.

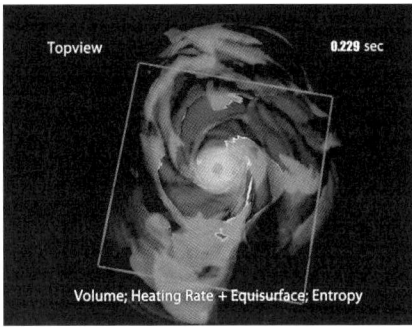
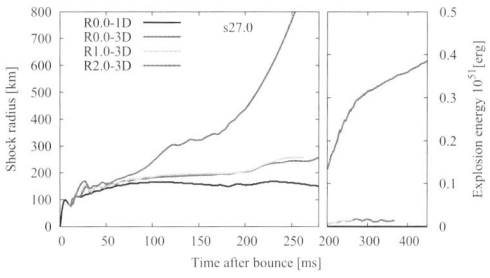

Figure 3. 3D image (left panel) shows a snapshot (at 229 ms after bounce) of the 3D rotating model ($\Omega_0 = 2$ rad/s, 27 $M_\odot$ star (Woosley *et al.* (2002))) seen from the direction parallel to the spin axis. Due to the growth of non-axisymmetric instabilities, the explosion develops preferentially toward the equatorial direction. Right panel shows the evolution of the (average) shock radii and the diagnostic explosion energy for the 1D or 3D models without rotation or with rotation, with R0.0, R1.0, and R2.0 representing $\Omega_0 = 0$ (non-rotating), 1, and 2 rad/s, respectively.

explosion energy may suggest that the diagnostic energy could barely reach 10^{51} erg ($\equiv$ 1 Bethe) in most of the 2D models. It is true that further global simulation, taking account of gravitational energy of an envelope and nuclear energy released via recombination process behind the shock, is necessary to determine the final explosion energy, but there might be some missing ingredients in 2D models to account for the 1 Bethe.

3. Seeking Ingredients to Foster Explosions : Impacts of Rotation

Here we focus on the roles of rotation and report results from a series of 3D rotational core-collapse simulations with spectral neutrino transport for a $27M_\odot$ star. The initially constant angular frequency of $\Omega_0 = 1$ or 2 rad/s is imposed inside the iron core with a cut-off ($\propto r^{-2}$) outside. We assume such rapid rotation to clearly see the impacts of rotation in this study.

As depicted in the left panel of Figure 3, we find a new type of rotation-assisted explosion for the $27M_\odot$ model, which otherwise fails to work when the precollapse core has no angular momentum (right panel of Figure 3). The unique feature was not captured in previous 2D self-consistent rotational models (e.g., Suwa *et al.*(2010)) because the growth of *non-axisymmetric* instabilities is the key to foster the explosion by enhancing the energy transport from the PNS to the gain region (see Takiwaki, Kotake, Suwa (2015) for more details). By systematically changing the precollapse rotation rates, we furthermore point out that rotation has also negative effects on the shock-revival, which was not treated accurately in previous 3D rotating models with fixed core neutrino luminosity or excision inside the PNS (e.g., Nakamura *et al.*(2014)). The SASI-modulated neutrino emission is significantly affected by the inclusion of rotation and the interesting correlation with the gravitational-wave emission is also found. Detailed coherent detector network analysis (e.g., Hayama *et al.*(2015)) of the correlated signals will be submitted elsewhere very soon.

References

Kitaura, F. S., Janka, H.-T., & Hillebrandt, W. 2006, A&A, 450, 345
Mezzacappa, A., Bruenn, S. W., Lentz, E. J., *et al.* 2015, arXiv:1501.01688

Foglizzo, T., Kazeroni, R., Guilet, J., *et al.* 2015, arXiv:1501.01334
Janka, H.-T., 2012, *Annual Review of Nuclear and Particle Science*, 62, 407
Kotake, K., Sumiyoshi, K., Yamada, S., *et al.* 2012, Progress of Theoretical and Experimental Physics, 2012, 01A301
Hanke, F., Müller, B., Wongwathanarat, A., Marek, A., & Janka, H.-T. 2013, *ApJ*, 770, 66
Nakamura, K., Takiwaki, T., Kuroda, T., & Kotake, K. 2014, arXiv:1406.2415, accepted to PASJ
Woosley, S. E., Heger, A., & Weaver, T. A. 2002, *Reviews of Modern Physics*, 74, 1015
Bruenn, S. W., Mezzacappa, A., Hix, W. R., *et al.* 2013, ApJL, 767, L6
Tanaka, M., Kawabata, K. S., Maeda, K., *et al.* 2009, *ApJ*, 699, 1119
Takiwaki, T., Kotake, K., Suwa, Y, in preparation
O'Connor, E. & Ott, C. D. 2011, *ApJ*, 730, 70
Ugliano, M., Janka, H.-T., Marek, A., & Arcones, A. 2012, *ApJ*, 757, 69
Suwa, Y., Kotake, K., Takiwaki, T., *et al.* 2010, PASJ, 62, L49
Nakamura, K., Kuroda, T., Takiwaki, T., & Kotake, K. 2014, *ApJ*, 793, 45
Hayama, K., Kuroda, T., Kotake, K., & Takiwaki, T. 2015, arXiv:1501.00966

MACHETE: A transit Imaging Atmospheric Cherenkov Telescope to survey half of the Very High Energy γ-ray sky

Rubén López-Coto, Juan Cortina and Abelardo Moralejo

Institut de Fisica dAltes Energies (IFAE), The Barcelona Institute of Science and Technology,
Campus UAB, 08193 Bellaterra (Barcelona) Spain
email: rlopez@ifae.es

Abstract. Current Cherenkov Telescopes for VHE gamma ray astrophysics are pointing instruments with a field of view up to a few tens of deg^2. We propose to build an array of two non-steerable telescopes with a FoV of 5×60 deg^2 oriented along the meridian. Roughly half of the sky drifts through this FoV in a year. We have performed a MC simulation to estimate the performance of this instrument, which we dub MACHETE. The sensitivity that MACHETE would achieve after 5 years of operation for every source in this half of the sky is comparable to the sensitivity that a current IACT achieves for a specific source after a 50 h devoted observation. The analysis energy threshold would be 150 GeV and the angular resolution 0.1 deg. For astronomical objects that transit over MACHETE for a specific night, it would achieve an integral sensitivity of 12% of Crab in a night. This makes MACHETE a powerful tool to trigger observations of variable sources at VHE or any other wavelengths.

Keywords. gamma rays, Imaging Atmospheric Cherenkov telescopes

1. Introduction

Very High Energy gamma rays are detected using space-based or ground-based detectors. From space Fermi-LAT is performing the deepest survey to date of the γ-ray sky from 20 MeV up to energies in excess of 100 GeV, although with limited sensitivity above 10 GeV due to its relatively small collection area (0.8 m^2). From the ground IACTs, such as the MAGIC, H.E.S.S. or VERITAS arrays, detect gamma rays with energies above 50 GeV and have collection areas of more than 105 m^2. They are pointing instruments with a Field of View (FoV) on the order of tens of deg^2. The 12 m diameter Medium-Sized Telescopes in the Cherenkov Telescope Array (CTA), currently under design, have a FoV of around 60 deg^2. On the other hand air-shower instruments such as Milagro, Tibet and HAWC detect gamma rays at higher energies, have a comparable collection area of 80000 m^2, but with a much larger FoV of 5000 deg^2 and high duty cycle. They are non-tracking instruments. Unfortunately they are not as efficient as IACTs in eliminating the cosmic ray background, so they suffer from a lower sensitivity and they have poorer angular or spectral resolutions. We propose to build an array of two non-steerable IACTs with a wide FoV of 300 deg^2. We call this array Meridian Atmospheric CHErenkov TElescope (MACHETE). The reader may find a detailed description of the instrument, its performance and its physics goals in Cortina et al. 2016. Here we focus mainly in the optics, review the performance that is estimated using a full Monte Carlo simulation and go through some of its possible applications.

2. Wide FoV telescopes

A Schmidt telescope is a well-known solution to achieve a large FoV with a small focal ratio. The optical components are an easy-to-manufacture spherical primary mirror, and an aspherical correcting lens, known as a Schmidt corrector plate, located at the center of curvature of the primary mirror. The corrector plate reduces optical aberrations and at the same time acts as a stop which defines the aperture of the telescope. Our concept is inspired by a Schmidt telescope, but we have aimed at simplifying it so that it is easy and cheap to implement:

- Like in the original Schmidt telescope, the shape of the primary mirror is spherical. The nominal focal length is half of the radius of curvature.
- Also like a Schmidt telescope, the shape of the focal plane is spherical and concentric with the mirror.
- An IACT is not as stringent as an optical telescope in terms of mirror Point Spread Function (PSF). A PSF on the order of 0.05 deg is good enough. We shall remove the corrector plate and achieve an acceptable PSF by increasing the focal ratio.
- However if we eliminate the corrector plate we are not only worsening the optical performance of the instrument, but also eliminating the stop. We must find an alternative way to limit the aperture. Compared to optical telescopes IACTs are in fact peculiar because each pixel is typically implemented as a light concentrator followed by the actual photodetector. In a natural way light concentrators can be used to define the section of the mirror which is viewed by each pixel and effectively the aperture.

The details in the optical design of MACHETE can be found in Cortina *et al.* 2016. Here we will only state that MACHETE will be composed of two instruments to improve the stereoscopic reconstruction of the γ-ray images. Each of them will have a section of spherical mirror (34 m curvature) covering a region of $5° \times 60°$ of sky.

3. Expected performance

We performed a Monte Carlo simulation of the MACHETE system to evaluate its performance. In this proceeding we only describe its results, for further details on design and performance, refer to Cortina *et al.* 2016. Here you can find a summary of the main results we obtained for the performance of the instrument:

- The sensitivity of MACHETE is significantly better than the sensitivity of HAWC for the same observation time and energies below 5 TeV. In addition, even one year of MACHETE has better sensitivity than 5 years of HAWC below 2 TeV. In addition, the angular and spectral resolution would be significantly better than those of HAWC.
- We have taken into account the change of acceptance as a function of position of the source in the FoV. The best integral sensitivity of MACHETE is reached at 500 GeV and it is 0.77% of the Crab flux at 30 deg declination after 5 years of operation.
- The sensitivity of MACHETE can also be compared to the sensitivity of the planned extragalactic scan of CTA. For the full CTA-South array and an observation of half of the sky for 1000 hours, the expected integral sensitivity of the survey is 0.6% of the Crab flux above 125 GeV. MACHETE achieves a similar sensitivity in 5 years of operation at a slightly higher energy but for the same fraction of the sky.
- MACHETE would achieve an integral sensitivity of 12% of the Crab flux in a single night for all sources in the fraction of the sky that is observable in that specific night.

References

Cortina, J., López-Coto, R., & Moralejo, A. 2016, *APh*, 72, 46-54

Atomic and Molecular Databases, VAMDC

M. L. Dubernet[1], C. M. Zwölf[1,], N. Moreau[1,], Y. A. Ba[1] and VAMDC Consortium[2]

[1]LERMA, Observatoire de Paris, PSL Research University, CNRS, UMR8112, 5 Place Janssen, 92195 Meudon, France
email: marie-lise.dubernet@obspm.fr

[2] http://www.vamdc.eu

Abstract. The VAMDC Consortium is a worldwide consortium which federates Atomic and Molecular databases through an e-science infrastructure and a political organisation. About 90% of the inter-connected databases handle data that are used for the interpretation of spectra and for the modeling of media of many fields of astrophysics. This paper presents how the VAMDC Consortium is organised in order to publish atomic and molecular data for astrophysics.

Keywords. databases, atomic data, molecular data

1. Introduction

The VAMDC Consortium originates from two European funded projects: the VAMDC (http://www.vamdc-project.vamdc.eu/)(Dubernet *et al.* 2010) and the SUP@VAMDC (http://www.sup-vamdc.vamdc.org/)(Zwölf *et al.*2014) projects. The main scientific outcomes of those two projects are: 1) an e-science infrastructure that interconnects about thirty databases (http://www.vamdc.eu/activities/research); 2) a political and technical organisation: "the VAMDC Consortium" that was launched on the 1rst November 2014 through the signature of a Memorandum of Understanding (MoU) between 15 partners. This structure ensures the organisation and the sustainability of the VAMDC activities. This MoU is complemented by an "Internal Regulations" Document describing in detail the implementation of the MoU and a Roadmap that provides the general strategy of the VAMDC Consortium †. The VAMDC Consortium activities cover 4 domains: research which is the most developed domain, education and industry that are currently in their early development stage, and outreach activities.

2. The Research Services

The Research Services are organised towards offering a common entry point to all databases thanks to the VAMDC portal (http://portal.vamdc.eu), towards offering the possibility to include new data and new databases within the VAMDC e-infrastructure, towards providing software librairies and modules that can be included into customers software, towards providing standalone users oriented software that retrieve and handle atomic and molecular data. The VAMDC e-infrastructure has evolved over the years through successive releases (Rixon *et al.*(2011), Doronin *et al.*(2012), Dubernet *et al.*(2015)) in order to arrive to the current version of standards and software (release version v12.07).

The current VAMDC e-infrastructure includes databases related to atomic and molecular spectroscopy and to heavy particle collisional processes, and is appropriate to the type of currently accessible data. Any producer of data can join the VAMDC infrastructure

† http://www.vamdc.eu - see section "About us"/How to join us

through different means: 1) they may include their data in existing atomic and molecular databases that are partners of VAMDC; 2) they may create a new database hosted by a partner of VAMDC; 3) they may create a new node in the VAMDC e-infrastructure. In cases 1 and 2, the data producers can contact the databases managers directly, while the general support system (support@vamdc.eu) should be contacted in the third case. Furthermore VAMDC aims to provide atomic and molecular data providers and compilers with a large dissemination platform for their work through its communication tools (website, forum) and through its citation policy. Currently all products related to VAMDC, portal and tools, explicitly warn that the VAMDC users should cite both the original papers where the data have been published and the relevant databases.

The librairies, software modules and software can be found on the VAMDC website (http://www.vamdc.eu/software). The integration of those librairies are documented, supported via tutorials and illustrated in scientific use cases (Dubernet *et al.* 2014). Users might want to create new librairies and software, and we provide support for those activities (support@vamdc.eu). In addition the VAMDC Consortium can provide on-demand additional innovative tools for easily handling and processing results.

3. Conclusion

The VAMDC Consortium continuously welcomes new members and is opened to welcome new type of data. Two main motivations would be considered in order to extend the scope of VAMDC Consortium: 1) a new community of data provider is interested to benefit from our experience and from part of our software, 2) one of our user community needs different types of data to be combined with the set of data already available in the VAMDC e-infrastructure. The inclusion of new types of data would certainly impact some of the "VAMDC Consortium" members, therefore the "VAMDC Consortium" members supporting such changes should make a case showing that this community is strategic for reasons such as increase of visibility, new customers, new stakeholders leading to consolidation of sustainability.

Acknowledgements

The VAMDC Support has been provided through the VAMDC and the SUP@VAMDC projects funded under the "Combination of Collaborative Projects and Coordination and Support Actions" Funding Scheme of The Seventh Framework Program. Call topic: INFRA-2008-1.2.2 and INFRA-2012 Scientific Data Infrastructure. Grant Agreement numbers: 239108 and 313284.

References

Doronin, M., Dubernet, M. L., Walton, N., *et al.* 2012, *in Astr. Soc. of the Pacific Conf. Series, Vol. 461, Astronomical Data Analysis Software and Systems XXI*, 331

Dubernet, M. L., Aboudarham, J., Ba, Y. A., *et al.* 2014, *in SF2A-2014: Proceedings of the Annual meeting of the French Society of Astronomy and Astrophysics*, 17–23

Dubernet, M. L., Boudon, V., Culhane, J. L., *et al.* 2010, *J. Quant. Spect. Rad. Trans.*, 111, 2151

Dubernet, M. L., Rixon, G., & Doronin, M., VAMDC Collaboration 2015, *Highlights of Astronomy*, 16, 685

Rixon, G., Dubernet, M. L., Piskunov, N., *et al.* 2011, *in AIP Conf. Series, Vol. 1344*, ed. A. Bernotas, R. Karazija, & Z. Rudzikas, 107–115

Zwölf, C., Dubernet, M.-L., Ba, Y., & Moreau, N., VAMDC Consortium 2014, *in IEEE, IST-Africa Conference Proceedings*

FM12p: Focus Meeting 12 Poster Session

1. Poster Presentations:

FM12p.01 Electron impact excitation of Astrophysically Important C III Ion. Author(s): Kanti M Aggarwal, Francis P Keenan

FM12p.02 Dielectronic Satellite Lines of Fe XVII. Author(s): Peter Beiersdorfer, Gregory V Brown, Alexander Laska, Jaan K. Lepson

FM12p.03 Calculations of effective recombination coefficients for nebular astrophysics. Author(s): Xuan Fang, Xiaowei Liu, Pete J. Storey

FM12p.04 Atomic lines in multiwavelength spectral analysis of late-type stars. Author(s): Hugh Jones

FM12p.05 Charge Exchange Produced Emission of Carbon in the Iron M-shell Dominated 150-200 Extreme Ultraviolet Region. Author(s): Jaan K. Lepson, Peter Beiersdorfer, Manfred Bitter, A. Lane Roquemore, Robert Kaita

FM12p.06 The role of hydrogen collisions in non-LTE abundance analyses of aluminium. Author(s): Thomas Nordlander, Karin Lind

FM12p.07 Progress in Identifying Fe I Level Energies and Lines from Stellar Spectra. Author(s): Ruth Peterson

FM12p.08 Atomic Oscillator Strengths for Stellar Atmosphere Modeling. Author(s): Matthew Ruffoni, Juliet C Pickering

FM12p.09 Improved and Expanded Near-IR Oscillator Strengths for Fe-group Elements. Author(s): Michael Wood, Gillian Nave, Christopher Alan Sneden

FM12p.10 Improved wavelengths for Fe V and Ni V for analysis of spectra of white dwarf stellar stars. Author(s): Jacob Ward, Gillian Nave

FM12p.11 Wavelengths, energy levels and hyperfine structure of Mn II and Sc II. Author(s): Gillian Nave, Juliet C Pickering, Keeley I. M. Townley-Smith, . Hala

FM12p.12 New laboratory atomic data for neutral, singly and doubly ionised iron group elements for astrophysics applications. Author(s): Juliet C Pickering, Gillian Nave, Florence Liggins, Christian Clear, Matthew Ruffoni, Craig Sansonetti

FM12p.13 Report on the recent advances performed in the determination of radiative parameters for spectral lines of astrophysical interest in heavy elements. Author(s): Pascal Quinet

FM12p.14 A systematic and detailed investigation of radiative rates for forbidden transitions of astrophysical interest in doubly ionized iron peak elements. Author(s): Pascal Quinet, Vanessa Fivet, Manuel Bautista

FM12p.15 The Interstellar Abundance of Lead: Experimental Oscillator Strengths for Pb II $\lambda 1203$ and $\lambda 1433$ and New Detections of Pb II in the Interstellar Medium. Author(s): Adam Michael Ritchey, Negar Heidarian, Richard E. Irving, Steven R. Federman, David G. Ellis, Song Cheng, Larry J. Curtis, W. A. Furman

FM12p.16 High Energy Laboratory Astrophysics Experiments using electron beam ion traps and advanced light sources. Author(s): Gregory V Brown, Peter Beiersdorfer, Sven Bernitt, Sita Eberle, Natalie Hell, Caroline Kilbourne, Rich Kelley, Maurice Leutenegger, F. Scott Porter, Jan Rudolph, Rene Steinbrugge, Elmar Traebert, Jose R. Crespo-Lopez-Urritia

FM12p.17 High-resolution oscillator strength measurements of the v' = 0,1 bands of the B- X, C -X, and E- X systems in five isotopologues of carbon monoxide. Author(s): Glenn Stark, Alan Heays, James Lyons, Michelle Eidelsberg, Steve Federman, Jean Louis Lemaire, Nelson de Oliveira, Laurent Nahon

FM12p.18 Identifying New Molecules from Comparison of Herschel-HIFI Spectra with ab initio Computational Spectra. Author(s): Naseem Rangwala, Sean Colgan, Timothy J Lee, Xinchuan Huang, Ryan Fortenberry

FM12p.19 IR Line Lists of SO_2 and CO_2 Isotopologues for Atmospheric Modeling on Venus and Exoplanets. Author(s): Xinchuan Huang, David W Schwenke, Timothy J Lee

FM12p.21 Laser Induced Fluorescence Spectroscopy of Neutral and Ionized Polycyclic Aromatic Hydrocarbons in a Cosmic Simulation Chamber. Author(s): Salma Bejaoui, Farid Salama

FM12p.22 Infrared spectra of interstellar deuteronated PAHs. Author(s): Mridusmita Buragohain, Amit Pathak, Peter Sarre

FM12p.23 Synthesis of Pure and N-substituted Cyclic Hydrocarbons (e.g. Pyrimidine) via Gas-Phase Ion-Molecule Reactions. Author(s): Partha P Bera, Roberto Peverati, Martin Head-Gordon, Timothy J Lee

FM12p.24 Coronene and Pyrene (5,7)-member Ring Defects. Author(s): Silvia Oettl, Stefan Kimeswenger, Stefan E. Huber

FM12p.26 Structural Evolution of Interstellar Polycyclic Aromatic Hydrocarbons. Author(s): Mark Hammonds, Alessandra Candian, Tamami Mori, Fumihiko Usui, Takashi Onaka

FM12p.27 Laboratory Astrophysics Studies with the COSmIC Facility: Interstellar and Planetary Applications. Author(s): Farid Salama, Cesar S. Contreras, Ella Sciamma-O'Brien, Salma Bejaoui

FM12p.28 Interstellar Methanol from the Lab to Protoplanetary Disks. Author(s): Maria Drozdovskaya, Catherine Walsh, Ruud Visser, Daniel Harsono, Ewine van Dishoeck

FM12p.29 Organic molecules in ices and their release into the gas phase. Author(s): Edith Fayolle, Karin I Oberg, Robin Garrod, Ewine van Dishoeck, Mahesh Rajappan, Mathieu Bertin, Claire Romanzin, Xavier Michaut, Jean-Hugues Fillion

FM12p.30 Combining Laboratory and Observational Data to Elucidate the Pathway from Simple to Complex Chemistry. Author(s): Helen Jane Fraser, Aleksi Suutarinnen, Anita Dawes, Jennifer Noble

FM12p.31 Are interstellar ices porous, and how do the pores collapse? Author(s): Catherine Rachel Hill, Christian Mitterdorfer, Tristan G. A. Youngs, Natalia Pascual, Olivier Auriacombe, Thomas Loerting, Helen J. Fraser

FM12p.32 THz Time-Domain Spectroscopy of Interstellar Ice Analogs. Author(s): Sergio Ioppolo, Brett A McGuire, Xander de Vries, Brandon Carroll, Marco Allodi, Geoffrey Blake

FM12p.33 Evidence of amino acid precursors: C-N bond coupling in simulated interstellar CO_2/NH_3 ices. Author(s): Sasan Esmaili

FM12p.34 Untangling molecular signals of astrochemical ices in the THz: distinguishing amorphous, crystalline, and intramolecular modes with broadband THz spectroscopy. Author(s): Brett A McGuire, Sergio Ioppolo, Marco Allodi, Brandon Carroll, Geoffrey Blake

FM12p.36 The MAON model of Astronomical Unidentified Infrared Emission Bands. Author(s): Sun Kwok, Yong Zhang

FM12p.37 Molecules, Dust and Ices in Brown Dwarf Atmospheres. Author(s): Sandy Leggett, Mark Marley, Caroline Morley, Didier Saumon

FM12p.38 The survival of PAHs and hydrocarbon nanoparticles in H II regions: theory and observations. Author(s): Elisabetta Micelotta, Marco Bocchio, Aurelie Remy-Ruyer, Melanie Kohler, Nathalie Ysard, Anthony Peter Jones

FM12p.39 Laboratory polarization and permittivity measurements to interpret dust polarimetric observations and in-situ radar studies. Significance for Rosetta mission at 67P/Churyumov-Gerasimenko. Author(s): Anny-Chantal Levasseur-Regourd, Yann Brouet, Edith Hadamcik, Essam Heggy, Dean Hines, Jeremie Lasue, Jean-Baptiste Renard

FM12p.40 Line survey observations of irradiated protostars - photo-destruction and evaporation. Author(s): Johan E Lindberg, Steven B Charnley, Jes K Jorgensen, Yoshimasa Watanabe, Suzanne E Bisschop, Nami Sakai, Satoshi Yamamoto

FM12p.41 Amorphous Silica- and Carbon- rich nano-templated surfaces as model interstellar dust surfaces for laboratory astrochemistry. Author(s): Natalia Pascual, Anita Dawes, Fernando Gonzlez-Posada, Neil Thompson, Dinko Chakarov, Nigel J Mason, Helen Jane Fraser

FM12p.42 Better Alternatives to Astronomical Silicate: Laboratory-Based Optical Functions of Chondritic/Solar Abundance Glass With Application to HD 161796. Author(s): Angela K Speck, Karly Pitman, Anne Hofmeister

FM12p.45 Infrared spectroscopy in the C-H stretching region towards embedded highmass young stellar objects in the Large Magellanic Cloud. Author(s): Takashi Shimonishi, Emmanuel Dartois, Takashi Onaka, Franois Boulanger

FM12p.46 Laboratory Analysis of Comet Samples. Author(s): Don Brownlee

FM12p.47 Fireball data analysis: bridging the gap between small solar system bodies and meteorite studies. Author(s): Maria Gritsevich, Manuel Moreno-Ibez, Daria Kuznetsova, Alexis Bouquet, Josep Trigo-Rodrguez, Jouni Peltoniemi, Detlef Koschny

FM12p.48 Origin of the Earths's Moon and Neptune's Triton. Author(s): Fred Johnson

FM12p.49 The Evolution of the Sun and the Planetary System. Author(s): Fred Johnson

FM12p.50 The Origins of the Planetary Satellites. Author(s): Fred Johnson

FM12p.52 Cosmological Plasma Instabilities – Plasma Physics at the Boundary. Author(s): Avery E Broderick

FM12p.54 Laboratory experiments investigating magnetic field production via the Weibel instability in interpenetrating plasma flows. Author(s): Channing Huntington, Frederico Fiuza, James Steven Ross, Alex Zylstra, Brad Pollock, R. Paul Drake, Dustin Froula, Gianluca Gregori, Nathan Kugland, Carolyn Kuranz, Matthew Levy, Chikang Li, Jena Meinecke, Richard Petrasso, Bruce Remington, Dmitri Ryutov, Youichi Sakawa, Anatoly Spitkovsky, Hideke Takabe, David Turnbull, Hye-Sook Park

FM12p.55 High performance computing (HPC) simulations of laboratory experiments probing astrophysical processes. Author(s): Giovanni Lapenta

FM12p.56 Studying counterstreaming high velocity plasma flows relevant to astrophysical collisionless shock. Author(s): James Steven Ross, Peter Amendt, Laurent Divol, Brad Pollock, Bruce Remington, Dmitri Ryutov, Wojciech Rozmus, David Turnbull, Dustin Froula, taichi morita, Youichi Sakawa, Hideke Takabe, R. Paul Drake, Carolyn C Kuranz, Gianluca Gregori, Jena Meinecke, Michel Koenig, Anatoly Spitkovsky, Hye-Sook Park

FM12p.57 Laboratory formation of a scaled protostellar jet by coaligned poloidal magnetic field: recent results and new exeprimental studies. Author(s): Tommaso Vinci, Guilhem Revet, Drew Higginson, Jrome Bard, K. Burdonov, Sophia Chen, D. Khagani, B. Khiar, K. Naughton, S. Pikuz, Caterina Riconda, R. Riquier, A. Soloviev, O. Willi, O. Portugall, Henry Pepin, Andrea Ciardi, Julien Fuchs, Bruno Albertazzi

FM12p.58 Laboratory experiments on Radiative Shocks relevant to Stellar Accretion. Author(s): Uddhab Chaulagain

FM12p.60 Modeling the interaction of solar wind with a dipole magnetic field with intense lasers. Author(s): Jiayong Zhong

FM12p.61 Merged Beams Studies for Astrobiology. Author(s): Daniel Wolf Savin, Aodh P. O'Connor, Nathalie de Ruette, Kenneth Miller, Julia Stuetzel, Xavier Urbain

FM12p.62 Science Goals and Laboratory Astrophysics Needs of the James Webb Space Telescope. Author(s): George Sonneborn

FM12p.63 Entoto Twining Telescopes: first robotic facilities in East Africa for astronomy research. Author(s): Solomon Belay Tessema

FM 13:
Brightness variations of the Sun and Sun-like stars

Part 2
Brightness variations of the Sun and Sun-like stars

Properties of stellar activity cycles

Heidi Korhonen

Finnish Centre for Astronomy with ESO, University of Turku, Väisäläntie 20, FI-21500 Piikkiö, Finland
email: `heidi.h.korhonen@utu.fi`

Abstract. The current photometric datasets, that span decades, allow for studying long-term cycles on active stars. Complementary Ca H&K observations give information also on the cycles of normal solar-like stars, which have significantly smaller, and less easily detectable, spots. In the recent years, high precision space-based observations, for example from the Kepler satellite, have allowed also to study the sunspot-like spot sizes in other stars. Here I review what is known about the properties of the cyclic stellar activity in other stars than our Sun.

Keywords. stars: activity, chromospheres, late-type, spots

1. Introduction

The Sun exhibits well established 11-year spot cycle, and a 22-year magnetic cycle. Unfortunately, activity cycles are much more demanding to detect observationally in other stars. Still, long-term monitoring of numerous stars has given us information on the properties of the activity cycles also in other stars. Due to the time and page constraints this review concentrates on observed properties of stellar photospheric and chromospheric cycles. Coronal cycles have also been studied (recent papers include Lalitha & Schmitt 2013; Sanz-Forcada, Stelzer & Metcalfe 2013; Ayres 2015).

2. Methods

One of the best known methods to study activity cycles is to investigate the chromospheric emission in the cores of the Ca II H&K lines. This method is based on the fact that the active regions in the solar chromosphere give rise to emission in the cores of the Ca II H&K lines. Typically, so-called S-index is calculated by comparing the flux in the Ca II H&K line cores to the flux at a close-by continuum region.

Photospheric activity cycles can also be studied from broad-band photometry. In many active stars the starspots are so large that they cause brightness variations which can be few tens of percent from the mean light level, thus making them easily observable even from the ground. Even smaller spots are reachable from space-based photometry (detectable cycle lengths are limited by relatively short instrument lifetimes). Small ground-based automatic photospheric telescopes, that have been operational last two decades or longer, provide excellent data-sets for cycle studies.

Doppler imaging (see, e.g., Vogt *et al.* 1987; Piskunov *et al.* 1990) is a method that provides the best spatial resolution on the stellar surface. It uses high resolution, high signal-to-noise spectroscopic observations obtained at different rotational phases of the star. If the star has a non-uniform surface temperature, i.e., has starspots, the spectral lines show small distortions from the normal Gaussian shape. These distortions move in the line-profile when the position of the starspots on the surface changes, due to the change of line-of-sight velocity caused by the stellar rotation. The temperature maps, which are constructed by tracking the movement of these distortions, can be used for detailed studies of stellar spot configurations. If spectropolarimetric observations are obtained over the stellar rotation, similarly to what is done in Doppler imaging, the surface magnetic field of the star can mapped using Zeeman-Doppler imaging technique (Semel 1989).

3. Activity cycles in 'normal' stars

One cannot talk about stellar activity cycles without mentioning the Mt. Wilson H&K survey. The project was initiated in 1966 by Olin Wilson (Wilson 1968; Wilson 1978), and it originally monitored 91 stars with spectral types ranging from F to M. The monitoring was expanded to more stars and star types, and in total some 2000 stars were observed during the project's lifetime (Baliunas et al. 1995; Baliunas et al. 1998).

The project found many stars with variations in the chromospheric emission, and distinguished four different activity types: variable, cyclic, trend, and flat activity. From the core sample of 111 dwarf stars 60% of the targets show cyclic or apparently cyclic behaviour, 25% are variable and 15% exhibit flat activity (Baliunas et al. 1995). In the sample of 175 more evolved stars, 40% show cyclic and basically the same fraction variable activity (Baliunas et al. 1998). Meaning that the regularly cyclic activity seems to change to more chaotic behaviour when the star evolves.

For correlating the photospheric and chromospheric activity one can compare the chromospheric indicators from Ca II H&K and the photospheric spots from broadband photometry. Radick et al. (1998) compared these two different indicators. Their results show that the younger, more active stars tend to become fainter as their HK emission increases, whereas the older stars tend to become brighter as their HK emission increases. This means that the activity of young stars is spot dominated, and that of older stars is dominated by chromospheric plagues.

Brandenburg, Saar, & Turpin (1998) correlated the stellar activity with $\log \omega_{cyc}/\Omega$ and noticed that the stars concentrate on 'active' and 'inactive' branches, later also 'super saturated' branch was added (Saar & Brandenburg 1999). Stars move along the active branch as they age. Once reaching activity range -4.8 – -4.7 the star makes a rapid, maybe discontinuous, transition to the inactive branch (see also Saar & Brandenburg 2002). Several authors have correlated cycle length with the rotation period (e.g., Baliunas et al. 1996; Böhm-Vitense 2007; Oláh et al. 2009). Stars are seen to group again to the inactive and active branches, and longer rotation periods tend to produce longer cycles

The original Mt. Wilson sample does not include many M dwarfs. Gomes da Silva et al. (2012) studied 27 M0 – M5.5 dwarfs observed by HARPS planet search. Half of the sample showed significant cycles in the time range of few years. Surprisingly, there does not seem to be any correlation between cycle length and rotation period for M dwarfs (Savanov 2012).

4. Cycles in active stars

In many active stars the starspots are so large that they cause brightness variations which can be few tens of percent from the mean light level, thus making them easily observable even from the ground. Many studies on photospheric cycles in active stars have been carried out, e.g., Henry et al. (1995), Jetsu (1996), Oláh, Kolláth & Strassmeier (2000), Lanza et al. (2002), Berdyugina, Pelt & Tuominen (2002), Donati et al. (2003), Järvinen et al. 2005, Savanov (2009), Vida et al. (2010), Lehtinen et al. (2012), and Metcalfe et al. (2013). Here few papers are discussed in more detail as examples of typical results.

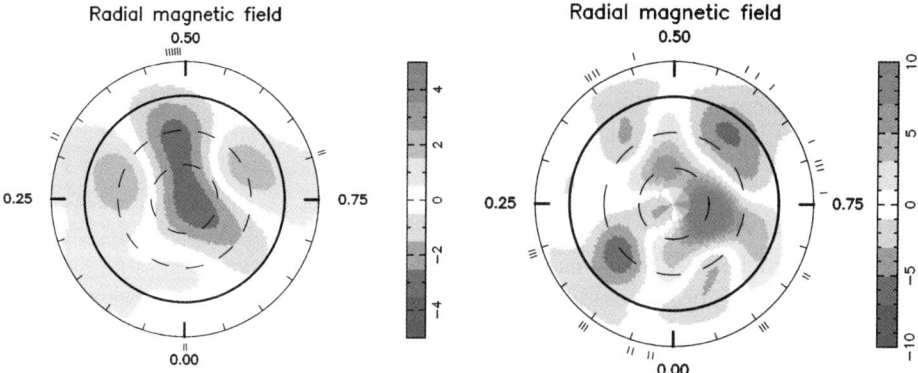

Figure 1. Radial magnetic field of τ Boo in June 2006 (left, from Catala *et al.* 2007) and in June-July 2007 (right, from Donati *et al.* 2008. The maps show the reversal of the polarities in the polar area.

Messina & Guinan (2002) monitored six K0 - G0 active dwarfs stars for more than a decade. Many of their main results are similar to the ones obtained from the more extended Mt. Wilson sample of stars with more 'normal' activity levels. Messina & Guinan (2002) find that the observed cycle lengths seems to converge with stellar age from a maximum dispersion around the Pleiades' age towards the solar cycle value at the Sun's age, and that the overall short- and long-term photometric variability increases with inverse Rossby number.

The cycles active stars show are not as regular and cyclic as many of their more normal counterparts have. Oláh, Kolláth & Strassmeier (2000) already show that many active stars exhibit multiple cycle lengths simultaneously. The more detailed analysis using short-term Fourier transforms shows that the cycle lengths in active stars are also often variable (Oláh *et al.* 2009). The positive relation between the rotational period and cycle length is also seen from the Oláh *et al.* (2009) sample.

Active stars also show more exotic cycles, e.g., the so-called flip-flop phenomenon. This effect was first discovered in an active single giant, FK Com, in the early 1990's (Jetsu *et al.* 1993). In this phenomenon the activity concentrates on two permanent active longitudes, and flips between the two every few years. Since the discovery this effect has been reported in binaries (e.g., Berdyugina & Tuominen 1998), young active stars (e.g., Järvinen *et al.* 2005), and even in the Sun (e.g., Berdyugina & Usoskin 2003). Recent studies show that flip-flops are not as regular and cyclic as at some point thought, and that there are also smaller 'phase jumps' (e.g., Korhonen, Berdyugina & Tuominen 2002; Oláh *et al.* 2006; Hackman *et al.* 2013).

5. Butterfly diagrams

It would be very important to obtain information on the spot latitudes during activity cycles in other stars In the Sun we know that the first spots of the cycle appear at higher latitudes (±30°) and migrate towards the equator as the cycle advances. We assume the same behaviour in the other stars, but we have not really produced proper 'butterfly diagrams' of distant stars. This would be best done using Doppler imaging, where we get detailed information on the spot latitudes. Still, it is difficult to obtain frequent enough long-term data for these studies.

Period variations have been detected in many active stars (e.g. Messina & Guinan 2003; Vida *et al.* 2014). These result can be explained by surface differential rotation and changing spot latitudes. Still, from these photometric studies we do not obtain accurate latitude information, and cannot tell which direction the spots have migrated. Berdyugina & Henry (2007) recovered the butterfly diagram of active RS CVn binary HR 1099 from photometry. They used the known differential rotation law, obtained from previous Doppler imaging work, in their reconstruction.

6. Magnetic cycles

Spectropolarimetric observations of stars and mapping the detailed configurations of the surface field has been possible soon for two decades. Still, the weak signal from the magnetic field demands superb observations, and the technique of Zeeman Doppler imaging requires many observations over a rotational phase for accurate surface reconstruction. Therefore, the observations of the stellar magnetic cycles have become possible only recently.

The first polarity revels in another star than the Sun was observed in the planet hosting F7 dwarf τ Boo (Donati et al. 2008). This was established by comparing the magnetic surface maps from June 2006 (Catala et al. 2007) and June-July 2007 (Donati et al. 2008), which are shown in Fig. 1. In 2008 the Second reversal in τ Boo was seen by Fares et al. (2009), implying a magnetic cycle of about 2 years.

Petit et al. (2009) discovered a polarity reversal in rapidly rotating solar-like star HD190771. But so far no cycle has been detected. Other long-term monitoring of the magnetic field configurations include e.g., K2 dwarf ϵ Eri (Jeffers et al. 2014), rapidly rotating solar-like star ζ Boo (Morgenthaler et al. 2012), and a sample of solar-like stars (Morgenthaler et al. 2011). No clear periodic reversals have been found in these studies, but we are collecting a dataset of long-term observation which will enable studying magnetic cycles in other stars than the Sun.

7. Final remarks

Since the discovery of activity cycles, first in our own Sun and later in other stars, we have identified many characteristics of these cycles.

Stellar activity cycles are seen in a large number of stars in the lower main sequence, meaning stars that have masses approximately solar mass and less. Several studies have also shown that the stellar rotation period and the cycle length correlate, i.e., stars with longer rotation period produce longer cycles. Therefore also the cycle length changes with stellar age, while the stellar rotation rate decreases due to the magnetic breaking. Very active stars, with huge starspots, can have multiple and changing activity cycles, and also exhibit 'exotic' behaviour: active longitudes, phase jumps and flip-flops.

Longer spectropolarimetric monitoring and Zeeman-Doppler imaging techniques are providing the opportunity to also probe the full magnetic field configuration changes during stellar cycles. So far stellar magnetic cycle has been seen in one stars, tau Boo, but the polarity change in the polar regions has seen in few others.

With dedicated telescopes and improved observing techniques we are starting to probe the detailed properties of stellar activity cycles. This is an important development as the cyclic magnetic activity is produced by Dynamo action operating in the star, and by studying the properties of the cycles, we can also give constraints to the Dynamo operation itself.

Acknowledgments The author acknowledges the travel support from Emil Aaltonen foundation, Turku University Foundation, and IAU to participate the IAU General Assembly 2015.

References

Ayres, T. R. 2015, *AJ*, 149, 58
Baliunas, S. L., Donahue, R. A., Soon, W. H., et al. 1995, *ApJ* 438, 269
Baliunas, S. L., Nesme-Ribes, E., Sokoloff D., & Soon, W. H. 1996, *ApJ* 460, 848
Baliunas, S. L., Donahue, R. A., Soon, W., & Henry G. W. 1998, *Cool stars, Stellar Systems, and the Sun*, ASP Conference series 154, p. 153
Berdyugina, S. V. & Tuominen, I. 1998, *A&A* (Letters), 336, L25
Berdyugina, S. V. & Usoskin, I. G. 2003, *A&A* 405, 1121
Berdyugina, S. V. & Henry, G. W. 2007, *ApJ* (Letters) 659, L157
Berdyugina, S. V., Pelt, J., & Tuominen, I. 2002, *A&A* 394, 505
Böhm-Vitense, E. 2007, *ApJ* 657, 486
Brandenburg, A., Saar, S. H., & Turpin, C. R. 1998, *ApJ* (Letters) 498, L51
Catala, C., Donati, J.-F., Shkolnik, E., Bohlender, D., & Alecian, E. 2007, *MNRAS* 374, L42
Donati, J.-F., Cameron, A. C., Semel, M., et al. 2003, *MNRAS*, 345, 1145
Donati, J.-F., Moutou, C., Farès, R., et al. 2008, *MNRAS*, 385, 1179
Gomes da Silva, J., Santos, N. C., Bonfils, X., et al. 2012, *A&A*, 541, A9
Hackman, T., Pelt, J., Mantere, M. J., et al. 2013, *A&A* 553, A40
Henry, G. W., Eaton, J. A., Hamer, J., & Hall, D. S. 1995, *ApJS* 97, 513
Järvinen, S. P., Berdyugina, S. V., Tuominen, I., Cutispoto, G., & Bos, M. 2005, *A&A* 432, 657
Jeffers, S. V., Petit, P., Marsden, S. C., et al. 2014, *A&A* 569, A79
Jetsu, L. 1996, *A&A* 314, 153
Jetsu, L., Pelt, J., & Tuominen, I. 1993, *A&A*, 278, 449
Korhonen, H., Berdyugina, S. V., & Tuominen, I. 2002, *A&A*, 390, 179
Lalitha, S. & Schmitt J. H. M. M. 2013, *A&A*, 559, A119
Lanza, A. F., Catalano, S., Rodonó, M., et al. 2002, *A&A* 386, 583
Lehtinen, J., Jetsu, L., Hackman, T., Kajatkari, P., & Henry, G. W. 2012, *A&A*, 542, A38
Messina S. & Guinan E. F. 2002, *A&A* 393, 225
Messina, S. & Guinan E. F. 2003, *A&A* 409, 1017
Metcalfe, T. S., Buccino, A. P., Brown, B. P., et al. 2013, *ApJ* (Letters) 763, L26
Morgenthaler, A., Petit, P., Morin J., et al. 2011, *AN* 332, 866
Morgenthaler, A., Petit, P., Saar, S., et al. 2012, *A&A* 540, A138
Oláh, K., Kolláth, Z., Strassmeier, K.G. 2000,*A&A* 356, 643
Oláh, K., Korhonen, H., Kővári, Zs., Forgács-Dajka, E., & Strassmeier, K. G. 2006, *A&A*, 452,303
Oláh, K., Kolláth, Z., Granzer, T., et al. 2009, *A&A*, 501, 703
Petit, P., Dintrans, B., Morgenthaler, A., et al. 2009, *A&A* (Letters) 508, L9
Piskunov, N. E., Tuominen, I., & Vilhu, O. 1990, *A&A* 230, 363
Radick, R. R., Lockwood G. W., Skiff, B. A., & Baliunas, S. L. 1998, *ApJSS* 118, 239
Saar, S. H. & Brandenburg, A. 1999, *ApJ* 524, 295
Saar, S. H. & Brandenburg, A. 2002, *AN* 323, 357
Sanz-Forcada, J., Stelzer, B., & Metcalfe T. S. 2013, *A&A* (Letters) 553, L6
Savanov I. S. 2009, *ARep* 53, 950
Savanov I. S. 2012, *ARep* 56, 716
Semel, M. 1989, *A&A*, 225, 456
Vida, K., Oláh, K., Kővári, Zs., et al. 2010, *AN* 331, 250
Vida, K., Oláh, K., & Szabó, R. 2014, *MNRAS* 441, 2744
Vogt, S. S., Penrod, G. D., & Hatzes, A. P. 1987, *ApJ* 321, 496
Wilson, O. C. 1968, *ApJ* 153, 221
Wilson, O. C. 1978, *ApJ* 226, 379

Sun-like Stars: magnetic fields, cycles and exoplanets

Rim Fares

INAF - Osservatorio Astrofisico di Catania, Via Santa Sofia 78, 95123 Catania, Italy
email: rfares@oact.inaf.it

Abstract. In Sun-like stars, magnetic fields are generated in the outer convective layers. They shape the stellar environment, from the photosphere to planetary orbits. Studying the large-scale magnetic field of those stars enlightens our understanding of the field properties and gives us observational constraints for field generation dynamo models. It also sheds light on how "normal" the Sun is among Sun-like stars. In this contribution, I will review the field properties of Sun-like stars, focusing on solar twins and planet hosting stars. I will discuss the observed large-scale magnetic cycles, compare them to stellar activity cycles, and link that to what we know about the Sun. I will also discuss the effect of large-scale stellar fields on exoplanets, exoplanetary emissions (e.g. radio), and habitability.

Keywords. stars: magnetic fields, stars: activity, technique: polarimetry

1. Introduction

More than a century after the discovery of the magnetic nature of sunspots (Hale 1908), and after achieving milestones in our understanding of solar magnetism, many questions remain unanswered. The Sun, despite having high spatial and temporal resolution, is nonetheless a single case study. Investigating the magnetism of Sun-like stars offers us the opportunity to place the Sun among its analogs, in order to have a wider view. The knowledge we have from solar studies regarding small and large-scale fields, as well as solar cycles, can guide, on the other hand, our exploration of magnetism in stars.

The solar magnetic field evolves in time. The latitude of emergence of sunspots, those small-scale magnetic features of thousands of Gauss, migrate from middle values toward the equator over a timescale of 11 years. This gives the butterfly diagram, showing a cyclic behaviour that I will refer to as the activity cycle. The large-scale field, which has a mean value that is much weaker (few Gauss), switches polarity every 11 years, giving a cycle length of about 22 years (see, e.g. Solanki *et al.*(2000) and Mordvinov *et al.*(2012) for a discussion about radial and azimuthal fields' evolution). During solar minimum, the large-scale field has a simple dipolar configuration, that becomes more complex during the solar maximum.

Stellar activity cycles are observed by studying activity proxies such as Calcium H&K and X-rays, see e.g. Baliunas *et al.*(1995), Metcalfe *et al.*(2010), Sanz-Forcada *et al.*(2013) and Berdyugina(2005). Solar-like stars have their magnetic fields generated by dynamo mechanisms in their outer convective layers. Studying their magnetism can thus give insights and observational constraints on dynamo generation theories. In this review, I will present the results of large-scale magnetic field studies of solar-like stars with masses between 0.7-1.5 $M_\odot$. The results presented here are obtained through the MagIcS initiative (a Magnetic Investigation of various classes of Stars), and in particular from the

Star-Planet Interaction (http://lamwws.oamp.fr/exo/starplanetinteractions/SPIScience) and the Bcool (http://bcool.ast.obs-mip.fr/) threads of this consortium.

2. Methodology: mapping of the magnetic field

Magnetic fields, when present in stellar photospheres, can cause Zeeman splitting and polarisation of spectral lines. Zeeman broadening is hard to detect in sun-like stars at optical wavelength (Anderson *et al.* 2010). The polarisation, on the other hand, can be detected. Its properties depend on the relative position of the observer to the orientation of the magnetic field. Circular polarisation, for example, is sensitive to the line-of-sight component of the field (see Landi Degl'Innocenti & Landolfi 2004).

To map stellar magnetic fields, one can use tomographic imaging technique, called Zeeman-Doppler Imaging (ZDI). This technique relies on the fact that when the star rotates, we observe different parts of the stellar disc. If the magnetic field distribution is not homogeneous on the stellar surface, for each rotational phase, a different polarisation signature can be observed. ZDI consists of inverting circular polarisation profiles into the magnetic topology (strength, distribution, polarity) of the field that produces such signatures (Semel 1989). The reconstruction is an ill-posed problem, a regularisation technique should be used to reconstruct a unique map (Maximum entropy, e.g. Brown *et al.*(1991) and Hussain *et al.*(2000); Tikhonov regularisation, e.g. Piskunov & Kochukhov(2002)). Results shown in this review are obtained using a maximum entropy regularisation, and describing the field by its poloidal and toroidal components, all described using spherical harmonics expansions (Donati *et al.* 2006a). We note however that ZDI does not recover small-scale field, since the spatial resolution is limited, and the signatures of small-scale fields can cancel out in some field geometries.

Polarisation spectra can be collected using spectropolarimeters, such as ESPaDOnS at CFHT, its twin instrument NARVAL at TBL, or HARPSpol at La Silla (Donati *et al.* 2006b, Aurière 2003, Piskunov *et al.* 2011). The signal in these spectra is extremely small, usually within the noise level. In order to improve the Signal-to-Noice (S/N), a multi-line technique dubbed Least-Square Deconvolution (LSD, Donati *et al.*(1997), Kochukhov *et al.*(2010))is used. It improves the S/N by a factor ~ 30 relative to single lines.

3. Magnetic properties and cycles of Sun-like stars

I will review, in this section, the current picture we have about the cool-stars' magnetism. But before that, I will discuss the magnetic properties of two stars with different stellar properties.

HD 76151 has been investigated in the Bcool program. It is a G3 star, with a mass of 1.24 ± 0.12 $M_\odot$ and a $v \sin i$ of 1.2 ± 0.5 km s^{-1}. It was observed in the beginning of 2007. Fig. 1 show its reconstructed magnetic map. Radial, azimuthal and meridional magnetic field are reconstructed. The mean magnetic energy is of 5.6 ± 2 Gauss (Petit *et al.* 2008), the field is mainly a simple dipolar radial field. HD 189733, on the other hand, is a cooler active star (K2) that hosts a transiting hot Jupiter. It was observed under the SPI program. It has a mass of 0.82 ± 0.03 $M_\odot$ and a $v \sin i$ of 2.97 ± 0.22 km s^{-1} (Moutou *et al.* 2007). The reconstructed map for July 2008 (Fares *et al.* 2010) shows a mean magnetic field of 36 Gauss, with a dominant azimuthal toroidal component as shown in fig.1.

These two examples show that the magnetic field of different solar-like stars can have different strengths and topologies. All the stars mentioned in this review are solar-like,

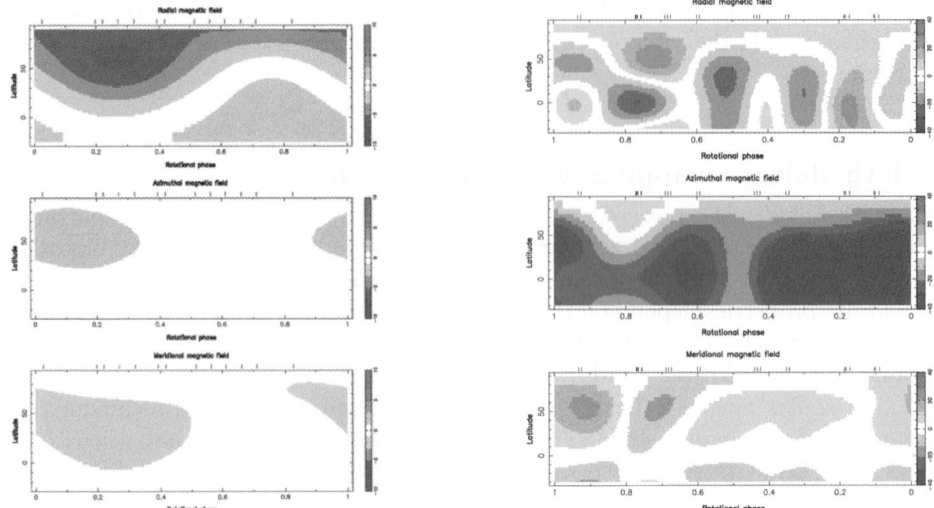

Figure 1. Left panel: The three components of the field in spherical coordinates system are presented for HD 76151 (from Petit et al. 2008). Right panel: The magnetic map of HD 189733, adapted from Fares et al.(2010). Please not that the colour scale is not the same for the left and the right maps.

with an outer convective envelope. We compare their properties in a mass-rotation diagram, since both the mass and the rotation are important in this context. The strength of the field, the percentage of poloidal energy and how axisymmetric the poloidal field is are considered, because they are important ingredients for dynamo. Data were compiled from Moutou et al. (2007); Donati et al. (2003, 2008); Petit et al. (2008, 2009); Jeffers & Donati(2008); Donati & Landstreet(2009); Fares et al.(2009,2010,2012,2013); Morgenthaler et al.(2011,2012) and Marsden et al.(2011). Fig. 2 shows the magnetic properties in the mass-rotation plane. As mentioned before, different properties appear in this diagram. However, they follow a trend that is different if the star's Rossby number is greater or lower than one. Stars with Rossby > 1, including the Sun, have smaller, dominantly axisymmetric poloidal field. The Sun, in this perspective, is not different from its analogs.

This is the general picture we have currently for magnetism in those cool stars. But the question one can ask is: what about magnetic cycles?

As mentioned in the introduction, stars exhibits activity cycles. A spectropolarimetric follow-up has been done for some targets. The first star to show a large-scale magnetic cycle with polarity switches was τ Boo, a massive planet hosting F-star (Donati et al.(2008), Fares et al.(2009), Fares et al.(2013)). The rotation of the star is synchronised with the planet. The magnetic field of τ Boo has a mean value of about 5 Gauss. A polarity flip was observed between June 2006 and June 2007, and another flip was observed between June 2007 and June-July 2008, indicating a fast cycle of 2 years (or less). The field was mainly poloidal for all these epochs. Observed in January 2008 between two plarity flips, the field showed an important toroidal component, indicating a change in configuration between the flips. It was first suggested that the planet might have played a role synchronising the outer convective layers of the star, and thus affecting the dynamo. But other stars (without hot Jupiters) have shown fast polarity reversals, indicating that fast cycles exist in cool Stars (e.g. F star (HD78366), and G dwarf (HD190771)). Note that for some of these stars, unlike the solar case, reported activity cycles are longer than the observed magnetic cycles. Most of the stars with polarity flips are fast rotators,

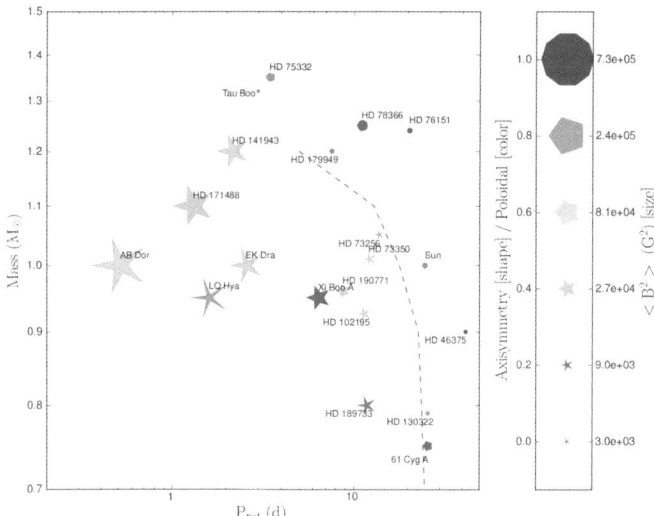

Figure 2. Mass-rotation diagram of reconstructed magnetic fields of Sun-like stars. The dashed line shows Rossby number of unity. The size of the symbol represents the field strength, its color the contribution of the poloidal component to the field, and its shape the degree axisymmetry of the poloidal component.

however, a polarity flip is observed on the slow-rotator 61 Cyg (Boro Saikia et al. in prep). This new result indicates that it is possible to detect magnetic cycles on slow rotators, now that we have long time-span of observations.

4. Exoplanet's environment

Stellar active regions can mimic the signal of an exoplanet. In order to detect earth-like planets, stellar activity should be well understood and modelled. Characterising exoplanets, however, is not limited to determining their masses, radii, and orbits. It also consists of studying their environment, their emissions (e.g. radio) and their interactions with their surrounding (i.e. stellar wind). It is possible to model more realistically the stellar corona when the large-scale magnetic field is reconstructed based on spectropolarimetric observations. The magnetic map can be used as a boundary condition for modelling the corona. This allows us to better understand the magnetic environment around the planet. Such environment can be responsible for bow shock formation (Vidotto et al. 2011) and radio emission (e.g., Zarka 2007, Jardine & Collier Cameron 2008). If the bow shock or the radio emission is observed simultaneously with spectropolarimetric observations of the star, one can indirectly infer the planetary magnetic field. A planetary magnetic field can act as a shield protecting the planet from the stellar wind. Earth-mass planets around M dwarfs, the focus of future missions, are subject to "aggressive" environment because of the intense stellar fields. Modelling the corona and the planetary orbit's environment is needed to understand better the conditions of habitability around active stars.

5. Conclusions

In this review, I presented the technique and results used to study the large-scale magnetic fields of solar-like stars. Spectropolarimetry, the study of polarisation in the spectral lines, and Zeeman-Doppler Imaging allow the reconstruction of the large-scale field maps. These techniques delivered, in the last 25 years, great knowledge on stellar magnetism. Magnetic fields of cool stars have different properties related to the strength of the field, the percentage of the energy in the poloidal to toroidal components, and how axisymmetric the field is. Stars with Rossby number greater than one seem to have similar properties while stars with a Rossby lower than one have different properties when compared to the first group, while they share between them similar properties. Magnetic cycles have been detected in Sun-like stars, mostly around fast rotators. Different types of cycles have been found, regular and complex cycles. The magnetic field polarity flips every year for at least one of the stars observed, much faster than the solar case. The length of the activity cycle to the large-scale cycle can be different from the solar case (11 vs 22 years for the Sun). Being able to access stellar activity data to review activity cycle length in the shadow of large-scale cycle finding is an important step to understand better stellar magnetism. Collecting more data to study activity, whether in the calcium H&K, other activity tracers, X-rays (corona) is important to understand directly the link between small-scale, large-scale, and even larger-scale (corona) activity cycles. This will help understand field dynamo generation and improve our theories.

References

Anderson, R. I., Reiners, A., & Solanki, S. K. 2010, A&A, 522, A81
Aurière, M. 2003, EAS Publications Series, 9, 105
Baliunas, S. L., Donahue, R. A., Soon, W. H., et al. 1995, ApJ, 438, 269
Berdyugina, S. V. 2005, Living Reviews in Solar Physics, 2, 8
Brown, S. F., Donati, J.-F., Rees, D. E., & Semel, M. 1991, A&A, 250, 463
oro-Saikia et al., in preparation
Donati, J.-F., Semel, M., Carter, B. D., Rees, D. E., & Collier Cameron, A. 1997, MNRAS, 291, 658
Donati, J.-F., Collier Cameron, A., Semel, M., et al. 2003, MNRAS, 345, 1145
Donati, J.-F., Howarth, I. D., Jardine, M. M., et al. 2006a, MNRAS, 370, 629
Donati, J.-F., Catala, C., Landstreet, J. D., & Petit, P. 2006b, Astronomical Society of the Pacific Conference Series, 358, 362
Donati, J.-F., Moutou, C., Farès, R., et al. 2008, MNRAS, 385, 1179
Donati, J.-F. & Landstreet, J. D. 2009, ARAA, 47, 333
Fares, R., Donati, J.-F., Moutou, C., et al. 2009, MNRAS, 398, 1383
Fares, R., Donati, J.-F., Moutou, C., et al. 2010, MNRAS, 406, 409
Fares, R., Donati, J.-F., Moutou, C., et al. 2012, MNRAS, 423, 1006
Fares, R., Moutou, C., Donati, J.-F., et al. 2013, MNRAS, 2010
Hale, G. E. 1908, ApJ, 28, 315
Hussain, G. A. J., Donati, J.-F., Collier Cameron, A., & Barnes, J. R. 2000, MNRAS, 318, 961
Jardine, M. & Collier Cameron, A. 2008, A&A, 490, 843
Jeffers, S. V. & Donati, J.-F. 2008, MNRAS, 390, 635
Kochukhov, O., Makaganiuk, V., & Piskunov, N. 2010, A&A, 524, A5
Landi Degl'Innocenti, E. & Landolfi, M. 2004, Astrophysics and Space Science Library, 307
Marsden, S. C., Jardine, M. M., Ramírez Vélez, J. C., et al. 2011, MNRAS, 413, 1922
Metcalfe, T. S., Basu, S., Henry, T. J., et al. 2010, ApJ(Letters), 723, L213
Mordvinov, A. V., Grigoryev, V. M., & Peshcherov, V. S. 2012, Solar Physics, 280, 379
Morgenthaler, A., Petit, P., Morin, J., et al. 2011, AN, 332, 866
Morgenthaler, A., Petit, P., Saar, S., et al. 2012, ApJ, 540, A138
Moutou, C., Donati, J.-F., Savalle, R., et al. 2007, A&A, 473, 651
Petit, P., Dintrans, B., Solanki, S. K., et al. 2008, MNRAS, 388, 80
Petit, P., Dintrans, B., Morgenthaler, A., et al. 2009, ApJ, 508, L9
Piskunov, N. & Kochukhov, O. 2002, A&A, 381, 736
Piskunov, N., Snik, F., Dolgopolov, A., et al. 2011, The Messenger, 143, 7
Sanz-Forcada, J., Stelzer, B., & Metcalfe, T. S. 2013, A&A, 553, L6
Semel, M. 1989, A&A, 225, 456
Solanki, S. K., Schüssler, M., & Fligge, M. 2000, Nature, 408, 445
Vidotto, A. A., Jardine, M., & Helling, C. 2011, MNRAS, 414, 1573
Zarka, P. 2007, P&SS, 55, 598

The Photometric Variability of Solar-Type Stars

Mark S. Giampapa

National Solar Observatory,
950 N. Cherry Ave., POB 26732, Tucson, Arizona USA
email: giampapa@nso.edu

Abstract. The joint variability of chromospheric emission with the integrated flux in the *Kepler* visible band for the Sun as a star is examined. No correlation between our Ca II K line parameter and the *Kepler* passband is seen, suggesting that visible-band variability in solar-like stars is mostly independent of solar-like chromospheric activity. However, the K-line parameter time series and the total solar flux in the infrared K band appear weakly correlated, reflecting the wavelength dependence of the relationship between magnetic activity and broadband variability. We then apply a schematic, three-component model as a framework for the discussion of stellar photometric variability as observed by *Kepler*. The model confirms that spots tend to dominate stellar photometric variability in the visible though interesting cases do emerge where the facular disk coverage may become important in determining the amplitude of broadband variability.

Keywords. Variability, spots, faculae, plage, *Kepler*

1. Introduction

The Sun and solar-type stars exhibit photometric variability on short and long time scales that is associated with magnetic structures. The amplitude of the brightness changes is wavelength-dependent according to the nature of the magnetic field-related component that is the dominant contributor to the variability. Prior to the *Kepler* mission, the most extensive, long-term study of brightness changes in solar-type stars utilizing high-precision, ground-based differential photometry is summarized by Lockwood *et al.* (2007). Hall *et al.* (2009) discuss an extension of this effort to a larger sample of more nearly sun-like stars.

In this invited paper, we carry out an initial exploration of the applicability of the solar paradigm to the interpretation of stellar photometric variability. We conclude with suggestions for achieving a more comprehensive picture of the origins of magnetic field-related variability, which is a topic of renewed importance given that magnetic fields in solar-type stars modulate the radiative and energetic particle environments in which exoplanetary systems form and evolve.

2. The Sun as a Star

A comparison of stellar variability with that of the Sun-as-a-star becomes appropriate given that (1) solar variability can inform our interpretation of stellar variability, (2) the amplitude of solar irradiance variability in the visible at the 1 – 3 mmag level overlaps with that seen in solar-type stars in the *Kepler* field sample, and (3) near-simultaneous, space - and ground-based data of superb quality are available for the Sun as a star.

We utilize the parameter time series for the 1 Å bandpass centered at the Ca II K line (hereafter referred to as K-line) at 3933.68 Å as derived from high resolution spectra ($R \simeq 300\,000$) obtained daily since 2006 by the Integrated Sunlight Spectrometer (ISS),

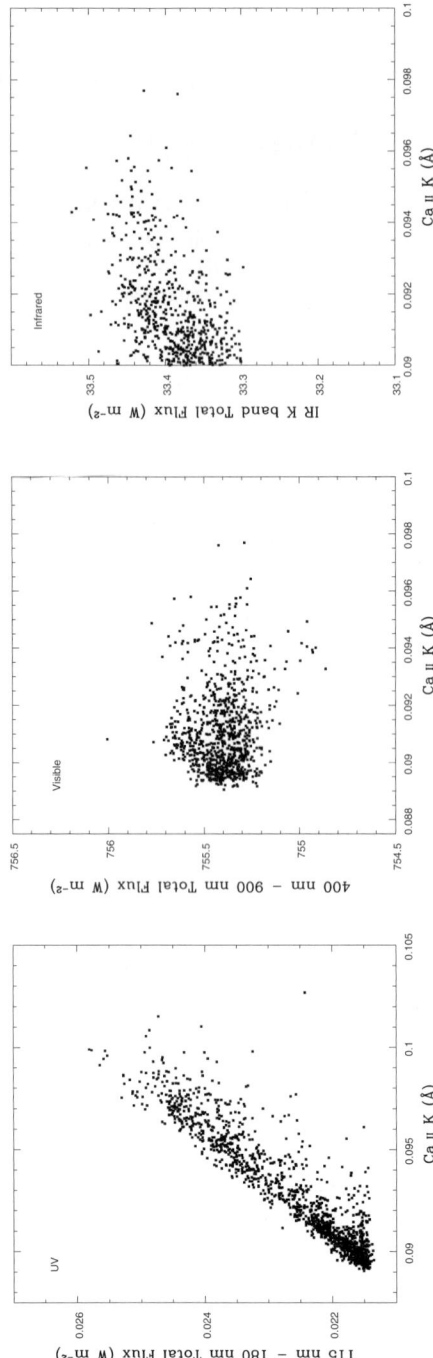

Figure 1. The daily variation of total flux with the Ca II K 1 Å parameter in the Sun as a star. The selected bandpasses include the far UV (*bottom*), the visible *Kepler* bandpass (*middle*) and the infrared K band (*top*). The accuracy of the daily monochromatic flux measurements is 2% while a typical measurement error in the K-line parameter is $\sim 0.001\%$. The correlation between the K-line and the far UV bandpass confirms that each has qualitatively similar origins in magnetic active regions. Variability in the visible tends to be spot dominated with little or no correlation with K-line emission. In the infrared the spot contrast has declined significantly accompanied by an emerging weak correlation with the K-line.

an instrument of the NSO SOLIS facility (see Keller *et al.* 2003). We compare the relative strength of the K-line core with solar monochromatic absolute flux measurements recorded by the Spectral Irradiance Monitor (SIM) on board the SORCE satellite (McClintock *et al.* 2000; Rottman *et al.* 2006). We utilized those SIM data that overlapped with the *Kepler* visible bandpass of approximately 400 nm – 900 nm.

SIM flux measurements and SOLIS/ISS K-line spectra obtained for the same Julian Day number, respectively, yield the plots displayed in Fig. 1. There is no, or very little correlation, of the relative flux in the core of the K-line with the solar flux measured by the SIM instrument in the *Kepler* visible bandpass. Hence, the solar flux in the *Kepler* bandpass is effectively independent of chromospheric activity in the Sun at primarily quiescent solar levels. We then extended this comparison to the Johnson K band in the infrared, which is centered at approximately 2.2μm. Interestingly, some correlation appears to be present, especially in contrast to what we see in the visible band. The result for the infrared K band is reminiscent of some early work in the solar infrared that revealed substantial intensity structure in a deep photospheric band at 1.64 μm correlated with the chromospheric emission network as seen in the Mg I feature at 1.72 μm (Worden 1975). We also show in Fig. 1 the expected correlation between the K-line parameter and far UV emission, thereby completing the sequence illustrating the wavelength dependence of the behavior of broadband flux with narrow band chromospheric emission.

In brief summary of the variability of the Sun-as-a-star, we find that (a) chromospheric activity and photometric variability are uncorrelated in the *Kepler* visible (400 nm – 900 nm) band *at solar-like levels of activity*; (b) there may be an emerging correlation of the K-line with IR photometric bands in the Sun-as-a-star; (c) chromospheric emission levels can be a somewhat ambiguous guide to the predicted amplitude of photometric light-curve variations in the visible; and, (d) low-amplitude photometric variability is not necessarily an indication of only quiet chromospheric activity. Though over the $\sim$ 11-year solar cycle time scale the spot and facular disk coverage are directly correlated, the daily spot and facular disk coverage do not appear to be correlated. The variability in the visible passband is dominated by sunspots on rotational time scales. The Ca II K line does not provide much information about the filling factor of spots on the solar disk and, therefore, it is not really informative about variability in the visible band on daily time scales.

3. The Applicability of the Solar Paradigm

In Fig. 2 we display the normalized photometric variability of a solar-type star in the open cluster M35 (age $\sim$ 150 Myr; Meibom *et al.* 2010) as observed with the repurposed *Kepler* mission, *K*2, along with that of the Sun-as-a-star as seen in the same *Kepler/K*2 visible band. The relatively high-amplitude, sinusoidal variations in the active M35 star are in vivid contrast to the comparatively flat "continuum" of solar variability, punctuated by the disk passage of sunspots. It was a similar comparison that prompted H. Hudson (Hudson 2015) to observe, *"The photometric behavior of these two stars could hardly be more different."*

To begin to gain some quantitative insight on the stellar photometric variability illustrated in Fig. 2, we adopt a simple three-component model—or, really, a schematic representation—of the normalized broad band flux. The three components are assumed to consist of the "immaculate" stellar photosphere (i.e., pure photosphere without any magnetic structures), cool spots analogous to sunspots, and faculae. Recall that faculae are bright spots in the solar photosphere that are often associated with concentrations of magnetic field lines and a higher ionization fraction of neutrals. Plage is the

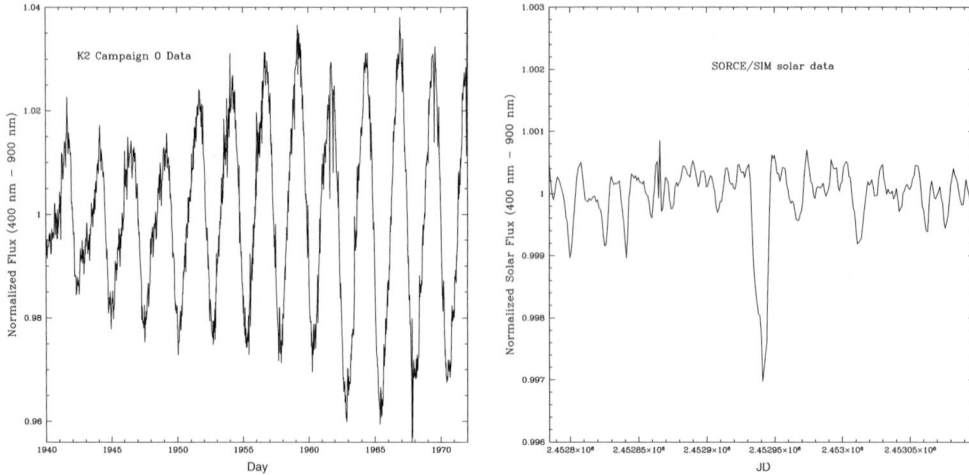

Figure 2. Rotational modulation of the light curves in (*left*) an active solar-type star in M35 and (*right*) the Sun, each over ten respective rotations as seen in the visible bandpass.

chromospheric counterpart of a facular region. The ratio of the observed flux to the undisturbed or normal stellar photosphere is given by

$$F/F_{star} = 1 - f_{spot}\left(1 - F_{spot}/F_{star}\right) + f_{fac}\left(F_{fac}/F_{star} - 1\right). \quad (1)$$

Since (1) involves the products of spot and facular filling factors (i.e., fractional area coverages) and a function of their respective contrasts, it is difficult to obtain unique solutions with single band photometry alone: a large, warm spot is equivalent to a cooler, small spot. We know from models that $F_{spot}/F_{star} \sim 10^{-3} - 10^{-1}$ and from solar observations that $F_{fac}/F_{star} \approx 1$. In these approximations, (1) simplifies to $F/F_{star} = 1 - f_{spot}$, meaning that the photometric modulation could be entirely attributed to cool spots. But since we typically see enhanced K-line emission in active stars with strong spot modulation as well as rotational modulation of photospheric lines in the Sun-as-a-star (Hall & Lockwood 2000), we retain a potential contribution due to faculae and plages so that (1) can be approximately expressed as

$$F/F_{star} \simeq 1 - f_{spot} + \epsilon f_{fac},$$

where ϵ is the facular contrast. The average contrast of solar magnetic elements is 3.7% in the quiet Sun (Kobel *et al.* 2011). Therefore, the above approximation suggests that the modulation of the photometric light curve is controlled primarily by spots *unless* the spot filling factor is only ∼ a few percent (which is still an order of magnitude higher than in the Sun) *and* the fractional area coverage of faculae/plage-like regions is near unity. In such a case the facular contrast competes with spots in the modulation of the light curve. We explore the implications of this perspective in the following where we assume that the values at the maximum and minimum phases in the photometric light curve are excursions from a mean value representing pure stellar photosphere.

We utilize equation (1) to find the range of facular and spot filling factors at maximum and minimum light in a light curve such as that for the M35 star (Fig. 2), which exhibits excursions of roughly ±2.5%. We adopt $T_{eff} = 5600$ K (∼ G5–6 V) and a solar-like $T_{spot}/T_{eff} = 0.70$. We estimate spot-to-star flux ratios from PHOENIX model atmospheres (Husser *et al.* 2013) yielding $F_{spot}/F_{star} = 0.074$ in the 400 nm – 900 nm *Kepler* visible bandpass. The results are graphically displayed in Fig. 3.

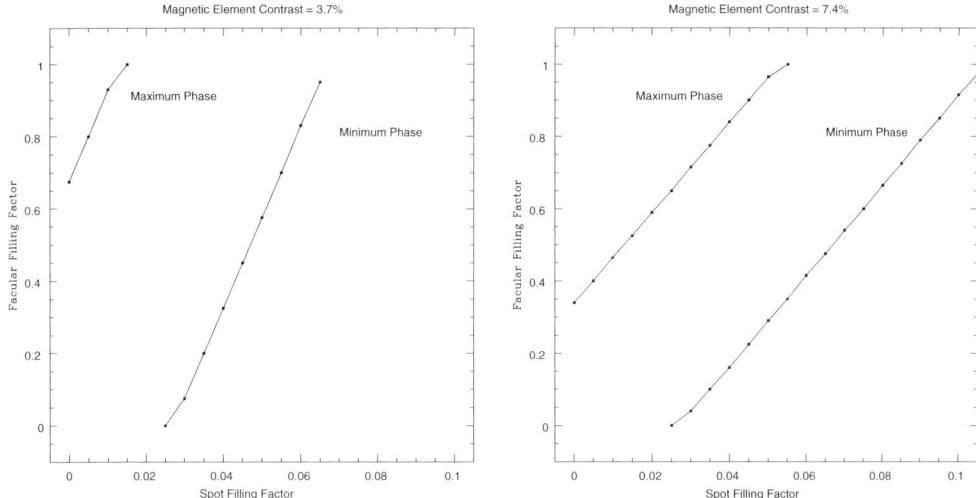

Figure 3. Relative photometric maximum and minimum for solar (*left*) and non-solar (*right*) values of facular contrast value in the spot—facular filling factor plane.

For a solar-like, 3.7% facular contrast we find that the possible range of facular filling factors is highly restrictive. At a 1% spot filling factor, the facular filling factor must be about 100% at maximum light (Fig. 3). In order to avoid unobserved, high-amplitudes in the rotational modulation of the K-line emission, facular filling factors of about 68% with spot filling factors of about $\sim 6\%$ account for maximum and minimum light, respectively. But this is for 0% spots on the bright side. In other words, this would imply a strong longitudinal asymmetry in at least the spot filling factor. As seen in Fig. 3 for a non-solar, 7.4% facular contrast, there is a broad range of possible facular filling factors though with a minimum facular coverage of 35%. At these levels, spot filling factors are at 6% (for the same facular coverage). Again, there is a strong asymmetry in the spot coverage. But with this high facular contrast, we avoid the possibility of modulation of the K-line emission at amplitudes that are not usually seen in active stars.

In brief summary, this interpretation of non-solar light curves in solar-type stars seems to drive us in the direction of high facular contrasts that may be non-solar combined with strong longitudinal asymmetry in the spot surface distribution. Could these non-solar facular/plage contrasts arise from high magnetic energy densities that, in turn, are the ultimate origin of "superflares"? Whether this is the case cannot be confirmed without flux-calibrated light curves. As pointed out in the discussion at this FM, the facular contrast may simply elevate the "DC signal" in the light curve with spots as the primary source of modulation. We explore this further in the next section where we focus on the range of photometric variability.

3.1. The Range of Photometric Variability

In the context of our schematic model, the range of the observed photometric variability from equation (1) is

$$\Delta F/F_{star} = (f_{smin} - f_{smax})(1 - F_{spot}/F_{star}) + \epsilon\,(f_{fmax} - f_{fmin}), \quad (2)$$

where ΔF is the range from maximum to minimum in the light curve, ϵ is the facular contrast, f_{smin} is the filling factor of cool spots at light-curve minimum, f_{smax} is the corresponding value at maximum, f_{fmax} is the filling factor of faculae at light-curve maximum and f_{fmin} is the fractional area coverage of faculae at minimum. The values of the facular filling factors range from zero to maximum values of $1-f_{smin}$ for f_{fmin} and $1-f_{smax}$ for f_{fmax}, respectively.

For the special case where the facular filling factors attain their maximum values in the hemispheres visible to us at light-curve maximum and minimum, respectively, we have that

$$\Delta F/F_{star} = (f_{smin} - f_{smax})(1 + \Delta),$$

where $\Delta = \epsilon - F_{spot}/F_{star}$, that is, the difference between the facular and spot contrasts. We see that the value of Δ ranges from a minimum of $-F_{spot}/F_{star}$ to a maximum of ϵ, i.e., the facular contrast. Normally, we would expect that $\Delta \ll 1$ in the visible band so that the range in the photometric light curve remains dominated by the difference in spot filling factors between maximum and minimum, *unless* stellar faculae have an unusually strong, non-solar contrast.

The final case we consider is where the filling factor of faculae at maximum (minimum) in the light curve dominates over the filling factor at minimum (maximum), suggesting a strongly asymmetric surface distribution of faculae and spots on the stellar surface. In this case we have

$$\Delta F/F_{star} = (f_{smin} - f_{smax})(1 - F_{spot}/F_{star}) + \begin{bmatrix} f_{fmax} \\ -f_{fmin} \end{bmatrix} \epsilon, \quad (3)$$

where the term in brackets takes on the value of either f_{fmax} or $-f_{fmin}$, whichever most appropriately represents the dominant facular surface distribution. In contrast to the previous cases where the photometric variability is governed primarily by spots, the contribution by faculae now plays a critical role in determining the range of photometric variability in equation (3). The photometric range is amplified if $f_{fmax} \gg f_{fmin}$. However, if faculae are concentrated mainly on the side of the star with the greatest fractional coverage of spots, that is, when $f_{fmin} \gg f_{fmax}$, and f_{fmin} is near its maximum value of $1 - f_{smin}$, then the amplitude of photometric variability is actually reduced. In this case, the second term is approximately $-\epsilon$. Given that the magnitude of the facular contrast can be similar to the difference in spot filling factors, i.e., $\epsilon \sim (f_{smin} - f_{smax})$, then we would see only a low amplitude of photometric variability with $\Delta F/F_{star} \ll 1$.

This counterintuitive result is due to the offsetting effects of a high filling factor of relatively brighter faculae in the presence of cool spots. Does such a special case exist in reality? Perhaps there is evidence for its occurrence in the developing work of Bastien *et al.* (2015, in preparation). These investigators find in their Ca II survey of solar-type stars in the *Kepler* field a population of objects with normalized chromospheric emission $\sim 3-6$ times that of the mean Sun but with a range in photometric variability < 1 mmag. This combination of relatively high chromospheric activity with very low photometric variability could equally arise from a homogeneous surface distribution of spots and plages. In contrast to the homogeneous case, however, the asymmetric surface distribution represented in equation (3) when $f_{fmin} \gg f_{fmax}$ (and the second term becomes approximately equal to $-\epsilon$) would give rise to strong rotational modulation of the Ca II H & K lines but in the presence of only a small amplitude of broad band photometric variability. In our differential analysis, it was, of course, not necessary to invoke non-solar facular contrasts to account for the range of photometric variability in active, solar-type stars.

4. Summary

Our simple three-component model suggests that photometric variability in the visible band is dominated by spots (consistent with what we see in the Sun-as-a-star) characterized by a strong longitudinal asymmetry. However, special cases exist where the facular contrast can play a critical role in determining the range of photometric variability. In the case of active solar-type stars, our schematic representation of the normalized variability indicates the possible presence of high facular contrasts that may be non-solar. But this interpretation assumes that the observed, uncalibrated light-curves represent an excursion from a relatively quiet photosphere in their minimum–maximum range of variation.

In this preliminary analysis, we did not consider inclination effects nor did we include the wavelength dependence of facular contrast in the visible band. Shapiro *et al.* (2015) find that the viewing angle with respect to the stellar rotation axis can determine whether the observed variability is spot-dominated or faculae-dominated on rotational time scales, at least at solar-like activity levels. The relative amplitude of the rotational modulation also declines at low inclinations due to foreshortening effects on the spot coverage as well as reduced spot contrasts as seen toward the limb (Shapiro *et al.* 2015). Further progress will require calibrated, multi-color photometry in conjunction with spectroscopic observations of key diagnostics, such as the Ca II resonance lines, the G band, and the He I lines at 5876 Å and 10830 Å, respectively (Andretta *et al.* 2015), which are sensitive to bright magnetic elements extending from the photosphere to the high chromosphere.

Acknowledgements

The author acknowledges interesting discussions with Hugh Hudson, Sasha Shapiro and Charlie Lindsey concerning the interpretation of stellar variability in a solar context. The National Solar Observatory is operated by AURA under a cooperative agreement with the National Science Foundation.

References

Andretta, V., Giampapa, M. S., Covino, E. *et al.* 2015, in preparation
Bastien, F. A., *et al.* 2015, in preparation
Hall, J. C., Henry, G. W., Lockwood, G. W., Skiff, B. A., & Saar, S. H. 2009, *AJ*, 138, 312
Hall, J. C. & Lockwood, G. W. 2000, *ApJ*, 541, 436
Hudson, H. S. 2015, *Journal of Physics Conference Series*, 632, 012058
Husser, T.-O., Wende-von Berg, S., Dreizler, S., *et al.* 2013, *A&A*, 553, A6
Keller, C. U., Harvey, J. W., & Giampapa, M. S. 2003, *Proc. SPIE*, 4853, 194
Lockwood, G. W., Skiff, B. A., Henry, G. W., *et al.* 2007, *ApJS*, 171, 260
McClintock, W. E., Rottman, G. J., & Woods, T. N. 2000, *Proc. SPIE*, 4135, 225
Meibom, S., Mathieu, R. D., & Stassun, K. G. 2009, *ApJ*, 695, 679
Rottman, G. J., Woods, T. N., & McClintock, W. 2006, Advances in Space Research, 37, 201
Shapiro, A. I., Solanki, S. K., Krivova, N. A., Yeo, K. L., & Schmutz, W. K. 2015, *A&A*, in press. doi: 10.1051/0004-63611201527527
Worden, S. P. 1975, Ph.D. Dissertation, University of Arizona, Tucson

Solar influence on Earth's climate

Rémi Thiéblemont[1] and Katja Matthes[1,2]

[1] GEOMAR - Helmholtz Institute for Ocean Research Kiel,
Düsternbrooker Weg 20, 24105, Kiel, Germany
email: rthieblemont@geomar.de & kmatthes@geomar.de

[2] Christian-Albrechts Universität zu Kiel, Kiel, Germany

Abstract. Understanding the influence of solar variability on the Earth's climate requires knowledge of solar variability, solar-terrestrial interactions and observations, as well as mechanisms determining the response of the Earth's climate system. A summary of our current understanding from observational and modeling studies is presented with special focus on the "top-down" stratospheric UV and the "bottom-up" air-sea coupling mechanisms linking solar forcing and natural climate variability.

Keywords. solar forcing, climate dynamics, stratosphere, ozone, climate modelling

1. Introduction

The thermal structure and the composition of the atmosphere strongly depend on solar irradiance. Therefore, fluctuations of the incoming radiation are expected to modulate the Earth's climate and its variability. Although changes in total solar irradiance have been shown to account for a small part of the recently observed global mean temperature changes (IPCC (2013)), there is increasing evidence that variations in solar irradiance are an important source of regional climate variability (Gray *et al.* (2010), Zhou & Tung (2010)). For instance, several observational and modeling studies suggested a link between the 11-year solar cycle and surface climate in North America and Europe (Matthes *et al.* (2006), Ineson *et al.* (2011)). These results could therefore have significant societal and economical impacts since they would imply an increase of seasonal-to-decadal climate predictability. The attribution and quantification of the climate response to solar variability are however still strongly debated, and require further modeling and observational efforts.

Here we aim at providing a brief overview of the current understanding of the solar irradiance-climate connections on decadal timescales. In section 2, we review the physical mechanisms that are proposed to describe the influence of total and spectral solar irradiances (TSI and SSI) variations on climate. Section 3 presents the most recent sun-climate modeling outcomes, which have been produced in the framework of the Coupled Model Intercomparison Project Phase 5 (CMIP5). Suggestions for future work are given in section 4.

2. Mechanisms of solar influences on climate

The mechanisms proposed to explain the projection of solar irradiance variability onto regional climate are represented in Fig. 1. They are separated into two categories: the "bottom-up" mechanism (box 1), arising from the absorption at the surface of VIS/IR energy variations (yellow arrow), and the "top-down" mechanism (box 2-4), arising from the absorption of UV energy variations in the stratosphere (red arrow).

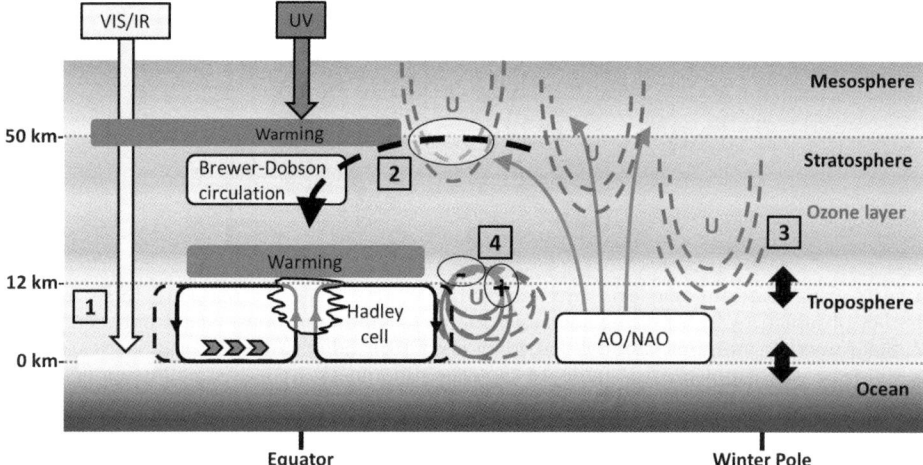

Figure 1. Sketch of the "bottom-up" and "top-down" concepts. See text for details.

Over the course of a solar cycle, the TSI varies by ~ 1 Wm^{-2}, corresponding to a relative variation of less than 0.1%. Despite the small initial perturbation, amplifying dynamical and thermodynamical "bottom-up" mechanisms have been proposed to describe some regional solar effects. It has been suggested that in response to the increased surface heating in subtropical cloud-free regions during solar maxima, the oceans release more water vapor into the atmosphere, which is then transported toward the Inter Tropical Convergence Zone (blue arrows). Water vapor carries latent heat that is released upon condensation, typically in deep convective clouds in the tropics. This release of energy fuels the tropical zonal (Walker) and meridional (Hadley) overturning tropical circulations. The latter is characterized by an ascending branch near the equator and subsidence in the subtropics (black filled stream contours). During solar maxima, a strengthening of the Hadley circulation has thus been proposed (van Loon et al. (2007)). Furthermore, increased subsidence induces even fewer clouds in subtropical regions and more incoming solar radiation, which provide a positive feedback and an amplification of the initial circulation perturbation (Meehl et al. (2008)). The clear solar signals attribution and details of the "bottom-up" mechanism are still under debate, however.

The second possible source of solar-induced climate variability is the fluctuation of UV radiation ($\lambda \sim 200\text{-}300$ nm), which typically varies by $\sim 6\%$ from solar minima to maxima. Although the relative irradiance variability is much stronger in the UV than in the VIS/IR spectral domain, the UV radiation is absorbed in the upper atmosphere and therefore no direct effect is expected on regional climate. Indirect "top-down" mechanisms are required to explain the transfer of the solar signal from the upper stratosphere to the surface. The starting point of the "top-down" mechanisms is the additional warming in the upper tropical stratosphere under solar maximum phases (red band), due to increased ozone absorption ($\lambda \sim 200\text{-}300$ nm), and more ozone production through oxygen photolysis ($\lambda < 242$ nm). The warming anomaly in the stratopause region results in a strengthening of the poleward temperature gradient, which in turn leads to a strengthening of the subtropical westerly jet through thermal wind balance (dashed green contours). These changes in the zonal background flow alter planetary wave propagation (brown filled arrows) and their interaction with the mean flow (transparent ellipse), so that (i) the stratospheric overturning circulation decelerates (black dashed arrow, box 2), and (ii) the initial westerly wind anomalies amplify and propagate poleward and downward through

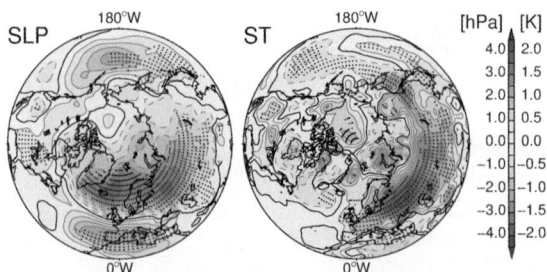

Figure 2. ERA-Interim sea level pressure (SLP) and surface temperature (ST) wintertime anomalies obtained by subtracting solar maximum to minimum phase years between 1958 and 2012. Solar maxima (minima) are defined when the F10.7 index value in winter is greater (lower) than 160 (85) sfu. Stippled areas indicate that the differences are statistically significant.

the winter season (evolution of green contours). The latter phenomenon results in a strengthening of the stratospheric polar vortex (box 3) which ultimately influences the regional climate at mid and high-latitudes through stratosphere-troposphere coupling processes (e.g. Gerber et al. (2012)). This version of the "top-down" mechanism is also known as the "polar route" (Kodera & Kuroda (2002)).

A "tropical route" has also been suggested, initiated by the secondary warming maximum signal observed in the lower tropical stratosphere (red band). This secondary warming would arise from adiabatic warming combined with increased ozone heating due to the deceleration of the stratospheric overturning circulation. Modeling studies have shown that this temperature anomaly in the lower tropical stratosphere can cause a weakening and a poleward shift of the upper troposphere/lower stratosphere subtropical jet (dashed green contours), as well as an expansion of the Hadley cell (Haigh et al. (2005)).

Both "top-down" approaches have been used to explain the observed surface climate response to solar variability at mid and high latitudes. In particular, they are retained as probable drivers of the Artic Oscillation (AO) and/or North Atlantic Oscillation (NAO) at quasi-decadal timescales, which are the most important modes of variability in the Northern Hemisphere atmospheric circulation in winter (Hurrell et al. (2001)). Fig. 2 shows the observed winter sea level pressure and surface temperature anomalies associated with the 11-year solar cycle. The sea level pressure anomalies are characterized by a decrease of ∼3 hPa over the North Atlantic which extends farther over Northern Europe and an increase of ∼2 hPa over Euro-Atlantic mid-latitudes region. The associated atmospheric circulation modulation results in warm (cold) anomalies of ∼2 K (∼-0.5 K) over Northern Europe (Mediterranean basin). In fact, this pattern strongly resembles the positive phase of the NAO, which induces mild and wet winters in Northern Europe.

The various mechanisms described in this section have been mostly proposed based on observations and/or climate model experiments of reduced complexity in order to minimize the computational costs, e.g., without a complete representation of the stratosphere, or by turning off the interactive ocean-atmosphere coupling. The perception of these mechanisms has however progressively evolved in time. Studies have proposed that the "top-down" may actually act in concert with the "bottom-up" mechanism (Meehl et al. 2009), and that the combination of both ocean-atmosphere couplings and stratospheric processes is crucial to transfer solar irradiance variability into climate variability.

3. Solar signals in CMIP5 simulations

Recent CMIP5 historical simulations provided an unprecedented opportunity to assess the ability of the latest generation of climate models to simulate solar-climate interactions. Analyses of solar signals in CMIP5 historical simulations have been performed by the Solar Model Inter-comparison Project (SolarMIP) as part of the WCRP/ SPARC-SOLARIS-HEPPA project. SolarMIP addressed (1) the role of ocean-atmosphere coupling (Misios et al. (2015)), (2) the stratospheric dynamical response and its influence on the surface (Mitchell et al. (2015)), and (3) the importance of the ozone solar signal and its feedback on stratospheric dynamics (Hood et al. (2015)). Among all CMIP5 models that conducted historical simulations, 31 models were first selected after excluding those where it was not possible to obtain information on the source of the solar forcing. Among the 31 models, 13 were classified as "high-top", i.e., including a full resolution of the stratosphere, and finally 6 out of the 13 models defined a subgroup which includes interactive ozone chemistry.

The results revealed a globally warmer surface ($\sim$0.07 K) and troposphere in response to solar maxima, that was a robust feature reproduced in the majority of models. The signal maxima showed a lag of $\sim$2 years consistently with observations. This lag was found to be induced by the extra energy absorbed over solar maxima that is primarily stored in the upper layer ($\sim$50-100 m) of the ocean (Misios et al. (2015)). Overall, models reproduced a globally average response to solar variability that is similar.

Analysis of solar signals at regional scales depicted less clear results, however. The two regions where solar signals are the most pronounced (i.e. Equatorial Pacific, and the Northern Hemisphere mid and high-latitudes) were examined in detail. Despite a large spread between the different models, evidence of a warmer equatorial central Pacific (with a lag of $\sim$2 years) and a warmer Arctic were found in response to solar maxima (Misios et al. (2015)). These features showed weak sensitivity to the accuracy of the model stratospheric representation, suggesting the prevalence of the "bottom-up" effect. Conversely, it has been suggested that a good representation of the stratosphere may help to simulate the observed North Atlantic signal, although the magnitude of the signals was weak and the agreement between models relatively low (Mitchell et al. (2015)).

To better understand the discrepancies in the regional response to solar irradiance variability, the stratospheric dynamical response was further investigated. The direct temperature response to increased UV was found to be robust between the 13 different "high-top" models but weaker than in the reanalysis. The resulting "top-down" mechanism was also not consistent across all models, although some models showed better performance. The latter are those which include a high spectral resolution radiative scheme in the shortwave region (Mitchell et al. (2015)). Interestingly, it was noted that models which include interactive ozone chemistry showed particularly relevant dynamical signals in the stratosphere (Hood et al. (2015)).

In summary, the three studies revealed that the models compare favorably with each other and with the observations/reanalysis at global scale. Conversely, the regional climate and stratospheric responses to the solar cycle were generally found challenging. In the next session, we propose future directions of research which may help to clarify the regional climate and stratospheric aspects.

4. Outlooks

Here, we develop three important issues which should be addressed in the near future:
• *Model formulation*. A rigorous representation of the impact of UV forcing on the atmospheric thermal structure is obtained by using a radiative transfer module which adequately resolves spectral solar variability and by including effects of solar induced ozone variations. The latter can be achieved by (i) prescribing the ozone changes or (ii) interactively calculating the ozone chemistry. While the first option is usually preferred

as it requires significantly less computational ressources, the second option remains the most accurate way to simulate ozone feedbacks on atmospheric dynamics and radiation which are important for the representation of the solar signal. Note that out of 31 CMIP5 models, only 3 fullfill these characteristics. More modeling sensitivity studies should be conducted to assess the relevance of implementing interactive ozone chemistry.

- *SSI forcing.* All the CMIP5 models were forced with the NRLSSI (Naval Research Laboratory Solar Spectral Irradiance) conservative solar spectral irradiance model, which represents the lower limit of the magnitude of SSI solar cycle variation among all models and measurements available (Ermolli *et al.* (2013)). This could partly explain the weaker upper stratosphere solar temperature anomalies compared to reanalyses, and the weak representation of the "top-down" mechanism. For instance, modeling studies using stronger UV forcing than NRLSSI revealed promising stratospheric and regional climate solar signal (Ineson *et al.* (2011), Scaife *et al.* (2013)). In this context, it is crucial to assess the best estimate of SSI variability.

- *Signals attribution.* There is increasing evidence that some other variability sources may alias the solar signals, leading to wrong signals attribution. Chiodo *et al.* (2014) inferred that the supposed solar response in tropical lower-stratospheric ozone and temperature (Fig. 1) may partly result from aliasing of the solar cycle with the major volcanic eruptions El Chichón and Pinatubo, which by chance occurred following solar maxima in 1982 and 1991. Recently, Thiéblemont *et al.* (2015) showed that the coupled ocean-atmosphere system can itself generate intrinsic quasi-decadal climate oscillations, i.e. without influence of any external forcing, and that solar variability may synchronize this internal variability. These examples highlight that more model sensitivity studies are necessary to precisely identify the role of the various internal or external variability sources in their interaction with the Earth's climate variability.

References

Chiodo, G., *et al.* 2014, *Atmos. Chem. Phys.*, 14, 5251-5269
Ermolli, I., *et al.* 2013, *Atmos. Chem. Phys.*, 13, 3945-3977
Gerber, E. P., *et al.* 2012, *Bull. Am. Meteor. Soc.*, 93, 845-859
Gray, L. J., *et al.* 2010, *Rev. Geophys.*, 48, RG4001
Haigh, J. D., Blackburn, M., & Day, R. 2005, *J. Climate*, 18, 3672-3685
Hood, L. L., *et al.* 2015, *Q.J.R. Meteor. Soc.*, doi:10.1002/qj.2553
Hurrell, J. W., Kushnir, Y., & Visbeck, M. 2001, *Science*, 291, 603-605
IPCC 2013, *Fifth Assessment Report of the Intergovernmental Panel on Climate Change*, Cambridge University Press, 1535 pp, doi:10.1017/CBO9781107415324
Ineson, S., *et al.* 2011, *Nat. Geosci.*, 4, 753-757
Kodera, K. & Kuroda, Y. 2002, *J. Geophys. Res.*, 107, 4749
Matthes, K., *et al.* 2006, *J. Geophys. Res.*, 111, D06108
Meehl, G. A., *et al.* 2008, *J. Climate*, 21, 2883-2897
Meehl, G. A., *et al.* 2009, *Science*, 325, 1114
Misios, S., *et al.* 2015, *Q.J.R. Meteor. Soc.*, accepted
Mitchell, D. M., *et al.* 2015, *Q.J.R. Meteor. Soc.*, 141, 2390-2403
Scaife, A. A., *et al.* 2014, *Geophys. Res. Lett.*, 40, 434-439
Thiéblemont, R., *et al.* 2015, *Nat. Commun.*, 6:8268
van Loon, H., Meehl, G. A., & Shea, D. 2007, *J. Geophys. Res.*, 112, D02108
Zhou, J. & Tung, K. K. 2010, *J. Climate*, 23, 3234-3248

FM 20:
Astronomy for Development

Preface - Focus Meeting 20: Astronomy for Development

Kevin Govender

IAU Office of Astronomy for Development, Cape Town, South Africa
email: kg@astro4dev.org

1. Overview

This Focus Meeting was about the global developmental impact that all aspects related to astronomy can deliver. The interdisciplinary nature of the meeting made it relevant to all IAU Divisions and the professional astronomy community in general. The manner in which the strategic plan has been designed and the way in which OAD implements it allows for input and innovation from the professional community both to develop the astronomy field globally and to stimulate the developmental benefits arising from the astronomy field. IAU members have played a key role in every stage of implementation of the strategic plan, from its ratification, through to strong participation in its implementation. This meeting served to report back to them in terms of progress, as well as seek input from them in terms of shaping the way forward.

2. Structure

The programme was organised to try to cover the wide variety of topics related to the OAD within a very limited amount of time allocated. The first session was for overview talks covering the history, achievements and activities of the OAD. This was followed by three sessions for the Task Forces (Universities and Research, Children and Schools, Public Outreach). We then had contributions from the Regional Offices and Language Expertise Centres, followed by a Poster Plug Session (where poster contributors could say a few words to attract people to their posters). The final session comprised a Panel discussion with Representatives from IAU Divisions, and an unconference session which sought topics from participants during the General Assembly in order to adapt to popular demand. These Proceedings reflect this general structure with a table summary at the end listing poster contributors' topics and authors. Presentations from the Focus Meeting, and any related information can be found on the OAD website www.astro4dev.org.

3. Acknowledgements

We are grateful to the IAU for the opportunity to conduct this Focus Meeting at the 2015 General Assembly and to the SOC for their voluntary efforts in driving the event before, during and after. A special word of thanks to the OAD staff, Visiting Fellows and Interns who handle all the logistics and detail required behind the scenes to ensure smooth implementation of such activities, including assistance with compiling these proceedings. Lastly, and most importantly, we wish to acknowledge and appreciate the contributions from our many collaborators (both old and new) from around the world. Without your Oral and Poster contributions, this Focus Meeting would not have happened at all, let alone with the incredible amount of enthusiasm, energy and positive feedback we experienced before, during and after the event.

The IAU Strategic Plan

George Miley
Leiden Observatory, Leiden University, The Netherlands

Abstract. I shall review the content of the IAU Strategic Plan (SP) to use astronomy as a tool for stimulating development globally during the decade 2010 - 2020. Considerable progress has been made in its implementation since the last General Assembly.

1. Introduction

The IAU Strategic Plan 2010 – 2020: "Astronomy for Development: Building from IYA 2009" was ratified by the IAU General Assembly in August 2009. It is based on the unique power of astronomy as a tool for furthering human and technological capacity building throughout the world. Astronomy is a fundamental science itself and a gateway to physics, chemistry, biology and mathematics. Astronomy has been an important driver for the development of the most sophisticated technology and provides a link to our deepest cultural roots and origin.

2. The IAU Strategy

The Plan http://www.iau.org/static/education/strategicplan_2010-2020.pdf is a blueprint for using astronomy as a tool for development. The vision is a global one, namely that eventually all countries should participate at some level in astronomical research and that all children throughout the world will be exposed to knowledge about astronomy and the Universe. The strategy of the SP has several components:

1. A strategic phased integrated approach, including primary, secondary and tertiary education, research and public outreach. This will be based on the potential for astronomy research and education in each country, using objective data augmented by advice from experts in the region.

2. Regional involvement. A bottom-up approach, as pioneered during the International Year of Astronomy (IYA2009), involves regional input, including designation of regional institute nodes to coordinate development efforts throughout their region.

3. Special attention to Sub-Saharan Africa. Because of its relative underdevelopment, sub-Saharan Africa is receiving special attention.

4. Using IYA2009 as a springboard. Several IYA "cornerstones" are being continued and supported (e.g. UNAWE, GTTP) and the network of IYA contacts exploited.

5. Enlarging the number of active volunteers. We have recruited volunteers from amongst members, doctoral students, postdoctoral trainees, non-member experts on education and outreach and amateur astronomers. Expatriates are particularly important.

6. Exploiting innovative techniques. Innovative approaches to education and development, are being explored, including distance-learning, archives, robotic telescope networks and mobile delivery via astro-buses.

7. Creation of a small global "Office of Astronomy for Development "(OAD). Mobilising large number of volunteers, implementing new programs and inputting strategic information need professional coordination. Setting up of the IAU OAD was crucial to the success of the SP.

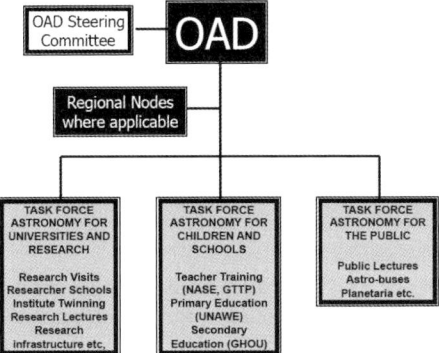

Figure 1. Structure of Astronomy for Development (AfD) activities envisaged in the IAU Strategic Plan (SP). Three task forces are coordinated by the IAU Office of Astronomy for Development (OAD) and regional input ensures a portfolio of demand-driven activities.

Figure 1 shows the implementation structure envisaged by the SP and now implemented. Three Task Forces of experts coordinated by the OAD manage the various Astronomy for Development (AfD) activities. The OAD injects a strategic component and safeguards accountability and transparency. Input from the regions ensures that the AfD portfolio matches local needs.

3. Implementation of the IAU Plan

The SP covers the ten-year period 2010 – 2020 and is being implemented gradually, to match the available funding. The first step was to set up the IAU Office of Astronomy for Development (OAD) to coordinate activities. Following a competitive call for proposals, and the selection of a host organisation by the IAU Executive Committee, the OAD started operation in March 2011. The IAU OAD is a joint venture between the IAU and the South African National Research Foundation (NRF), with Kevin Govender as its first Director. The official launch of the OAD took place on 16th April 2011 by the South African Minister of Science and Technology and the IAU President. By the last GA in Beijing, the OAD had set up the three Task Forces envisaged in the Strategic Plan and two regional nodes (ROADs) had been inaugurated. Enormous progress has been made in implementing the IAU Strategic Plan since then. By the end of the present GA, there will be eight functioning IAU ROADs. Earlier this year, the OAD received an oustanding report by a high-level independent review committee appointed by the IAU and the NRF. This has resulted in the intention by both the IAU and the NRF to renew the agreement between them to operate the OAD until at least 2021 and the resolution to approve this extension that we will be asked to approve later in this GA. During this period the NRF will fund an additional OAD position and the IAU will support a fund raising campaign to enable the implementation of additional activities.

During recent years it has become increasingly important for the scientific communities to justify the large expenditure on the infrastructures and facilities needed to carry out fundamental science. The IAU Strategic Plan and its implementation by the OAD is a unique endeavour for a scientific community. It demonstrates that astronomy is not only one of the most fundamental and exciting sciences, but that it can be of great benefit to society as a whole.

In the next talk Kevin Govender will next tell you more about the recent accomplishments of the OAD.

Towards "Astronomy for Development"

Kevin Govender

IAU Office of Astronomy for Development, Cape Town, South Africa
email: kg@astro4dev.org

Abstract. The ambition of the IAU's decadal strategic plan is to use astronomy to stimulate development globally. The Office of Astronomy for Development was established in 2011 to implement this visionary plan. This talk will reflect on the past, present and future activities of the office, and describe the status of implementation of the plan at this halfway point in the 2010- decade.

Keywords. OAD, astronomy, development

1. Background

The International Astronomical Union (IAU), together with the South African government, established the global Office of Astronomy for Development (OAD) in 2011 in order to fulfill the IAU's 10 year strategic plan, which aims to realise the global developmental benefits of astronomy. We interpret astronomy's role in achieving sustainable development in several ways including (i) social benefits (common humanity, scientific engagement & discourse); (ii) human capital development (education, skills, career choices); (iii) economic growth (knowledge economy, innovation); and (iv) human welfare (all of the above, technology transfer).

The OAD is tasked with establishing and strategically coordinating Regional Nodes (ROADs) and Language Expertise Centres (LOADs) across the world as well as three Task Forces, namely (i) Astronomy for Universities and Research, (ii) Astronomy for Children and Schools, and (iii) Astronomy for the Public. The bulk of the implementation of projects is carried out by volunteers, coordinated by the Task Forces with the support of the OAD and its regional offices. In the first 5 years of its existence, 68 projects were funded through an open annual call for proposals, 9 agreements were concluded for the hosting of regional offices and language expertise centres, and memoranda of understanding were signed with 9 partner organisations. Over 500 volunteers registered their skills with the OAD and are a valuable source of project proposals and support. This first phase of the existence of the OAD managed to achieve a transition from previous IAU activities in the area, a substantial amount of network building and strong international credibility for the concept.

2. The OAD within the IAU landscape

The IAU landscape can sometimes appear confusing since there are several activities that seem to deal with similar objectives. The three that most often stimulate queries for clarification are the existence of the OAD, the IAU Office for Astronomy Outreach (OAO), and the IAU's Division C (Education, Outreach and Heritage). We aimed during this talk to clarify the roles with three respective words: Development, Access and Knowledge. Astronomical knowledge resides within the IAU and its Divisions. In order to provide access to this knowledge the OAO (based at the National Astronomical Observatory of Japan) performs its global networking and outreach work, sustaining the

Table 1. 2015 Strategic goals and status.

Goal	Status
a. Active regional nodes on all populated continents, collectively covering at least two thirds of world population in terms of target regions.	By August 2015, there were 9 regional offices established which meet this target.
b. Active Task Forces with an established annual strategic planning process and funding procedure (yearly cycle) for Astronomy-for-Development (AfD) activities.	Successfully implemented and ongoing.
c. Secured funding and plan for the continuation of the OAD in the 2016 to 2020 period.	This has been achieved as of September 2015
d. Sustainable volunteer programme including efficient registration of both volunteers and opportunities for volunteers.	Registration exists although more can still be done to engage volunteers
e. Total accumulated funding commitments from external sources exceeding the annual investments made through the IAU-NRF agreement.	Funds raised or committed in the first 3 years (including regional offices and partnerships) already amount to €2,217,595, far exceeding the total 5-year summed contributions from IAU and NRF/DST of €1,320,000.

momentum from the International Year of Astronomy 2009. The specific role of the OAD then is to look at the issue of development. As such, the OAD is responsible for using the knowledge within the IAU/astronomy community, along with the access provided through the OAO, to stimulate global development. To achieve each mandate there obviously needs to be synergy between all structures in this landscape, which we believe we have achieved.

3. External Review of the OAD

In February 2015, an external review was conducted of the OAD. The review panel was very complimentary about the OAD activities, and recommended the continuation of the OAD in South Africa until 2021. This recommendation was accepted by all parties (IAU, NRF, DST). A new agreement is expected to be signed in September 2015 with revised financial commitments to the OAD. Specific recommendations of the review were: (i) continuation of the OAD till 2021; (ii) there should be a resolution of the IAU General Assembly for the continuation of the OAD/Strategic Plan beyond 2020; (iii) there should be an increase in funding and staff (fundraiser + astronomer); (iv) simpler oversight structures; (v) consolidation of activities; (vi) annual high level meetings between the partners (IAU and South African National Research Foundation); (vii) there should be simpler visa processes to get more skills to the OAD in South Africa easier. The details of the review are publicly available on the OAD website.

4. Strategic Goals

During the first 5 years the OAD worked according to a business plan that listed long term (5 year) strategic goals for the OAD up to December 2015. These goals and the state of achievement are reflected in the following table:

In the second half of the decade of the Strategic Plan, and leading up to the 2021 General Assembly, the OAD will build on its experience thus far and strive to position itself as the global reference point for best practice and evidence on astronomy related human capital development, education and outreach interventions leading the community to maximally realise the potential of astronomy for development. The possible long terms goals which the OAD will strive to achieve by 2021 (to be approved by the new OAD Steering Committee) are:

(a) A user friendly impact cycle for projects containing a library of best practice resources and evidence on what works and what doesn't work. This cycle should include support for rigorous evaluation of impact as well as a project monitoring system for OAD funded projects.

(b) Synergised regional leadership providing input to the OAD in a systematic way as well as local coordination for the implementation of evidence-based pilot projects and for taking successful projects to scale.

(c) The OAD should host and coordinate externally funded programmes, working with strategic partners, to develop and evaluate potentially high impact pilot projects that meet specific needs or gaps not currently addressed by other organisations, and take to scale those projects shown to work. External funding would be used to fund additional staff to manage the development process for each project.

(d) The OAD model should be adopted by other scientific bodies for their respective fields, with the International Council for Science (ICSU) possibly driving a global science for development initiative based on the OAD model.

(e) The OAD volunteer management system should provide a user friendly platform to allow for the flow of skills from the astronomy community to other regions and fields.

In summary, at this point in time, halfway through the decade of the IAU Strategic Plan, the OAD is well positioned to making a significant difference, by 2020, in this growing area of astronomy (and more generally, science) for development.

An Evidence-Based Framework to Optimise Social Impact in Astronomy for Development

Eli Grant

IAU Office of Astronomy for Development, Cape Town, South Africa
email: eg@astro4dev.org

Abstract. Astronomy for development projects conceive of development in very broad terms and seek to affect a wide range of social outcomes. The histories of education, development economics and science communication research indicate that positive social impacts are often difficult to achieve. Without a scientific approach, astronomy's potential as a tool for development may never be realised nor recognised. Evidence-informed project design increases the chances of a project's success and likely impact while reducing the risk of unintended negative outcomes. The IAU Office of Astronomy for Development (OAD) Impact Cycle is presented here as a possible framework for integrating evaluation and evidence-based practice in global Astronomy outreach and education delivery. The suggested framework offers a way to gradually accumulate knowledge about which approaches are effective and which are not, enabling the astronomy community to gradually increase its social impact by building on its successes.

Keywords. outreach, science communication, education, development, evaluation, evidence-based practice

1. Background

The International Astronomical Union's (IAU) Strategic Plan 2010–2020 *Astronomy for Development* (2012) set out the IAU's vision of how astronomy could be used as a tool to achieve a range of development objectives and resulted in the establishment of the IAU Office of Astronomy for Development (OAD) in 2011. In these proceedings, Govender describes the work done by the OAD to expand the reach and capacity of astronomy for development projects. Since the OAD's establishment, thousands of students and members of the public have been exposed to Astronomy through a diverse range of projects seeking to realise the plan's aims.

The ability to reach people is a prerequisite for social impact. Exposure and impact are not, however, synonymous. Some interventions are more effective than others. Not all individuals who have been exposed to astronomy have been effectively engaged or learned what the project intended them to learn. From a more sceptical, scientific point of view each project (and the IAU Strategic Plan itself) is an idea, essentially a hypothesis, about a specific way in which astronomy might be used for positive social impact. This is how the IAU's Strategic Plan was deliberately conceived, allowing for grassroots innovation and experimentation. Unless we can test these hypotheses, however, we do not know which projects have been most and least effective. Measuring project impacts would help to ensure that limited resources are directed to the most cost-effective projects. that is, activities that produce the largest impacts for the largest number of people for the least amount of money.

2. Good Intentions and Unintended Consequences

Human beings are complex and embedded in complex systems. For some natural scientists, social complexity suggests that scientific measurement and testing is not possible. However, methodological and computing advances across multiple disciplines over the past century have provided valuable tools for detecting signals amidst the randomness and noise. There are also regularities in how human beings tend to respond. Indeed, human and social complexity is the very reason that it is imperative to test project impact†. A few examples of unintended consequences from astronomy for development projects include loss of interest in astronomy from those previously interested; loss of self-confidence and interest in science when educational activities are mismatched in pace or content to some students' baseline skills and knowledge; and the perpetuation of negative social norms in projects that bring together disadvantaged youth. The probability of negative unintended consequences can be reduced by learning from other fields working to effect positive social impact and adopting a more scientific, evidence-informed approach.

3. Methods for Measuring Project Impact

There are three key challenges for determining which projects 'work', which do not and which are most likely to work in a new context. These are: measuring target outcomes, measuring a project's causal effects and then predicting likely effects in a new context. Not all social impacts are measurable. Many astronomy for development projects, for example, focus on inspiring their audiences. Inspiration is not readily measurable. However, observable changes must occur at some point for a project to be considered useful for development. Current global development priorities can be found in the Sustainable Development Goals (SDGs), agreed by the United Nations this year after extensive surveys and consultations with millions of citizens, experts, governments and development organisations across the world. These goals are the outcome of a long consensus-building process focused on identifying which aspects of development matter most. Each goal is associated with a list of targets or indicators: these are observable and measurable‡. Examined in relation to the SDGs, inspiration might be seen as a key stepping stone towards achieving a multitude of development goals (such as improvements in education quality) rather than a final goal. Many astronomy for development projects target multiple, observable, development-related outcomes alongside invisible or internal psychological processes. Examples of observable and measurable outcomes typically targeted by astronomy for development projects include improved or increased learning outcomes; participation in science classes; scientific literacy; selection of science majors by undergraduates; enrolment in science, physics or astronomy courses or post-graduate studies; and research output.

Measuring the causal effects of project activities on observable outcomes is the second challenge. Drawing from medical science and economics, a project's causal impacts and/or key components can be tested using experimental designs. In these applications, experimental designs are known as randomised controlled trials. Participants are randomly assigned to the experimental condition (e.g. an innovative astronomy outreach activity)

† See: Macintyre, S. & Petticrew, M. (2000) Good intentions and received wisdom are not enough. *Journal of Epidemiology and Community Health*, 54, 802-803 and Dishion, T. J., McCord, J. & Poulin, F. (1999) When interventions harm: Peer groups and problem behavior. *American Psychologist*, 54, 755-764.

‡ See the International Council of Science Union's (ICSU) report *Review of Targets for the Sustainable Development Goals: A Science Perspective*, available from http://www.icsu.org/publications/

or to an alternative (e.g. no activity or a standard astronomy outreach activity). Because assignment to one condition or the other is random, it is statistically independent of outcome: the two samples are drawn from the same population, with intervention exposure known to be the only thing distinguishing them. The average outcome in the comparison condition is thus known to offer a measure of what the average outcome for the experimental sample would have been. In the absence of random assignment, it is often impossible to know what caused what. For example, if a high proportion of students who attend a summer school enrol in astronomy postgraduate courses, is that because the school motivated and engaged them? Or is that because only those students who were most interested (and already intending to continue with astronomy) were the ones who were most likely to apply and be best-prepared for the school? Would they have otherwise enrolled in a different areas of science? Or did the summer school actually increase the total number of science post-graduates in the target population? Because they are often over-subscribed, summer schools and workshops are particularly well-suited to impact evaluation using random assignment.

A particularly promising and often low-cost approach for helping projects to learn from one another is to randomly assign participants to experience different *versions* of a project. An excellent example of this approach is offered by Buck (2013), who randomly assigned post-secondary students to visualisations of dark matter in a planetarium shown in three different colours. She found that those exposed to the 'standard' visualisation (dark matter depicted in white so that it stands out from a dark blue background), were four times *less* likely to correctly identify dark matter in a post-exposure test than participants who were exposed to visualisations where dark matter was darker than the background†. Findings from this simple experiment can now be used to improve learning outcomes amongst planetarium attendees.

Readers interested in learning more about research and project evaluation methodologies are referred to (1) the MIT Poverty Action Lab (JPAL), who conduct randomised controlled trials of international development projects including those in primary and secondary eduction and provide both data from these experiments and extensive resources on evaluation methods themselves: www.povertyactionlab.org/education (2) Nobel prize-winning economist Angus Deaton's more in-depth explanation of the statistical theory underpinning randomised controlled trials and ongoing debates as to whether project mechanisms rather than whole projects should be tested: Deaton, A. (2009) Instruments of Development: Randomization in the Tropics, and the Search for the Elusive Keys to Economic Development. *NBER Working Paper*,14690. For examples of how evidence and evaluation research have been used to inform the design and implementation of projects in related fields see Yeager and Walton (2011) Social-Psychological Interventions in Education: Theyre Not Magic. *Review of Educational Research*, 267-301.

4. The OAD Impact Cycle to Support Astronomy-for-Development

Simple experiments like these can advance our understanding of how to achieve social impact more effectively. Through co-ordination and collaboration across the education, outreach and development communities, the OAD hopes that more experiments will be conducted and fed into a shared database that can be used to make evidence-based decisions about 'what works'. The aim is to create a positive feedback cycle through which projects can draw on research that has already been conducted to inform their project

† Buck, Z. (2013) The Effect of Color Choice on Learner Interpretation of a Cosmology Visualization. *Astronomy Education Review*, 12, 1, 010104.

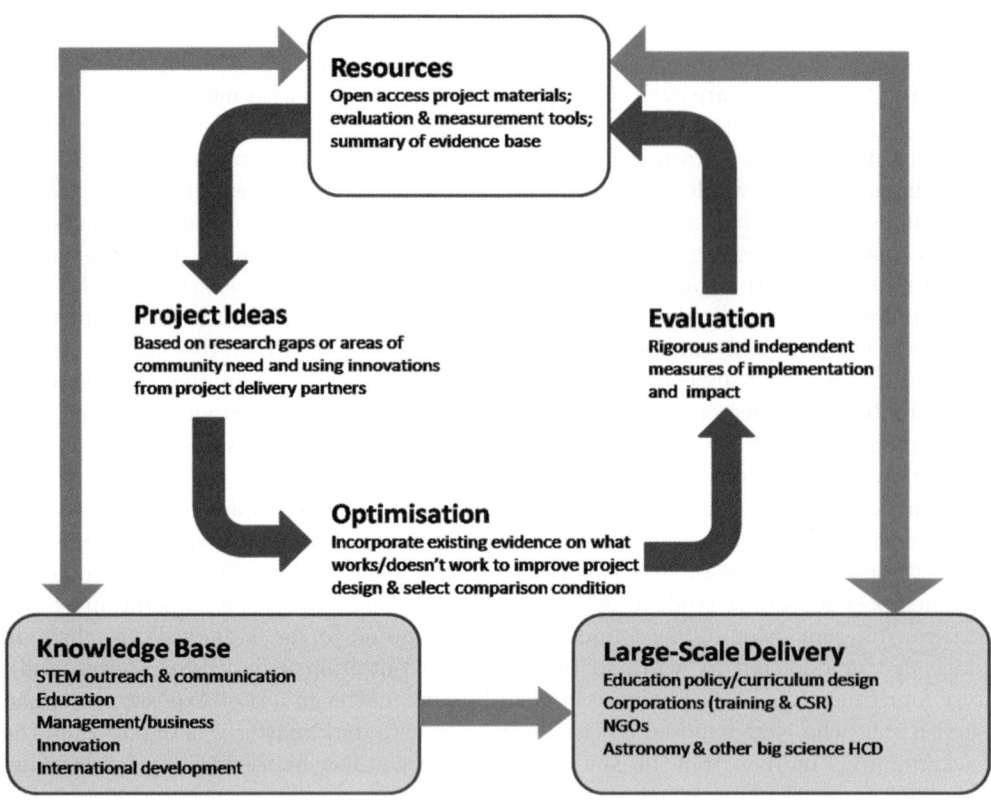

Figure 1. OAD Impact Cycle for Project Evaluation and Optimisation

designs; test creative new ideas, innovations and adaptations, including implementation in a new context or country; and then contribute their own results to a gradually accumulating knowledge base. This framework is referred to as the OAD 'Impact Cycle' (see Figure 1).

Resources will include systematic reviews† of relevant evidence (e.g. educational techniques and international development programmes); best-practice guidelines; standardised and validated outcome measures (e.g. of scientific and statistical literacy) and evaluation tools to assist in the design and implementation of project evaluations and outcome measurement. The *Ideas* stage refers to support for creative innovations or projects that members of the astronomy community come up with for how astronomy might be used for development. Project ideas will ideally be informed by Resources or developed in response to specific development-related problems or needs. *Optimisation* refers to a stage of refinement and improvement of project ideas, informed by evidence provided in the Resources of what has already been tried, what works and what does not. The Ideas and Optimisation phases provide mechanisms through which innovations and ideas can more easily build upon, rather than reinvent, previous work. Drawing from evidence will allow new ideas to maximise their likelihood of success and reduce their risk of unintended negative consequences. Finally the *Evaluation* phase involves embedding impact evaluation designs of entire projects or of specific innovative components within projects,

† See for example Greenhalgh, T. (1997) How to read a paper. Papers that summarise other papers (systematic reviews and meta-analyses). BMJ, 13, 315(7109):672-5.

to complete the cycle and contribute new evidence of 'what works' to the communities' resources.

5. Conclusion

The OAD Impact Cycle offers a framework for accumulating scientific evidence from across contexts and countries to support efforts to understand the variability and localisation of specific types of projects' impacts. The Impact Cycle seeks to support the astronomy community in drawing from resources from other disciplines, including science communication, education and development efforts across fields of science, technology, engineering and mathematics. The Impact Cycle also offers a way through which the astronomy community can contribute to global development efforts by participating in cross-disciplinary research on what works.

Astronomy for a Better World: IAU Office of Astronomy for Development Activities to Grow and Advance Astronomy Education and Research at Universities in the Developing World

Edward F. Guinan[1] and Katrien Kolenberg[2,3,4]

[1]Astronomy & Astrophysics Department, Villanova University,
800 Lancaster Ave, Villanova PA 19085, USA, email: edward.guinan@villanova.edu

[2]Physics Department, University of Antwerp, Groenenborgerlaan 171, 2020 Antwerp, Belgium
[3]Harvard-Smithsonian Center for Astrophysics, 60 Garden Street, Cambridge MA 02138, USA
email: kkolenberg@cfa.harvard.edu
[4]Institute of Astronomy, KU Leuven, Celestijnenlaan 200D, 3001 Heverlee, Belgium

Abstract. In 2012, the International Astronomical Union (IAU), through its Office of Astronomy for Development (OAD), established the three Task Forces which drive global activities using astronomy as a tool to stimulate development. These Task Forces are: (i) Astronomy for Universities and Research; (ii) Astronomy for Children and Schools; and (iii) Astronomy for the Public.

Keywords. astronomy, development, strategic plan

1. Astronomy for Development at the Universities & Research level

Task Force One (TF-1) on Astronomy for Universities & Research was established in 2012 as part of the IAU Office of Astronomy for Development (OAD). This Task Force drives activities related to astronomy education and research at universities mainly focused on the developing world. Astronomy is used to stimulate research and education in STEM fields and to develop and promote astronomy in regions of the world where there is little or no astronomy. There is also potential for developing research in the historical and cultural aspects of astronomy which may prove important for stimulating an interest in the subject in communities where there is yet no established interest in the science. Since the establishment of the OAD, twenty-six TF-1 programs have been funded to support a wide variety of interesting and innovative astronomy programs in Africa, Asia, South-East Asia, the Middle-East, and in South & Central America. Nearly every aspect of development has been supported. These programs include supporting: regional astronomy training schools, specialized workshops, research visits, exchange visits, university twinning programs, distance learning projects, university astronomy curriculum development, as well as small telescope and equipment grants.

Also supported within the TF-1 program framework are Visiting Astronomer & Exchange programs; Sabbatical Visits; funds for travel of astronomer to developing country-airfare / small stipend; Development of distance learning packages for University Astronomy Courses; Regional Specialized /Focused Workshops & Schools, Start-up funds for Association of ISYA Alumni Technology Internships e.g. instrument specialists to consult; Telescope adoption programs in which existing observatories "adopt" a facility in a developing region; Collaborations with institutions where a percentage of research grants are spent on development; Open source text book for astronomy education at all levels and many others that related to the development of Astronomy education and research at the university level.

As examples, the TF-1 programs approved for 2015 (Ten programs approved for support/ partial support) are given below. Because of the limited amount of funds, there are many valuable programs that could not be funded. Fund raising is being carried out to support these other highly ranked programs. The over subscription rate for TF-1 proposals over the last three years is about 4/1.

2015 (Ten programs approved for support/ partial support)

- Visiting Program at the Harvard-Smithsonian Center for Astrophysics (USA-SA)
- Time Variability in Modern Astrophysics (Thailand)
- Latin American School of Observational Astronomy
- Summer School on Statistical Data Analysis and Data Mining in Astronomy
- Optical camera for a 14-inch telescope in Namibia (Namibia)
- Astronomy for Africa: Student Support - Astronomy Modules via Distance Learning
- National School on Astrophysical Simulation
- Introducing Data Analysis in the University System
- TARA, Fergusson College Node, Pune, India
- Guatemalan School of Astrophysics 2015
- Andean Cosmology School

In addition, a new multi-year program - Starlab - was introduced by TF1-members, J-P De Greve and Michele Gerbaldi, and proven successful through initial funds by OAD in 2013 and 2014 to bring "starlight" into the class room. In this program, students carry out and reduce CCD photometry secured by them using remotely controlled telescopes around the world. The primary goal is to develop the Astrolab program into a standard tool for astronomy education, useful for training in institutions and locations without easy telescope access. Pilot programs have already been carried out in Uganda, Rwanda and Ethiopia.

The following proposal selections criteria are considered in the selection of new projects: Alignment with the goals of IAU's 10 year strategic plan; Quality and relevance of content. Innovativeness and creativity of approach; Clear timeline indicators; Cost effectiveness of the project including a detailed budget; Potential to raise funding from other sources; Range of stakeholder participation (e.g. partnerships universities / industry / government); Degree of impact of the project with potential for sustainability.

Information and advice are provided about applying for support in the future. Additional information is available on the OAD website at www.astro4dev.org/. The next OAD call for proposals will be during mid-2016 for programs to be carried out in 2017. Formal submission of proposals is typically due during September and the proposal evaluation outcomes are released during December. The OAD is continuously improving and expanding these programs to bring astronomy education and research to a greater number of people and indeed to use Astronomy for a Better World.

References

IAU 2012, http://www.iau.org/static/education/strategicplan_2010-2020.pdf

On formation and activity of IASS

A. S. Hojaev

Ulugh Beg Astronomical Institute, Uzbek Academy of Sciences
Astronomicheskaya, 33, Tashkent, 100052, Uzbekistan
email: hojaev@yahoo.com

Abstract. A brief description of International Aerospace School (IASS) is given.

Keywords. sociology of astronomy, history and philosophy of astronomy, miscellaneous

---

Central Asia has a remarkably wealthy and deep intellectual tradition in the sciences and was the creative intellectual capital of the world between 800 and 1600, which achieved signal breakthroughs in many fields including astronomy. They gave algebra its name, calculated the earth's diameter with unprecedented precision, wrote the books that later defined European medicine, and penned some of the world's greatest poetry (see, e.g., Starr 2015).

Present Uzbek astronomers still play a key role in scientific research collaborating with many foreign colleagues. International Aerospace School (IASS) is a unique project in the region held annually since 1989(twice a year during a few first years), i.e. active for more than quarter of a century. Academician S.Vakhidov and famous cosmonaut, general V. Djanibekov were founders of IASS . The main support is provided by Vakhidov's Youth AeroSpace Fund. At present the summer camp gathers about 50 teenage and undergraduate students over different regions of Uzbekistan, Russia, Kazakhstan, Kyrgyzstan, Ukraine, Tajikistan, Azerbaijan, Georgia, Pakistan, Malaysia, Turkey, Poland and France (since 2005), etc. They are selected based on tests of astronomy and space science, and on qualifying competition. Working languages are Uzbek, Russian, French, English. The qualified simultaneous translation of lectures and instructions is provided. During two weeks of IASS camp the invited physicists and astronomers, cosmonauts and astronauts, pilots and instructors of the airlines as well as other specialists give lectures and engage in practical exercises with IASS students in astronomy, including daily observations of the Sun and night sky observations with meniscus telescope, space research and exploration, aerospace modelling, visiting the special technical facilities like unique Large Solar Furnance in Parkent, preparation and presentation of original projects. Lectures and practice in astronomy are carried out with the generous assistance of Uranoscope Association (Paris, France). This is important that IASS gives not theoretical basics only but also practically train the students, and the hands-on training are among or within the major aims of IASS. IASS alumni become students of physics, technology, aviation universities and colleges, pilots of cosmonaut corps and national airlines,design engineers in aerospace, scientists of research institutes and teachers at universities. The 26-th IASS season was held in July-August 2015, the next one is scheduled for the same time in 2016. IASS invites students as well as lecturers and instructors from abroad to participate in the camp and take part in other IASS activities. It will appreciate any sponsor contributions. Contact e-mail is presented in affiliation line above.

References

Starr, S. F. 2015, *Lost Enlightenment: Central Asia's Golden Age from the Arab Conquest to Tamerlane* (Princeton University Press), 680 pp.

The Role of Astronomy in Development: The Case of Uganda

Edward Jurua

Department of Physics, Mbarara University of Science and Technology, Uganda
email: ejurua@gmail.com

Abstract. Science and technology play a key role in economic development; and Universities have a direct stake in this process. A knowledge-based economy requires scientific and technological expertise that is strongly influenced by the strength of training in science and technology. However, in Uganda not many students opt for science subject at higher levels, and subsequently in the University. Therefore, there is need to encourage and motivate students to study science subjects in order for this to be successful. This can be achieved through introduction of stimulating subjects such as astronomy in the university curriculum. Astronomy is considered as the most appealing subject and an excellent tool for conveying scientific knowledge to young students. In this paper, the role that astronomy has played to motivate and interest students to study physics in Mbarara University of Science and Technology, in Uganda, is discussed.

Keywords. Astronomy, Ecomony, Development.

1. Introduction

The major economic activity in Uganda is Agriculture, and most of the Agricultural produce is exported in the raw form, thus generating low income for the government. As a result, the Government of Uganda embarked on the expansion of her industrial base and mechanization of Agriculture in order to generate more income. However, the current challenge is the unavailability of skilled, dedicated, and well-motivated human resource that is required to make such a programme successful. This resulted in government policies that aim to popularize and develop science and technology in order to provide the knowledge, skills, and professional and technical expertise that is required for the expansion of the industrial base and mechanization of Agriculture. However, very few students opt for science subjects and this has a direct effect on students admitted for sciences in tertiary institutions. Therefore, there is need to encourage and motivate students to study science subjects. This is discussed in the following section.

2. Motivation of Students into Science

To get the expertise in Science and Technology needed for economic development, young students should be motivated and attracted to the science subjects. Introducing subjects that motivate and attract young people into mathematics and science can do this. At this point, it is important to recognize the power of astronomy in attracting students into science subjects, which is vital for scientific and technological development. Fierro (1999) described Astronomy as the most appealing subject and an excellent tool for conveying scientific knowledge to young students. Besides, astronomical discoveries virtually interest everyone on the globe since astronomy is a frontier science (Anguma & Jurua (2009)). Unlike other natural sciences, it takes the universe as its laboratory, in which physical laws and theories are applied, tested and refined at temperatures, pressures and scales unobtainable in terrestrial laboratories (Stobie (1995)).

In order to promote science and technology in Uganda, Mbarara University of Science and Technology (MUST) introduced Astronomy as a course in the Physics department in 2003. The introduction of astronomy has not only created awareness among students, but also cultivated a lot of interest in pursuing further careers in astronomy. This has resulted in increased number of students opting for science, particularly in physics, in Mbarara University of Science and Technology. To make this a success, there is need to develop human resource in astronomy through postgraduate training. The plan for human resource development is discussed in the following section.

3. Human Resource Development in Astronomy

In 2012, the Department of Physics, MUST was invited to apply for the International Science Programme (ISP) support for human resource capacity development in astrophysics and space science. An application was submitted and the support was awarded for a period of three years (2014 - 2015); and this is renewed after every three years depending on the output of the project. The main aim is to develop human resource capacity in astronomy and space science. As a result of this project, the number of students that have cultivated interest in pursuing further careers in astronomy and space science in particular and physics in general has drastically increased. For instance, over 30 students (down from an average number of 8) applied for admission into Master of Science in Physics programme in Mbarara University of Science and Technology for the academic year 2015/2016. This is an increase by more that 300%.

During these first three years of ISP support, a number of students were supported directly and indirectly. In particular, 5 students graduated with Master of Science during the February 2015 graduation. Four (4) students are expected to complete their MSc and graduate during the next graduation in 2016. All these students have expressed interest to continue with PhD in astronomy. Two (2) students were recruited and supported for PhD programme; and they are expected to graduate during the next graduation ceremony in 2016. Many of our undergraduate students have also expressed interest to pursue further studies in Astronomy. However, there is still much more to be done facilitate the training in order to attain an optimum capacity.

4. Conclusion

The introduction of astronomy has resulted in increased number of students opting for science. However, there is still a critical need for continued support from both the government of Uganda and development partners in building research infrastructure and human resource capacity in Astronomy in Uganda. These will promote and generate new knowledge and continuously facilitate training in this discipline. This will in turn produce the skilled, dedicated, and well-motivated human resource required for technological development.

References

Anguma, K. S. & Jurua, E. 2009 *The Role of Astronomy in Society and Culture, Proceedings IAU Symposium*, 260, 2009; D. Valls-Gabaud & A. Boksenberg, eds.
Fierro, J. 1999, *Teaching of Astronomy in Asia Pacific Region Bulletin*, 15, 16
Stobie, R. S. 1995 *Astrophysics and Space Science*, 230, 9

West African International Summer School for Young Astronomers

Linda E. Strubbe[1] and Bonaventure Okere[2]

[1]Science Teaching & Learning Fellow, University of British Columbia,
6224 Agricultural Road, Vancouver, British Columbia, V6T 1Z1, Canada
email: `linda@phas.ubc.ca`

[2]NASRDA Centre for Basic Space Science,
Nsukka, Nigeria
email: `ibokere2001@yahoo.com`

The West African International Summer School for Young Astronomers (WAISSYA) is a week-long program for university science students and teachers from West Africa to develop their interest in astronomy. The first summer school was held in Abuja, Nigeria, in 2013; the second Summer School was held in Nsukka, Nigeria, in July 2015. West Africa has a large number of students interested in science, but a paucity of facilities or interest from funding bodies in developing West African astronomy. Our broad goals for the WAISSYA program are: (1) to introduce West African students to astronomy; (2) to exchange ideas about teaching and learning in West Africa and abroad; and (3) to continue building a sustained astronomy partnership between West Africa and Canada. We now briefly describe three defining aspects of WAISSYA 2015.

• We designed our curriculum from evidence-based teaching strategies, emphasizing active learning and scientific thinking (e.g., Wieman 2014, NRC 2012, Chinn & Malhotra 2002). The curriculum included a two-day hands-on inquiry activity on the Cosmic Distance Ladder, interactive lectures with group discussions and multiple-choice voting, and a project where students designed their own astronomy teaching module to teach at their home institutes after the Summer School.

• The instructional team comprised astronomers from Canada, Nigeria, Germany, and Gabon. We held a workshop for instructors before the school, with goals of sharing evidence-based teaching strategies, and building collaboration and sustainability in the team. Instructors formed international pairs who collaborated to design their lessons using the teaching strategies we modeled and discussed in the instructor workshop. This paradigm builds on research on paired teaching for professional development in other contexts (e.g., Henderson *et al.* 2009, Stang & Strubbe 2015).

• We are evaluating outcomes of the Summer School in several ways. Students took the Colorado Learning Attitudes about Science Survey (Adams *et al.* 2006) and a set of astronomy concept inventory questions (e.g., Hufnagel 2002) at the beginning and end of the Summer School. Students wrote daily reflections about their learning, and also completed a school evaluation survey. We collected reflections from the instructors as well. All of these data are currently being analyzed.

The next WAISSYA will be held in Accra, Ghana, in 2017.
We are pleased to acknowledge the WAISSYA 2015 teaching team: R. Eze, W. Kerzendorf, V. Murray, T.D.C. Nguyen, F. Odoh, P. Okouma, E. Sudum, J. Zhang, and the authors Linda E. Strubbe and Bonaventure Okere. We are grateful for funding and support

from: the IAU Office of Astronomy for Development, the Dunlap Institute for Astronomy and Astrophysics, the Canadian Institute for Theoretical Astrophysics, the NASRDA Centre for Basic Space Science, the European Southern Observatory, and the University of British Columbia.

Web-based Teaching Radio Interferometer for Africa

Claude Carignan[1,2] and Yannick Libert[1]

[1] Department of Astronomy, University of Cape Town, P.B. X3, Rondebosch 7701, South Africa
[2] Observatoire d'Astrophysique de l'Université de Ouagadougou, Burkina Faso
email: ccarignan@ast.uct.ac.za

Abstract. This presentation describes the web-based Teaching Radio Interferometer being built on the campus of the University of Cape Town, in South Africa, to train the future users of the African VLBI (Very Long Baseline Interferometry) Network (AVN).

Keywords. instrumentation: interferometers, techniques: interferometric

1. Overview

Practical training for the future use of the African VLBI Network (AVN) or any VLBI experiment starts by understanding the basic principles of radio observations and radio interferometry. The aim of this project is to build a basic interferometer that could be used remotely via a web interface from any country on the African continent. This should turn out as a much less expensive and much more efficient way to train AVN researchers from SKA partner countries to the principles of radio astronomy and to interferometric data analysis. The idea is based on the EUHOU (European Hands-On Universe) project, already very successful in Europe. The former EUHOU manager, Dr Yannick Libert, arrived for a 3 years postdoc with Prof Claude Carignan at the University of Cape Town to implement the same project on the African continent (AHI: African Hands-on Interferometry). Besides the use of AHI for the AVN researchers, this web-based system could be used be any undergraduate program on radio astronomical techniques across the African continent as the EUHOU is used all across Europe.

Our goal is to build a robust instrument that can be used remotely in classrooms on the African continent for years to come. The dishes will be two 2.3m parabolas hooked up to equatorial mounts. The mounts of the telescopes are computerised and the signal will be sent over a coaxial cable to the receiver box. This box will be linked to a fast fiber optic switch so that it can be accessible by a server installed in the Astronomy department that will provide access to the instrument through the Web. The receivers will be designed for 21cm observations and coupled with a fast processor that will allow interferometric acquisition between the 2 antennas. A server will be installed for storing the data and sharing the control of the instrument over the Web. This project is funded by the AVN training programme. It aims at training students to radio observations techniques all over Africa. The teaching interferometer received R420 000 to be developed.

With the advent of SKA (Square Kilometre Array) and its precursors KAT-7 and MeerKAT, and the development of the AVN, the field of radio astronomy is undergoing a massive development. To sustain this growth, we need to develop efficient training programmes. Successful realisation, demonstration and fruitful usage of the facility on the University of Cape Town premises during 2015 can lead to realisation of similar facilities in the Universities of SKA partner countries. Botswana and Mauritius already showed interest in replicating this project.

Space Awareness: Inspiring A New Generation of Space Explorers

Pedro Russo[1]

(on behalf of the Space Awareness Project Consortium)
[1] Leiden Observatory / Leiden University, the Netherlands
email: russo@strw.leidenuniv.nl

1. Overview

Space Awareness (EUSPACE-AWE) will use the excitement of space to attract young people into science and technology and stimulate European and global citizenship. Our main goal is to increase the number of young people that choose space-related careers.

EUSPACE-AWE targets diverse groups that are influential in the complex processes that lead to career decisions. The project will show teenagers the opportunities offered by space science and engineering and inspire primary-school children when their curiosity is high, their value systems are being formed and seeds of future aspirations are being sown.

A carefully crafted portfolio of EUSPACE-AWE activities will: 1. Acquaint young people with topical cutting-edge research and "role-model" engineers, 2. Demonstrate to teachers the power of space as a motivational tool and the opportunities offered by space careers, 3. Provide a repository of innovative peer-reviewed educational resources, including toolkits highlighting seductive aspects of Galileo and Copernicus and 4. Set up a "space career hub" and challenging contest that will appeal to teenagers. Attention will be paid to stimulating interest amongst girls and ethnic minorities and reaching children in underprivileged communities, where most talent is wasted. Targeting policy makers via high-impact events will help ensure sustainability and demonstrate the social value of the space programme.

EUSPACE-AWE shall maximise cost effectiveness of the activities by joining with and supplementing existing space teacher training networks and courses and exploiting and expanding infrastructures of the proven FP7-Space projects, EU Universe Awareness for young children and Odysseus for teenagers.

EUSPACE-AWE will join with and complement existing space-education programs and be coordinated closely with the European Space Agency (ESA). We shall reach European teachers, schools and national curricula through the national host organisations of ESA's European Space Education Resource Offices (ESEROs) and the extensive networks of European Schoolnet, (Scientix) and Universe Awareness (UNAWE). Designated EUSPACE-AWE nodes will provide curriculum and resource localisation and test beds for academic evaluation. A partnership with the IAU Office of Astronomy for Development in Cape Town (South Africa) will ensure global reach.

More information: www.space-awareness.org

GalileoMobile, sharing astronomy with students and teachers around the world

Sandra Benitez-Herrera and Patricia F. Spinelli

Museum of Astronomy of Rio de Janeiro, Rua General Bruce, 586, Rio de Janeiro, Brazil
email: sandraherrera@mast.br

GalileoMobile is a non-profit, itinerant, science outreach initiative that brings Astronomy closer to young people in areas with little or no access to outreach programs. We perform astronomy-related activities in schools and communities we visit and encourage follow-up activities through teacher training workshops and the donation of telescopes and other educational resources. GalileoMobile also extends its impact to a worldwide audience through deliverable products. Our work is shared worldwide through the production of documentaries, books and a wide range of Internet resources (OfficialWebsite - www.galileo-mobile.org - and Blog, Facebook page, Google+,Twitter, Youtube and Vimeo). GalileoMobile is an unprecedented initiative promoting science knowledge and the interaction beyond borders through Astronomy while raising awareness for the diversity of human cultures, conveying the message of "unity under the same sky". We take advantage of the local astronomical culture of the visited communities to establish a dialogue between different ways of understanding the world and to share different types of knowledge (historic, scientific, anthropological...), encouraging a process of mutual learning.

GalileoMobile is composed of 13 volunteer team members and more than 30 collaborators from different countries. Several of our expeditions were initially motivated by the desire of one of our volunteers to bring GalileoMobile to their country of origin, in communities where they had witnessed first-hand the lack of access to science outreach programs. Since its creation in 2008, we have organised expeditions in Chile, Bolivia and Peru (2009), Bolivia (2012), India (2012) and Uganda (2013), Brazil and Bolivia (2014), Colombia (2014) and extended actions in Portugal (2012, 2013), Nepal (2013), Guatemala (2013), Dominican Republic (2013), the United States (2013) and Haiti (2014) reaching over 12,400 students; 1,300 teachers and 1,700 community persons. Our efforts and activities have been shared with the public in over 80 conferences and talks, including a TEDx talk. Today, we continue our efforts with the support of Universe Awareness (UNAWE) and the collaboration of Galileo Teacher Training Program (GTTP) and A Touch of Universe (ATU).

The Brazil-Bolivia (BraBo) expedition preceded the celebration of the International Year of Light (IYL2015) and thus the activities performed were also oriented to the understanding of astronomical observations as a manifestation of cosmic light. To support these efforts, the documentary "Light-Year" was recorded during the expedition which had both the aim to portray the encounter of the scientific and traditional interpretation and the presence of light through a beautiful manifestation. The BraBo expedition performed the activities in schools and communities developed during previous expeditions, and included four main innovations: (1) the implementation of GalileoMobile activities for visually impaired people; (2) the organisation of roundtable debates and Astronomy training to undergraduate students of local universities in the field of Natural Sciences, and (3) the visit to indigenous schools. Our new initiative for 2015 is the Constellation Project (www.constellationproject.org), which aims to establish a South American network of schools committed to the long-term organisation of astronomical outreach activities amongst their pupils and local communities. This project has been supported by the Cosmic Light Project of the International Astronomical Union (IAU) and is being partially funded by the Office for Astronomy Development.

Ten years of RELEA: Achievements and challenges for astronomy education development

P. S. Bretones[1], L. C. Jafelice[2] and J. E. Horvath[3]

[1] Departamento de Metodologia de Ensino, Universidade Federal de São Carlos, São Carlos, Brazil

[2] Departamento de Fsica, Universidade Federal do Rio Grande do Norte, Natal, Brazil

[3] Instituto de Astronomia, Geofísica e Ciências Atmosféricas, Universidade de São Paulo, São Paulo, Brazil

When an area of education, and more particularly the research within this area, is aimed to development, one of the basic requirements is the existence of a regular publication that accounts for the scientific production in that area. This study aims to analyze 10 years of Latin-American Journal of Astronomy Education (RELEA) [http://www.relea.ufscar.br/].

The 75 articles published in 18 editions were analyzed. The acceptance rate of the articles is 60.2%, while the refereeing time average is about 14 days, the annual average is 7.5 articles and the issue average is 4.2 articles. A large majority of the articles originated from Brazil (61-81.3% of the published). The rest of the works were contributed by Argentina (6-8.0%), Uruguay (2-2.7%), USA (3-4.0%), Spain (2-2.7%), and New Zealand (1-1.3%). Concerning the school grade level or public outreach addressed in the papers, the largest percentage were related to university education (28.0%) and to high school (28.0%), while 25.3% of works deal with unspecified school grade level. Concerning the focus of study on education there is a predominance of papers on learning and teaching of astronomy education (34.7%), followed by studies of students understanding (17.3%) and development and discussion of teaching materials (13.3%). Other works were found related to: Teacher's education (8.0%), studies on teachers understanding (8.0%), curricular discussions/programs in astronomy (6.7%), studies of history and philosophy of science (5.3%), non-school programs (4.0%). With respect to the contents, most studies do not deal with specific topics in astronomy and we labeled these as General (33.3%). Studies of Sun-Earth-Moon System (26.7%) and Solar System (18.7%) are the second and the third largest groups. Less frequent are studies which include history of astronomy, positional astronomy and constellations as the main. These results present trends and shortcomings of the production.

This journal is now consolidated but the number of articles is still relatively small probably because of the lack of publishing tradition in the area. Its future challenges include how to increase the number of published articles; specially from Latin American countries, and how to bring in the issues and subjects not addressed until now; training of the community for a better quality of the submitted articles and the recent restructure of the Editorial Board. It is also considered the possibility of encouraging graduate studies, new lines of research in astronomy education, and dissemination of material in universities and schools for teachers and students.

Finally, existing future possibilities are given by the IAU development programs. For example, more article submission from Portuguese-speaking countries with the support of Regional Nodes and Language Expertise Centers, advertising by the OAD and opportunities for volunteer IAU members and global projects for the development of astronomy education.

Task Force 3 Discussions

Sze-leung Cheung

IAU Office for Astronomy Outreach, Mitaka, Tokyo, Japan
email: cheungszeleung@iau.org

Abstract. Task Force 3 is focused on public outreach. A number of projects have been funded by the OAD in the past, and in this meeting, the opportunity was provided for both funded projects as well as other non-funded projects with similar objectives.Three projects were presented during the session which sparked discussion in this area.The following outlines the talks.

Keywords. OAD, astronomy, development

1. Star parties in Mexico, extended to Colombia and China

by Silvia Torres-Peimbert, Jose Franco Institution: Universidad Nacional Autonoma de Mexico

Sparked by the enthusiasm of the International Year of Astronomy, a set of simultaneous star parties have been held since 2008 in several cities in Mexico. These star parties have raised big expectations among the population and they have been repeated at least yearly. The activity has increased in size and participating sites, which was very successful. The most recent one took place on November 29th 2014, and it included 55 locations across Mexico as well as 5 in Colombia and one in China. To organize this activity a Mexican National Committee was created formed by several universities, the French Embassy, related industries and astronomical societies.

Overall, this was a very successful initiative and triggered a great public interest in astronomy.

2. Reflections on a Multi-stakeholder National Campaign in India around Comet ISON

By Prajval Shastri Institution: Indian Institute of Astrophysics

Astronomy has been repeatedly demonstrated to be an effective vehicle to promote learning science by doing and to propagate a scientific temper among the public. In this spirit, the efforts undertaken in India during the international Year of Astronomy enabled the building of collaborations between professional astrophysicists, amateur astronomers, science teachers, science activists, theatre artistes and artist-designers. In the six years since, these networks have managed to not only sustain, but expand in reach, and have built focused attention of school-going children around astronomical events. The activities that were part of the build-up to both the Transit of Venus (2012) and the arrival of Comet ISON (2013) will be described.

A somewhat informal consortium led the campaign around the comet. The campaigners were enabled by open-source digitally based resource material available in both English and regional languages so that volunteers could reproduce the materials locally and to reach schools and tens of thousands of students across India. Since it involved so many schools and lack of monitoring and quality control was the major issues.

3. Dark Skies Africa: a Prototype Project with the IAU Office of Astronomy for Development

by Constance Elaine Walker, Daniel Tellez, Stephen M. Pompea Institution: National Optical Astronomy Observatory

The IAU's Office of Astronomy for Development (OAD) awarded the National Optical Astronomy Observatory (NOAO) with a grant to deliver a "Dark Skies Outreach to Sub-Saharan Africa" program to institutions in 12 African countries during 2013: Algeria, Nigeria, Rwanda, Tanzania, Ghana, Zambia, South Africa, Ethiopia, Gabon, Kenya, Namibia and Senegal. The program helped students identify wasteful and inefficient lighting and provided ways to reduce consumption and to keep energy costs in check. The goal was to inspire students to be responsible stewards in helping their community safeguard one of Africa's natural resources - a dark night sky. Thirteen kits made by the NOAO Education and Public Outreach group were sent to coordinators at university, science center and planetarium-type institutions in the 12 countries and to the IAU OAD. The program's kit included complete instructional guides and supplies for six hands-on activities (e.g., on the importance of shielding lights and using energy efficient bulbs) and a project on energy conservation and responsible lighting (through energy audits). The activities were taught to the coordinators in a series of six Google+ Hangout sessions scheduled from June to mid-November. The coordinators at the institutions in turn trained local teachers in junior and senior high schools. The Google+ Hangout sessions also included instruction on carrying out evaluations. From the end of November until mid-December students from the different African countries shared final class projects (such as posters or powerpoints) on the program's website. The entire program was designed to help coordinators and educators work with students, parents and the community to identify dark sky resource, lighting and energy issues and to assess their status, efficiency and effectiveness.

Although precautions were taken to minimize difficulties with Customs, challenges arose with half of the kits, even though official invoices, letters specifying the kits were for educational purposes only and not for resale, and detailed content lists and prices were included in each kit.

Attendance by all at the Google+ Hangout sessions has also been a challenge. To accommodate schedules, 3 different dates were offered for the 1st Google+ Hangout session and 2 different dates for the subsequent Google+ Hangouts. For those unable to attend, the sessions were all recorded and placed on-line.

The third major problem was the communication and poor internet facilities across some sites.

4. Concluding remarks

Although these were just a few of the many projects evisaged under Task Force 3 (Astronomy for the Public) it gave the audience a taste of the type of activity in this area. The OAD continues to work closely with the IAU Office for Astronomy Outreach in order to further explore and expand on these types of projects and more.

Star Parties in Mexico extended to Colombia and China

Silvia Torres-Peimbert and José Franco

Instituto de Astronomía, Universidad Nacional Autónoma de México, México

The preparation for The International Year of Astronomy 2009 stirred our interest in preparing star parties in Mexico. The lunar eclipse of February 20th 2008 was the perfect event for the first massive observation in Mexico City that attracted over 25,000 people. To accompany this event there were additional attractions: a massive astronomical lecture, more than 100 telescopes were set up for people to watch the sky, exhibits of astronomical images, children hands-on projects, rock concert, dance performance, and chats with astronomers. Already in 2009 a collective program was organized to involve more than 30 sites in Mexico to hold star parties at the same time once a year. These star parties were more in the spirit of science fairs, that include lectures, astronomy exhibits, children projects, as well as concerts and other cultural displays. The scope of each one of them depended on the local support from volunteers and from the local authorities. After the International Year of Astronomy the group that organized these star parties decided to continue its activities. The main attraction in these fairs has been the opportunity to see the Moon, Jupiter and Saturn (if observable) through a telescope. For this program the presence of the amateur astronomers has been crucial. They have brought their instruments to the sites and have generously taught the public how to look through the telescopes and pointed out to the interesting features on the sky.

Since then, the program *Noche de las Estrellas* has grown to include a coordinated project of many sites in Mexico to hold star parties on the same night. This effort has been very successful, and been held once every year. From 2011 to 2014 the number of sites grew in number from 41 to 57 to include most of the large cities in Mexico, and some of the smaller ones. There has been an army of volunteers that have made this program possible. The total number of volunteers (or staff) has been approximately 5,000 each time. The number of telescopes involved on average has been around 1,500. And the average number of people that have attended these science fairs has been around 160,000 each time. For the last 2 years the organizing group has extended its membership to include the participation of sites beyond our country. The most recent star party, in November 2014 included Bogota, Colombia and Beijing, China. In Bogota the planetarium hosted this activity, and it was extended to visits with telescopes to 5 schools in the city; while in Beijing the planetarium received 5,200 visitors and had lectures and additional activities.

Reflections on a Multi-stakeholder National Campaign in India around Comet ISON

Prajval Shastri[1] **(on behalf of the *Eyes on ISON* campaign team)**

[1]Indian Institute of Astrophysics, Bengaluru 560034 India
email: pshastri@iiap.res.in

Abstract. A country-wide awareness campaign was conducted in India on the trail of Comet ISON, which built upon the networks created during IYA2009, thus maximising its reach.

A striking trend in modern-day India is the salience of technology at all socio-economic levels. There has not been a concomitant rise in scientific temper, however, and superstitions are prevalent even among the educated or privileged classes. A parallel trend in most schools is that learning by doing and discovering is on the back burner, with an emphasis on rote learning instead. This is despite government policy on education embodied in the National Curriculum Framework 2005 that aims at equality of outcome regardless of socio-economic background, which perforce requires activity-based learning. In this context, a national awareness campaign around Comet ISON was designed, particularly aimed at teachers and students. The campaign built heavily upon the networks and collaborations created preparatory to IYA2009 spearheaded by the Indian Institute of Astrophysics (IIA Newsletters 2009, 2010). The hallmarks of the campaign were:

- It was co-ordinated by a loose consortium of practitioners from professional astrophysics and science communication, amateur astronomy and peoples' science movements,
- It was built around a cascading series of training workshops with a common open-source curriculum, a workshop primer and resource material generated by the campaign,
- The resource material (activity books, posters, multi-media presentations) was generated in several regional languages on the web (and continues to be a long-term resource),
- It leveraged the world-wide web to make the open-source multi-lingual high-resolution material digitised and downloadable, for poster printing and multi-media use,
- Its workshop primer emphaisized inclusive organization and participation.

Details of the team, training workshops, curricula and resource materials are on the Eyes on ISON campaign website (http://ison.metastudio.org)†; also *cf.* on-line presentation.

Outputs & Outcomes. The curriculum for the training workshops included talks on the solar system, comets and the history of Indian astronomy and astrology (along with multi-media presentations); activities on day-time astronomy (along with videos), comet craft, astronomy role playing and night-sky watching (along with activity books); discussion activities on frequently asked questions by the public on astronomy and astrology. The three national-level workshops trained a total of 200 trainers. The 38 state-level workshops conducted in regional languages inturn trained over 2000 train-

---

† HBCSE-TIFR hosted the brainstorming meeting that kicked off the campaign. *Vigyan Prasar*, Dept. of Science & Technology, Govt. of India, funded this meeting and the national-level workshops. The state-level workshops were largely funded by DST-NCSTC. Funds for several state-level and all district-level workshops as well as other events were locally generated.

ers, who also constitute a network which is a resource for future campaigns. While the events of the campaign were heavily under-reported, at least 130 district-level workshops trained over 5000 communicators country-wide, which includes a large fraction of school teachers.

Dark Skies Africa

Constance E Walker, Daniel Tellez and Stephen M Pompea

National Optical Astronomy Observatory
950 N Cherry Ave, Tucson, AZ 85719, USA
email: cwalker@noao.edu

Abstract. The first IAU Office of Astronomy for Development Task Force 3 project on light pollution is described along with evaluations and recommendations for future projects.

Keywords. Light Pollution, Energy Conservation, Quality Lighting, Education Outreach

1. Overview

The IAU Office of Astronomy for Development (OAD) awarded the National Optical Astronomy Observatory (NOAO) with a grant to deliver a "Dark Skies Outreach to Sub-Saharan Africa" program to institutions in 12 African countries during 2013. The program's 1st goal was to help students identify wasteful and inefficient lighting and provide ways to reduce consumption and to keep energy costs in check. The 2nd goal was to inspire students to be responsible stewards in helping their community safeguard one of Africa's natural resources - a dark night sky. The 12 countries chosen were based on 3 criteria: 1) coordinators were English-speaking and willing to train teachers, 2) coordinators, teachers, and students had to have some computer and internet access, and 3) the countries should for the most part be in sub-Saharan Africa. During the 1st half of 2013, 13 kits were designed and produced by the NOAO Education and Public Outreach group and sent to the 12 coordinators (who were located at universities, science centers, and a planetarium-type institution) and to the IAU OAD. The program's kit included complete instructional guides and supplies for 6 activities and a project on energy conservation and responsible lighting. From June through November, the 6 activities and project were taught to the coordinators and some of the teachers in a series of 6 Google+ Hangout sessions. One Google+ Hangout session included instruction on carrying out evaluations. All Google+ Hangout sessions were recorded for future viewing at any time. During the same period, the 12 coordinators trained local teachers in junior and senior high schools. From November until the following February, students from the different African countries undertook final class projects and shared them on the program's PBWorks website. Also shared on the program's website is every document connected to the program. Everyone in the program will continue to have access to the web site.

2. Findings from the External Evaluator

Findings fro the external evaluator include: 1) Activities and supporting materials (e.g. kits and Google Hangout sessions) were implemented successfully in almost 60% of partnering countries. 2) Coordinators trained by NOAO, and the teachers they trained, were equipped with the knowledge and skills needed to provide access to quality STEM learning experiences for students, regardless of their geographical location, academic ability and socioeconomic status. Overall, members from seven countries were able to perform key activities of the project. 3) There were substantial student outcomes, including increases in target knowledge, and engagement as a result of their participation in the project. Additionally, active students created excellent project outputs demonstrating their increased knowledge and engagements. 4) There was a spectrum of participation in the project by active partner countries. Some were incredibly active, despite barriers, and some did not engage in the project at all. Those who did participate were highly successful, and 5) There were many challenges and lessons learned through the course of the project that will be useful in future projects working with diverse and distant populations.

3. Recommendations for Future Projects

Recommendations for future projects with diverse participants include: 1) Use short applications to ensure that coordinators are interested and committed to the project. 2) Be aware of issues with customs when mailing kits out to different countries so ensure efficient delivery of materials. 3) Use pre-recorded tutorial videos of activities and use Google hangout time to answer questions. 4) Provide multiple ways for coordinators to get information online, and 5) Use a variety of techniques to gather data to measure success of the project including online surveys, paper surveys, and analysis of artifacts submitted by coordinators from each country.

The East Asian Regional Office of Astronomy for Development

Richard de Grijs[1], Ziping Zhang[2] and Jinhua He[3]

[1] Kavli Institute for Astronomy & Astrophysics, Peking University, Yi He Yuan Lu 5, Hai Dian District, Beijing 100871, China
email: grijs@pku.edu.cn
[2] Beijing Planetarium, 138 Xizhimenwai Dajie, Xicheng District, Beijing 100044, China
[3] Yunnan Observatories, Chinese Academy of Sciences, Phoenix Mountain, East District, Kunming 650011, China

Abstract. At the 2012 General Assembly of the International Astronomical Union (IAU), the Office of Astronomy for Development announced a number of exciting new partnerships to assist with the IAU's decadal strategic plan (2010–2020). These landmark decisions included establishing a new coordinating centre that aims at using astronomy as a tool for development in East Asia. The agreement covers two important functions. One is known as a Regional Node, which entails the coordination of astronomy-for-development activities in countries within the general geographical region of East Asia. The other is known as a Language Expertise Centre which deals with all aspects relating to (mainly) the Chinese language and culture. The impact of the latter may obviously spread well beyond the geographical region to other parts of the world. Here we provide an update of the achievements and aims of the East Asian Office of Astronomy for Development.

Keywords. miscellaneous, sociology of astronomy

1. Facilitation of Astronomy for Development in the far East

An important component of the International Astronomical Union's (IAU) Strategic Plan 2010–2020 is the adoption of a "bottom-up" approach, with a substantial degree of decentralisation. This involves the appointment of regional development coordinators and the designation of regional nodes. The first of these pioneering agreements, concerning a coordinating centre established in the East Asian region (in China), was signed on Tuesday, 21 August, 2012. The main institutes involved in the East Asian Regional Office of Astronomy for Development (EA-ROAD) are the Kavli Institute for Astronomy and Astrophysics (KIAA, Peking University) in Beijing, Beijing Planetarium and Yunnan Observatories, Chinese Academy of Sciences (YNAO). Our institutes are supported in our efforts by a number of important partners, including the National Astronomical Observatories of the Chinese Academy of Sciences (NAOC), the East Asian Observatory (EAO), the National University of Mongolia (NUM) and Pyongyang Astronomical Observatory (PAO).

Our key geographical focus area of interest comprises mainland China, Mongolia and the Democratic People's Republic of Korea (DPRK). Through the EAO, we have indirect links to Japan (and the IAU Office of Public Outreach at the National Astronomical Observatories of Japan; NAOJ), Taiwan (Academia Sinica Institute of Astronomy and Astrophysics), and the Republic of Korea (Korean Astronomy and Space Science Institute). This area covers of order a quarter of the world's population, and given the small size of the EA-ROAD's steering committee, our role is mainly one of facilitation and information exchange rather than that of a major driver of practical implementation. Nevertheless, we do not consider the boundaries to our main focus area to imply exclusion of activities in neighbouring countries if and when the opportunity arises to make a difference there in the context of our "Astronomy for Development" remit.

The majority of EA-ROAD facilitation and active engagement during the first triennium of its operation occurred in China. Astronomy in China is concentrated in a number of large institutions under the umbrella of the Chinese Academy of Sciences (CAS) and in a few university departments (predominantly in Beijing, Nanjing, and Xiamen). CAS institutes include, among others, NAOC, YNAO, Shanghai Astronomical Observatory (SHAO), Purple Mountain Observatory (Nanjing), the Nanjing Institute of Astronomical Optical Technology, and Xinjiang Astronomical Observatory. We have invited representatives from SHAO and NUM to join the EA-ROAD's steering committee. We have also established contacts with the team running the new FAST radio telescope, under development in Guizhou province, aiming at setting up a teacher training event in collaboration with a number of tertiary education colleges we incorporated as associated partners, thanks to significant efforts by our YNAO-based steering committee member.

Despite this focus on China, we have also engaged positively with our colleagues in both Mongolia and the DPRK. We organised a summer school and the first ever international conference aimed at professional astrophysicists at the NUM in September 2014. Although Mongolian astronomy dates back thousands of years, the country has only been an interim member of the IAU since 2006. The small Mongolian astronomical community has interacted with foreign organisations through several workshops and schools, most prominently the 2008 Astronomical Summer School, held in Ulaanbaatar, and also through interaction with NAOJ. The Mongolian astronomical community is keen to develop their research abilities and exposure to attain an internationally competitive level.

The long-term aims of the DPRK astronomical community are to increase the level of astronomy in all areas of research, education and public outreach to an internationally acceptable high level. This includes conducting cutting-edge professional research and astronomy education of the general public to a sufficient level to secure the benefit of astronomy as a tool of national capacity building. To support this aim, we hosted several researchers from PAO for 6 months each at NAOC's Huairou Solar Observatory near Beijing. In addition, two of their colleagues spent 6 months at Leiden University (Netherlands) thanks to the great efforts of Prof. George Miley.

We have also facilitated Master's and PhD student exchanges with Mongolia, Nepal, Indonesia, Bangladesh and other regional centres. In addition, we have actively worked on establishing links with external partners, including the UK's Institute of Physics, with whom we support monthly science days in underresourced schools for children from so-called migrant families in Beijing, and with the European Union's EURAXESS networking organisation, bringing together Chinese and international researchers.

The EA-ROAD team is very keen to involve a larger fraction of the astronomical community in the wider region. For more details of our achievements and ambitions, we refer the reader to the EA-ROAD website, http://eastasia.astro4dev.org. A full description of the EA-ROAD, in Chinese, and a call for participation was published in *The Amateur Astronomer* (published by Beijing Planetarium; also read by many professional astronomers around the country) of October 2012 (pp. 40–43); an English version is available at http://astro-expat.info/ROAD-EastAsia.pdf. Further details can be obtained by email, eastasia@astro4dev.org or chinese@astro4dev.org (Chinese or English). We will redouble our efforts to engage the community during the annual meeting of the Chinese Astronomical Society, held on the campus of Peking University in October 2015.

Strategies for Astronomy Development in the Southeast Asia

Boonrucksar Soonthornthum

National Astronomical Research Institute of Thailand,
Ministry of Science and Technology

Astronomy is the science in which we study about everything in the universe, in which we derive the properties of celestial objects, planets, stars, galaxies and so on. These properties deduce the laws by which the universe operates. Astronomy is a remarkable subject which has had a profound effect on the human curiosity and inspiration. So, Astronomy can be used as powerful tool for the development of human resources and human capacity buildings on Science and Technology.

During the past few decades, Astronomy in Southeast Asia has developed significantly. Major astronomical facilities such as optical and radio telescopes have been provided and human resources in Astronomy have been developed and increased. Several countries in Southeast Asia have established their national institutes in Astronomy, Space Science and related disciplines, namely, LAPAN in Indonesia, ANGKASA in Malaysia, PAGASA in Philippines and NARIT in Thailand.

One of the important strategies for developing Astronomy is to use the national institute to drive the national policy in promoting astronomy in the country. The national institute can be a center for the coordination of astronomical activities in schools, universities and research institutes. Fund raising for supporting astronomical activities and the construction of the national astronomical infrastructures can be done through the national institute.

National and international collaborations are also a crucial strategy for the development of Astronomy in Southeast Asia. National institutes can create a national network in Astronomy with universities for the cooperation on research, supporting education and public outreach activities. University staff and researchers can cooperate on some mutual interest research topics with joint publications. University students can access the national astronomical facilities and work under co-supervision of researchers of the national institute. Moreover, the national institute can also get the support from universities and schools in the country for the astronomical outreach activities throughout the country.

International collaboration on Astronomy in Southeast Asia is operated through the Southeast Asia Astronomy Network (SEAAN). SEAAN is a powerful mechanism in nurturing the development of Astronomy in Southeast Asia. Several activities have been done through the SEAAN network especially the promotion of human resource development and human capacity building in Astronomy in Southeast Asia through training. Each year, several schools and workshops in Astronomy on different levels were organized for participants in the Southeast Asian countries. Support and cooperation, including experts and funding, came from many well-established astronomical institutes and observatories outside Southeast Asia.

Recently, the Southeast Asian Regional Office of Astronomy for Development (SEA ROAD) was established in Thailand through the appointment of the Office of Astronomy for Development (OAD) under the International Astronomical Union, (IAU). The operation of SEA ROAD, under the support of the OAD, would be a substantial mechanism

Figure 1. Astronomical infrastructures at the national institutes in Southeast Asia

Figure 2. The Southeast Asia Astronomy Network (SEAAN) Meeting

for materializing the cooperation to promote and proliferate Astronomy in Southeast Asia.

East African ROAD

Kelali Tekle

East African Regional Office of Astronomy for Development, Addis Ababa, Ethiopia

Abstract. In the developing world astronomy had been treated as the science of elites. As a result of this overwhelming perception, astronomy compared with other applied sciences has got less attention and its role in development has been insignificant. However, the IAU General Assembly decision in 2009 opened new opportunity for countries and professionals to deeply look into Astronomy and its role in development. Then, the subsequent establishment of regional offices in the developing world is helping countries to integrate astronomy with other earth and space based sciences so as to progressively promote its scientific and development importance. Gradually nations have come to know that space is the frontier of tomorrow and the urgency of preeminence on space frontier starts at primary school and ascends to tertiary education. For this to happen, member nations in east African region have placed STEM education at the center of their education system. For instance, Ethiopian has changed University enrollment strategy to be in favor of science and engineering subjects, i.e. every year seventy percent of new University entrants join science and engineering fields while thirty percent social science and humanities. Such bold actions truly promote astronomy to be conceived as gateway to science and technology. To promote the concept of astronomy for development the East African regional office has actually aligned it activities to be in line with the focus areas identified by the IAU strategy (2010 to 2020).

1. Universities and Research

In the triple helix approach Universities are epicenter of qualified workforce production. Universities incorporate what we call the *three pillars of successful learning: education, perspective and inspiration. Education includes understanding how science uses observation and logical thinking to learn about the world around us; perspective shows students how what they are learning can affect their views on their own lives and on humanity's place in the universe; and inspiration encourages students' want to learn more, and to think about how they personally can help make their world a better place for human kind.* Such knowledge transaction places Universities to be sources of wisdom in Research and Development. For the sake of expanding the knowledge horizon in astronomy, the regional office is closely working with Universities and research institutions. Up until recently it was difficult to offer astronomy and space science courses in universities. Now the situation is progressing very fast: many public Universities have started offering PhD and MSc Programs in astronomy and space science and technology which is a big step towards furtherance of astronomy for development. Entoto Observatory and Research Center (EORC) which is serving as the research center of the regional office is teaching 18 students in PhD and 8 students in MSc. In close consultation with EORC, the regional office has secured two scholarships for East African students in Astronomy.

2. School and Children

The regional office believes that the easiest way to explore global knowledge in astronomy at ease is to open access to education for boys and girls through robust STEM curricula. Since we put more emphasis on STEM subjects, the number of schools involved in space science clubs increased at a faster rate; more and more students are participating in astronomy program activities, parents and teachers are continuously expressing their interest to support space science programs. This is one of the promising activities of the regional office in advancing the science of astronomy for development because future school of astronomy will entirely depend on member countries performance in STEM education at all levels. In all the awareness programs, conference and workshops organized and hosted by the ROAD, students, teachers, academicians, researchers and volunteers have participated, and the office is committed to broaden this experience in member countries.

3. Public Outreach

For the last one year, the regional office has spent much of its time in promoting public awareness towards astronomy for development. Member countries' policy decision-makers, academic and research institutions, private sector and the general public have been the focus in its awareness promoting programs. Specifically:

(a) the regional office and EORC jointly organized and hosted East African Astronomical Society and ROAD granted US$ 5000 for the conference to be successful;

(b) the regional office was invited to participated in the 3rd international conference on "Financing for Development" held in Addis Ababa, Ethiopia and to present a paper under the sub-theme African-European Radio Astronomy platform. ROAD was among the 14 sectors selected by the national exhibition committee to exhibit its program activities along with Ethiopia Space Science Society. The president of the Federal Democratic Republic of Ethiopia who was the guest of honor expressed his hearth felt appreciations and pledged to help the space program to achieve its set out objectives;

(c) The oversight committee decided to approach member countries to own and commit resources for the regional office to play its role in promoting astronomy for development in particular and the space science and technology in general. Accordingly a delegation led by the former deputy prime ministry of Ethiopia and board chair of Entoto observatory and research center visited Uganda, Rwanda and Tanzania. The delegation met with respective ministers in charge of space science and technology in each member state and has agreed to promote space science and technology to be a driving force for sustainable development. The unwavering commitment of member states to advance importance of space science and technology encouraged the delegation to visit the other member states. The visit to other member countries will be made before the end of this year. Onces the visit is over there will be a regional conference to be hosted here tentatively end of January 2016.

(d) The regional office made experience sharing visit to Republic of Korea, USA and has scheduled to visit India this coming November.

The office, under the guidance of the oversight committee, is working to broaden the mandate of reaching member countries to promote astronomy for development in a pragmatic way.

Southern African Office of Astronomy for Development: A New Hub for Astronomy for Development

Moola S. Mutondo[1] and Prospery Simpemba[2]

[1]Department of Biological Science, The Copperbelt University,
P.O. Box 21692, Kitwe, Zambia
email: `moola.mutondo@cbu.ac.zm`

[2]Department of Physics, The Copperbelt University,
P.O. Box 21692 Kitwe, Zambia
email: `prospery.simpemba@cbu.ac.zm`

Abstract. A new Astronomy for Development hub needs innovative tools and programs. SAROAD is developing exciting tools integrating Raspberry Pi technology to bring cost-effective astronomy content to learning centres. SAROAD would also like to report achievements in realizing the IAU's strategic plan. In order to manage, evaluate and coordinate regional IAU (International Astronomical Union) capacity building programmes, including the recruitment and mobilization of volunteers, SAROAD has built an intranet that is accessible to regional members upon request. Using this resource, regional members can see and participate in regional activities. SAROAD has commenced with projects in the three Task Force areas of Universities and Research, Children and Schools and Public Outreach. Under the three Task Force areas, a total of seven projects have commenced in Zambia (some supported by funds from IAU Annual Call for proposals).

Keywords. General

1. Introduction

In January 2008, the IAU started developing a strategic plan for the decade entitled "Astronomy for the Developing World". This plan aimed to use astronomy in stimulating development at all levels, such as education, research and public outreach. A central coordinating "Office of Astronomy for Development" (OAD), hosted in South Africa at the South African Astronomical Observatory (SAAO) was created as a step towards implementation. In August 2012, at the General Assembly in Beijing , an updated version of the strategic plan was released including an updated title, "Astronomy for Development".

Table 1. Overview of SAROAD Current and Future Members

Current Members	Need Representation	LOAD/SAROAD Members	SKA
Zambia	Swaziland	Angola	Ghana
Namibia	Seychelles	Mozambique	Kenya
Botswana	Lesotho		
South Africa	Congo DR		
Mauritius	Malawi		
Zimbabwe			
Madagascar			

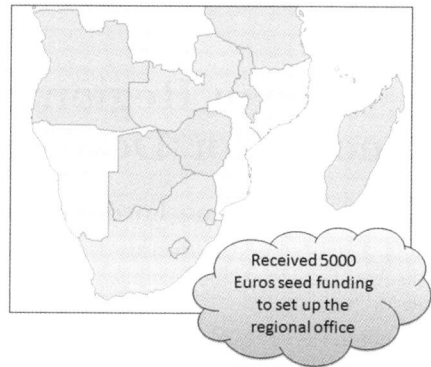

Figure 1. SAROAD Region and Member Countries.

Table 2. Connecting the Astronomy Community in SAROAD Countries

Providing information: INTRANET	Connecting the Community: Website
http://southernafrica.astro4dev.org/	http://sites.google.com/saroadoffice

Table 3. Some Projects under Task Forces

Task Force: Children/Schools	Task Force: Universities / Research	Task Force: Public Outreach
Raspberry Pi Project for schools	Astrolab project (2013-2014)cont.	Solar Astronomy
Science Opera (IYL)	Visiting expert Prof. J. Murphy	SAROAD Road show
	Traditional Astronomical Knowledge	

2. SAROAD hosted at the Copperbelt University (CBU), Zambia

The CBU and the IAU signed an important agreement to host a Southern African regional node of the IAU Office of Astronomy for Development on 8th August 2014. This is the second regional node to be established on the African continent and is an important part of the IAU's decadal strategic plan for astronomy to benefit society. The Governance of SAROAD rests on direction from the Steering committee[F. Tailoka: Chairman (CBU); K. Govender (OAD); P. Whitelock (SAAO); H. Mweene (UNZA); M. Backes (UNAM); L. Tonga (MSEVTEE, Zambia); T. Nemaungani (SKA)]. SAROAD spans the African Southern sub-region (Fig. 1) including the countries presented in Table 1 .

3. SAROAD Activities and Future Outlook

SAROAD has been active for a year and has achieved the results outlined in Table 2 and Table 3. SAROAD needs to work hard to support astronomy at all levels: amateur, professional and outreach. It is now imperative to build on the current good will expressed across the region concerning Astronomy for Development. SAROAD needs to increase and establish strong support offices at the country level in order to achieve its mandate of facilitating Astronomy to contribute to national and regional development.

The new Andean Regional Office of Astronomy for Development

Farid Char[1] and Jaime Forero-Romero[2]

[1] Unidad de Astronomía, Facultad de Cs. Básicas, Universidad de Antofagasta
Avenida U. de Antofagasta 02800, Antofagasta 1270300, Chile
tel.: +56 55 2637596, email: `farid.char@uantof.cl`

[2] Departamento de Física, Universidad de los Andes
Carrera 1 18-10, Bloque Ip., Bogotá, Colombia
tel.: +571 3394949-5183, email: `je.forero@uniandes.edu.co`

Abstract. The Andean Regional Office of Astronomy for Development (ROAD) is a new effort in South America to serve several goals in astronomical development. The six countries in the Andean ROAD (Bolivia, Colombia, Chile, Ecuador, Peru and Venezuela) represent a common language block in the region. They work together to develop strategies to strengthen the professional research, education and popularization of astronomy. Our current Working Structure comprises a ROAD Coordinator and one Coordinators in each Task Force. Here we describe the main points of the ROAD's current action plan.

Keywords. andean, office, development

1. Introduction

The countries in the Andean Region (Bolivia, Colombia, Chile, Ecuador, Peru and Venezuela) represent a common language block in South America, sharing a common vision about the importance of cultural development that can be achieved through regional cooperation and institutional commitment. The Andean Regional Office of Astronomy for Development (ROAD) serves this goal - strengthening ongoing collaboration efforts, creating channels of communication and developing new strategies to exchange knowledge and human resources. The participating institutions of the Andean ROAD will share research, educational and outreach visions and activities. We expect to increase the engagement with astronomy in all the Andean ROAD countries, leveraging the special role of Chile with its great conditions for astronomy and the involvement of many professional observatories, universities and astronomical institutions.

2. Overview of Task Forces

The Andean ROAD is under supervision of the Office of Astronomy for Development (OAD) of the IAU. The main goal of the Andean ROAD is to guarantee and strengthen effective methods of communication between the representatives and coordinators of global, regional and local projects implemented in the Andean countries, through the three Task Forces established by the OAD : (i) Astronomy for Universities and Research, (ii) Astronomy for Children and Schools and (iii) Astronomy for the Public. The ROAD Main Coordinator is Jaime Forero-Romero, PhD, from the Universidad de los Andes (Bogotá, Colombia). The Andean ROAD's plans include several projects and ideas for each Task Forces, under supervision of a single representative, as follows:

Task Force 1. *Coordinator: Germán Chaparro, PhD (Universidad ECCI, Bogotá, Colombia).*

- Andean School on Astronomy/Astrophysics. The main objective is to organize a school for advanced undergraduate and graduate students. The first version of this school was held in Quito (Ecuador) during December 2014 with great success, and we expect to develop a new version during 2016, to be held in Lima (Perú), as a joint effort between Perú and Chile.
- Andean Graduate Program. The main objective is to determine the feasibility of creating and funding an Andean graduate program. This could be a great opportunity to provide powerful experiences in astronomy for those countries without formal programs.
- Massive Open Online Courses. The main objective is to create a MOOC at the introductory undergraduate level and general public. This MOOC could also have a version for the general public, and can be considered as a TF3 project as well.

Task Force 2. *Coordinator: Ángela Pérez (Parque Explora, Medellín, Colombia)*.
- Virtual Training Sessions. The main objective is to create virtual training and activities open to teachers and students, by developing a virtual platform to hold virtual workshops on information technologies applied to astronomy and space sciences.
- Special teachings. Design teaching material to work with visually impaired students. The first step will be a comprehensive research on this kind of material and current experiences. Later on will be defined the best way to produce and distribute these materials.
- Annual TF2 meeting. An annual meeting to gather all the collaborators in the Task Force, looking for new collaborations and alliances. The venue will change every year, with a different key topic.

Task Force 3. *Coordinator: Farid Char (Sociedad Chilena de Astronomía, Chile)*.
- Development for planetariums. The main objective is to develop special shows for planetariums in the region, with the aim of highlighting ancestral traditions and the particular cosmovision from many communities in South America.
- Communicating Astronomy with the Public. The main objective is to organize a periodic meeting to gather all the collaborators in the Task Force. This should be a very powerful platform for all people involved in communicating astronomy to the public. The first meeting will take place at the same time as the International CAP meeting to be held in Medellín (16-20 May 2016), only a few months before the XV Latin American Regional IAU Meeting, in Cartagena de Indias (3-7 October 2016).

3. The role of Chile and future plans

Due to the large size of the Chilean professional astronomical community and its special role in global astronomy, the interaction to coordinate activities in Chile will be done through the Sociedad Chilena de Astronomía (SOCHIAS). SOCHIAS has designated specific representatives per Task Force, and a main coordinator of the Chilean branch. The physical address of the Chilean branch (formally Oficina Nacional de Coordinación) is located in the Unidad de Astronomía of the Universidad de Antofagasta.

We expect that the Andean ROAD will improve the overall state of astronomy development in our region following the vision of the IAU Strategic Plan 2010-2020 International Astronomical Union (IAU, 2012).

More information about our future plans, can be accessed through the official website of the Andean ROAD http://andean-road.uniandes.edu.co.

References

International Astronomical Union 2012, *Astronomy for Development Strategic Plan 2010-2020*.

Portuguese Language Expertise Center for the OAD

Rosa Doran[1], Lina Canas[1], Sara Anjos[1], Thilina Heenatigala[1], João Retrê[2], José Afonso[2] and Ana Alves[2]

[1]NUCLIO - Núcleo Interactivo de Astronomia,
Largo dos Topázios, 48, 3 Frt, PT2785-817 S. D. Rana, Portugal
email: rosa.doran@nuclio.pt

[2]Instituto de Astrofísica e Ciências do Espaço,
Observatório Astronómico de Lisboa,
Tapada da Ajuda, PT1349-018 Lisboa, Portugal

Abstract. Supporting the use of astronomy as a tool for development in specific regions and languages, the International Astronomical Union's (IAU) Office of Astronomy for Development (OAD) has established a Portuguese 'Language Expertise Centre for the OAD' (PLOAD), hosted at Núcleo Interactivo de Astronomia (NUCLIO), in collaboration with the Institute of Astrophysics and Space Sciences (IA) in Portugal. The centre is one of the new coordinating offices announced at the IAU General Assembly in Honolulu, Hawaii on 13 August 2015.

Keywords. Miscellaneous, sociology of astronomy

"The office in Portugal will be the first of the OAD's regional coordination offices to focus exclusively on a particular language. The Portuguese language region is an important global community with a strong connection across several borders - the LOAD will maximise on this in order to realise astronomy-for-development ambitions" says Kevin Govender, director of OAD. The president of NUCLIO, Rosa Doran, adds "The LOAD will be essential to build bridges between the diverse stakeholders in the different member countries. Members already committed are very hopeful that this will be a strong partnership that will open a variety of opportunities for country members and a privileged place to exchange best practices. NUCLIO will be the first coordinator and we hope to design an effective road map to help strengthen communication among countries and set the example for other similar initiatives in the future." João Retrê of IA adds "This is an excellent opportunity for Portuguese speaking countries around the world to work together developing resources, sharing ideas despite the borders and the cultural differences. The Instituto de Astrofísica e Ciências do Espaço will support the LOAD to fulfil its goals and vision."

The Portuguese-speaking countries are the home for more than 240 million people located across the globe, having cultural similarities and a shared history. Many children in some of these countries are out-of-school youth and the overall communities are scarce in technological resources, particularly in Africa and South America. Despite the natural wealth of some of these nations there is an urgent need for capacity building and improvement of competence profiles.

The vision of the PLOAD is to establish a strong collaborative network across the Portuguese speaking community, from countries where Portuguese is the national language and from other Portuguese speaking communities established across the world. The main objective is to empower these countries, communities and groups with the necessary tools to build their own local support structures and development goals on Astronomy and Space.

The PLOAD mission will be to fulfil its vision by following the guidelines of the IAU strategic plan and build on existing challenges and opportunities towards strong collaborative and active structures. This mission will be accomplished by several steps, starting from a careful research of resources and existing needs and design an effective implementation strategy of aid and support.

The following steps are foreseen:

• Establish solid online connections in each Portuguese speaking countries' teaching communities;

• Make curricula assessment and recommendations for improvements, understanding the already established importance of astronomy on each country in a comprehensive guide and promote science cross-curricular linkage to other fields of learning such as relevant cultural references and cultural importance of the sky in language and history;

• Promote online teacher training workshops supported and enhanced by the already existing networks of international non-profit organizations such as Galileo Teacher Training Program (www.galileoteachers.org) and Universe Awareness (www.unawe.org) as well as to support face-to-face teacher trainings;

• Promote scientific literacy by exploring online free tools and resources and in close collaboration with each of the local contact points understand the material needs of the communities and find appropriate funding to provide technological resources to these communities;

• Support student's exchange between universities within the participant countries according to existing expertise and needs;

• Build-in partnership with Portuguese speaking universities via online learning opportunities by exploring the MOOC concept;

• Promote job opportunities between researchers and research centers and local industries, by fostering internships in the institutions;

• Start collaboration with local entities for the development of research infrastructures, using national and international congress, meetings and other alike opportunities to make local stakeholders aware of the potential of astronomy for the development of the communities;

• Promote global initiatives to engage the general public in astronomy outreach activities and to promote intercultural exchanges between the different Portuguese speaking countries using global initiatives such as Global Astronomy Month, the Earth Hour, the Portuguese Language Day (celebrated in 5th May annually), "Bons Raios te Meçam" (a proposal to celebrate each year the solstices and equinoxes by reproducing Eratosthenes measurements), among others.

The creation of the PLOAD will open a new door to cross country collaboration in the field of science and education. All partners already involved in the effort are highly motivated and with high expectations for the power of this new initiative. The PLOAD will build the perfect stage of collaboration in all levels starting from early school education to innovation in research and industrial development. This new alliance will build on IAU's vision and move beyond its challenges bringing equal opportunities by bridging the existing gaps and offering a rich collaboration and support environment. Equality and no borders will be our moto.

IAU South West Asian ROAD

Areg Mickaelian, Naira Azatyan, Sona Farmanyan and Gor Mikayelyan

Byurakan Astrophysical Observatory (BAO), Byurakan 0213, Aragatzotn Province, Armenia
email: aregmick@yahoo.com

Abstract. Armenia is hosting the IAU South West Asian (SWA) Regional Office of Astronomy for Development (ROAD). It is a county of ancient astronomy and is also rich in modern astronomical facilities and infrastructures, hence may successfully serve as a regional center for various activities. Byurakan Astrophysical Observatory (BAO) has 2.6m and 1m Schmidt, as well as a number of smaller telescopes that are an observational basis for joint projects and collaborations. Armenian Virtual Observatory (ArVO) is hosting astronomical databases, such as the Digitized First Byurakan Survey (DFBS) and may also serve as a basis for development of VO structures in this region. Recently we have conducted a number of new activities; a meeting on "Relation of Astronomy to other Sciences, Culture and Society" (RASCS) was organized by BAO and Armenian Astronomical Society (ArAS) in Oct 2014 in Byurakan. Activities related to Archaeoastronomy and Astronomy in Culture (AAC) were initiated as well. Discussions on future Armenian-Iranian collaboration in astronomy were carried out, including an Armenian-Iranian Astronomical Workshop held in Oct 2015 in Byurakan. Similar workshops have been carried out between BAO and Abastumani Astronomical Observatory (AbAO, Georgia) since 1974.

Keywords. telescopes, astronomical data bases: miscellaneous, miscellaneous

---

Armenia will be hosting an IAU Regional Office of Astronomy for Development (ROAD), namely in the South West Asian (SWA) region. The agreement was signed between IAU and Byurakan Astrophysical Observatory (BAO) during the IAU General Assembly in August 2015 in Honolulu, Hawaii, USA.

The history of astronomy in Armenia goes back to very old ages. Since ancient times Armenians accumulated astronomical knowledge and have left this heritage in the forms of rock art, ancient observatories, calendars and chronology, historical records of astronomical events, medieval sky maps, astronomical terms, etc.

Nowadays Armenia has developed astronomy as well. Being a small country by its territory and population, it has high activity in astronomy at all levels: professional, educational and public. Armenia is situated in a region, where efforts are needed to develop and promote astronomical education and knowledge. Armenia is a reliable centre for astronomy development in the Southwest Asian region, involving the South Caucasus countries (Armenia, Georgia and Azerbaijan), Iran, Turkey and Israel, where efforts are being made to develop astronomy. In addition, Armenia also is one of the former Soviet Union republics and has tight relations to other such countries, including South Caucasus, Russia, Ukraine and Central Asian states. The latter may also join SWA ROAD. At last, Armenia is regarded by many European countries as a link for astronomical contacts between Europe and East in frame of the European Eastern Partnership (EaP) program.

The modern Armenian astronomy has an international recognition due to a number of reasons:

- Byurakan Astrophysical Observatory (BAO) is one of the important astronomical centres in the Middle East,

- discoveries and achievements by outstanding scientist, former IAU and ICSU President Viktor Ambartsumian and his famous colleagues are well known,
- the largest in the region 2.6m and one of the largest in the world 1m Schmidt telescopes,
- many important international meetings, including six IAU ones and the European JENAM-2007,
- recently established series of Byurakan International Summer Schools (BISS), where the regional students train and get experience,
- active international collaboration with a number of countries, such as France, Germany, Italy, Russia, USA, and others,
- international PhD program that has awarded scientific degrees to astronomers from Hungary, Bulgaria, Georgia, Azerbaijan, Uzbekistan, Jordan, etc.,
- famous Byurakan surveys and one of the largest astronomical spectroscopic databases, which is included in UNESCO "Memory of the World" international register,
- Armenian Virtual Observatory, the only such project in the Middle East and one of 20 members of IVOA,
- one of the major international prizes in astronomy (USD 500,000), Viktor Ambartsumian International Prize,
- Galileo Teachers Training Program (GTTP) and successful participation of Armenian pupils in International Astronomical Olympiads.

Recently we have conducted a number of new activities related to astronomy for development. A meeting on "Relation of Astronomy to other Sciences, Culture and Society" (RASCS) was organized by BAO and Armenian Astronomical Society (ArAS) in October 2014 in Byurakan. Astronomers, philosophers, biologists, historians, archaeologists, philologists, linguists, artists, and other specialists took part in the meeting. The meeting was important from the point of view of increasing the visibility of astronomy as a leader in interdisciplinary and multidisciplinary sciences.

Activities related to Archaeoastronomy and Astronomy in Culture (AAC), as encouraged by a number of international organizations (UNESCO, IAU, ISAAC, SEAC, etc.), were initiated as well. Armenia is especially rich in astronomical heritage issues and this area may strongly support the ROAD project.

We have already put efforts on strengthening the collaboration with the neighboring countries. A number of research projects have been accomplished between Armenian and Georgian astronomers and many Georgian astronomers have defended their Ph.D. theses in BAO. Armenian-Georgian astronomical colloquia have been organized since 1974 and in total 14 meetings were held, both in Byurakan and Abastumani observatories. In 2013, the last meeting was organized in Byurakan.

The first Armenian-Iranian Astronomical Workshop was organized in October 2015 to strengthen our scientific relations and establish new collaboration. We plan to organize such workshops on annual basis, succeeding each other one in Armenia and one in Iran. The collaboration will include several forms: collaborative research projects; stays of Armenian scientists at Iranian institutions and stays of Iranian scientists at Armenian institutions for joint research work; organization of joint meetings; BISS with invitation of Iranian students; observations on joint projects; collaboration between VOs; joint archaeoastronomical and cultural studies.

More countries may be involved in our ROAD as well, such as Turkey, Israel, Pakistan, Afghanistan, and Central Asian states. This will be of mutual benefit both for our ROAD and these countries.

Divisions Panel Discussion: Astronomy for Development

Kevin Govender[1], Mary Kay Hemenway[2], Anna Wolter[3], Nader Haghighipour[4], Yihua Yan[5], E.F. van Dishoeck[6], David Silva[7] and Edward Guinan[8]

[1] Facilitator of Panel Discussion, IAU Office of Astronomy for Development, Cape Town, South Africa
[2] University of Texas at Austin, USA
[3] INAF-Osservatorio Astronomico di Brera, Via Brera 28, 20121, Milano, Italy
[4] Institute for Astronomy, University of Hawaii-Manoa
[5] National Astronomical Observatories, Chinese Academy of Sciences, A20 Datun Road, Chaoyang District, Beijing 100012, China
[6] van Dishoeck, Leiden Observatory, Leiden University, the Netherlands
[7] National Optical Astronomy Observatory, 950 North Cherry Avenue, Tucson, AZ, 85719 USA
[8] Villanova University, 800 Lancaster Avenue, Villanova, PA 19085 USA

Abstract. The main purpose of this panel discussion was to encourage conversation around potential collaborations between the IAU Office of Astronomy for Development (OAD) and IAU Divisions. The discussion was facilitated by the OAD and the conversation revolved mainly around two questions: (i) What should the OAD be doing to enhance the work of the Divisions? (ii) What could the Divisions (both members and respective scientific discipline in general) contribute towards the implementation of the IAU strategic plan?

Keywords. OAD, Divisions, astronomy, development

1. Format and Participants

The discussion followed a relatively open format with conversation around potential collaborations between the OAD and respective IAU Divisions. Due to the large number of panellists there were only two rounds of "going-around-the-table" in turn and the rest were voluntary contributions as points arose. The participants in the discussion (representing their respective divisions) are listed in Table 1 below:

2. Summary of discussions

The discussion was in general quite dynamic and included input and question from the audience. Some important points are captured here and include input received by panellists by email before and after the discussion.

(i). Each Division (or area of astronomy) may produce specific resources that relate to education and development - the OAD could serve to collect, compare and disseminate these e.g. tutorials, lecture notes, assessments, outreach materials, etc.

(ii). The OAD has 9 regional offices and a large network of collaborators - this can be used to disseminate information and resources related to each Division's activities.

(iii). Divisions could consider the establishment of a small committee or single point of contact to be the interface with the OAD.

Table 1: List of IAU Divisions and Representatives

Division	Representative
Division A Fundamental Astronomy	Anne Lemaitre
Division B Facilities, Technologies and Data Science	David Silva
Division C Education, Outreach and Heritage	Mary Kay Hemenway
Division D High Energy Phenomena and Fundamental Physics	Anna Wolter
Division E Sun and Heliosphere	Yihua Yan
Division F Planetary Systems and Bioastronomy	Nader Haghighipour
Division G Stars and Stellar Physics	Ed Guinan (also represented Division J)
Division H Interstellar Matter and Local Universe	Ewine van Dishoeck
Division J Galaxies and Cosmology	Ed Guinan (also represented Division G)

(iv). The OAD could develop guidelines for symposia to enhance the regional developmental benefits that could result e.g. how to maximise the impact of an outreach component, addressing issues of inclusion, increase the international participation, etc.

(v). The OAD could coordinate the development of assessment materials in order to effectively measure the impact of training activities such as schools/workshops/symposia.

(vi). There could be a collection of talks online, similar to TED talks, which are made available to growing astronomy communities around the world (it was noted by the audience though that online material could be problematic in some parts of the world)

(vii). A long discussion was held around evaluation and measurement of impact. While it was generally agreed that this was important there were concerns raised about the cost and whether some goals would be realistically achievable. This is a challenge for the OAD to apply its mind to.

(viii). OAD and the IAU Office for Astronomy Outreach (OAO) should consider Division C as a "Think Tank" for education and outreach, noting that many research based projects indicate the results of extensive evaluation. Summative evaluation is very expensive, but many aspects of projects can benefit even by small evaluation efforts. Many Division C members have experience with these efforts that they would be pleased to share.

(ix). The OAD and the projects it supports should be using best practices, and not reinventing all the time. Better communication within the Astronomy Education and Outreach communities would make it easier to identify these best practices.

(x). As the OAD compiles a list of resources it should note that resources should include more than just activities, but also include management advice on things such as how to recruit teachers, how to form teams at a workshop, how to pace a workshop, how to do a needs assessment, how to evaluate the project, and how to do follow-up so that participants feel like part of a community. (Not all of these are appropriate for every project, but for those working with teachers and/or students, or development teams, they are.)

(xi). The OAD should foster use of Internet for mentoring new practitioners, with recognition that the local person has a better understanding of the local culture.

(xii). In addition to the expert advice from Commission C.C1 (Astronomy Education and Development) and Commission C.C2 (Communicating Astronomy with the Public), the other Commissions in Division C (C.C3 on History of Astronomy and C.C4 on World Heritage and Astronomy) can provide input into helping with the historic and/or cultural links to the projects done within the OAD.

(xiii). Division E provides a forum for observers, theoreticians, modellers and instrumentalists studying a wide range of phenomena related to the structure, radiation and activity of the Sun, the dynamic magnetized solar wind that shapes the heliosphere, and their combined impact on the multitude of bodies within the solar system, including the Earth. The phenomena of solar eclipses also provide a unique chance for professionals, amateurs and the public in general to understand the natural process by providing information on, and facilitating involvement in eclipse research.

(xiv). Division D hosts a Commission on Gravitational Wave Astrophysics, an Inter-Division commission on Galaxy Spectral Energy Distribution, one on Supermassive Black Holes, Feedback and Galaxy Evolution and a Supernova Working Group. The strongest assets of Division D which could be of interest of OAD are in general: technology; statistics; x-rays in general (they are well known to the general public even if for different purposes); programming and visualization tools; analytic skills; inspiration. One suggestion would be to create a repository for activities that have been already performed - something similar to astroEDU, but more general, with link to useful sites.

(xv). A difficulty is to involve the single individuals of the Division to contribute to OAD and how to differentiate this from the general outreach activities. How do we collect the inputs from everyone?

(xvi). The description material of OAD is large and well written, however it seems to be lacking of a short and easy description that can be shared with the members to encourage them to participate. This is for the OAD to address.

(xvii). If most of the work has to be done on a voluntary basis, then we need some kind of recognition to give to participants, otherwise it could be difficult to recruit them.

(xviii). A good way of advertising the OAD in the Divisions is through Division Bulletins where applicable.

(xix). Every Division could have a designated liaison with the OAD so that information flows back and forth. A telecon with these Division representatives, say, 2 times per year could keep the links active and information/ideas flowing.

3. Concluding remarks

The session proved to be very useful for exploring synergies between the OAD and IAU Divisions. It was clear that greater time was needed to continue the conversation and that the OAD should remain in close communication with points of contact within each Division. The session also seemed to be very useful from the audience perspective with many people expressing appreciation for the brief summaries of what each Division does. There were a number of points that need to be followed up on and from the OAD perspective, we look forward to expanding on these discussions into the future. It was clear that this is an initiative that should be implemented at each General Assembly.

Unconference session at the IAU General Assembly 2015

Tibisay Sankatsing Nava, Ramasamy Venugopal and Silvia Verdolini

IAU Office of Astronomy for Development, Cape Town, South Africa

Abstract. The Astronomy For Development Focus Meeting 20 at the IAU General Assembly encompassed an 'Unconference' session as part of the proceedings. Unstructured conferences, with their potential to unleash innovative ideas, are gaining traction in various conferences and symposia. Astronomy For Development is a field that is applicable to the entire Astronomy community (and even beyond) and hence an unconference inviting ideas and fostering frank dialogue is very pertinent.

Officially one of the final sessions of the the 2015 General Assembly, the unconference session was intended to provide a balanced platform for a diverse set of participants and act as an informal setting to promote open discussion on topics of relevance to Astronomy for Development.

Keywords. miscellaneous, sociology of astronomy, Unconference, Inclusion, Diversity, Evidence, Evaluation, Translation

1. Introduction

An unconference is a non-traditional conference that is loosely structured, allowing for an informal exchange of ideas. One was held for the first time at the IAU General Assembly 2015. True to the nature of such a gathering, there were no pre-defined topics. Rather, for a week before the session, members of the IAU and other conference participants were encouraged to suggest issues of concern to Astronomy for Development, which were then displayed publicly at the Office of Astronomy for Development (OAD) booth. The topics were loosely assembled under different themes such as "Inclusion & Diversity", "Education", "Outreach", "Evidence & Evaluation", "Light pollution", "Translation & Localisation", "Volunteering", and "Interdisciplinary & Partnerships". Visitors then engaged in anonymous polling to choose the most interesting topics for discussion at the unconference.

2. Themes and Topics

A brief summary of each of the eight themes and the corresponding topics proposed follows.

2.1. Inclusion & Diversity

A subject of significant importance, inclusion and diversity are hot topics in and out of the Astronomy world. Topics that were suggested addressed the issue of diversity in the Astronomy community and the contribution of Astronomy-for-Development in furthering the inclusion goal.

2.2. Education

Education has a direct connection to the Astronomy for Development agenda. Conference participants were keen to discuss transfer of practical skills to communities using Astronomy as well as running specialized workshops for teachers.

2.3. *Outreach*

Outreach activities, very popular in Astronomy and related areas, have the potential to be crowd-pleasers as well as serious instruments in using Astronomy for Development. Proposed topics focused on the differing objectives of such activities and the measurable benefits of outreach to all parties involved.

2.4. *Evidence & Evaluation*

Evidence collection and evaluation are essential to learn the best methods and practices. The need to tap into the growing body of evidence from other fields of research was highligted to be paramount in Astronomy for Development.

2.5. *Light pollution*

Light pollution is a major hurdle for Astronomy observations. But protecting our shrinking dark skies need not necessarily hold back development. The main issue tackled how combating light pollution goes hand in hand with Astronomy for Development.?

2.6. *Translation & Localisation*

Suggested topics included the difference in approach to Astronomy for Development work necessitated by the contexts of language, culture and region and translation of resources for different regions.

2.7. *Volunteering*

Managing and engaging volunteers and running internship programs for Astronomy for Development were the major topics of concern.

2.8. *Interdisciplinary & Partnerships*

The proposed topics tackled collaboration with development agencies and using insights from other fields such as cognitive sciences.

3. Unconference session

A total of three topics were chosen according to the votes coupled with the discretion of the organizers who took into account relevance and degree of importance of the topic. Participants at the unconference each selected one of the topics below, moderated by experts in the respective fields.

Evidence and Evaluation: Moderator(s) - Eli Grant, Linda Strubbe
Inclusion and Diversity: Moderator(s) - Beatriz Garcia
Translation and Localisation: Moderator(s) - Sze-Leung Cheung

A summary description of the discussion in the three groups follows.

Evidence and Evaluation
The importance of conducting assessment and evaluation of Astronomy for Development activities was highlighted. It helps identify the most effective and efficient processes to reach a certain outcome. Astronomy for Development is as much a domain of Astronomy as development. So, it is necessary to engage and collaborate with the development community as well as utilize research findings from other fields.

Inclusion and Diversity
Diversification has been an ongoing exercise in all areas of science and technology and has been given importance by the IAU. Discussions revolved around the potential of

Astronomy-for-Development to target and inspire a wider section of society to participate in technological innovation.

Translation and Localisation

The Astronomy for Development movement has spread across the globe with nine regional offices established on four continents. A global presence demands translation of resources, activities, best practices and news across languages. The discussion group touched on the merits and demerits of free translation services such as Google Translate and the urgent need for better software. In addition, volunteer translators need to be used to improve performance of such software with an appropriate reward system in place. The case of scientific terminology translating to exotic or obscure terms in certain languages is an example where knowledgeable human translators are required. The group recommended the creation of a website or other platform that could be used to send and receive translation requests as the foundation for further efforts in this area.

4. Conclusion

The first Astronomy for Development unconference session was reasonably well-attended. Recommendations from the participants included organizing a longer unconference session with more lead time for discussions.

List of Posters for Focus Meeting 20: Astronomy for Development

The following table contains a list of all posters and first authors that were registered to a participant during the the IAU General Assembly under Focus Meeting 20: Astronomy for Development. Due to the limited amount of time allocated for talks at the Focus Meeting, several Oral proposals had to be converted to Posters. We hope this table will be useful to illustrate the diversity of contributions to the Focus Meeting

Table 1: Titles and Authors for Poster Session at Focus Meeting 20: Astronomy for Development

Titles	Author
Looking for Partners to Engage the Global Community in Connecting to the Sky through International Observe the Moon Night	Sanlyn Buxner, Teaching, Learning, and Sociocultural Studies, University of Arizona, Tucson, Arizona, UNITED STATES
Universe Awareness: a global educational programme	Tibisay Sankatsing Nava, Leiden Observatory, Leiden University, Leiden, NETHERLANDS
How Big is Earth?	Bonnie Thurber, Northwestern University, Deerfield, Illinois, UNITED STATES
Astronomy and development: a multidisciplinary project in the Mexican countryside	Hector Bravo Alfaro, Departamento de Astronomia, Universidad de Guanajuato, Guanajuato, Guanajuato, MEXICO
Development Programs and Activities for Southeast Asia Regional Office of Astronomy for Development	Wichan Insiri, Foreign Affairs Department, National Astronomical Research Institute of Thailand, Chiangmai, THAILAND
Astronomy for Extremely ill or Traumatically Injured Children and Their Families - IAU OAD Grant	Donald Lubowich, Physics and Astronomy, Hofstra University, Hempstead, New York, UNITED STATES
3D Printing Meets Astrophysics: A New Way to Visualize and Communicate Science	Thomas Madura, NASA Goddard Space Flight Center, Greenbelt, Maryland, UNITED STATES
Astronomy Outreach In Parana state/Brazil	Marcelo Emilio, Geociências, Ponta Grossa State University, Ponta Grossa, Paraná, BRAZIL
"Astronomy for a Better World": IAU/OAD Task Force One Activities to Develop Astronomy Education and Research at Universities in the Developing World	Edward Guinan, Astrophysics & Planetary Science, Villanova University, Villanova, Pennsylvania, UNITED STATES
Current state of Czech astronomy popularization and its potential for enhancing science career interest	Radek Kříček, Astronomical Institute, Charles University in Prague, Prague, CZECH REPUBLIC

Table 1: *Continued.*

Titles	Authors
Peer-review Platform for Astronomy Education Activities	Thilina Heenatigala, Leiden Observatory, Leiden University, Leiden, NETHERLANDS
Astronomy in the DPR Korea	George Miley, Leiden University, Leiden, NETHERLANDS
Leiden University astronomy for development projects	George Miley, Leiden University, Leiden, NETHERLANDS
A Pilot Astronomy Outreach Project in Bangladesh	Dipen Bhattacharya, Dept. of Natural Sciences, Moreno Valley College, Moreno Valley, California, UNITED STATES
The Sharjah Center for Astronomy and Space Sciences (SCASS 2015): Concept and Resources	Hamid Al Naimiy, Office of the Chancellor, University of Sharjah, Sharjah, Sharjah, UNITED ARAB EMIRATES
AstroNavigation: Freely-available Online Instruction for Performing a Sight Reduction	Susan Stewart, Astronomical Applications, U.S. Naval Observatory, Nashville, Tennessee, UNITED STATES
Principles of Celestial Navigation: An Online Resource for Introducing Practical Astronomy to the Public	Sean Urban, US Naval Observatory, Washington, District of Columbia, UNITED STATES
Journey of Ethiopia Astronomy	Solomon Tessema, Astronomy and Astrophysics Research Division, Entoto Observatory and Research Center, Addis Ababa, ETHIOPIA
Recent activities in Armenia related to IAU ROAD and strategic plan	Areg Mickaelian, Armenian Virtual Observatory (ArVO), Byurakan Astrophysical Observatory (BAO), Yerevan, ARMENIA
Visiting nursery, kindergarten and after-school day care as astronomy for development	Akihiko Tomita, Wakayama University, Wakayama, JAPAN
Design, transport, and installation of autonomous Cherenkov detectors at high altitude	Mario Calderón Cueva, Escuela Politécnica Nacional, Quito, Pichincha, ECUADOR
Results from pre- and post- tests of professional development astronomy workshops for teachers in Peru	Susana Deustua, STSCI, Baltimore, Maryland, UNITED STATES
ISODEX: An entry point for developing countries into space activities	Mark Skinner, Boering Research & Technology, The Boeing Company, Albuquerque, New Mexico, UNITED STATES
Exploring new possibilities of astronomy education and outreach	Kodai Fukushima, Hosei University, Koganei, JAPAN
eGTTP - bridging distances with online training	Rosa Doran, Executive Council, NUCLIO - Nucleo Interativo de Astronomia, Sao Domingos de Rana, PORTUGAL
NASE 2015: Implementation of a Management Quality System	Rosa Ros, Applied Mathematics 4, Technical University of Catalonia, Barcelona, SPAIN

FM 21:
Mitigating Threats of Light Pollution & Radio Frequency Interference

Session 21.1 – Observations, Advances in LED Technology, and Dark Sky Protection

Dan M. Duriscoe

National Park Service, Natural Resources Stewardship and Science Directorate,
351 Pacu Lane, Bishop, CA, 93514, United States
email: dan_duriscoe@nps.gov

Abstract. The importance of dark sky protection, potential threats to further degradation from LED technology, the announcement of a new world atlas of artificial night sky brightness, and the use of color images from the orbiting International Space Station for monitoring potential sources of light pollution were discussed in the six talks of this session. It was clear from the presentations that the work of professional astronomy depends upon continued restraint in the use of outdoor lighting, especially new LED technology, which relies upon blue-rich sources to support the advantages of high luminous efficacy and resulting energy savings.

Keywords. techniques: photometric, atmospheric effects, radiative transfer

1. The search near Near Earth Objects — why dark skies are so important

AUTHOR: Richard Wainscoat (Institute for Astronomy, University of Hawaii, Honolulu, Hawaii, United States)

The Pan-STARRS telescopes located on Haleakala, Maui, Hawaii, and the Catalina Sky Survey, located near Tucson, Arizona, continuously search for Near Earth Objects (NEOs). Such objects may present a threat to Earth, especially if undetected before entering the atmosphere. These objects are very faint, therefore early detection requires broad passband detectors to maximize the amount of light and signal to noise. The addition of any amount of artificial sky glow between 420 and 820 nm will hinder the search. ue In addition to the Maui and Arizona search telescope locations, follow up telescopes, spread across the planet, play a crucial role in the verifying and determination of accurate positions to establish and orbit. Only then can it be determined whether or not the object poses a threat to Earth. The majority of these followup telescopes are at locations that are impacted by light pollution, and this seriously impacts their ability to secure additional observations.

The evidence of past NEO impacts on the earth's surface, such as Meteor Crater in Arizona (a relatively small object) and the much older Chicxulub crater near the Yucatán Peninsula, estimated at 66 million years ago, are reminders of the potential consequences of these objects. Effects include a cloud of super-heated dust, ash and steam, dust covering the whole Earth, global earthquakes and volcanic activity, and possible mass extinctions.

The Pan-STARRS telescopes, two 1.8-meter diameter instruments at Haleakala Observatory, are the largest digital cameras in the world, surveying large blocks of sky with a sequence of images separated by about 20 minutes (Figures 1a, b, c). The gigapixel cameras collect 3 terabytes of data per night which is subsequently processed looking for moving objects among the stars. To date, almost 13,000 Near Earth Asteroids have been discovered by both Pan-STARRS and the Catalina Sky Survey, of which almost 1600 are classified as potentially hazardous asteroids.

 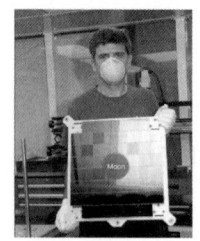 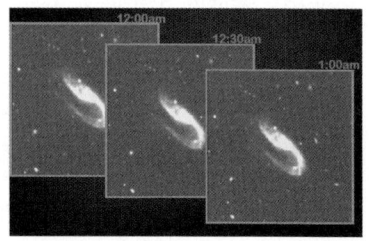

Figure 1: (a) The PanSTARRS telescope atop Haleakala, Maui, Hawaii, (b) the sensor of the gigapixel camera used on the 1.8 meter telescope, and (c) sample sequential images used to detect Near Earth Objects.

Once an object is determined to exist, follow up observations by telescopes across the globe contribute to efforts of orbit determination. Many are affected by light pollution, decreasing the effectiveness of observing programs. In addition, cloudy weather at professional observatory sites may necessitate the observations being taken by amateur instruments from more urban locations. Continuing strong efforts to protect the dark night sky from light pollution are essential for the search for Near Earth Objects, and for mitigating the threat that they pose.

2. LEDs/ALAN—Working to be Good Neighbors

AUTHOR: Robert Adams (CW-Energy Solutions, Paradise Valley, Arizona, United States)

ALAN (Artificial Light at Night) and LEDs have recently become major discussion topics in the areas of astronomy, light pollution, endangered species and human health to mention but a few. In years past, MH, LPS and HPS dominated night lighting with LPS and its associated narrow spectrum as the preferred source around observatories and shorelines. LEDs offer the ability to modify the spectrum, realize substantial energy savings and other associated benefits while meeting the requirements of the astronomy community. The primary concern of the different groups relates to blue light content of the LED. For astronomers, the molecular (Raleigh) scattering related to the blue light interferes with certain portions of the spectrum used for deep space studies. Blue light in the environment also has an impact on sensitive species, such as leatherback turtles and bats, and on the human circadian rhythms.

The spectral power distribution (SPD) of various light sources and the CIE color chart (X,Y coordinates) adequately describe the quality of light emitted and identify potential concerns. The luminous efficacy of the source in lumens/watt is also important. Figure 2 illustrates various types of LED lights including filtered 3000K warm white, both warm white and traditional cool white LEDs with various filters attached, phosphor coated (PC) amber LED, and narrow beam amber, which show promise for meeting outdoor lighting needs while protecting night skies and the environment. It is evident that the metric CCT is not adequate for specifying the new LED solutions with the modified spectra. Percent blue, percent blue and green, XY coordinates, or scotopic to photopic (S/P) ratio are better metrics.

The status of the technology indicates that it is mature enough for wide application, 10,000 new lights are being installed on the roadways of the County of Hawaii over the next 18 months. The efficiency of the lighting has improved 80% over the past three years from 55 lumens per watt to 100. In order to be competitive, efficiency should be

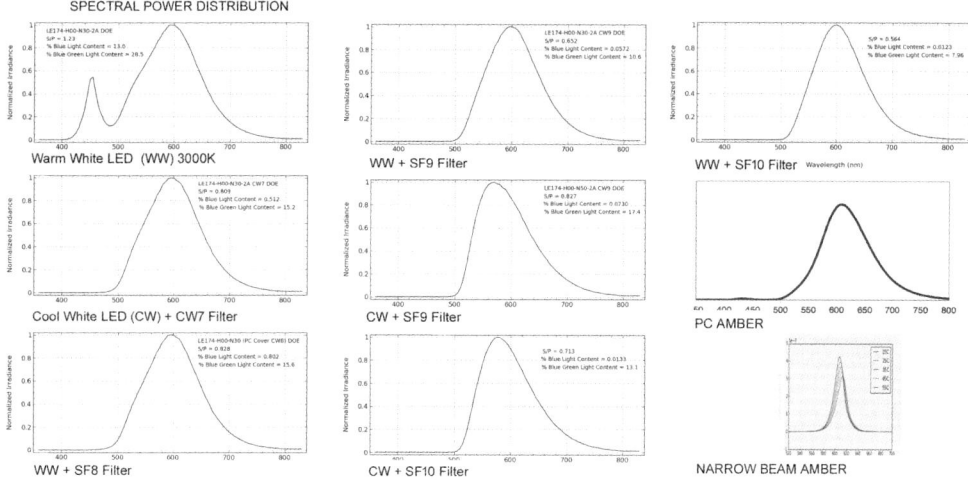

Figure 2: Spectral power distribution of selected LED lamps.

maintained at 70 to 80 lumens per watt. Three new families of filters have been developed and one family qualified for use in the Canary Islands.

Today, lighting plans and implementation are all too often based on opinions and limited data. The ensuing problems and repercussions make it imperative to collect accurate and thorough information. Data collection is now ongoing using a variety of techniques analyzing the "before" and "after" lighting results from the County of Hawaii LED streetlight conversion. The studies will focus on any quantifiable impact LEDs may have on such topics as light pollution, endangered animals, astronomy and most importantly, the citizens of our local communities.

3. Light emitting diodes and astronomy — a change for restoration of the dark night sky or for further loss

AUTHOR: Richard Wainscoat (Institute for Astronomy, University of Hawaii, Honolulu, Hawaii, United States)

Across the planet, conventional light sources such as high pressure sodium, are rapidly being replaced by light emitting diodes (LEDs). As light fixtures are being replaced, there is a tremendous opportunity for restoration of dark night skies through replacement of poorly shielded fixtures by fully shielded fixtures. Also, the huge advantage that LEDs offer is that the light from and LED is much easier to direct, allowing the use of less lumens for a given task. However, it is critically important to limit the amount of blue light from the LEDs.

Sales people are strongly promoting LEDs with high correlated color temperature (CCT), such as 5000K. They are promoting them on energy efficiency grounds - higher energy efficiency is easier to sell. These LEDs have tremendous amounts of blue light near 450 nm. The photopic human eye is relatively insensitive to this blue light, but the dark adapted scotopic eye is much more sensitive, and CCDs are also very sensitive to this wavelength of light. As a consequence, both professional and amateur astronomers are very seriously impacted by high CCT LED lighting. The sodium lighting that the LEDs are replacing has relatively little blue light. Blue light is strongly scattered by air molecules in the atmosphere. There is little reason to install LEDs with high blue light content now because there is only a 5-7% difference in energy efficiency between 3000 K

Figure 3: Skyline of the city of Honolulu, Hawaii, at night.

and 5000K white LEDs. Use of high CCT LED lighting will cause further deterioration of night sky quality.

In contrast, use of LED lighting with low CCT (e.g., 2400K or 2700K), or use of filters to remove the blue light, can restore the dark night sky. LED lighting is much easier to direct, meaning that an area such as a roadway can be lit with many less lumens with LEDs compared to conventional lights such as high pressure sodium. And use of fully shielded fixtures will eliminate direct uplighting. Figure 3 illustrates part of Honolulu as it is currently lit at night, with an estimated 30 million lumens directed upwards resulting in a waste of approximately $300,000 per year in energy costs.

4. Advanced strategies for outdoor LED lighting applications and technologies to curtail regional light pollution effects

AUTHOR: Christian Monrad (Monrad Engineering, Tucson, Arizona, United States)

LED lighting systems for outdoor lighting applications continue to evolve as do strategies to mitigate related effects upon regional astronomical and ecological assets. The improving availability and relative lumen-per- watt efficiencies of blue-suppressed low correlated color temperature emitters, narrow band amber, phosphor converted amber, and various combinations of broadband emitters and sub-550NM and sub-500NM filters allow for a wide palette of choices to be assessed to suit site-specific and task-specific lighting needs. In addition to static spectral content options, readily available luminaire designs also include precise geometric beam shape selections and adaptive controls to include dimming, dynamic spectral shifting, motion detection, and dynamic beam shaping to minimize total environmental lumen emissions throughout the course of the nighttime hours.

Major motivators for large scale replacement of metal halide, high pressure sodium, and low pressure sodium lamps with LEDs are reduced maintenance, energy savings, and many HID lighting systems reaching the end of their life. This, and the factors listed above, makes now a good time to modernize. There is an efficient, cost effective and properly shielded LED replacement for almost every legacy lighting system.

Regional and international light pollution mitigation regulations are inconsistent. For example, in Chile near major observatories a maximum output of 15% of the total in

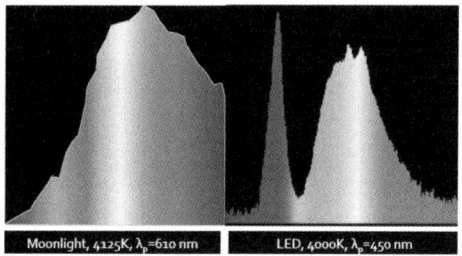 

Figure 4: Comparison of the spectrum of moonlight (left) and 4000K LED lamp (right).

Figure 5: Scene illumination from 3000K LED lamps at Tucson International Airport.

sub-500 nm wavelengths is prescribed, in Cohise County, Arizona, USA, at specification of maximum CCT of 3000K is employed, in Pima County, Arizona this value is 3500K, while in the Canary Islands a maximum of 25% sub-550 nm is used. Of these, the use of CCT is archaic and inappropriate for many LEDs, since it is based upon a Plankian blackbody radiator. Scotopic to photopic (S/P) ratio is a better metric. Some advertisers claim that 4000K is "natural moonlight" but with the spectral power distribution of most LED lamps rated at this CCT, the percent blue and S/P ratio are quite different (see Figure 4), and natural moonlight is close to a 2200K CCT LED lamp. Not only is S/P ratio available as part of the CIE/IES testing format, it can be field verified with a handheld spectrometer.

Of the LED technologies available today, narrow band amber LED has the lowest S/P ratio at 0.3 (see bottom illustration in Figure 2). An LED with a CCT of 3000K typically will result in and S/P ratio of 1.3, with cooler (higher CCTs) colors producing even greater ratios. Filtered LEDs provide a method of reducing the S/P ratio to below 1.0. For comparison, low pressure sodium and high pressure sodium have typical S/P ratios of 0.23 and 0.54, respectively. With the improved aiming characteristics of LED luminaires, often very significant reductions in the amount of light compared to HID systems may be realized in a re-design. For example, at Tucson, Arizona, USA international airport, airline terminal parking area outdoor lighting was replaced, converting from 22 megalumens (primarily HPS) to 6 megalumens 3000K LED (Figure 5). Even after accounting for the increase in S/P ratio, there was a 26% reduction in downward looking luminance. This installation is expected to result in skyglow reduction, energy savings, and lowered maintenance costs.

5. The new world atlas of artificial sky brightness

AUTHORS: Fabio Falchi, Pierantonio Cinzano (ISTIL—Light Pollution Science and Technology Institute, Thiene, Italy), C.C. Kyba (Leibniz-Institute of Freshwater Ecology and Inland Fisheries, Berlin, Germany), B. A. Portnov (Department of Natural Resources & Environmental Management, Universtiy of Haifa, Haifa, Israel)

The main steps toward the completion of the new World Atlas of Artificial Sky Brightness (WA II) have been realized, and a whole earth zenith sky brightness prediction has been computed. The upward radiance data used are those from Visible Infrared Imaging Radiometer Suite (VIIRS) Day-Night Band (DNB) on board the Suomi NPP satellite. The use of this newly available radiance data allows for an increased real resolution, even while maintaining the same 30×30 latitude and longitude pixel size. The computational technique has been updated, in comparison to the first World Atlas, to take into account

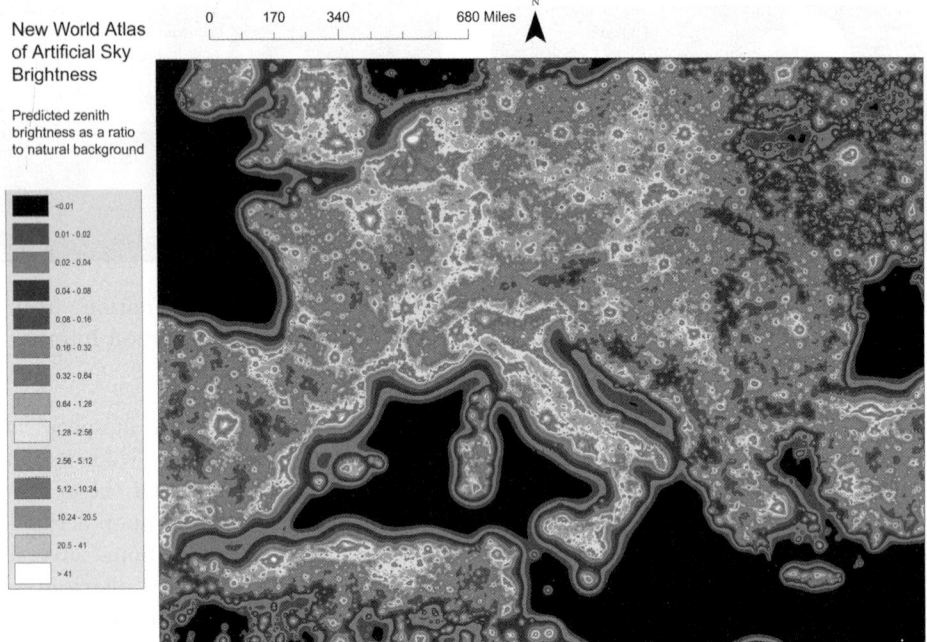

Figure 6: Artificial sky brightness at zenith prediction for south-central Europe from the new World Atlas.

both sources and sites elevation. The elevation data are from USGS GTOPO30 global digital elevation model, with the same pixel size as the WA II maps.

The upward emission function used to compute the Atlas is a three parameter function, representing three different upward patterns: Lambertian reflection, high zenith angle emission typical of direct emission from "cobra head" streetlights, and an emission pattern at moderate zenith angles. The parameters are then constrained to Earth based night sky brightness measurements, primarily from SQM-L devices, using a database of tens of thousands of measures from sources world-wide. In this way we use a global fitting upward function for the final map's calibration.

We maintained constant atmosphere parameters over the entire Earth, identical to those used for the first Atlas (Garstang atmospheric clarity coefficient k=1, equivalent to a vertical extinction at sea level of 0.33 magnitude in the V band). This was done in order to avoid introducing a local bias due to different conditions that may confound the light pollution propagation effects.

The VIIRS DNB data used for the input data were chosen from the months ranging from May to September 2014 in order to avoid introducing bias from the variable snow coverage in mid to high northern latitudes. In the southern hemisphere this problem is far less pronounced.

The final maps demonstrate greater detail and resolution than the first world atlas. Figure 6 shows a close-up of south-central Europe. The predicted artificial sky brightness is shown as a ratio to the natural background, which is assigned at 22.0 V-magnitudes per square arc second or 172 micro-candela per meter squared. The completion of the project will include estimates of the percentage of the population of the earth residing in each of five separate classes of night sky degradation. The atlas should provide a useful tool for urban planners, professional and amateur astronomers, protected area managers, and the general public seeking locations from which to enjoy as natural a night sky as possible.

6. ISS images for observatory protection

AUTHORS: A. Sánchez de Miguel (CEGEP de Sherbrooke, Sherbrooke, Québec, Canada), J. Zamorano (Astrofísica y CC. de la Átmosfera, Universidad Complutense de Madrid, Madrid, Spain), M. Aubé, J. Gallego.

Light pollution is the main factor of degradation of the astronomical quality of the sky along the history. Astronomical observatories have been monitoring how the brightness of the sky varies using photometric measures of the night sky brightness mainly at zenith. Since the sky brightness depends in other factors such as sky glow, aerosols, solar activity and the presence of celestial objects, the continuous increase of light pollution in these enclaves is difficult to trace except when it is too late. The use of direct light detection from above the earth's surface is up to 14 times more sensitive than sky brightness measurements due to the natural variations. Also, tracking trends in the artificial light sources is important to night sky protection efforts.

Using models of light dispersion on the atmosphere one can determine which light pollution sources are increasing the sky brightness at the observatories. The input satellite data has been provided by DMSP/OLS and SNPP/VIIRS. Unfortunately, their sensors are in panchromatic bands which have little or no sensitivity to blue wavelengths. Therefore they are not useful to detect potential increases in upward emissions due to the dramatic change produced by the irruption of blue-rich LED technology in outdoor lighting. The only instrument in the space that is able to distinguish between the various lighting technologies are the DSLR cameras used by the astronauts onboard the ISS. This is because their sensors contain three channels, R, G, and B.

The objectives of this work are to contribute to knowledge of upward emissions for use in sky brightness models, develop methods to calibrate satellite measures of upward radiance to an absolute scale, and identify trends in spatial, temporal, and spectral variation of light pollution sources and their relationship with the sky brightness variations. Through the use of both remote sensing and ground based observations, and the identification of associated relationships between the two classes of observations, a more comprehensive evaluation of world-wide impact from light pollution is obtained.

The examination and evaluation of the remote sensing data, from earth observation satellites and ISS DSLR images, revealed that while it is possible to detect the presence of artificial light sources, they were not designed primarily for measuring radiances. The main contribution of this study include: 1) Calibration of the DMSP/OLS data from 1992-2010, 2) Evolution of energy consumption in street lighting with satellite data, 3) Absolute photometric calibration of nighttime images of the Earth from the International Space Station (ISS/D3S), and 4) The cataloguing and geo-referencing of the ISS image archive. Examples of upward radiance images of the city of Madrid, Spain, are shown in Figures 7a, b, and c, with sources DMPS/OLS, VIIRS/DNB, and ISS/D3S/RGB, respectively.

The RGB data from the ISS camera allows for the interpretation of the type of light source observed in each image. The amount of light detected by R, G, and B channels from each type of lighting is computed, and the ratio of combinations of the channels in the observation images may be used to infer the lamp type (Figures 8 and 9).

Photometric calibration of the green channel of the ISS camera with reference stars allows for comparison with VIIRS/DNB data. Often such a comparison necessitates allowing for atmospheric extinction and tilt, transmission of the window of the ISS, and

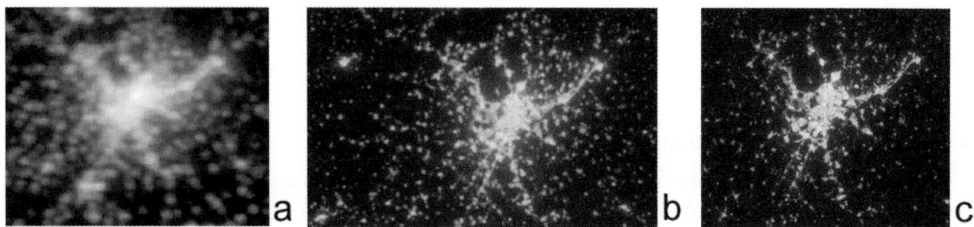

Figure 7: Images of Madrid, Spain from space, (a) DMSP, (b) VIIRS, and (c) ISS.

Figure 8: High resolution ISS image of Madrid showing RGB combined and separate channels (inset).

camera linearity. After these corrections are made, a good fit between the two sensors is achieved (Figure 10).

Near an observatory, emitting sources may be identified in ISS images and their spectra characterized approximately. The manner in which these sources may affect the night sky spectrum may be predicted. Ultimately the prediction of such effects will lead to wise choices in outdoor lighting. We are planning to send an official request to NASA with a plan to get images for the most important astronomical observatories. We ask support for this proposal by the astronomical community and especially by the US-based researchers. See www.citiesatnight.org.

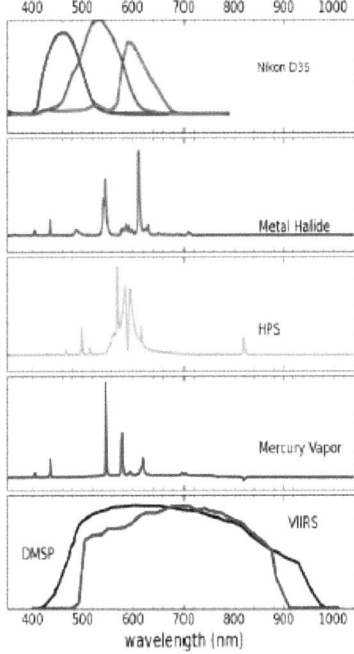

Figure 9: Comparison of senor sensitivity and emission from light sources.

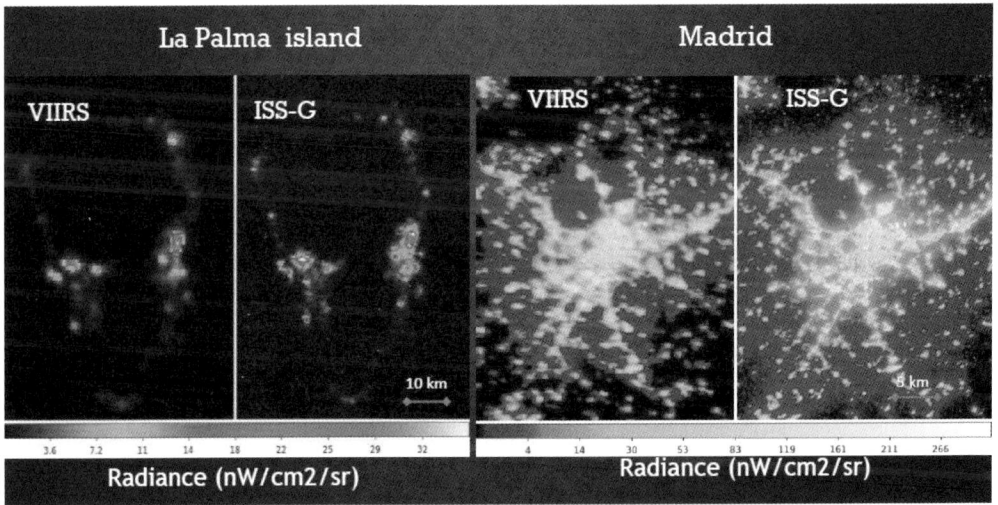

Figure 10: Comparison after calibration of VIIRS and ISS calibrated upward radiance measures.

Session 21.2 – Measurement of Light at Night

Richard J. Wainscoat

University of Hawaii, Institute for Astronomy
email: rjw@IfA.Hawaii.Edu

Abstract. The introduction of the mercury vapor lamp for general lighting in the 1930s probably marked the beginning of significant light pollution. Lighting levels have increased slowly, year-to-year, with sky brightness levels increasing only slowly on timescales of a year; no measurement protocols or instruments existed to quantify this increase. However, on timescales of 10–20 years, or on generational timescales, the increases in night sky levels, particularly in urban areas, have been dramatic. Younger people speak with their parents or grandparents who remark how beautiful the sky used to be, and how many stars they could see when they when they were younger. Older people can themselves remember how many stars were visible in the sky when they were younger. Whole generations of children now grow up without ever seeing the Milky Way.

Society has not had tools to easily measure sky brightness, and monitoring from space has only recently become available. A subtle increase of 10% sky brightness per year, for example, is not noticeable to the human eye on the time scale of a year, and has been tolerated by society. But such an increase compounds to an increase of a factor 2.6 in 10 years, 6.7 in 20 years, and a factor 45 in 40 years, corresponding to a dramatic increase in sky brightness, an almost complete loss in ability to see faint objects in the night sky, and rendering the sky unusable for most forms of astronomy. The most striking examples are the urban observatories found in many major cities that can no longer be used.

Session 2 was primarily focused on measurement of light at night, with an emphasis on measurement of light pollution. It comprised of 6 papers that are summarized below. Over the last decade, our ability to measure light pollution has grown tremendously, and the instrumentation needed to produce reliable quantitative measurements has become much more affordable, and now includes consumer grade digital cameras and even smart phones. During this same time period, light pollution has continued to grow. The widespread changes from mostly high-pressure sodium lighting to LED lighting that are now occurring make continued monitoring and measurement of light pollution particularly important into the future.

Complete presentations may be viewed at: http://www.noao.edu/education/IAUGA2015FM21

Keywords. light pollution, measurement

---

1. Derivation of sky quality indicators from photometrically calibrated all-sky image mosaics

Authors: D.M. Duriscoe, C.A. Moore, Night Skies Program, U.S. National Park Service, Bishop, California, United States; C.B. Luginbuhl, Dark Sky Partners, L.L.C., Tucson, Arizona, United States

For many years, the US National Park Service Night Skies Program has been acquiring carefully calibrated mosaic images of the night sky. A large database of high resolution all-sky measurements of V-band night sky brightness at sites in U.S. National Parks and astronomical observatories, including a few international locations, has been amassed. This database can be utilized to describe sky quality over a wide geographic area.

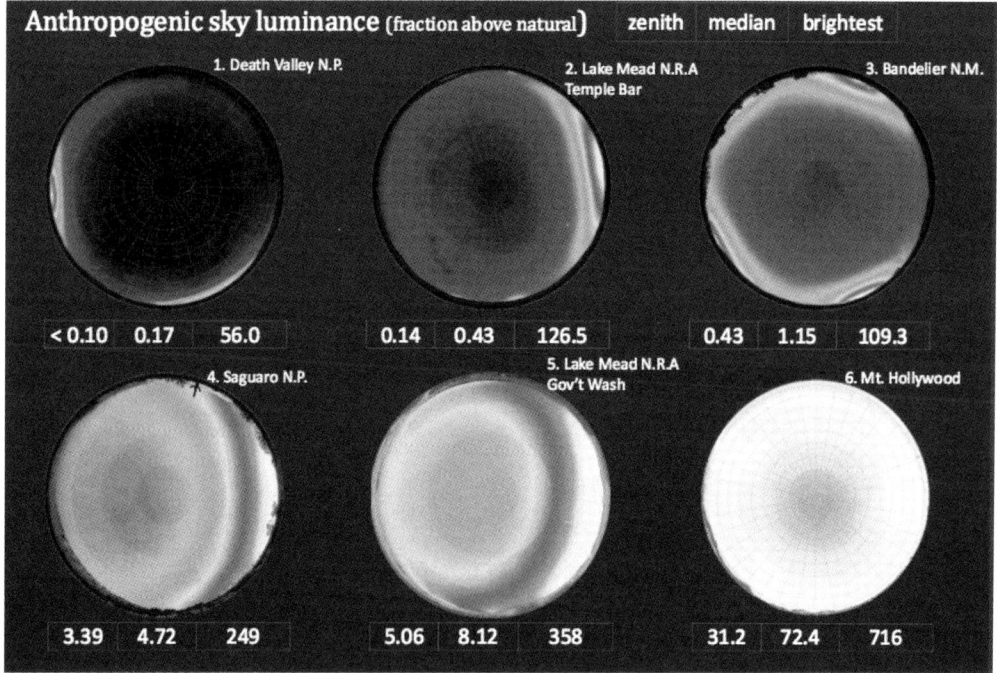

Figure 1: Artificial Sky Luminance

Mosaics of photometrically calibrated V-band imagery are processed with a semi-automated procedure to reveal the effects of artificial sky glow through graphical presentation and numeric indicators of artificial sky brightness (Figure 1). Comparison with simpler methods such as the use of the Unihedron SQM and naked eye limiting magnitude reveal that areas near the horizon, which are not typically captured with single-channel measurements, contribute significantly to the indicators maximum vertical illuminance, maximum sky luminance, and average all-sky luminance.

Artificial skyglow at the zenith only is difficult to measure accurately because of variations in the natural airglow, and because artificial skyglow is typically lowest at the zenith.

Distant sources of skyglow may represent future threats to areas of the sky nearer the zenith. The brightest area of the sky is a very sensitive indicator of visual intrusion and points to areas of the sky for continuous monitoring where trends can be easily detected. Timely identification and quantification of these threats may allow mitigating strategies to be implemented.

Astronomers should establish a median all sky limiting value for light pollution at research-grade observatories. A value of no more than 20% above natural is suggested. A smaller value, such as no more than 10% above natural is highly desirable for the best astronomical observatory sites where 4-meter or larger telescopes are located.

Expansion of this work beyond the V filter will be needed in the future to properly quantify the effects of changes of lighting to Light Emitting Diodes (LEDs). White LEDs emit blue light that is particularly damaging both to astronomy and to the dark adapted scotopic human eye's view of the night sky, and this blue emission, centered at 450 nm, falls outside the transmission curve of the V filter.

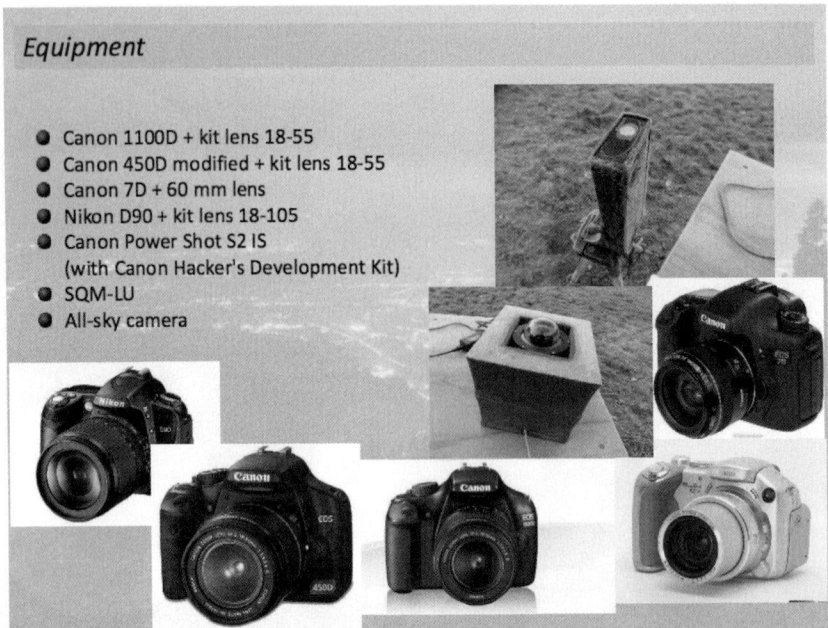

Figure 2: DSLR cameras used in measuring night sky brightness.

2. Night sky photometry with amateur-grade digital cameras

Authors: T. Mrozek, M. Steslicki, Solar Physics Division, Space Research Centre PAS, Wroclaw, Poland, T. Mrozek, D. Gronkiewicz, S. Kolomanski, Astronomical Institute, University of Wroclaw, Wroclaw, Poland

Measurements of night sky brightness can give us valuable information on light pollution. The more measurements we have, the better is our knowledge on the spatial distribution of the pollution on local and global scales.

High accuracy professional photometry of night sky can be performed with dedicated instruments. The main drawbacks of this method are high price and low mobility. The apparatus used by the National Park Service in Section 1 above, for example, consists of an astronomy grade cooled CCD, heavy mount, computer and portable power supply. It is expensive and not easily portable, and beyond the reach of most people.

This limits the amount of observers and observations and therefore the amount of photometric data that can be collected. In order to overcome the problem of the limited amount of data, we can involve amateur astronomers in photometry of the night sky. However, to achieve this goal we need a method that utilizes equipment that is usually used by amateur astronomers—digital cameras.

Digital cameras have improved tremendously over the last decade, with both professional and amateur grade digital cameras now having very low noise performance. The noise performance of these digital cameras is temperature dependent, and their noise is lower at colder temperatures (albeit with poorer battery performance at these colder temperatures).

We propose a method that enables good accuracy photometry of night sky with a use of digital compact or digital single lens reflex (DSLR) cameras. (See Figure 2.) Reduction of observations and standardization to BVR system are performed.

We tested several cameras, including popular compact digital cameras and amateur grade DSLRs in the Izera Dark Sky Park in Poland. We compared our results to Sky Quality Meter (SQM) measurements. The overall consistency for results is within 0.2 mag. Our method is very simple but is quantitative and uses inexpensive equipment that is already owned by many people including amateur astronomers.

We tested our methodology with secondary school students, a group of teachers, and a group of amateur astronomers. This allowed us to simplify the method and to identify the most problematic issues that could lead to degradation of results. We believe that this method can be deployed in a widespread manner to obtain a substantial number of light pollution measurements over wide areas and spanning long periods of time to show long terms trends in the sky brightness.

3. Evaluation of the night sky quality at El Leoncito and LEO++ in Argentina

Authors: M. Aubé, N. Fortin, S. Turcotte, Physics, Cégep de Sherbrooke, Sherbrooke, Quebec, Canada, B. Garcia, A. Mancilla, J. Maya, CNEA-CONICET-UNSAM, Instituto de Tecnologas en Deteccion y Astropartculas, Mendoza, Argentina

Light pollution is a growing concern at many levels, but especially for the astronomical community. Artificial lighting veils celestial objects and disturbs the measurement of night time atmospheric phenomena. This was the motivation for this sky brightness measurement experiment in Argentina. The goal was to determine the quality of two Argentinian observation sites: LEO++ and El Leoncito. (See Figure 3.) Both sites were candidates to host the Cherenkov Telescope Array (CTA). This project consists of an arrangement of many telescopes that can measure high-energy gamma ray emissions via their Cherenkov radiation produced when entering Earth's atmosphere. A very dark sky is essential for this experiment.

Even if the two Argentinian sites had been excluded from the final CTA site competition, they are still of great interest for other astronomical projects. Especially the El Leoncito site which already hosts the Leoncito Astronomical Complex (CASLEO). Both sites are in western Argentina, close to the Chilean border. The closest large town is San Juan.

The measurement methods used to determine the sky quality were described in detail. These included two different techniques: Sky Quality Meters (SQM) and the Spectrometer for Aerosol Night Detection (SAND). Seasonal variations in sky brightnesses at El Leoncito were clearly observed, with the fall season having an average sky brightness 0.3 magnitudes brighter than average. The darkest skies were in the summer. The variation in sky brightness is related to seasonal variations in aerosol content in the atmosphere.

The results were compared to different renowned astronomical sites (Kitt Peak, Arizona, USA, and Mont-Mégantic, Canada). The two Argentina sites studied have similar sky brightness levels to Kitt Peak and Mont Mégantic. We found that LEO++ is a high quality site, however there are a lot of aerosols that can interfere with astronomical measurements. El Leoncito shows very low sky brightness levels, which are optimal for low light level detection, and has the highest potential for future development as an astronomical site. It shows low radiance in the 569 nm sodium line, low aerosol content (resulting in lower extinction) and is surrounded by mountains that suppress the impact of the surrounding luminous halos.

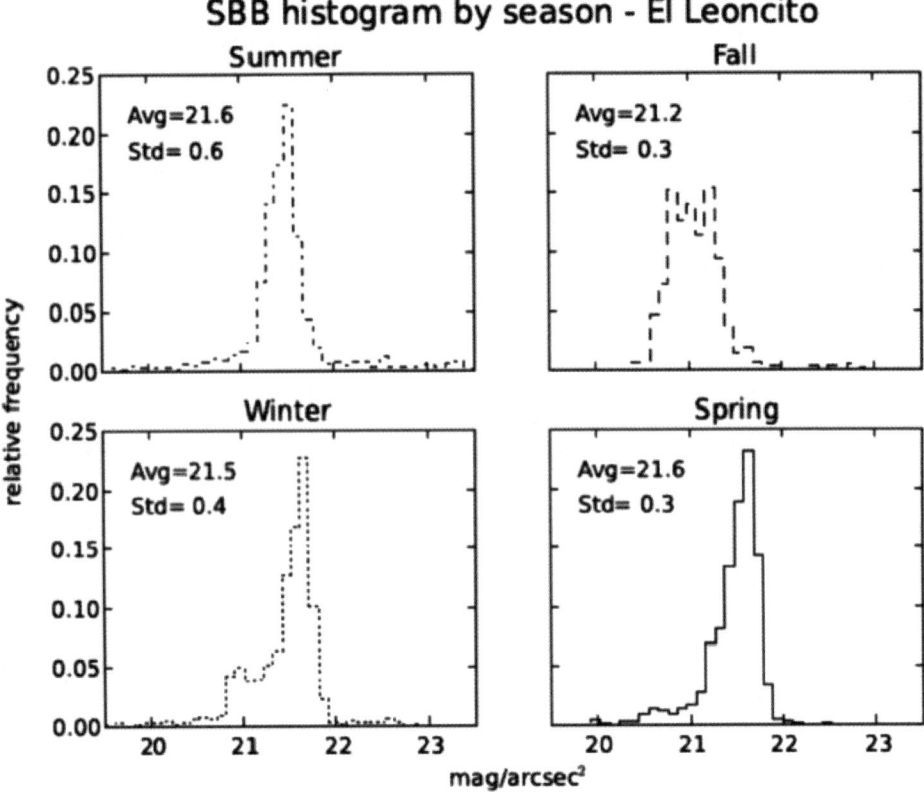

Figure 3: Seasonal behavior of the night sky brightness at El Leoncito.

4. Globe at Night - Sky Brightness Monitoring Network

Authors: S. Cheung, IAU Office for Astronomy Outreach, Mitaka, Tokyo, Japan, S. Cheung, Y. Shibata, H. Agata, National Astronomical Observatory of Japan, Tokyo, Japan, J.C. Pun, C. So, The University of Hong Kong, Hong Kong, Hong Kong, C.E. Walker, National Optical Astronomy Observatory, Tucson, Arizona, United States

The Globe at Night - Sky Brightness Monitoring Network (GaN-MN) is an international project for long-term monitoring of night sky conditions around the world. The GaN-MN consists of fixed monitoring stations each equipped with a Sky Quality Meter - Lensed Ethernet (SQM-LE), which is a specialized light sensor for night sky brightness (NSB) measurement. NSB data are continuously collected at high sampling frequency throughout the night, and these data are instantly made available to the general public to provide a real-time snapshot of the global light pollution condition. A single data collection methodology, including data sampling frequency, data selection criteria, device design and calibration, and schemes for data quality control, was adopted to ensure uniformity in the data collected. This is essential for a systematic and global study of the level of light pollution. The data collected also provide the scientific backbone in our efforts to contribute to dark sky conservation through education to the general public and policy makers. The GaN-MN project is endorsed by the IAU IYL Executive Committee Working Group as a major Cosmic Light program in the International Year of Light.

The data are collected using the Unihedron lens-ethernet sky quality meters. These provide sky brightnesses in units of mag arcsec^{-2}, with an accuracy of ± 0.1 mag arcsec^{-2}.

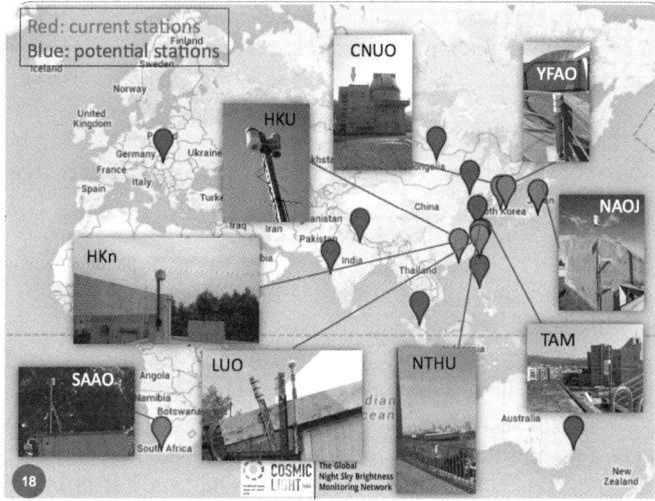

Figure 4: Current and potential sky brightness monitoring station locations worldwide

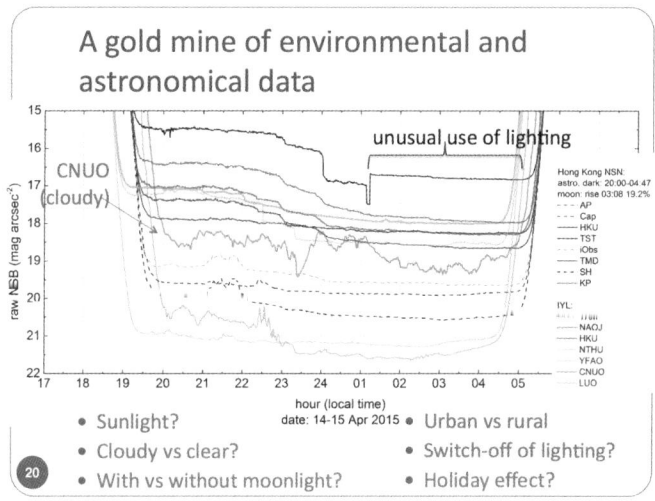

Figure 5: The Sky Quality Meter measurements provide a plethora of environmental and astronomical data.

The SQM-LE has a spectral response function centered at 540 nm, with a full-width half-maximum of 240 nm, roughly corresponding to the spectral sensitivity of human photopic vision. On the sky, the SQM-LE samples a circular area with full-width half-maximum of 20°.

The SQM-LE was chosen for its modest cost ($250) and robustness. A weatherproof housing provided by the manufacurer allows standardized installation and data reduction. A short (30-second) sampling interval has been adopted to record short-term sky brightness variations. Date from the SQM-LE is sent directly to a cloud-based database. Infrastructure requirements are minimal (power, internet and mount); maintenance is minimal—regular calibration is needed to maintain data quality.

Data are available in real-time via a public interface, and have been integrated into the Google Map platform. Existing stations include locations in Taiwan, Japan, Hong Kong,

South Korea and South Africa. Many other potential sites exist, and new installations are strongly encouraged with the aim of producing a live light pollution map of the world. (See Figures 4 and 5.)

5. Dark Sky Meter - International Year of Light 2015 app

Authors: P. Russo, T. Sankatsing Nava, Leiden Observatory, Leiden University, Leiden, Netherlands

Many people now have smart phones, and the intent of the Dark Sky Meter app is to provide everyone who has a smart phone an easy way to measure the brightness of the night sky. The cameras in smart phones have improved dramatically in their capabilities over the last few models of phone, including better optics and lower noise levels.

The Dark Sky Meter IYL 2015 edition app measures the night sky brightness using a smart phone. The app is presently limited to the Apple iPhone. The principal reason for not expanding the app to the larger Android phone market is related to hardware. The iPhone has a single manufacturer (Apple), whereas Android phones are made by many different companies. As a result, there are only a few variants of the camera in iPhones (corresponding to different models of iPhone), but there is a very diverse range of cameras in Android phones. This made it impossible, within the scope of the current app, to expand usage to the cameras in Android phones. This, of course, limits the app to only iPhone owners. However, the iPhone presently has a market share of approximately 40%, so the app can be used by many people.

The night sky brightness measured by the Dark Sky Meter app was compared to measurements from the Unihedron Sky Quality Meter (SQM) over a range of sky brightness spanning 15.5–21.5 mag arcsec^{-2}. Some systematic differences are evident, but these can be corrected in future versions of the app. (See Figure 6.)

From the reviews of the app, it is clear that some users are trying to use the app under street lamps or under a full moon. This produces inconsistent and unusable results, and also produces user frustration, producing the illusion that the app is not useful and flawed.

Plans for the future (version 2.0) include user interface improvement and better instructions, analysis of the data, finding funding the keep the app free, developing educational material, and demonstrating the relationship between sky brightness and energy waste. An observing campaign is planned for September 2015 during the International Year of Light 100 hours of light.

6. in situ Measures of LED Installations: Results of Air and Ground Surveys

Authors: E.R. Craine, B.L. Craine, Western Research Company, Tucson, Arizona, United States, E.R. Craine, B.L. Craine, STEM Laboratory, Inc., Tucson, Arizona, United States

Light Emitting Diode (LED) outdoor light fixtures of different types are rapidly proliferating in many communities, particularly in the form of continuous roadway, work, and parking lot lights. These lights offer a wide range of benefits, but many in the astronomical community have expressed various concerns about their impact on local observatory facilities. We have spent several years developing complementary ground-based and aerial techniques of measuring light installations in the field. Unfortunately, large community retrofits of lighting preclude comprehensive measurement of the changes that result unless

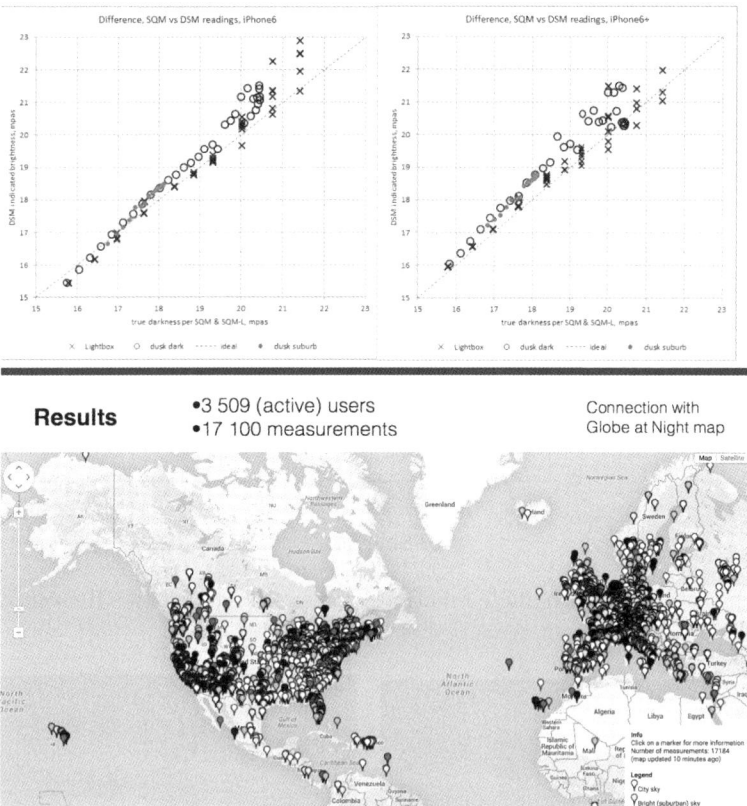

Figure 6: The Dark Sky Meter app performance compared with Sky Quality Meters' plus a world map of over 17,000 measurements

baseline data have been collected prior to completion of the new installations. Because of the rapidity of conversion to LEDs, it is increasingly difficult to conduct informative before and after surveys.

Our LED measurement projects include the Rosemount Copper Mine, Tucson International Airport, and Sierra Vista (all in Arizona, US), and Waikoloa Village and Kawaihae Harbor (both in Hawaii, US). The measurement protocols include static ground surveys, mobile ground surveys, airborne surveys, and satellite data surveys. The Hawaii locations are undergoing changes from low-pressure sodium lighting to filtered LED lighting.

The survey data show, for example, that LED lighting is typically much more uniform than high-intensity discharge lighting (e.g., high-pressure sodium), and can also highlight installation or design errors.

As a point of interest to astronomers, we offer examples of some in situ measurements of LED installations, compare those measurements to results for older light fixtures, and discuss some of the implications for astronomy. These objective data may be helpful in reaching an informed perspective on how LED lights perform in typical settings.

Quantitative *in situ* measurements are pivotal to understanding the impact of lighting strategies on artificial light at night. Although models are interesting, real observational data lead to the truth. LEDs are the lighting of the future because of their economics. But the spectral energy distribution of LEDs is less desirable from the dark sky perspective.

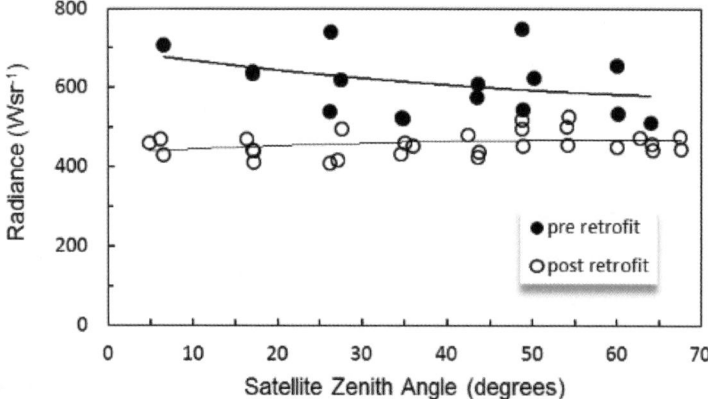

Figure 7: The satellite zenith angle function versus radiance for the Tucson International Airport (TIA) "LED Retrofit Area".

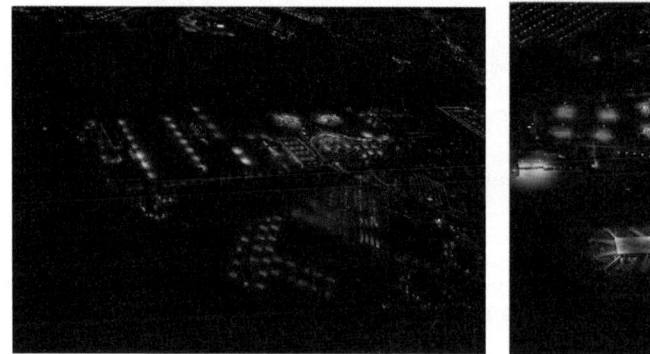

Figure 8: Seeing the task oriented footprints on the ground of the parking lot area of TIA from the Airborne Survey.

Figure 9: Seeing the task oriented footprints on the ground of the terminalarea of TIA from the Airborne Survey.

LEDs can reduce overall brightness levels. A full cutoff design is not the same as a full cutoff installation. Multimodal quantitative measurement of installed fixture removes the ambiguity about installations, provides feedback for improvement, and allows monitoring to validate the status of light installations.

Session 21.3 – Radio and Optical Site Protection

Ramotholo Sefako

South African Astronomical Observatory,
PO Box 9, Observatory, 7935, South Africa
email: rrs@saao.ac.za

Abstract. Advancement in radio technology means that radio astronomy has to share the radio spectrum with many other non-astronomical activities, majority of which increase radio frequency interference (RFI), and therefore detrimentally affecting the radio observations at the observatory sites. Major radio facilities such as the SKA, in both South Africa and Australia, and the Five-hundred-meter Aperture Spherical radio Telescope (FAST) in China will be very sensitive, and therefore require protection against RFI.

In the case of optical astronomy, the growing urbanisation and industrialisation led to optical astronomy becoming impossible near major cities due to light and dust pollution. Major optical and IR observatories are forced to be far away in remote areas, where light pollution is not yet extreme. The same is true for radio observatories, which have to be sited away from highly RFI affected areas near populated regions and major cities.

In this review, based on the Focus Meeting 21 (FM21) oral presentations at the IAU General Assembly on 11 August 2015†, we give an overview of the mechanisms that have evolved to provide statutory protection for radio astronomy observing, successes (e.g at 21 cm HI line), defeats and challenges at other parts of the spectrum. We discuss the available legislative initiatives to protect the radio astronomy sites for large projects like SKA (in Australia and South Africa), and FAST against the RFI. For optical protection, we look at light pollution with examples of its effect at Xinglong observing station of the National Astronomical Observatories of China (NAOC), Ali Observatory in Tibet, and Asiago Observatory in Italy, as well as the effect of conversion from low pressure sodium lighting to LEDs in the County of Hawaii.

Keywords. Radio astronomy, optical astronomy, RFI, light pollution, spectrum protection

1. Introduction and Background

Both radio and optical astronomy face a number of challenges relating to radio frequency interference (RFI) for radio astronomy, and light and dust pollution for optical and InfraRed (IR) astronomy. Both radio and optical sites require protection to ensure that observations are not detrimentally affected at observing sites. The differences between Radio and Optical with regard to protection strategies are small. Just like optical astronomy a few decades earlier, radio astronomy is no longer able to use radio spectrum everywhere, even for dedicated radio astronomy spectrum. It is therefore necessary for radio astronomy facilities to be installed in remote, radio quite sites far away from densely populated areas.

In some of the cases, the sites suitable for radio astronomy are also suitable for optical and IR astronomy, and therefore it is not uncommon to have both radio and optical observing facilities around the same area, examples include Kitt Peak, La Silla, Atacama Desert and Mauna Kea.

† Focus Meeting 21, Session 3 (FM21.3) presentations are available at http://www.noao.edu/education/IAUGA2015FM21

1.1. *Radio astronomy*

Radio astronomy observations from the ground are quite sensitive to radio interference from radio communications, aviation and satellites, and therefore radio sites need to be optimised for radio astronomy and should have explicit protection areas reserved as radio quiet zones (RQZs) with restrictions and regulations for the use of the spectrum. The use of the spectrum should and is managed in what is called spectrum management. Spectrum management is the politics of access to electromagnetic spectrum from 0 to 3000 GHz, mainly by a UN organ in Geneva called the International Telecommunication Union (ITU) Radiocummunication sector (ITU-R).

According to the IUT-R, a RQZ is meant to be any recognised geographic area within which the usual spectrum management procedures are modified for the specific purpose of reducing or avoiding interference to radio telescopes, thereby maintaining the required standards for quality and availability of observational data (Report ITU-R RA.2259, 2012).

1.2. *Light pollution*

For optical astronomy, light pollution is the most important challenge that need to be addressed and major observatories need to be protected from light pollution using either legislation and/or education. Observatory sites like Xinglong station of the NAOC, Ali Observatory and Asiago Observatory continue to invest in monitoring and looking at ways of reducing light pollution at their sites to ensure that astronomy continues to be viable at those sites. They continue to work on possible regulations that could be used to control light pollution at their sites.

Light pollution is a major threat to large telescopes around the world and astronomical observatories continue to engage their local governments to help set up legislations that protect optical astronomy facilities. Observatories are also involved in outreach and educational programs to teach people about the importance of preserving the night sky and protecting it from light pollution, not only for astronomy but for sky gazers and for economic savings related to reduced and efficient use of night lighting.

2. Radio astronomy protection

2.1. *Spectrum protection for radio astronomy: details, successes, failures, challenges and convergence*

Spectrum management is a big business with far more spectrum managers than astronomers†. It is however an important mechanism that provides statutory protection for radio astronomy observing sites. There have been successes, failures and challenges in the protection of radio astronomy radio quiet zones (RQZs). Successes include the protection of H I at 21 cm and OH at 18 cm, which are currently ONLY observable because radio astronomers having succeeded to protect their spectrum. There have been challenges, e.g. with GLONASS (satellite), which used to obscure 1612 MHz band, but eventually, after protracted negotiations with IUCAF [the Scientific Committee on the Allocation of Frequencies for Radio Astronomy and Space Science], left the 1612 MHz band.

Without care, the 21 cm H I will likely have very restricted availability once wireless broadband spectrum is expanded, and this will have negative effects even to using 21 cm H I as a teaching tool without heavy filtering (Figure 1). Many administrations are extremely cynical about protecting the 21 cm band. Transmitters are also moving to

† From presentation by H.S. Liszt, NRAO, Charlottesville, Virginia, USA

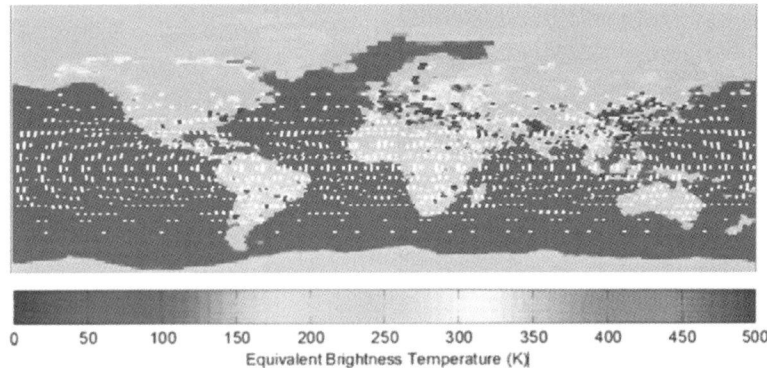

Figure 1: Radio Frequency Interference at the 21cm (or 1400-1427 MHz) HI line.

higher frequencies and broader bandwidths just when our own bandwidths are opening up, and therefore making it even more crucial to have spectrum management.

Spectrum management happens in parallel but partially overlapping sequences, which include **International allocations** at the ITU-R, the highest level, and **National allocations**, which largely track the Radio Regulations. Allocations are only the outline, the rules are the implementation, the details and the devil lives there. The ITU-R does not write the rules, it only provides some guidelines. National rules govern whether the radio allocations are usable, including

- Permitted power levels in shared bands;
- Unwanted emission levels for adjacent/nearby bands, and
- Limitations on operations in the vicinity of radio telescopes (around radio quiet and coordination zones).

Some of the allocated radio bands are unusable in some countries due to their rules, e.g. the 608 - 614 MHz in the US. Usability of 3 - 4 mm spectrum is subject to new radars, 7/8 allocated since 2000 but no rules until now. There is a global struggle to be carried out, country by country. Many of these battles are already lost.

There are a number of functioning spectrum management bodies world-wide, regionally and nationally, including IUCAF (global), CRAF (Europe and South Africa), RACAP (Asia-Pacific) and CORF (USA)†.

2.2. Radio Quiet Protection at the Australian Square Kilometre array site

The Australian Square Kilometre Array (SKA) and its pathfinders site is at the Murchison Radio-Astronomy Observatory (MRO), 350 km from the nearest populated centre. It has a large Radio Quiet Zone (RQZ) that is managed under a range of legislative agreements‡ (Figure 2). Murchison is sparsely populated with a population of about 120 people. The MRO site for SKA covers an area of about 50 km radius centred on the Australian SKA Pathfinder (ASKAP) centre.

There are still a range of challenges regarding the protection of the SKA site in Australia, including the challenge of having to protect a large frequency range, 70 MHz - 25.25 GHz, most of which is unprotected bands. The physical reach of the telescopes (long baselines), the time frames (50 years or more), as well as technical, social and

† See H.S. Liszt presentation for websites of these organisations
‡ From presentation by L. Harvey-Smith, CSIRO, Epping, New South Wales, Australia

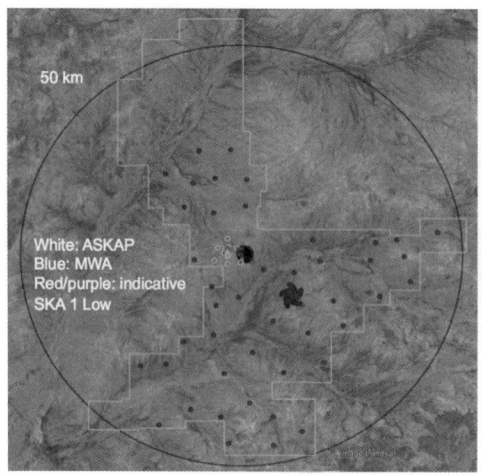 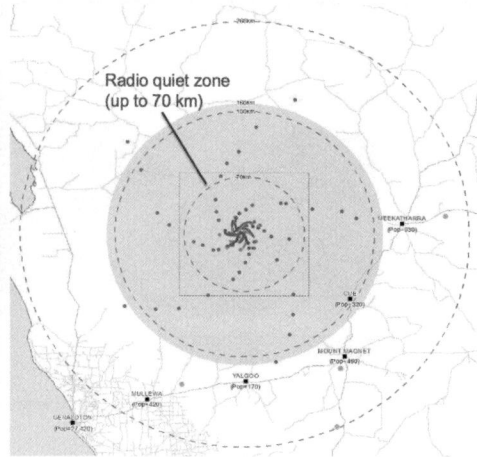

Figure 2: Radio Quiet Zone for the Murchison Radio-Astronomy Observatory.

Figure 3: Radio Quiet Zone for the Australian Square Kilometer Array.

economic implications of restricting radio usage, are some of the challenges faced in trying to protect the spectrum for the SKA. Radio astronomers need to appreciate that other uses of the spectrum are also vital.

The site protection of the Australian SKA include the radio quiet zone up to 70 km radius from the centre of the array where radio astronomy is the primary user of the spectrum, and the coordinated zone, which extends up to 260 km (Figure 3). Between the 70 km and the coordination radius, applicants must assess power spectral density at the centre and power over 50 km radius. Within the 70 km radius, radio communication transmitters are secondary to radio astronomy with regard to spectrum use and control of radio interference. Spectrum licences held by mobile network carriers exclude a region in the RQZ. Users of radio devices within 70 km radius, once notified, must not cause interference to radioastronomy. Activities such as mining within the 70 km radius of the RQZ centre are to submit a radio emission management plan.

Policies are in place for limits on self-generated emissions by the telescopes as well. Every radio frequency transmitter (9 kHz to 275 GHz) must be licensed in Australia. The RQZ, however, does not guarantee absence of radio inference to radioastronomy observations. Current regulations do not cover frequencies below 70 MHz, aircraft and satellite transmissions, and transmitters beyond 260 km from the centre of the observatory.

2.3. On the Development of Radio Astronomy and Protected Astronomy Reserves in South Africa

Recent initiatives to take advantage of various geographic locations in South Africa that exhibit excellent conditions for astronomical observations (optical and radio) have resulted in the establishment of a number of world class astronomical facilities, including the 10-m class Southern African Large Telescope (SALT), the 64-dish MeerKAT radio telescope (under construction), and the future Square Kilometre Array†.

The Northern Cape Province (NCP) in South Africa, which is over 1000 km across, and has very low demand on wireless services due to low population density, was identified as the natural geographic advantage area for both optical and radio astronomy. It is high

† From presentation by A. Tiplady, SKA South Africa, Johannesburg, Gauteng, South Africa

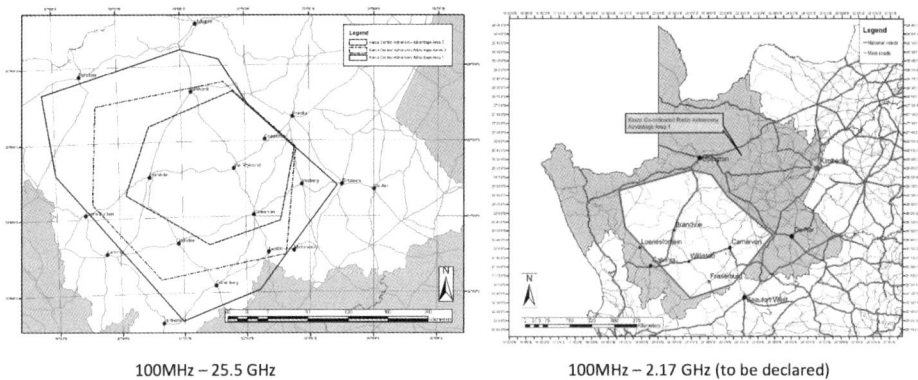

Figure 4: Declared Astronomy Advantage Areas

and dry, and its topographical features provide a shield to radio interference. It is remote, but also has a bulk of infrastructure already available.

To preserve these natural astronomical advantages in the NCP, a unique legislation, the Astronomy Geographic Advantage Act, No. 21 of 2007 (or the AGA Act), was promulgated to establish astronomy reserves. These reserves are protected through a unique set of regulations that enable protection of astronomical facilities located in declared areas from any current, and future, sources of potential interference. The legislation empowers the South African minister for Science and Technology (DST) to declare protected areas around strategic astronomy sites (radio, optical and other multi-wavelength astronomy). The legislation also allocates all the spectrum to radio astronomy services.

The legislation gives three tiers of protected areas, including:
- *Core area* – the physical area of the observatory / instrument;
- *Central area* – surrounds the core area. Minister prohibits certain activities / categories of activities in this area;
- *Coordination area* – the Minister sets standards which activities must comply with.

The AGA Act prevails over existing Electronic Communications Act where protection of radio astronomy is concerned. Protected areas apply to both existing and new activities. The Astronomy Management Authority within the DST is responsible for enforcement of the AGA Act related regulations.

The Central Areas for the radio astronomy have been declared and the 100 MHz to 25.5 GHz bands are protected under the AGA Act regulations (Figure 4). A protection standard, which gives the protection levels that should not be exceeded, has been adopted. There are, however, some drawbacks: Protection levels are inadequate for the next generation surveys, and the narrow band is no longer the only risk, but the broadband EMI (often self-generated) is also becoming a major risk, which can be more detrimental due to complete contamination of the RF spectrum.

Implementation requires buy-in from different stakeholders, such as telecommunication operators and government. Protection requirements are implemented through all government policies. Working with telecommunication operators to optimize existing coverage and reduce impact, as well as development of alternative means of delivery of telecommunication services, are vital to ensure increased spectrum efficiency.

Management policies require that everything brought on to the site is measured, characterized and issued a permit with conditions for use to ensure protection from self-generated EMI.

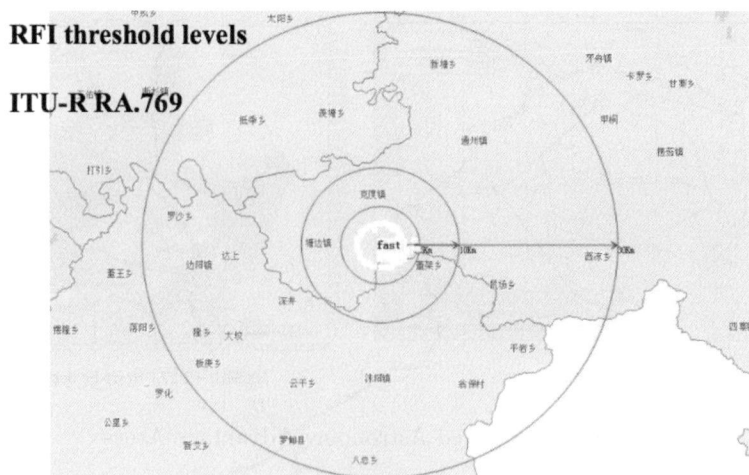

Figure 5: The outer circle represents the 30km radius of the FAST Radio Quiet Zone.

2.4. *RFI Mitigation for FAST*

The Five-hundred-meter Aperture Spherical radio Telescope (FAST) is a Chinese megascience project to build the largest single dish radio telescope in the world. The construction was officially commenced in March ,2011, and the first light of FAST is expected in 2016†.

Due to the high sensitivity of FAST, RFI mitigation for the telescope is required to ensure the realization of the FAST scientific goals. In order to protect the radio environment around FAST efficiently, the local government has established a radio quiet zone with 30 km radius around the site (Figure 5). Moreover, Electromagnetic Compatibility (EMC) designs and measurements for FAST have also been carried out and some examples, such as EMC designs for actuator and focus cabin, have been introduced.

3. Light pollution and protection measures

3.1. *Light pollution and site protecting of the Xinglong Station, NAOC*

The Xinglong station of National Astronomical Observatories of China (NAOC) is one of the most important astronomical sites in China. The site has 9 optical telescopes at the station, including LAMOST, a 2.16 meter reflecting telescope and several other optical/IR telescopes‡.

The results of urbanisation has had negative effects on the night sky brightness at Xinglong observing station. Using Walker (1977) Population-Distance relationship model, $P \sim D^{2.5}$ shows that Xinglong observing station is at same level of light pollution as the United States was in the 1970s.

Sky brightness measurements carried out in the past 15 years between 1996 and 2011 show that Xinglong station's brightness has increased by 0.5 magnitude in i-band (Figure 6). Measurements using SQM-LE meters in February 2011 give night sky brightness of about 19.9 to 20.0 in V. Monitoring of the night sky brightness using SQM is done

---

† From presentation by H. Zhang, National Astronomical Observatories of CAS, Key Laboratory of Radio Astronomy of CAS, Beijing, China

‡ From presentation by Y. Zhao, National Astronomical Observatories of China, Beijing, China

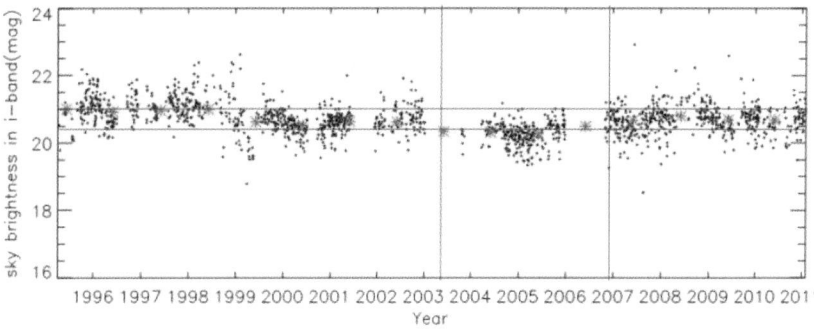

Figure 6: The Beijing-Arizona-Taipei-Connecticut (BATC) survey shows Sky Quality Meter data from 1996-2011. The i-band shows an increase of 0.5 magnitudes in sky brightness during the 15 years.

regularly to ensure that changes in observational conditions at Xinglong station are observed and the efforts to protect the site are implemented.

The protection area around Xinglong station is limited to within 5 km radius of the observing site. There are also some efforts to control dust levels around the observing station, as well as educational awareness on the effects of light pollution in the neighbouring villages.

3.2. Site Protection Program and Progress Report of Ali Observatory, Tibet

The Ali Observatory, Tibet, is a promising new site identified through ten year site survey over western China†. It is of significance in establishing guidelines for site protection during site development of the Ali Observatory.

The site protection program is done by looking at five aspects, including site monitoring, technical support, local government support, specific organization, and public education. The long-term sky brightness monitoring is ready with site testing instruments and basic light pollution measurements. The monitoring also includes directions of main light sources, providing periodic reports and suggestions for coordinating meetings. The technical supports with institutes and manufacturers help to publish lighting standards and replace light fixtures. The research done pays special attention to the blue-rich sources of light, which has serious impact on light pollution.

Ali Government has established an official group that leads development and protection of astronomical resources. One of the group's tasks is to issue regulations against light pollution, including special restrictions on airports, mines, and winter heating, as well as supervise lighting inspection and rectification.

A site protection office under the official group and local astronomical society is organized by Ali observatory. The office will coordinate activities at government levels and promote related activities. A specific website operated by the protection office releases activity programs, evaluation results, and technical comparisons with other observatories.

Both the site protection office and Ali Observatory take responsibility for public education, including popular science lectures, light pollution and energy conservation education. Ali Night Sky Park has been constructed and was opened in 2014. The Park provides a popular place and observational experience.

The establishment of Ali Observatory and Night Sky Park has brought unexpected

† From presentation by Y. Yao, National Astronomical Observatories, Chinese Academy of Sciences, Beijing, China

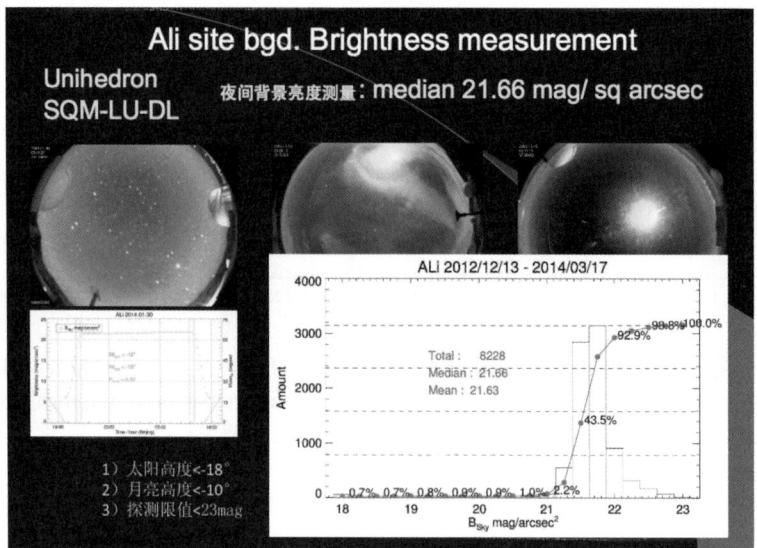

Figure 7: SQM data logger measurements of the night sky brightness of Ali Observatory averaged over a 15 month period (Dec 2012 to March 2014) give a median of 21.66 mag/sq arcsec.

social influence, and the starry sky trip to Ali has become a new form of culture-oriented travels in China. The related news reports and network programs have drawn attention of national top leadership to instruct further investigation into the national support policies.

A number of telescopes are planned to be sited at Ali Observatory, including two 1-m LCOGT telescopes, which are under construction, and a 1.0m infrared solar telescope. Average night sky brightness measurements using SQM-LU-DL during 2012-2014 is 21.7 mag/arcsec^2, which makes the Ali Observatory one of the potentially good observatories for optical astronomy in terms of limited light pollution (Figure 7).

3.3. *County of Hawaii - A Unique LED Street Light Conversion*

In 2010 the County of Hawaii was paying \$0.40/kW-Hr for electricity, \$1.5 million annual bill for 8,500 street lights. Over the past 20 years, costs have increased on an annual average of 7%. Inventory maintenance frequency for the 8,500 lights was 35%, which meant 3,000 visits per year for maintenance. The current Low Pressure Sodium (LPS) street lights were nearing 20 years of service and a complete replacement was imminent, a significant cost for the County of Hawaii and its 185,000 citizens†.

The astronomy community impact was identified early on and discussions conducted for an acceptable conversion path. Key concerns centred on the blue light content of the LED and reflected light.

A demo project with Federal ARRA funds installed 1,000 LED full cut off fixtures, achieving an energy savings of \$200,000 annually. The results were extremely successful and were loudly applauded by both the general public and the Astronomy Institute. Hence, the Traffic Division recommended to the County administration changing the remaining lights, now numbering 9,000, to new LED lights. The County administration approved the change to the LED lights and an upgrade to the outdoor lighting ordinance.

The remainder of the conversion, amounting to \$6 million for materials and labor, is

† From presentation by R.L. Thiel, Public Works, County of Hawaii, Hilo, Hawaii, USA

expected to yield an energy savings of approximately $800,000 annually with a 5 year recovery of costs that includes both energy savings and maintenance reduction.

Additional benefits achieved from using full cutoff fixtures include reduction in glare for drivers, pedestrians, and elimination of trespass light onto neighbouring residences.

Benefits achieved by using a filtered LED includes reducing blue light to <1%, diffusing the harshness of the direct LED light and the ability to use the most energy efficient lumen producing fixture to achieve in excess of 63% reduction in energy costs.

Additional aspects of this conversion will include steps to gather quantitative data showing reduction in light pollution, aerial and satellite surveys for gathering before and after ground level brightness plots along the roadways, and interpretive spectral analysis on skyward impacts.

3.4. *Checking the light pollution sources at Asiago Astrophysical Observatory from photometric and spectroscopic observations. Results from a unique experiment*

The results of recent sky brightness measurements at Asiago Observatory, in Italy, with the goal of understanding the sources and the propagation of light pollution are presented†, and show some interesting results.

The Asiago Observatory, which has four telescopes in two locations (Pennar and Ekar), is found north-east of Veneto region, and has a population of 4.9 million inhabitants. Although the region's population has been increasing slowly for the last century (by a factor <2; compare with LA with increase by a factor of about 20), its luminance has been increasing at an average of 6% per year. The region appears like a big city at night, with substantial amount of light pollution.

Light pollution regional protection law was developed between 1997 and 2009, with a committee that included astronomers, tasked with checking the effects of the law and reporting to the regional administration. From 2011, three SQMs were installed to monitor the night sky brightness.

There are actually two main models for the light pollution: one is based on a dominant Lambert diffusion. A recent model, instead, includes a cavity effect in the urban centers, limiting the low angles horizontal propagation. The effects of the local vs. distant light centers on the brightness of the night sky at the observatories are different in the two models and the regulations required to limit the light pollution are also different.

A unique experiment was carried out at Asiago Observatory in order to clarify this ambiguity, turning off more than 5000 public street lights in a 500 km^2 area around the telescopes, in a clear, moonless night. The sources of the emission lines in the spectra and their evolution were investigated.

In 2009, the Veneto regional law passed two innovative concepts, one being *upon request from the Astronomical Observatories, in connection with specific events, the local administrations agree to turn off completely (or reduce) the public outdoor illumination, up to 3 nights per year*. The 5000 street lights were indeed turned off during photometric night of 28 March 2014, in a population of about 20,000 people covering about 700 km^2. Photometric measurements were carried out using four SQMs, and observations were obtained at 3 telescopes. The main villages were in silent spectral darkness dominated by shining stars around the zenith.

The sky brightness differences between the March 28 and 30 (another photometric night), 2014, at Ekar and Pennar show some interesting results with a considerably higher gain at the nearby Pennar (50%) than at the relatively far Ekar (30%). There was

† From presentation by S. Ortolani, Dipartimento di Fisica e Astronomia, University of Padova, Padova, Italy

no tight correlation on the gain increase through the night. During the 'dark' night, both sites had the same night sky brightness.

From the results of this unique experiment, it is concluded that (i) distant lights (50-200 km) significantly affect the sky brightness over observatories, and (ii) the contribution of the private illumination, including parking lots, sport activities, bus or train stations, gardens, etc. is relevant and can account for about 50% of the total light pollution, and therefore should be controlled.

4. Summary

Both radio and optical astronomy are affected by man-made activities at ground based observing facilities. Radio observations are sensitive to RFI, and therefore require spectrum management and national legislations to ensure that activities that are detrimental to radio astronomy are regulated. Both South African and Australian governments have promulgated legislations that protect radio astronomy sites for the SKA in those countries. There is also a need to continue to engage with the ITU-R to ensure that standards are set and selected spectra for radio astronomy are reserved only for radio astronomical use.

Radio astronomy in Australia and in South Africa shares the spectrum with many other radio systems. The Australian SKA is naturally radio quiet and it is protected by a range of state and federal legislation and emissions management protocols. In South Africa, the radio quiet astronomy reserves are protected by the AGA Act of 2007, which also protects the optical astronomy instruments such as SALT in the Northern Cape Province.

Optical astronomy at observatory sites affected by light pollution need to engage with their communities and work together with relevant authorities to ensure that suggested mitigations against light pollution are implemented. Only gathering data and confirming that a site is indeed suffering from light pollution is not enough.

References

Mitigating Threats of Light Pollution & Radio Frequency Interference, http://www.noao.edu/education/IAUGA2015FM21, [FM21.3: Radio and Optical Site Protection]

Report ITU-R RA.2259 2012, http://www.itu.int/pub/R-REP-RA.2259-2012

The Astronomy Geographic Advantage Act, 2007 (Act No. 21 of 2007), *Government Gazette* No. 31157, 17 June 2007

Walker 1977, *PASP*, 89, 405

Session 21.4 – World Heritage and the Protection of Working Observatory Sites

Clive Ruggles

School of Archaeology and Ancient History
University of Leicester, Leicester LE8 0PJ, United Kingdom
email: rug@le.ac.uk

Abstract. This joint session between FM21 and FM2 ("Astronomical Heritage: Progressing the UNESCO–IAU Initiative") focused upon the need to preserve the dark skies necessary for the continued functioning of the world's leading optical observatories and whether, if some of the sites concerned could be inscribed on UNESCO's World Heritage List, this could help achieve this objective. Among the main issues addressed were: is a WHL inscription feasible in the first place? how could the strongest case for inscription be made? what progress has been made towards doing this? and what other effects might a WHL inscription have and would they all be desirable to astronomers? Addressing such issues involves not only scientific but also heritage and political considerations.

Keywords. Astronomical heritage, World Heritage, Leading observatories, High-mountain observatories

1. Introduction

Dark skies cannot of themselves be recognised under the World Heritage Convention. However, light pollution not only affects night sky quality but also affects the integrity of other resources and, indeed, whole ecosystems. It is also linked to the issue of energy waste through lighting. These factors affect the sustainable management of both cultural and natural sites, including existing and potential World Heritage Sites.

This—the first of two joint sessions with FM2 ("Astronomical Heritage: Progressing the UNESCO–IAU Initiative") dealing with preserving dark skies and protecting against light pollution in a World Heritage framework—focused upon working observatory sites.

The first four presentations were all by contributors from Chile, which is playing a vital role in the "Windows to the Universe" project, one of the main nomination projects being advanced within the UNESCO–IAU Astronomy and World Heritage Initiative. The last of the Chilean contributions, together with a further three papers, described efforts to protect observatories from light pollution and/or radio frequency interference at various leading observatories. The remaining paper outlined, both from an astronomical and a heritage perspective, a practical approach to the recognition of Dark Sky places as possible World Heritage sites using the Pic du Midi Observatory as a case study.

2. The protection of dark skies in Chile

Malcolm Smith (***The AURA Observatory in Chile—part of the IAU/UNESCO Extended Case Study***) described the "Windows to the Universe" concept and how, as is hoped, it could serve as the basis for an international serial (multi-site) nomination for the inscription on UNESCO's World Heritage List of outstanding, high-mountain, ground-based, observatory sites developed over the period 1870–2000. An Extended Case Study

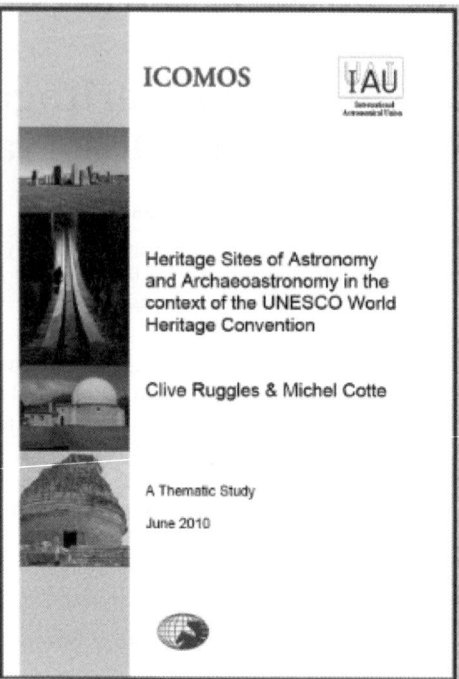

Figure 1: The publication of the UNESCO–IAU Portal to the Heritage of Astronomy

for AURA-O as a "Window to the Universe" has been prepared for inclusion in the second ICOMOS–IAU Thematic Study on astronomical heritage, to be published later this year (Ruggles & Cotte 2015), and has been published on the UNESCO–IAU Portal to the Heritage of Astronomy (http://www.astronomicalheritage.net/index.php/show-entity?identity=000059&idsubentity=005).

AURA–O includes the Cerro Tololo InterAmerican Observatory (CTIO) established in 1962 as the first of the major international observatories to be installed in Chile. The future of AURA–O now includes the Large Synoptic Survey Telescope (LSST). This Extended Case Study has provided the context for the development of possible initiatives to protect a variety of sites in Chile (e.g. Tololo, Pachón, La Silla, Las Campanas, Paranal, Armazones and Chajnantor) for their historical and scientific value to humanity. The dark skies and ideal weather patterns of northern Chile, along with its location in the southern hemisphere, have made this area of the world a major centre for astronomical facilities.

Following this presentation, Ambassador Gabriel Rodríguez G-H, Director for Energy, Science & Technology and Innovation in the Chilean Ministry of Foreign Affairs, was invited to give an example of a government's response to this initiative and its implementation in practice. His report is quoted here in detail:

Chile: an experience of government policy for the protection of "dark skies"

2.1. *The context*

Astronomy has been increasingly present in Chile during the last fifty years. As of today, almost 50% of all the optical and radio-astronomical observational capabilities on the entire planet are located in the arid northern regions of our country. Considering the new big astronomical projects recently inaugurated (ALMA), or planned to be built in

Chile in the next five to ten years (LSST, GMT, TAO, CCAT), by 2022–25 our country will host almost three quarters of the world's astronomical observation capabilities.

Just in the last decade, Chile has started to be fully aware of this exceptional situation and its consequences, in terms both of opportunities and responsibilities.

2.2. Why Chile?

For astronomers, it is pretty obvious that a clean and dry atmosphere, combined with high-altitude sites, sparsely populated areas, and dark skies, all of which exist practically year round, is an ideal mix of conditions for observing the universe. Adding the location in the southern hemisphere, where observations of our galaxy's center and the rest of the universe are better and easier, Chile's northern regions are a privileged place for optical and radio-astronomical observations.

In this sense, Chile has a "natural resource"—clean skies—no different from our resources of copper if we view them from an economical perspective. Clean and dark skies are a natural resource that we can rent, gaining a financial return, or "exploit", extracting some added value from it.

Since the 1960s, when international astronomical activities began in Chile, our country has facilitated the installation of scientific projects in the "Norte Chico" (350 km north of Santiago) and the Atacama desert. In that sense, and initially not being fully aware, the government has developed a "policy" for attracting astronomy projects to our country.

2.3. Chile as a "natural lab"

The Atacama desert represents for astronomy what Antarctica does for polar and subantarctic science, the Humbolt current for oceanography, the Cordillera de los Andes for vulcanology, and the geothermal resources and the geological Andean faults for seismology. Chile has unique natural conditions. We call them "natural labs" (a name coined by Chilean scientist José Miguel Aguilera), among which astronomy is the best known. This makes our country an exceptional place for scientific research.

All this represents a big opportunity—"an astronomy platform" for our country—and simultaneously a very serious responsibility: how to protect these "natural labs" for humankind's solace, curiosity and progress.

2.4. Government astronomy policy

Chile has a responsibility to preserve these exceptional sites for humanity, and we are determined to be world leaders in the challenge of protecting "dark skies" for the above-mentioned purposes.

With that in mind, the Chilean government has developed a four-tier policy which combines our national interests with global ones. We seek to answer the question: "how can we create virtuous connections between astronomy and national development?"

An obvious answer is **Astronomy** as a **Science**. We already have a community of more than a hundred Chilean astronomers, and around the same number of future scientists in the pipeline. We are supporting the development of science, and astronomy is a key scence. Today, around 10% of the papers published as a result of research carried out at the telescopes installed in Chile have Chilean scientists as authors.

The second pillar is **Astroengineering**. We can view all the astronomical observatories in Chile as scientific labs, but also as a gigantic engineering school for big-data technologies, software development, telecommunications, robotics, optics, electromechanical devices, anti-seismic engineering, electronics, new materials, and so on. Access to high-end technologies that can facilitate the identification of opportunities for development and for training have advanced human capital in key areas for our country, enabling it to jump ahead to reach development. Robotics today is central to unmanned mining, big data for genetics or geology, and optics for solar energy, just to mention a few examples.

Education and **Culture** is the third pillar. Astronomy is an attractive niche to entice young students into science, and science is central for development. There is no technological innovation and social progress without science. At the same time, astronomy is more than just science. It is culture. Astronomy has to do with understanding who we are and where we are. These are deep questions whose answers can illuminate personal daily lives and societies.

Finally, the **Involvement of Society** is crucial. Government can and must play a central role in identifying opportunities around astronomy, but in the end society itself is the key actor in taking this challenge in hand. Government has the duty to involve citizens in protecting the natural resource of dark and clean skies that is the basis for the human right to observe the stars with the naked eye. And it is part of the citizens' duty to protect skies not only as a natural lab for astronomy, but also as places for "astrotourism" and solace.

2.5. *Protecting "dark skies"*

The Chilean government is working in several areas to implement this astronomy policy.
- The Ministry of Foreign Affairs is a "one-stop" contact entry point for international astronomical consortia looking for a place in Chile.
- International Observatories can be granted "diplomatic status".
- Clearly regulated land property access and long-term concessions for scientific installations.
- Declaration of "scientific protected areas" and buffer zones to isolate the observatories from industrial (and particularly mining geothermal) activities, including wavelength interference.
- Infrastructure facilitation.
- Clear rules and technical norms for public lighting, and enforcement of legal capabilities.
- Environmental regulations for dust pollution.
- Energy supply. Energy generation and transmission in Chile is privately owned, but government can help create the best conditions for facilitating the energy supply to scientific observatories.

Because we are convinced that the natural labs we have are not permanent realities, protecting clean and dark skies, in the case of astronomy, is one the main objectives of this policy.

The UNESCO World Heritage Programme is looking to develop a new area of World Heritage: places that have universal value based on their outstanding scientific activities and discoveries.

Based on the definitions and objectives of our own astronomy policy and the opportunity that UNESCO opens up to us, Chile has taken the decision to pursue the declaration of World Heritage through UNESCO for its astronomical sites. There is a long road ahead but at the same time the whole process is an opportunity to involve society in this quest. Involvement is a two way process. On one side is a government responsibility to develop awareness between citizens of the value we already have in our country. On the other side is the responsibility of the Observatories to become involved in local realities and communities and to show the social value of what they are doing. Education and outreach are central to this.

There are many intermediate steps and possible paths. We can move ahead with a local declaration (involving only Chile) or a serial one, at an international level. We can select groups of astronomical sites based on their similar scientific characteristics or based on historical features. In all cases, however, it is the need to demonstrate Outstanding Universal Value that is the key element in determining the possibility of a World Heritage Declaration.

2.6. Conclusions

Chile's astronomy policy, the four pillars, is an example of how the government can support international astronomical research through protecting our privileged natural labs, while at the same time identifying how astronomy can be a key factor in pursuing development objectives that can have clear and effective social impact on our citizens.

Ambassador Rodríguez's report drew an enthusiastic response from several contributors in the ensuing discussion, who noted with approval the positive attitude and constructive approach taken by the Chilean government towards international astronomical research and maintaining the dark skies upon which it depends.

The following report by Chris Smith *et al.* (**Site protection efforts at the AURA Observatory in Chile**) described efforts over the past 20-30 years to highlight the importance of site protection at the AURA Observatory site in northern Chile through education and public outreach as well as through more recent promotion of IDA certifications in the region and support for the World Heritage initiatives. The site, which includes both Cerro Tololo and Cerro Pachón, has been operational for over 50 years, and hosts more than 20 operational telescopes, ranging from small projects with 0.4m telescopes to the Blanco 4m, the SOAR 4.1m, and the 8m Gemini-South telescopes, with construction of the Large Synoptic Survey Telescope (LSST) having recently begun.

The AURA–O faces a variety of challenges to its long-term future, particularly from mining and from lighting and development. One way forward has been through collaborative efforts with the Chilean government at local, regional, and national levels: in 2013 the Chilean government issued updated regulations for outdoor lighting in the region, which are beginning to be implemented. Another is to strengthen ties with the Chilean community through educational programs and outreach activities, in an effort both to get the public excited about the science and the wonders of the Universe and also to sensitize them to the fragile condition of the dark skies and the threats those skies face. Similar programs and activities are also being undertaken by other international observatories in Chile, including the European Southern Observatory, Carnegie Observatory, and the Giant Magellan Telescope.

Recently the AURA-O site was recognized as the first international Dark Sky Sanctuary by the International Dark-Sky Association, in recognition of the importance of the dark skies in northern Chile and the efforts to protect them. These efforts of international recognition are being furthered by the Chilean government through its initiative to propose key astronomical sites in northern Chile as UNESCO World Heritage sites. Such initiatives and collaborations aim to ensure that the skies of northern Chile remain dark for decades, if not centuries, to come.

3. Global efforts to protect observatories from light pollution and/or radio frequency interference

Pedro Sanhueza *et al.* (**Highlights of the new Emission Norm for the Regulation of Light Pollution in Northern Chile**) described efforts to address shortcomings in the first lighting regulations passed in Chile in 1998 (DS 686/1998) regarding scattering and overillumination, and to deal with the new menace of LEDs.

A new version of the regulations, developed by OPCC in collaboration with the Chilean

Ministry of Environment (MMA), was approved by Presidential decree (DS 043/2012) in May 2013. This new environmental standard includes the following main restrictions:

- A full-cut-off requirement for general lighting: a maximum of 0.49cd/KLm at 90 degrees (i.e., no light distribution above the horizontal).
- For sport and recreational activities, an allowed level of 10cd/Klumen at 90 degrees, together with a visor to cut upper-hemisphere emissions.
- Spectral restrictions divided into three regulated regions of the visible spectrum (as compared to the total light emission between 380 and 780nm):
  - not more than 15% of total light emission in the range 300 to 380nm;
  - not more than 15% in the range 380 to 499nm; and
  - not more than 50% in the range 781nm to 1000nm.
- Overillumination restricted to not more than 20% over the Chilean standard (NSEG 9 n71) for minimal levels in public lighting.
- Billboards with inner sources of illuminations (LED or large plasma screens) must emit no more than 50 cd/m^2 at night. No spectral restriction is applied.
- Externally illuminated billboards must have full cut-off and be installed horizontally.

This new lighting regulation has yet not come into force, owing to a delay in approving complementary technical protocols. Enforcement is also a critical issue, given that the institutional environmental framework in Chile is being modified.

For the future, the OPCC is working with both the Ministry of Public Works and also the Ministry of Housing, seeking to go beyond the new lighting regulation by applying a stronger approach in terms of spectral restriction, promoting the use of warm white LEDs with a CCT of 2700K and, in the case of outdoor illumination near professional observatories, monochromatic amber LEDs.

Richard Wainscoat (**Protection of Hawaii's observatories from light pollution and radio frequency interference**) explained that in the United States (as opposed to Chile or the Canary Islands), lighting ordinances are typically at the county or (occasionally) state level, meaning that astronomers have to deal with all levels of government: for example, military installations do not have to follow lighting ordinances.

The dark night sky over Mauna Kea—the largest collection of optical and infrared telescopes in the world—has been well protected by a strong lighting ordinance issued by Hawaii county, and remains very dark (Figure 2). Full shielding is now required for all new lights; there are now strict limits on blue light; and filtered LEDs are typically required for applications where white light is not required. The National Park Service night sky team visited Mauna Kea in July 2011, and found it to have a darker night sky than any of the US National Parks that they had visited. Astronomers have been working with Hawaii county's consultant on a revision of the Ordinance with, for example, widespread limitations on blue light. Enforcement is a key issue, with present enforcement being very lax, and the new draft includes new penalties for violations.

Haleakala Observatory on Maui—home to the Pan-STARRS telescopes, the Faulkes Telescope North, solar telescopes, and military telescopes—is more threatened, because Maui has a weaker lighting ordinance, and it is a smaller island, meaning that people live and work closer to the telescopes. Haleakala is also closer to Honolulu, and the urban glow from Honolulu contributes to an artificially bright sky towards the northwest (Figure 3).

Oahu has no lighting ordinance. The county is planning to replace all streetlights with 4000K LEDs, most but not all of which will be fully shielded, but the use of 4000K, rather a warmer colour, is troubling.

As regards radio frequency interference, on Mauna Kea fixed radio transmitters are banned, as is Wi-fi, and cellular phones must be turned off. On Haleakala powerful TV

Figure 2: A view from Mauna Kea of the light from Waimea and Waikoloa (on the island of Hawaii)

Figure 3: A close up of light from Waimea and Waikoloa and Honolulu on Oahu, Haleakala and the island of Maui.

transmitters have been removed from the summit area. In both areas weather radar is sector blanked. On Haleakala there are still FAA radios and other government transmitters in the summit area, and there is little chance that these will be removed; however it is not believed that these interfere with Pan-STARRS.

Richard Green et al. (***A Tale of Two Regions: Site Protection Experience and Updated Regulations in Arizona and the Canary Islands***) compared experiences at two of the world's largest concentrations of telescopes containing some of the world's largest telescopes. Active site protection efforts are underway in both regions; the common challenge is to stay ahead of the LED revolution in outdoor lighting.

In Arizona, observatories and space science institutions are working together for site protection and economic leverage (Figure 4). Initial activities included drafting a model outdoor lighting code and the creation of presentation materials for the public and policy makers. The current focus is upon working with localities on lighting code updates and enforcement issues. A successful statewide conference was held in 2014 to raise awareness among public officials about issues of light pollution for astronomy, safety, wildlife and public health (see www.keystone.org/darkskies). There has been progress in improving lighting codes in the City of Tucson, smaller communities in northern Arizona (e.g. Winslow) and Sierra Vista, but there State-level battle in 2012 over a legislative bill to extend lighted signs into previously protected areas: the outcome was that the Phoenix metropolitan area can now have bright, animated LED signs, but they are prohibited in much of the rest of the State. Other major challenges include the fact that the city of Phoenix has invited proposals for 90,000 replacement street lighting fixtures at a CCT of 4000K, having rejected a request for the use of 3000K (note the parallel to Honolulu). Another is that, as a result of current copper prices, the mining industry is seeking significant expansion. One contested development is at the foot of Mt. Hopkins (FL Whipple Obs. including MMTO) which would become the model lighting plant for outdoor mining operations, in violation of lighting protection zone limits; there will be a jurisdictional dispute.

In the Canary Islands, the 1988 "Sky Law" (31/1988) regulates light pollution, electromagnetic pollution, aircraft flight protected airspace, and air pollution. For the Oficina Técnica para la Protección de la Calidad del Cielo (OTPC), light pollution is the most time-consuming aspect: for example, the Canarian observatories are surrounded by National Parks and natural protected areas, so there are no air pollution problems. There is a range of legal lighting requirements, including no upward flux, switch-off or reduction

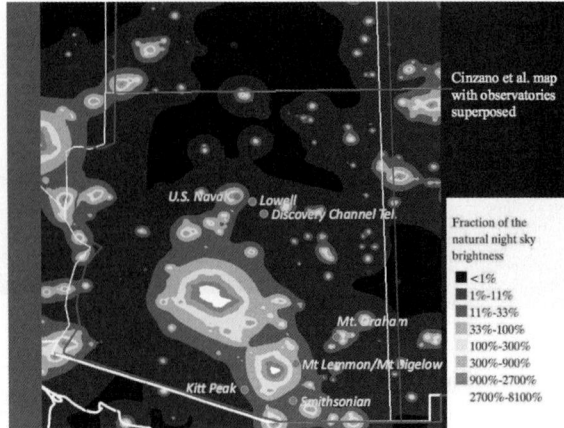

Figure 4: A map of Arizona from Cinzano et al.'s "World Atlas of the Artificial Night Sky Brightness" with observatory locations overlaid.

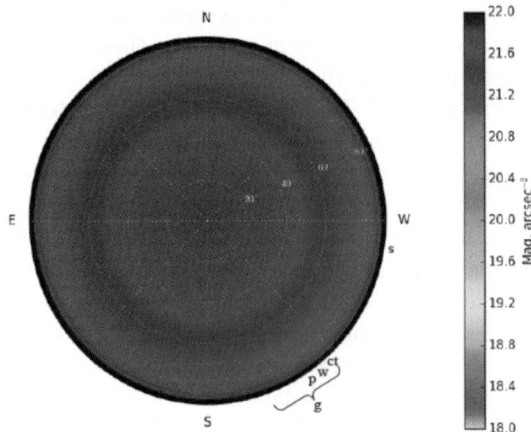

Figure 5: An all-sky image of night sky brightness in mag/sq arcsec. The "s", "ct", "w" and "p" denote the cities of Sutherland (14km from SAAO), Cape Town (360 km), Worcester (254 km), Paarl (304 km), respectively.

at midnight, and (generally) no emissions under 550nm. However, there are more flexible criteria for small pedestrian areas in gardens or surrounding buildings, and there are no colour restrictions on sport, advertisement and ornamental lighting. In the special case of La Palma, there are additional restrictions. The OTPC gives free technical advice, issues mandatory technical reports, certifies light devices, acts on complaints, collaborates with public bodies and associations, and undertakes popular outreach.

Ramotholo Sefako (***Protection of SAAO observing site against light and dust pollution***) described some issues faced by the South African Astronomical Observatory (SAAO) observing station near Sutherland, one of the darkest sites in the world for optical and IR astronomy (Figure 5). It hosts and operates several facilities, including the Southern African Large Telescope (SALT) and a number of international robotic telescopes.

In South Africa, the Astronomy Geographic Advantage (AGA) Act of 2007 empowers the Department of Science and Technology (DST) to regulate activities that pose a threat

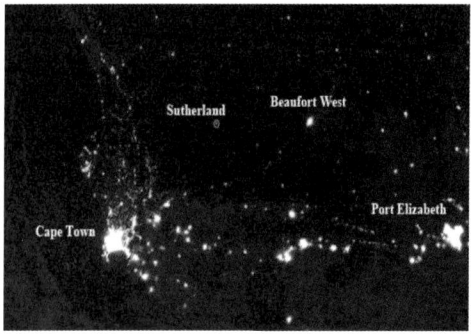 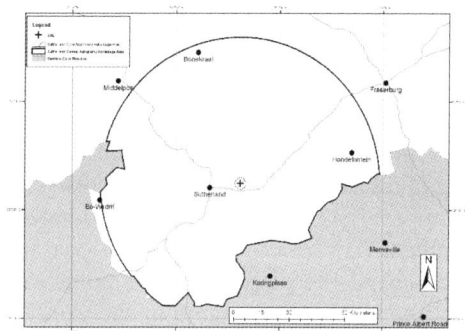

Figure 6: Nighttime image from orbit of Cape Town and cities near Sutherland. Cape Town is 360 km from Sutherland.

Figure 7: The Sutherland Central Astronomy Advantage Area (about 60 km radius). The Northern Cape and Western Cape provinces are shown in white and grey, respectively.

to optical and/or radio astronomy in areas declared Astronomy Advantage Areas (Karoo and Sutherland) (Figure 7). Such activities include mining or prospecting, outdoor lighting, power generation and "harmful industrial activities".

There are three tiers of protected areas:

- Core area—the physical area of the observatory or instrument;
- Central area—an area around the Core where the Minister can prohibit certain activities or categories of activity;
- Coordinated area—where the Minister sets standards with which activities must comply.

Challenges include planned developments around Sutherland, namely oil and gas exploration and exploitation (including fracking), mining, and wind farms. In particular, several wind farms are planned around Sutherland, within the central AAA; for example, a planned farm 20–25 km from the observatory proposes 50 wind turbines, lit, almost all of which will be visible from SALT/ Sutherland observatory. Astronomers continue to engage with different energy developers and local communities around the Sutherland Central AAA to find ways of protecting the observatory while at the same time permitting renewable energy developments.

Voluntary compliance continues to be the main strategy for protection of the observatory against light and dust pollution. This means promoting understanding and education: building good relationships with the public and the local community is an essential component of protecting South Africa's astronomical sites.

4. A practical, heritage-based approach to the recognition of Dark Sky places as possible World Heritage sites

Combining the perspectives of an astronomer and a heritage professional, Rémi Cabanac and Michel Cotte (*Toward a Serial International Approach of the High Mountain Observatories, within important Dark Sky Value*) considered how Outstanding Universal Value (OUV) might best be demonstrated in the case of a potential serial nomination of high-mountain observatories as "Windows to the Universe" that have maintained very important dark sky properties upon which their important astronomical functions depend.

The quality of the dark sky at a given place, and policies for its conservation, are very important, but insufficient in themselves to justify inscription on the World Heritage List. They must be related to important cultural or/and natural values, in other words significant cultural heritage features relating to astronomy and science and/or other exceptional natural attributes. The Dark Sky place must also demonstrate integrity/authenticity to an appropriate degree for today's tangible heritage of astronomy and have made a very significant contribution to international history of science and astronomy as an intangible attribute of the place. The last point must be demonstrated by a serious comparative analysis with similar places in the world and in the region. In the case of a serial nomination, each individual site must contribute significantly to the OUV of the global series.

The Pic du Midi Observatory in the French Pyrenees is likely to be an important component of such a nomination. It represents the early origin (at the end of the 19th century) of mountain scientific stations and observatories in Europe, with a long, continuous and important astronomical and scientific history lasting until the present with active programs of celestial and atmospheric observation.

5. Further information

Versions of the reports by M. Smith *et al.*, C. Smith *et al.*, and Sefako may be found in the FM2 pages in this volume. PDF versions of all the presentations in this session are available both on the FM2 pages on the UNESCO–IAU Portal to the Heritage of Astronomy (`www.astronomical heritage.net/index.php/community/news-events/focus-meeting-at-iau-general-assembly`) and on the FM21 pages on the NOAO website, `www.noao.edu/education/IAUGA2015FM21`.

Reference

Ruggles, C. & Cotte, M. (eds.) 2015, *Heritage sites of Astronomy and Archaeoastronomy in the Context of the UNESCO World Heritage Convention: Volume II* (Bognor Regis: Ocarina Books), in press

Session 21.5 – Light at Night and Protected Areas

Zhao Yongheng

National Astronomical Observatories
B313, 20A Datun Road, Chaoyang District, Beijing, China, 100012
email: yzhao@bao.ac.cn

1. The Ecological Implications of Light at Night (LAN)

(C. Henshaw, Health Studies and Training Centre, North West Armed Forces Prince Salman Hospital, Tabuk, SAUDI ARABIA)

Light at night (LAN) is now an established environmental problem, not only for astronomers but for the population at large. It has serious ecological effects that are wide ranging, and its environmental effects may be more serious than ever imagined. The ecological and environmental consequences are examined and emphasis is stressed on resolving the problem before it is too late.

Summary:

It has been demonstrated here that lighting has serious environmental implications, and is a contributor to climate change. One of the easiest ways to mitigate climate change is to eliminate unnecessary lighting.

It is a green issue, so environmental organisations in addition to astronomers need to campaign against it. Though light pollution cannot be eliminated entirely, technology is now available that can mitigate its worst effects (Fig. 1).

Public outreach campaigns by environmentalists need to be pro-active in creating awareness of the problem amongst the general public, the lighting industry, commerce, industry, municipalities, sporting and artistic communities and any other parties that may be tempted to abuse light.

Otherwise they will be considered as campaigns by a minor interest groups and may be construed as being a tyranny by a minority.

2. Ecological Impact of LAN: San Pedro Riparian National Conservation Area

(E.R. Craine, B.L. Craine, Western Research Company, Tucson, Arizona, UNITED STATES, STEM Laboratory, Inc., Tucson, Arizona, UNITED STATES)

The San Pedro River in Southeastern Arizona is home to nearly 45% of the 900 total species of birds in the United States; millions of songbirds migrate though this unique flyway every year. As the last undammed river in the Southwest, it has been called one of the "last great places" in the US. Human activity has had striking and highly visible impacts on the San Pedro River. As a result, and to help preserve and conserve the area, much of the region has been designated the San Pedro Riparian National Conservation Area (SPRNCA). Attention has been directed to impacts of population, water depletion, and border fence barriers on the riparian environment. To date, there has been little recognition that light at night (LAN), evolving with the increased local population, could have moderating influences on the area. STEM Laboratory has pioneered techniques of coordinated airborne and ground based measurements of light at night, and

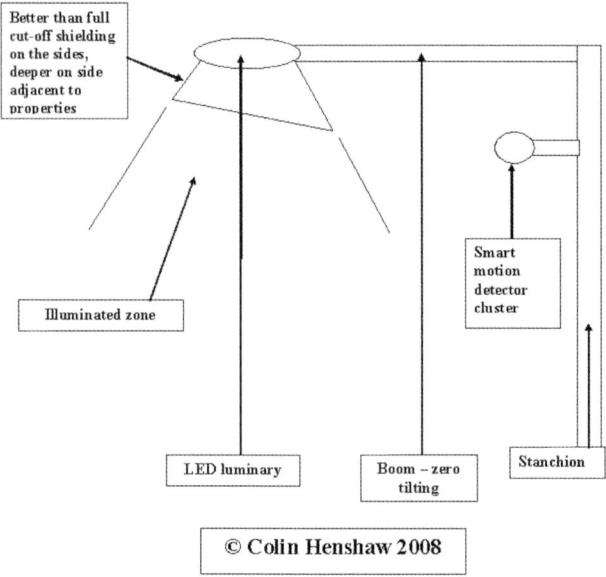

Figure 1. Anti-light pollution street light suitable for residential areas.

has undertaken a program of characterizing LAN in this region. We conducted the first aerial baseline surveys of sky brightness in 2012. Geographic Information Systems (GIS) shapefiles allow comparison and correlation of various biological databases with the LAN data. The goal is to better understand how increased dissemination of night time lighting impacts the distributions, behaviour, and life cycles of biota on this ecosystem. We discuss the baseline measurements, current data collection programs, and some of the implications for specific biological systems.

Summary:

LAN plays important, but largely unexplored, roles in wildlife ecology.

Such interactions can be extremely complex and may require years to characterize.

Comprehensive LAN databases are critical to the conduct of LAN/wildlife behavior research studies.

To understand impacts in specific regions, temporal monitoring is essential (Fig. 2).

3. Stars For Citizens With Urban Star Parks and Lighting Specialists

(Valentin Grigore, The Romanian Society for Meteors and Astronomy, Targoviste, Dambovita, ROMANIA)

General context

One hundred years ago, almost nobody imagine a life without stars every night even in the urban areas. Now, to see a starry sky is a special event for urban citizens.

It is possible to see the stars even inside cities? Yes, but for that we need star parks and lighting specialists as partners.

Educational aspect

The citizens must be able to identify the planets, constellations and other celestial objects in their urban residence. This is part of a basic education. The number of the

Light at Night and Protected Areas 475

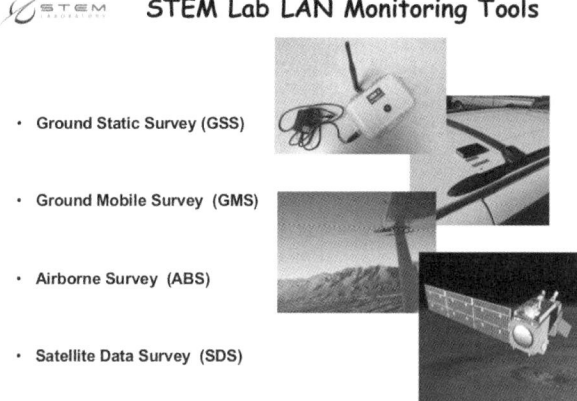

Figure 2. LAN monitoring tools.

Figure 3. Activities for urban community.

people living in the urban area who never see the main constellations or important stars increase every year. We must do something for our urban community (Fig. 3).

What is an urban star park?

An urban public park where we can see the main constellations can be considered an urban star park. There can be organized a lot of activities as practical lessons of astronomy, star parties, etc.

Classification of the urban star parks

A proposal for classification of the urban star parks taking in consideration the quality of the sky and the number of the city inhabitants,Two categories:
· city star parks for cities with < 100.000 inhabitants
· metropolis star parks for cities with > 100.000 inhabitants
Five levels of quality:
1* level = can see stars of at least 1 magnitude with the naked eyes
2* level = at least 2 mag
3* level = at least 3 mag
4* level= at least 4 mag
5* level = at least 5 mag

The urban star urban park structure and lighting system

A possible structure of a urban star park and sky-friend lighting including non-electric illumination are descripted.

The International Commission on Illumination

First: Ali Dark Sky Reserve

Ali Prefecture Government of Tibet, Ali Astronomical Observatory, and IDA Beijing, around the astronomical observation area, planed a 2,500 square km of Dark Sky Reserve.
The elevation is 4,500-6,000m, including the Core Area, the Buffer Area and the Peripheral Zone.

Figure 4. A Dark Sky Reserve around Ali astronomical observatory.

A description of this structure which has as members national commissions from all over the world.

Dark-sky activists - lighting specialists

National Commissions on Illumination organize courses of lighting specialist. Dark-sky activists can become lighting specialists. The author shows his experience in this aspect as a recent lighting specialist and his cooperation with the Romanian National Commission on Illumination working for a law of illumination in Romania and to implement the sky protection elements into the lighting specialist accreditation.

4. The New Progress of the Starry Sky Project of China

(Xiaohua Wang, Starry Sky Project of China, International Dark Sky Association Beijing Chapter, Beijing, CHINA)

Since the 28th General Assembly of IAU, the Starry Sky Project of China (SSPC) team made new progress:

1. Enhanced the function of the SSPC team. Established the contact with IAU C50, IUCN Dark Skies Advisory Group, AWB and IDA,and undertakes the work of the IDA Beijing Chapter. Got supports from China's National Astronomical Observatories, Beijing Planetarium, and Shanghai Science and Technology Museum. Signed cooperation agreements with Lighting Research Center, English Education Group and law Firm; formed the team force.

2. Put forward a proposal to national top institution The SSPC submitted the first proposal about dark sky protection to the Chinese People's Political Consultative Conference.

3. Introduced the Criteria and Guideline of dark sky protection The SSPC team translated 8 documents of IDA, and provided a reference basis for Chinese dark sky protection.

4. Actively establish dark sky places Plan a Dark Sky Reserve around Ali astronomical observatory (5,100m elevation) in Tibet (Fig. 4). China's Xinhua News Agency released the news. Combining with Hangcuo Lake, a National Natural Reserve and Scenic in Tibet, to plan and establish the Dark Sky Park. Cooperated with Shandong Longgang Tourism Group to construct the Dream Sky Theme Park in the suburbs of Jinan city.

In the IYL 2015, the SSPC is getting further development: First, make dark sky protection enter National Ecological Strategy of "Beautiful China". We call on: "Beautiful

China" needs "Beautiful Night Sky"; China should care the shared starry sky, and left this resource and heritage for children. Second, hold "Cosmic Light" exhibition in Shanghai Science and Technology Museum on August.Third, continue to establish Dark Sky Reserve, Park and Theme Park. We want to make these places become the bases of dark sky protection, astronomical education and ecological tourism, and develop into new cultural industry. Fourth, actively join international cooperation.

Now, "Blue Sky, White Cloud and Starry Sky" have become the common pursuit of Chinese society. In order to obtain this goal, the SSPC team would like to pay more efforts.

5. Measuring light pollution in Beijing and effects on Xinglong Station of National Astronomical Observatory

(Ligen LU, B. ZHANG, S. ZENG, Department of Astronomy, Beijing Normal University, Beijing, CHINA; M. AI, Liaoning Province Institute of Metrology, Shenyang, CHINA; J. LIU, National Institute of Metrology, Beijing, CHINA)

A light pollution survey in Beijing has been carried on to assess the quality of the night sky. To measure the absolute luminance of night sky directly, a portable night-sky luminance meter was developed specially for this survey. With a 2-degree field of view, the meter is sensitive only to a narrow cone of the sky and capable of detecting the minimum luminance of $10^{-6} cd/m^2$ (equivalent to 27.4 mag/arcsec2). The night-sky brightness was measured at seven sites, of which six are almost in line but with different distances from the city center. The Xinglong Station of National Astronomical Observatory was included to study the impacts of city lightings on an astronomical observatory. The survey shows that night skies at later time (from 0:00 to 3:00) keep mostly unchanged and are evidently darker than earlier time (e.g. the night-sky at 23:00 is about 40% brighter than midnight), which can be attributed to substantial artificial lightings for human activities being turned off after midnight. Moreover, zenith luminance of the night sky decreases with increasing distance from the city center. Compared with the night-sky luminance (21.50 mag/arcsec2) at Lingshan observation site which is closer to the city center, the night-sky brightness at Xinglong Station is a litter brighter (21.37 mag/arcsec2). This indicates that night sky at Xinglong Station has been brightened by outdoor lighting of the county town of Xinglong. The survey shows that either the luminance of zenith dark sky or the average luminance of skies at 45 degree altitude in all directions could be considered as a reasonable indicator of light pollution.

Summary:
A portable night-sky luminance meter (Fig. 5)
· Independent of the telescope.
· With narrow field of view (2 degrees) and high sensitivity ($10^{-5} cd/m^2$).
Measuring night-sky brightness in Beijing
· Seven observation sites with different distances from the city center, including Xinglong Station, a major optical site for astronomical observations in China.
· Potential risks to night sky.

6. Light pollution modelling the UK Highways Agency new environmental policy, inc. astronomical impact of blue- rich LED luminaires

(Christopher Baddiley, BAA Campaign for Dark skies, Nr. Malvern, UNITED KINGDOM)

Instrument – Night-sky luminance meter

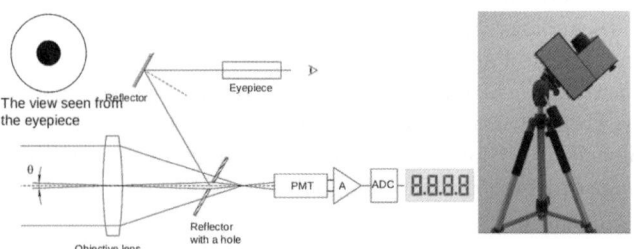

Table 1 Specifications of the Night Sky Luminance Meter.

Objective aperture	50 mm	Focal length	70 mm
Field of view	2 degrees	Sensitivity resolution	10^{-5} cd/m^2
Measurement range	0 ~ 19.99 cd/m^2	Measurement error	± 4.6% (± 0.05 mag/arcsec^2)

Figure 5. A portable night-sky luminance meter.

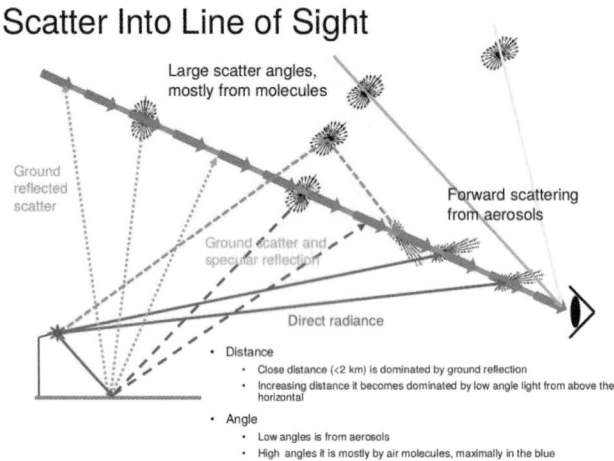

Figure 6. Light pollution model.

The Highways Agency are replacing their policy of full cut off class G6 road lighting specification on motorways (originally based on the author's work), and are adopting a categorised environmental impact based point system that can accommodate technical advances, such as LED lighting. The Skyglow component of this will be based on the modelling of skyglow versus cut-off angle, developed for determining the relative light pollution environmental impact of different streetlight designs, by the author (Fig. 6). Further modelling has been done concerning the effect of LED lighting, which potentially, has highly directional properties. But increasingly used blue rich colour temperatures may increase skyglow by 5 fold, compared to traditional lighting. This is due to enhanced reflection of vegetation and greatly increased atmospheric molecular Rayleigh scattering; a potential astronomical environmental disaster.

Prior to this, the author carried out a dark sky survey of the Malvern Hills area of outstanding natural beauty (AONB), relating it to the same light pollution model. The results confirm the general predictions of the model and also clearly illustrate the relative significance of different designs of light sources at different distances, to the dark sky environment.

The paper also briefly describes the results from the same model adapted to study the night-time environmental impact of a proposed very large sea based wind farm project in the English Channel, as a part of the planning process.

Summary:

The individual red tilted a aviation and navigation warning lights 15 km from landfall will be visible from the coast on the horizon at night. The brightness will be brighter than a bright Star, but dimmer than Jupiter.

The angular distribution is quite different from that caused by normal road lighting in a landscape environment. The reflectivity off the sea is also very low angle.

The skyglow from the navigation lights is limited to near the horizon, and on a clear night will be still less than those from that from external lighting in France.

This is due to their directionality, and being deep red, where scattering is low.

Session 21.6: Preserving Dark Skies and Protecting Against Light Pollution in a World Heritage Framework

Malcolm G. Smith

NOAO/CTIO,
Avenida Juan Cisternas 1600, La Serena, Chile
email: msmith@ctio.noao.edu

Abstract. This session opened with a crucial explanation by Michel Cotte of how astronomers first need to understand how to apply UNESCO World Heritage Criteria if they want to motivate their government(s) to make the case to UNESCO for World Heritage recognition. UNESCO World Heritage cannot be obtained just to protect dark skies.

Much more detail of this and the other presentations in this session, along with many images, can be found at the session website: http://www.noao.edu/education/IAUGA2015FM21.

The next speaker, John Hearnshaw, described the Aoraki Mackenzie International Dark Sky Reserve and the work it carries out . This was followed by a wide-ranging summary (by Dan Duriscoe and Nate Ament) of the U.S. National Park Service (NPS) Night Skies Program. The abstract of Cipriano's Marin's paper, "Developing Starlight connections with UNESCO sites through the Biosphere Smart" was shown in his absence. The final presentation (by Arkadiusz Berlicki, S. Kolomanksi and T. Mrozek) discussed the bi-national Izera Dark Sky Park.

Keywords. atmospheric effects, site testing

1. Overview

As adviser to ICOMOS and IAU, Michel Cotte opened the session by addressing the question of how to apply the UNESCO World Heritage Criteria to the "Windows to the Universe" Sites. He explained the terms of reference issued from the UNESCO World Heritage Convention and the vital details behind establishing the "Justification of the Outstanding Universal Value (OUV) of a given place or site". He then explained how the OUV of a site is expressed by the WH criteria - six of which are possible for cultural value and 4 for natural value. Dark sky is one of the natural attributes giving value to the place (Fig. 1). Unlike all the existing WH sites, sky itself could not be seen as a 'property' in a Human/Juridical (anthropocentric) sense. However, dark-sky quality could also be considered as a cultural attribute in the context of the history of the observatory place. The remarkable atmospheric quality for a given place such as High Mountain Observatories explains their settlement; their foundation as "Windows to the Universe" is partly justified by the dark sky quality. The human presence of the observers and astronomers gives meaning and life to the place, supporting important additional intangible value from its history of knowledge and human representations. The dark sky quality can be presented as a local environmental attribute. Studying a place from a World Heritage perspective starts by an inventory of its tangible attributes. Generally speaking, as mentioned earlier, a given place has a series of natural attributes and cultural attributes supporting and expressing its value. Clearly, intrinsic dark sky quality is a natural attribute of the place, among others. A discussion of dark sky quality must include an analysis of its properties (purity as a minimum of physical constraints). Visible

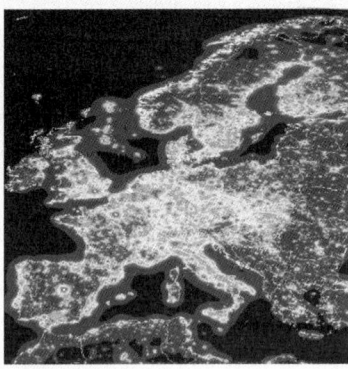 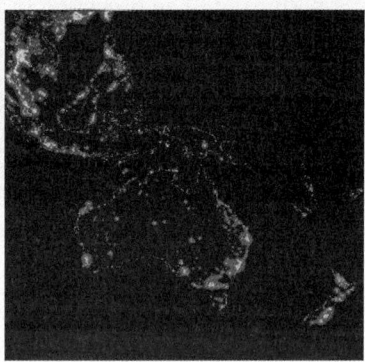

Figure 1: Dark Sky as one of the natural attributes giving value to the "Windows to the Universe" site

relationships between attributes give a landscape value. One can think of dark sky and the atmosphere below it as the "glass" in a "Window to the Universe", i.e. one of the physical attributes giving a specific value to the site. The site itself is then the "Frame" of the window and it has relevant properties in a WH context such as geography, atmosphere, architecture, landscape and nature. Turning to its history, the "Window" has come to be used by humankind with artifacts and instruments.

Thus, in presenting and discussing the OUV of individual sites in a World Heritage context in this session, Michel Cotte reminded us of the need to present Dark Skies among a cluster of natural attributes. We need to consider dark skies as a component of the global natural context of a given place. It belongs to a larger group of natural attributes of the site, forming its natural environmental components. One of the best ways to use Dark-Sky Value is among an ensemble of other remarkable attributes which all combine to constitute a generally remarkable landscape, remarkable at night and remarkable in daylight. Potential OUV is what can result from such a combination of remarkable attributes. Significance and beauty of the whole could be largely the result of the simple addition of individual attributes - which give a landscape specific value. World Heritage Criterion (vii) could be one to concentrate on - a monument of nature.

He also added the wisdom of including Dark Sky management for scientific efficiency and heritage preservation. Points to include are the international success and the importance of the Dark Sky reserves initiatives. Maintenance of the dark-sky quality could be presented as a strong management goal for an astronomical site and region. This leads naturally to the concepts of Dark Sky Reserve and WH Buffer Zones. Detailed examples of such places in different parts of the world provided the content for the rest of this Session.

2. Outstanding Sites & Evolution of Protection

John Hearnshaw spoke next and described the Aoraki Mackenzie International Dark Sky Reserve and light-pollution issues in New Zealand - and how the reserve is managed and promoted to the public to make them aware of light pollution issues and in order to promote star-gazing and astro-tourism. His full presentation can be found at http://www.noao.edu/education/files/IAU_AMIDSR_Hawaii_2015_short.pdf.

New Zealand has consistently been one of the world leaders. In 1981, a Lighting Ordinance was drawn up in the Mackenzie District Plan. It was enacted through the Town and Country Planning Act 1977. The Plan controls outdoor lighting, specifying types of light, full cut-off, limits emission below 440 nm, restricts times when outdoor recreational illumination is permitted. The objective of the ordinance is "Maintenance of the ability to undertake effective research at the Mt. John University Observatory and of the ability to view the quality of the night sky'. The lighting ordinance applies over a large area of the MacKenzie Basin, including all of Lakes Tekapo and Pukaki; this area stretches 60km EW and 100km NS.

John went on to discuss clearly why a Dark Sky Reserve was needed, given the existence of a lighting ordinance for over 30 years. He gave the following answer: (1) A dark-sky reserve is about (a) marketing and branding; (b) education (c) inspiration. (2) A reserve helps promote the romance of astronomy and allows people to connect spiritually with the universe (3) It also reinforces the message on the need to control light pollution.

These aims are very different from those of the lighting ordinance, which is strictly a legal document to protect the night sky, mainly for astronomical research, with public stargazing as a secondary aim, and the development of astro-tourism is not mentioned at all.

AMIDSR is the world's largest IDA International Dark Sky Reserve (4,367 km^2) and the first to obtain gold-tier status (recognized by IDA in June, 2012, following the application made in January of that year). It was the first IDSR in the southern hemisphere (Fig. 2). Views from space (Cinzano et al., Wold Atlas of Artificial Night Sky Brightness, 2001) demonstrate the dramatic contrast between the light (and wasted energy) going up into space from Europe versus that from New Zealand, particularly the western part of the South Island (Fig. 4). Viewed from Mt. John, there is still some light pollution from Tekapo, but the situation is generally good. Tekapo skies are still very dark.

John then mentioned that the World Heritage convention held in Christchurch in 2007 marked the beginning of a campaign to lobby for WH recognition of the dark sky above Mackenzie Basin. At the time this seemed reasonable, given that the Basin adjoins the WH site Te Wahipounamu (which was inscribed in 1990) (Fig. 3). It is an area of magnificent primeval vistas: snow-capped mountains, glaciers, forests, tussock grasslands, lake, rivers, wetlands and over 1,000km of wilderness coastline with OUV. His group completed short and full thematic case studies on the Aoraki Mackenzie site for the IAU-ICOMOS International WG on Astronomy and World Heritage.

As mentioned earlier by Michel Cotte, World Heritage Criteria require OUV (Outstanding Universal Value). John Hearnshaw recognized that two of the WH criteria for natural sites may be relevant to the Mackenzie Basin in New Zealand.

(vii) to contain superlative natural phenomena or areas of exceptional natural beauty and aesthetic importance;

(viii) to be outstanding examples representing major stages of earth's history, including the record of life, significant on-going geological processes in the development of landforms or significant geomorphologic or physiographic features.

In recent years it has become clear to John and to many others that an outstanding night sky free of light pollution is not recognized as OUV by the WHC, and is unlikely to be any time soon. John outlined New Zealand's rather disappointing record - it has only three WH sites and only one in he South Island. It already has 8 natural sites on its tentative list for consideration as WH sites, but there has been no movement on this list in the last 8 years.

As hinted above by John, a possible strategy is to apply for an extension of the adjoining Te Wahipounamu WH site so as to include the Mackenzie Basin, citing criteria (vii) and (viii).

Dan Duriscoe and Nate Ament authored the next presentation covering a huge effort aimed at Night Sky Protection and Restoration in U.S. National Parks. This has come such a long way from early, small attempts from Hawaii that was witnessed in the

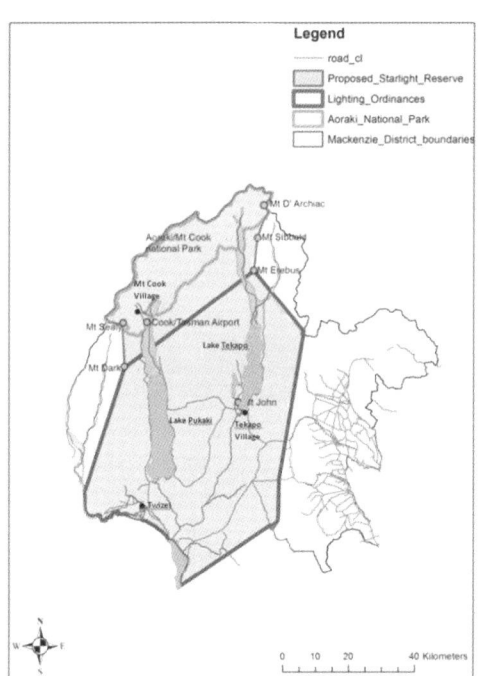
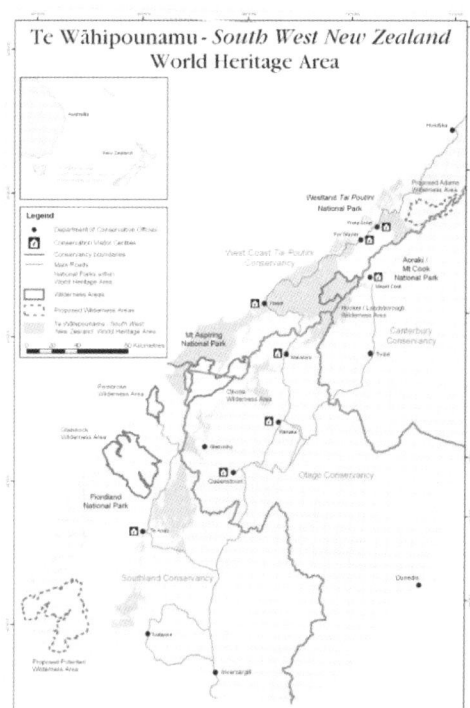

Figure 2: The Aoraki Mackenzie International Dark Sky Reserve, recognized by IDA in June 2012 as the first IDSR in the southern hemisphere.

Figure 3: Te Wahi Pounamu, the South Island of New Zealnd's only World Heritage site, has 2.6 million hectares of OUV wilderness and natural beauty.

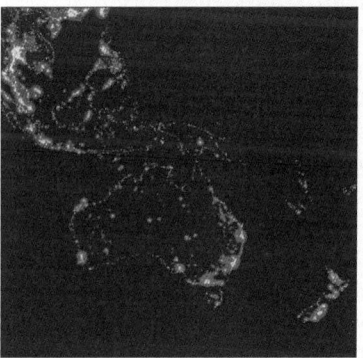

Figure 4: The light recorded in these satellite images represents wasted light and wasted energy going up into space. Image from Cinzano et al., 2001.

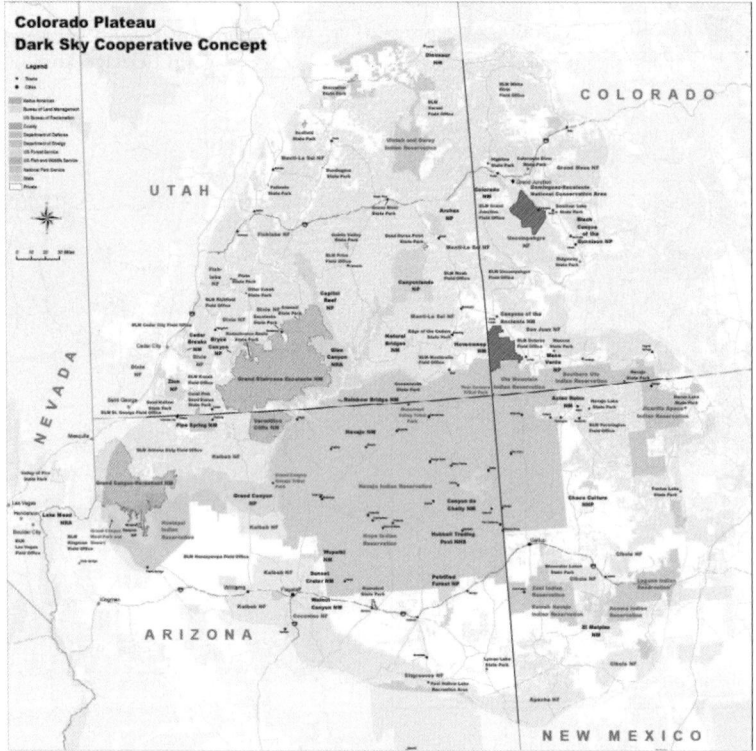

Figure 5: The Colorado Plateau Dark Sky Cooperative Concept

1980s - which were rejected by Washington as irrelevant to the mission of the NPS. The NPS now has a truly remarkable record of leadership in action, measurement, documentation, education and of course outreach on a huge, yet balanced scale that is an example to the world. (See http://www.noao.edu/education/IAUGA2015FM21/IAUGA2015FM21/2-6-04.)

Dan and Nate describe how the U.S. National Park Service (NPS) Night Skies Program contributes to the recognition of certain outstanding NPS lands as dark sky places. A combination of efforts including measuring resource condition, within-park outdoor lighting control, education outreach for visitors, and engagement with surrounding communities helps establish and maintain such places. In certain circumstances, communities and protected areas join forces in a cooperative effort to preserve the natural nocturnal environment of a region. One recent example, the Colorado Plateau Dark Sky Cooperative, is taking lighting, conservation, and educational steps to fulfill the mission of the NPS Call To Action- Starry Starry Night. This voluntary initiative forms America's first Dark Sky Cooperative, and links communities, tribes, businesses, state/federal agencies, and citizens in a collaborative effort to celebrate the view of the cosmos, minimize the impact of outdoor lighting, and ultimately restore natural darkness to the area. The authors presented progress and accomplishments of established dark sky parks and reserves in the western U.S., with particular emphasis on public response to the actions taken and the results achieved.

The presentation opened with a description of the "Colorado Plateau Dark Sky Cooperative", accompanied with associated images (Figure 5). The Colorado Plateau shows up clearly in images from space provided by the NPS Natural Sounds and Night Skies

Figure 6: International Dark Sky Parks/Places involved in the "Colorado Plateau Dark Sky Cooperative"

Division as one of the darkest areas of the continental United States. This Co-operative has the advantages of access to high elevation, low precipitation, a large number of clear days, clear, unpolluted air, low population density, dramatic landscape, worldwide visibility and a large amount of public land. Their presentation shows a map of the Plateau showing the Co-operative's involvement in regular stargazing sites and IDA Certified Dark Sky Places. For more detail search "Clear Air and Magnificent Skies (Utah)" and "Grand Canyon In Depth Night Sky". International Dark Sky Parks/Places include the Chaco Culture National Historical Park, Thunder Mountain Pootsee Nightsky - Astronomy with Kaibab Band of Paiutes, Capitol Reef National Park (an International Dark Sky Park), Flagstaff - the world's first international dark sky city, Parashant International Night Sky Province and the City of Sedona (Figure 6).

On the education side, there is an NPS booklet - Junior Ranger Night Explorer serving as an explorer's activeity guide for ages 5-12. In 2014 Bryce Canyon Night Sky Sanctuary led the number of night-sky-related park contacts (approx. 30,000). Others included Parashant (656), Black Canyon (6,356) Canyonlands (761) and Capitol Reef (3,766).

Regarding astrotourism and economic benefits, for Bryce Canyon (2012) astronomy-related attendance accounted for over 50,000 visits and US$2 million contributed to local economies. Tourism office campaigns have increased some park visitation by up to 30 percent. Regarding the importance of dark skies to NPS visitors, numbers indicating that Dark Skies are "Important" or "Very Important" over the period from the early 1990's to the late 2000's have increased five-fold. Spending per visiting party has increased from $90 to $270-390 (day vs. overnight). This change represents an additional $1.68 billion value added to Colorado Plateau economies over 10 years. The number of visitors has increased in the off-peak seasons and provide a longer, more sustained period of tourism activity.

In the area of civic engagement and partnerships, Flagstaff is the world's first International Dark Sky City; it has set standards and examples for comprehensive night-sky friendly lighting. Sedona has achieved a recent IDSC designation. Work is in progress with Moab, Monticello, Springdale and others.

Regarding plans for the future, these include extendion of the following current activities: (1) Develop Lightscape Management plans for Parks. (2) Inventory lights and plan for retrofits.(3) Work with partners to collect night sky quality data. (4) Collaborate with interpreters to develop night-sky resource educational materials (5) Engage with gateway communities and local businesses (6) Pursue International Dark Sky Park or similar designation.

Measurements of success include: (1) number of light fixtures retrofitted to be night-sky friendly (2) number of parks/communities designated as International Dark Sky Places. (3) number of guided tours that incorporate night-sky education and conservation (rafting trips, expeditions, OUV tours, astro tours). The website for this Focus Meeting Session contains the power points for most of these talk, including this one which presents a "Colorado Plateau Dark Sky Co-operative Measurements of Success'tracking spreadsheet.

Success stories include (1) Created night skies curriculum and training for Colorado Plateau outdoor guides and outfitters, reaching an estimated 40,000-50,000 trip participants each year. (2) 2014 - Partnered with Lowell Observatory and the Keystone Center to host the first Dark Skies and Emerging Technologies conference. (For me this transdisciplinary conference represented the first serious effort I have seen to start to engage all parties involved in, or affected by, the lighting of our planet). (3) Since 2013, 138 new media articles and videos featuring Colorado Plateau dark skies and the Cooperative. (4) 4 newInternational Dark Sky Parks since 2013 (PARA, CHCU, HOVE, CARE). (5) 3 new International Dark Sky Communities. (6) 9 National Parks, 3 State Parks and 3 communities are pending.

The slogan "Half the Park is after Dark" leaves a memorable impression as did the NPS-related presentation.

Cipriano Marin submitted the following abstract found at http://www.noao.edu/education/IAUGA2015FM21/2-6-05, "Developing Starlight connections with UNESCO sites through the Biosphere Smart". He pointed out that the large number of UNESCO Sites around the world, in outstanding places ranging from small islands to cities, makes it possible to build and share a comprehensive knowledge base on good practices and policies on the preservation of the night skies consistent with the protection of the associated scientific, natural and cultural values. In this context, the Starlight Initiative and other organizations such as IDA play a catalytic role in an essential international process to promote comprehensive, holistic approaches for dark sky preservation, astronomical observation, environmental protection, responsible lighting, sustainable energy, climate change and global sustainability.

Many of these places have the potential to become models of excellence to foster the recovery of the dark skies and its defence against light pollution, including some case studies mentioned in the Portal to the Heritage of Astronomy. Fighting light pollution and recovering starry sky are already elements of a new emerging culture in biosphere reserves and world heritage sites committed to acting on climate change and sustainable development. Over thirty territories, including biosphere reserves and world heritage sites, have developed successful initiatives to ensure night sky quality and promote sustainable lighting. Clear night skies also provide sustainable income opportunities as tourists and visitors are eagerly looking for sites with impressive night skies.

Taking into account the high visibility and the ability of UNESCO sites to replicate network experiences, the Starlight Initiative has launched an action In cooperation with Biosphere Smart, aimed at promoting the Benchmark sites. Biosphere Smart is a global observatory created in partnership with UNESCO MaB Programme to share good practices, and experiences among UNESCO sites. The Benchmark sites window allows access to all the information of the most relevant astronomical heritage sites, dark sky protected areas and other places committed to the preservation of the values associated with the

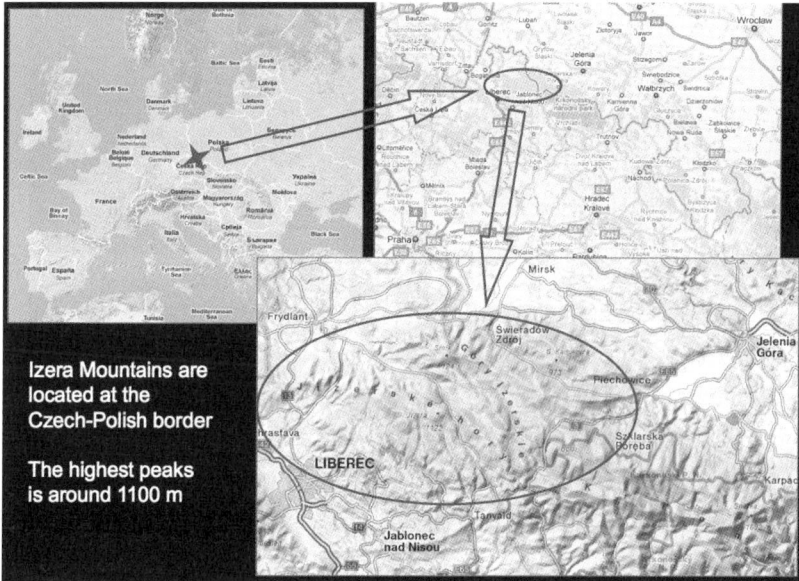

Figure 7: The location of the Izera Dark Sky Park in Europe and at the Czech-Polish border.

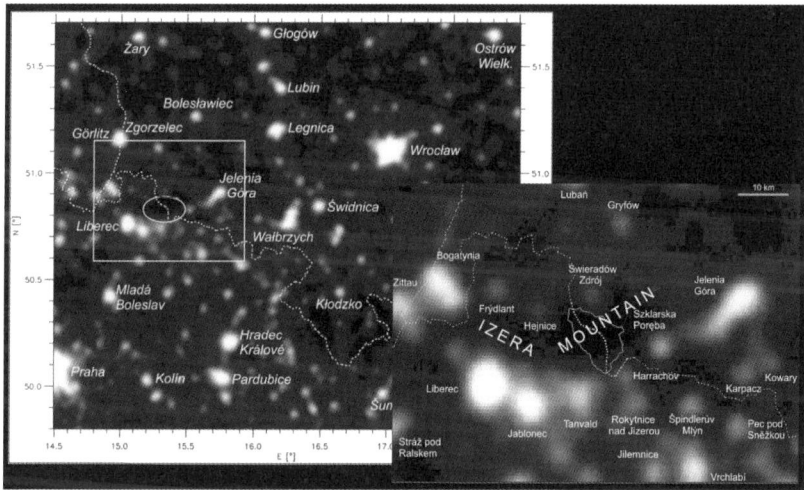

Figure 8: Southwest Poland/North Czech Republic at night – images are from the Defense Meteorological Satellite Program.

night sky. This is a new step ahead in our common task of protecting the starry skies at UNESCO sites.

The final presentation in the session was by Arkadiusz Berlicki, S. Kolomanksi and T. Mrozek (http://www.noao.edu/education/IAUGA2015FM21/2-6-07), who discussed dark-sky protection and education enabled at the Izera Dark-Sky Park (Figures 7 and 8). Their presentation emphasized that darkness of the night sky is a natural component of our environment which should be protected against negative effects of human activities. Darkness at night is necessary for balanced life of plants, animals and people. Unfortunately, development of human civilization and technology has led to substantial

Figure 9: Astronomical events for the public: Polish-Czech Astronomical Days, amateur astronomy meetings, etc

increase of the night-sky brightness and to a situation where nights are no longer dark in many areas of the world. This phenomenon - light pollution - can be ranked among such problems as chemical pollution of air, water and soil. Besides the environment, the light pollution can also affect e.g. the scientific activities of astronomers - many observatories built in the past began to find themselves within the glow of city lights making their (night-time) observations difficult, or even impossible. In order to protect the natural darkness of nights many so-called "dark sky parks" were established, where the darkness is preserved, similar to typical nature reserves. The role of these parks is not only conservation but also education, supporting the effort to make society aware of how serious the problem of the light pollution is.

The history of dark sky areas in Europe was initiated on November 4, 2009 in Jizerka, a small village situated in the Izera Mountains, when Izera Dark Sky Park (IDSP) was established. It was the first international, trans-boundary dark sky park in the world. The idea of establishing that dark sky park in the Izera Mountains originated from a need to give to the society in Poland and the Czech Republic knowledge about light pollution. Izera Dark Sky Park is a component of the astro-tourism project, "Astro Izery", that combines the tourist attraction of Izera Valley and astronomical education under the wonderful starry Izera sky. Besides the IDSP, the project, Astro Izery, consists of the set of simple astronomical instruments (gnomon, sundial), a naturel trail "Solar System Model", and astronomical events for the public (Figure 9). In addition, twice a year they organize a 3-4 day "Astronomy Workshop for Schools", where teachers and astronomers from the Astronomical Institute (University of Wroclaw) educate the young generations in the field of astronomy and other physical sciences.

3. Path Forward

Initiatives such as the above and continued networking between these and other similar groups will continue to provide ideas for protection of starlit sites across the world. They may also provide their governments with options for to explore further action with

UNESCO separately or together in the form of one or more serial nominations. It is becoming clear, from these joint sessions between FM2 and FM21, that the Chilean government is becoming increasingly interested in working with UNESCO and the IAU on these issues, as they apply to the optical, mountain-top observatories in northern Chile.

Session 21.7 – Education Programs Promoting Light Pollution Awareness and IYL2015

Constance E. Walker

National Optical Astronomy Observatory
950 N. Cherry Ave., Tucson AZ 85719 United States
email: cwalker@noao.edu

Abstract. By proclaiming the IYL2015, the United Nations recognized the importance of light and light based technology in the lives of the citizens of the world and for the development of global society on many levels. Light and application of light science and technology are vital for existing and future advances in many scientific areas and culture. Light is a key element in astronomy: as astronomers, it is what we study and makes our science possible, but it is also what threatens our observations when it is set-off from the ground (light pollution). The UN-designated year 2015 represented a magnificent and unique opportunity for the global astronomical community to disseminate these messages and raise the awareness of the importance and preservation of dark skies for heritage and the natural environment.

As such, the International Year of Light served as a launching pad for several projects during 2015. Two other projects with equally as impressive programs are highlighted and begin the narrative for this section on public education and outreach programs on light pollution issues and solutions.

Keywords. miscellaneous, stars: general, Galaxy: general, galaxies: general, solar system: general, ISM: general

---

1. Network for Light Pollution Education at Secondary Level

Beatriz García (ITeDA - UTN FRM, Godoy Cruz, Mendoza), Rosa Ros (Universidad Politécnica de Cataluña, Barcelona, Spain) and their colleagues at the Network for Astronomy School Education (NASE) offer a set of simple, low-cost and fun activities for secondary school educators and students. The activities focus around the three main types of light pollution (sky glow, light trespass and glare), the problems associated with over-consumption of energy and the effects on the astronomical observation (both visible and radio) and on human health. The activities include pin-point boxes (for use without a light, with an unshielded light, and then with a shielded light) (Figures 1 and 2), boxes to let students feel the difference in heat between incandescent lights, CFLs and LED lights and other simple instruments that you can make out of pieces of paper or use with cell phones. More details are at http://www.naseprogram.org.

2. Astronomy Education Under Dark Skies

Johanna Molenda-Zakowic and her colleagues at the Department of Physics and Astronomy at the University of Wroclaw in Poland have been providing professional support for the high school students and the astronomy teachers since 2007. Among their events and projects, like 'School Workshops on Astronomy' (SWA) and 'Wygasz', are dark skies awareness programs that count stars and measure the night sky brightness, familiarizing

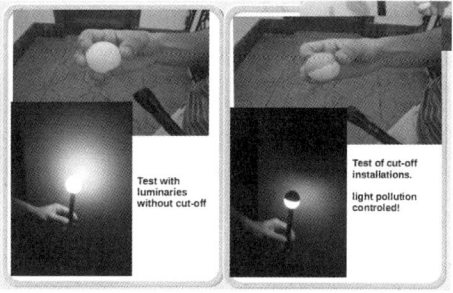
Figure 1: An activity on Sky Glow using a cardboard box with holes punched for a constellation, two 7" maglights with removable reflectors and two ping pong balls (one half painted) each with a small hole for the maglight bulb.

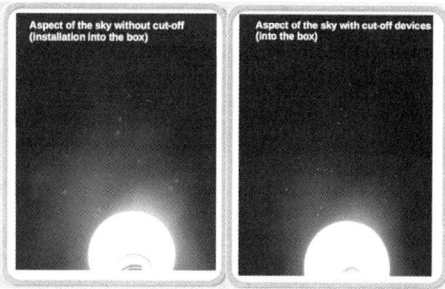
Figure 2: The Sky Glow activity in action. Which light fixture works better? With or without the top of the globe light blocked (e.g., the half-painted ping pong ball)?

teachers and students with the idea and the necessity of protecting the dark sky. All those activities are organized either in the Izera Dark-Sky Park in Poland or in other carefully selected locations in which the beauty of the dark night sky can be appreciated. Understanding dark skies awareness allows for enjoyment of other astronomy activities. Students work in groups on particular assignments such as learning about astrophotography, lenses, star maps, spectroscopy and even how to build a cloud chamber.

3. Dark Skies Rangers

Creating awareness about the importance of protecting our dark skies is the main goal of the Dark Skies Rangers project, a joint effort of the Galileo Teacher Training Program (Rosa Doran, Nucleo Interativo de Astronomia) and the National Optical Astronomy Observatory (Connie Walker). Hundreds of schools and thousands of students have been reached by this program. In particular, students in several municipalities in Portugal have conducted street light audits and produced suggestions on how to enhance (illumination) energy efficiency in specific urban areas. During the International Year of Light, efforts to export the successful Portuguese experience to other countries are being undertaken. The recipe is simple: train teachers, engage students, foster the participation of the local community and involve local authorities in the process.

Dark Skies Rangers activities focus on inquiry-based learning. The students are involved in making observations – when did the lights go on, when did they go off, what is the measure of light pollution, etc. Highlights of the program over the last two years involved an invitation to 1500 schools in Portugal to participate in a contest to create a student-designed calendar (Figure 5). Students have also designed special projects (Figures 3 and 4) and written letters to mayors with ideas for reducing light pollution (i.e. have sports games during the day to reduce need for lights at night). Students have designed three games: one for elementary schools, one for middle schools, and one for high school. The games take place on an imaginary island known for its biodiversity, but tourism wants to move in (Figure 6). The students deal with protecting biodiversity versus promoting commercialization. Between the IAU General Assembly and June 2016, there will be 120 workshops in more than 40 countries using Cosmic Light EDU kits that include 50 free light-based activities. The website has hangouts, resources, and

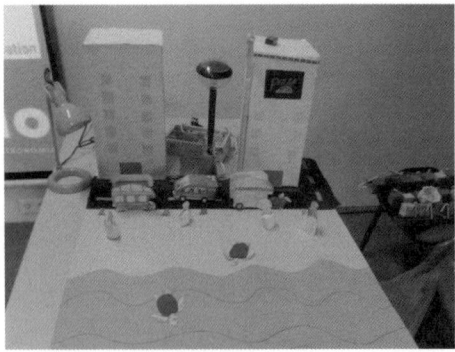

Figure 3: Student-made model of lighting and its effect on sea turtles.

Figure 4: Student-made model of good and bad lighting.

Figure 5: One of the winning, student-designed games from the Dark Skies Rangers Contest. The objective was to create awareness for the growing problem of light pollution and its effects on our lives.

Figure 6: Another student-designed game winner from the Dark Skies Rangers Contest.

activities: http://nuclio.org/cosmiclightedukit/. As a final comment, education isn't always from the top down. This project shows the power of using the energy of children bottom-up.

4. IYL2015 and "Cosmic Light" Message: Awareness and Dissemination in the UK

Global and National initiatives for the International Year of Light have been taking place during 2015 (and will beyond). As an IYL National Committee and Gold Sponsor of IYL2015, the Department of Physical Sciences at The Open University in London has been carrying out IYL2015 activities in the UK. Lucia Marchetti from that department has provided an overview here of what the activities are, how they developed their National Programme and how a long-lasting "Cosmic Light" communication strategy can be constructed to exploit the lessons learnt carrying out the IYL2015 UK year plan.

IYL2015 is cross-disciplinary educational and outreach program with more than 100 partners in 85 countries. The Institute of Physics is the coordinating body in the UK. Prince Andrew is the patron and has helped to promote its visibility. A special page on Cosmic Light has been created on the UK IYL webpage: www.light2015.org.uk. The

Figure 7: IYL2015 Global Open Lab Day promoted by CIE, May 9-25, any date, anytime, anywhere

Figure 8: The Skyglow Citizen-Science Campaign: Light Pollution and the UK's Changing Skies

Figure 9: The Skylight Opera: Students in 28 countries are creating a Science Opera about Cosmic Light, basing it on creative science learning and cross-border cooperation.

webpage provides a national calendar, but also is a way for people to participate in global projects, like Global Open Lab Days (Figure 7). One of the special projects is Skyglow, which compares the light at night from 1992 to the present (Figure 8). The website is building a page of education resources in 5 areas of the spectrum, delineated by age groups. Materials are being promoted through the Open University. There are Open University experiments online. An event has been organized for Parliament during British Science Week. Various exhibits have been organized including a traveling photography exhibit and a special event at a museum. A soapboxscience.org event is to be held. (Soapbox Science is a novel public outreach platform for promoting women scientists and the science they do.) A citizen science project is to measure the speed of light (www.speedoflight2015.co.uk). The solar eclipse in March 2015 was used engage the public. Many countries are involved in the Skylight opera (Figure 9). The Skylight opera web platform has been used to communicate with schools all around the world. The UK IYL team is also partnering with Lunar Mission One, which has a presence at the British Science Festival in September, and is doing projects with museums.

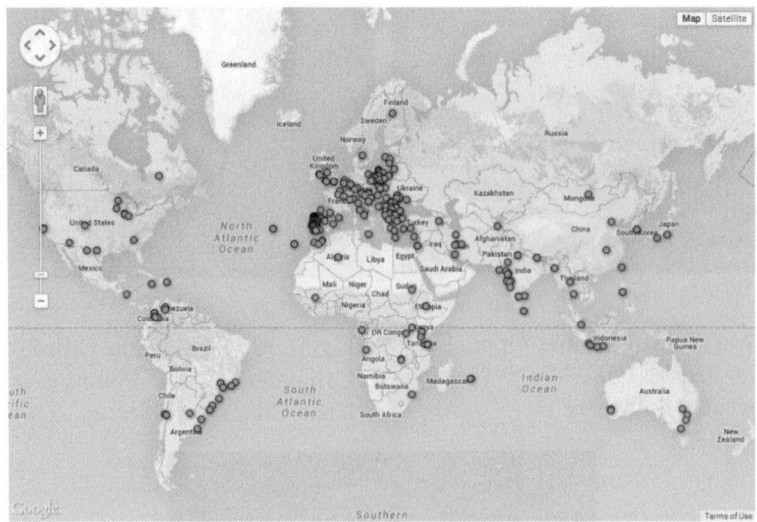

Figure 10: Countries worldwide registered for the Cosmic Light EDU kit and training.

5. Cosmic Light EDU Kit

The main aim of this project is to involve schools around the globe in awareness campaigns in the framework of the International Year of Light. The project is a joint collaboration between Rosa Doran (lead) and Thilina Heenatigala at Nucleo Interativo de Astronomia (NUCLIO), Lina Canas at the IAU Office of Astronomy Outreach (IAU OAD), Pedro Russo at the University of Leiden, and Connie Walker at the National Optical Astronomy Observatory (NOAO). The project is features a tool kit, the Cosmic Light Educational Kit, with simple tools to address the proposed thematic of the ephemeris, in particular, those related the nature of light, the impact of light on our knowledge of the Universe, and its importance for our existence. The richness of the networks involved in this project will also allow for a rich cultural interchange of experiences. The outlined campaign is reaching teachers and students all over the world. The kit contains components in material format and several digital tools and resources. Tutorials for its use and training is foreseen in the framework of this project. These will take diverse formats: face-to-face in some countries, asynchronous and synchronous formats, according to the possibilities and needs of each partner. Thematic hangouts, related to the main cornerstones of IYL2015 as well as to the components foreseen in this program, will be promoted into 2016. Accompanying training efforts will be implemented in order to empower teachers to the full exploitation of the kit and its different components.

The project aims to target diverse social and cultural audiences and has topics related to light in its rich variety and impact. Light as a source of life, light as a source of knowledge, light from the past, light from the future, light for inclusion. Colours we see and colours we don't see. A special component designed for children with visual impairment will be incorporated in the kit.

Over 120 workshops taking place from July 2015 through June 2016 are registered in the Cosmic Light EDU page coming from over 42 different countries (Figure 10). With the support of the Office of Astronomy Outreach, all workshop promoters registered until end of September will receive a package with a few resources donated by different institutions and resources in digital format of the kit on a pen drive. For more information on the Cosmic Light EDU kit, see http://nuclio.org/cosmiclightedukit.

6. The Quality Lighting Teaching Kit: Educating the Public about the Dark Side of IYL2015

Poor quality lighting not only impedes astronomy research and our right to see a starry night sky, but creates safety issues, affects human circadian sensitivities, disrupts ecosystems, and wastes billions of dollars/year in energy consumption. It also leads to excess carbon emissions. How do you change the mindset of society that is used to turning night into day? You educate the next generation on quality lighting.

The United Nations-sanctioned International Year of Light in 2015 (IYL2015) provided an opportunity to increase public awareness of dark skies preservation, quality lighting, and energy conservation. The Education and Public Outreach (EPO) group, namely Constance E. Walker (Connie), Stephen M. Pompea and Rebecca Levy, at the U.S. National Optical Astronomy Observatory (NOAO) received a grant through the International Astronomical Union (IAU) and the Optical Society of America (OSA) to produce official "Quality Lighting Teaching Kits" (QLT Kits) for the IYL2015 cornerstone theme, "Cosmic Light". The QLT kit allows students to do creative problem solving using quality lighting. The files for the QLT kit (Figure 11) can be found at www.noao.edu/education/qltkit.php.)

The concepts and practices of quality lighting are explored through problem-based learning, hands-on/minds-on activities, as well as formative assessment probes. The six activities use quality lighting to solve realistic cases on how light pollution affects wildlife, the night sky, our eyes, energy consumption, safety, and light trespass into buildings. The impact of the kits is amplified by providing professional development using a tutorial video created at NOAO and conducting question and answer sessions via Google+ Hangouts for program instructors. The Quality Lighting Teaching Kit program leverages NOAO EPO's work over the last ten years in lighting and optics education (e.g., "Hands on Optics", the International Year of Astronomy's "Dark Skies Rangers", the IAU "Dark Skies Africa", and Arizona Public Service's "Dark Skies Yuma" programs).

The premise of the activities is that the instructor is the mayor of a fictitious city in which the students live (inspired by the City of the Future Poster). The mayor has been receiving complaints from citizens of the city, which all have to do with the lights in the city (stated on the Issues Poster). The students have been assembled into 6 different task forces, to determine the underlying problems expressed in each of the 6 complaint categories, as well as to come up with feasible solutions to those problems.

The students start by reading the information presented in their group's poster. The "Now Try This!" section gives instructions for an experiment, game, or activity to complete in order to gain more understanding of the problems they face. They use the materials in their box and/or envelope to complete the activity. Using what they know along with help from the Problem Solving Poster, the students brainstorm solutions to their problem. The students then carefully consider the implications (both positive and negative) of their solutions as well as any exceptions where their solutions may not work. They determine if there is any other information they need to better understand the problem or have better solutions. This may involve using the links provided or key ideas from the poster to research more about their problem and possible solutions.

6.1. Example: Glare Poster and Activity

As one of the three main types of light pollution, glare is caused by an exposed light bulb. An overly bright bulb can severely impair vision, especially while driving at night. Glare is worse for older adults due to the presence of cataracts and loss of pupil control. In this activity, the students will explore glare from a "headlight" (a capless Mini-Maglite) at

Figure 11: The IYL2015 Quality Lighting Teaching Kit comes with materials for activities, handouts and posters on 6 different light pollution issues for use in student-led, problem-based learning.

night (in a darkened room). With an unshielded light source, students will see how glare affects their ability to read an eye chart 20 ft away. Layers of inkjet transparencies are used to simulate varying degrees of cataracts. The students then explore how cataracts (both with and without a glaring light) can impair their reading ability.

6.2. Example: Animals Poster and Activity

As a second example of the kit's activities, NOAO EPO staff designed a game for students to explore how light pollution affects animals, specifically birds. In the game they are Kirtland's Warblers, which migrate from the Bahamas to the Great Lakes region of the United States and back again. Along the way, they fly through many major cities. Each year, up to 1 billion birds are killed by crashing into buildings in North America alone. Lit buildings at night cause many of these deaths and injuries. Birds and other animals use the sun or stars to navigate, and the lights can confuse the animals causing them to circle the building and collapse from exhaustion. These issues are explored in the game. A great "Going Further" idea is to have students research and design a game centered on where they live and on an animal that is threatened by light pollution.

6.3. Capstone Presentations

A key component of problem-based learning is presenting methods and findings to an audience. After the students have completed their research involving one of the six activities, they present this information to the mayor of the city and other task groups. Presentations can take many forms, such as oral (e.g. Powerpoint) presentations, posters, videos, skits, songs, brochures, or pamphlets. After all groups have presented, the instructor leads a discussion in which the groups meld together their ideas. After the presentations

Figure 12: Earth Hour on March 28, 2015 kicked off an intensive year of astronomy outreach in Pittsburgh. This is before lights were turned off at 8:30pm.

Figure 13: Lights during Earth hour on March 28, 2015 went out from 8:30 to 9:30 pm in over fifty buildings downtown.

and discussion have concluded, the post-assessment is given, mainly to assess student understanding and growth during the project.

6.4. Project Partners

NOAO's partners are International Commission on Illumination (CIE), the International Dark-Sky Association (IDA), International Society for Optics and Photonics (SPIE), the Optical Society (OSA) and the IAU Office of Astronomy for Development, with sponsorship from IAU and OSA. This is the first time that all six stakeholders have partnered in educating the public on the importance of quality lighting and its effects on society. Starting in December 2015, the partners are disseminating the kits to audiences worldwide.

7. Our Pittsburgh Constellation

Riding on the Pittsburgh mayor's keen interest in astronomy and the ongoing change of 40,000 city lights from mercury and sodium vapor to shielded LEDs, Diane Turnshek from the Physics Department of Carnegie Mellon University in Pittsburgh, Pennsylvania, USA and her colleagues organized a series of city-wide celestial art projects to bring attention to the skies over Pittsburgh. Light pollution public talks were held at the University of Pittsburgh's Allegheny Observatory and other colleges. Earth Hour celebrations kicked off an intensive year of astronomy outreach in the city. Lights went out on March 28, 2015 from 8:30 to 9:30 pm in over fifty buildings downtown and in Oakland. Their art contest was announced at the De-Light Pittsburgh celebration at the Carnegie Science Center during Astronomy Weekend. "Our Pittsburgh Constellation" was an interactive Google map of all things astronomical in the city. Different colored stars marked locations of planetariums, star parties, classes, observatories, lecture series, museums, telescope manufacturers and participating art galleries. Contest entrants submitted artwork depicting their vision of the constellation figure that incorporated and connected all the "stars" in their custom city map. Throughout the year, over a dozen artists ran workshops on painting star clusters, galaxies, nebulae, comets, planets and aurorae with discussions of light pollution solutions and scientific explanations of what the patrons were painting, including demonstrations with emission tubes and diffraction

Figure 14: More than 23,000 Globe at Night campaign observations were measured from participants in 104 countries in 2016! A new world's record!)

grating glasses. The celestial art created in this International Year of Light was displayed at an art gallery as part of the City's Department of Innovation and Performance March 2016 Earth Hour gala. The organizers are thankful for the Astronomical Footprint grant from the Heinz Endowments, which allowed them to bring the worlds of science and art together to enact social change.

8. Globe at Night: From IYA2009 to the International Year of Light 2015 and Beyond

Celebrating its tenth year, Globe at Night (www.globeatnight.org), directed by Constance E. Walker with Stephen M. Pompea at the U.S. National Optical Astronomy Observatory, is an international citizen-science campaign to raise public awareness of the impact of light pollution by inviting citizen-scientists to measure and submit their night sky brightness observations via a "web app" on any smart device or computer. Globe at Night was invited to be an official citizen-science program for the International Year of Light in 2015. By the end of 2015, Globe at Night exceeded all records for the number of observations in a year, topping at 23,000 observations from 104 countries and all 50 US states (Figure 13). 56% of the observations were taken with mobile devices (smart phones and tablets) versus desktops; nearly three-quarters of the mobile device measurements were made by iOS devices versus Android devices; 3 out of 5 mobile device measurements were with the Dark Sky Meter app and 1 out of 5 mobile device measurements were with the Loss of the Night app. (The data from both apps feed into the Globe at Night data base.) Nearly 9000 measurements included readings from the handheld Sky Quality Meters devices.

The overall results for limiting magnitudes were distributed as follows: 2030 measurements at Limiting Magnitude 1, 3285 measurements at Limiting Magnitude 2, 4827 measurements at Limiting Magnitude 3, 4290 measurements at Limiting Magnitude 4, 3410 measurements at Limiting Magnitude 5, 2759 measurements at Limiting Magnitude 6 and 248 measurements at Limiting Magnitude 7.

Twenty-two countries qualified for the "Over 100" Club for Globe at Night. Five countries got over 1000 measurements: United States (8216), Croatia (2276), South Korea (1568), Uruguay (1455) and Germany (1363). The other "Over 100" Club countries are: Poland (987), Japan (808), Chile (739), United Kingdom (669), Spain (477), Macedonia

(FYROM) (423), France (404), Canada (368), Australia (324), Italy (232), Austria (179), Switzerland (159), Puerto Rico (156), Mexico (156), the Netherlands (133). Costa Rica (113) and Belgium (111).

Citizen-science is a rewardingly inclusive way to bring awareness to the public on important issues like the disappearing starry night sky, its cause and solutions. Citizen-science can also provide meaningful, hands-on "science process" experiences for students. Globe at Night will continue to do both for at least another ten years.

9. Summary

The ambitious projects launched during the International Year of Light have potential as legacy projects in the years to come. We urge you to use the projects in doing outreach to bring awareness to the public on issues and solutions surrounding light pollution. Many hands make light work.

Session 21.8 – Challenges and Solutions to Light Pollution, RFI and Implementing IAU Resolution 2009 B5

Richard Green

Steward Observatory, University of Arizona, 933 N. Cherry Ave., Tucson, AZ 85721 USA
email: rgreen@email.arizona.edu

Abstract. The closing session included a panel on the challenge of raising cultural awareness of the negative effects of light pollution and RFI, and a discussion about the means to implement the IAU Resolution on the Right to Starlight. The strongest arguments to the public are that light pollution wastes precious energy and adds greenhouse gases, and that artificial light at night can be damaging to human health and to the natural environment. As astronomers, our community is concerned that the world is blinding itself to the electromagnetic radiation connecting us to the Universe. An outcome of successful advocacy would be to create demand for commercial products that minimize blue light and upward radiation. Implementation of the resolution on the Right to Starlight has multiple aspects. The IAU, through its site protection commission, should provide a clear technical description of "astronomy friendly" lighting and specifications for protection of the near zones around optical observatories. In addition, the commission should provide reference materials for astronomers giving public presentations, provide a forum for those seeking stronger local or national regulation, seek IAU approval for endorsement of protected status of sites and regions, and support the process of gaining UNESCO World Heritage Status for observatories and their regions.

Keywords. site protection, light pollution, lighting standards

1. The Challenge of Changing Cultural Awareness

The Focus Meeting concluded with a session to address the cultural awareness challenge to stop the increase and reduce light pollution, as well as a plan of action for the IAU through its Site Protection commission. A panel of three passionate advocates for dark skies provided perspective on the challenge of culture change. Colin Henshaw, Elizabeth Griffin, and Audrey Fischer each offered their views on the advantages of quality outdoor lighting that should be most compelling to a thoughtful public.

Uniform lighting at the appropriate level of illumination provides superior nighttime visibility over situations with bright glaring sources producing high illumination and deep shadow. The deep cultural change challenge is in appealing to that rational result to overcome humans' innate fear of the dark and the consequent opinion that brighter is safer. The fear of crime as modern-day threat (as opposed to predatory animals) may be linked to that innate fear of the dark. There is a growing body of evidence, however, that criminal activity is highest during daylight, indoors, or when the moon is nearly full. (See, e.g., http://www.decodedscience.org/full-moons-crime-aka-lunar-effect-real-deal-pseudoscience/41881 for a compendium of lunar phase studies.) Collecting the results of research on low light vision, particularly with respect to aging eyes, and how that is best accommodated with outdoor lighting is an important aspect of presenting the facts to the public.

Figure 1. A major cultural challenge is changing the preference for strong artificial urban lighting to design that allows appreciation for darkness and the night sky. Photo of "Open Air Philly" public art display by James Ewing.

The presentation and discussion showed the divergence within the dark sky community on the issue of business motivation. Some feel, based on experience, that manufacturers have a vested interest in creating demand for more and brighter outdoor lighting, that they are willing to link the approach explicitly to nighttime safety, and that their field representatives can even resort to misrepresentation to bring home the sale. A broader view is that manufacturers have a vested interest in increasing sales, which they will do by both creating and trying to meet customer demand. Therefore, educating the public with the goal of creating demand for quality outdoor lighting can actually drive the market in a positive direction. Presentations by the lighting design engineers earlier in the session illustrated the growing availability of spectrally controlled LEDs and full cut-off fixturing in response to demand.

Light trespass is the most common complaint relative to outdoor lighting. Lack of directional control can send light from streetlights or outdoor advertising directly into residential windows, generating legitimate complaints. There is a wide range of response by localities and commercial interests to such complaints from region to region. A proposed goal is for the IAU and those working for the principle of the right to starlight to help assert the rights of "quiet enjoyment" for residents. Light trespass should be made comparable to noise trespass or any other external infringement on the living space of individuals. In many regions, particularly in dense urban areas, there is much work to be done to rebalance the approach to lighting, with protection against trespass and the legal means to enforce that protection as a critical beginning. The major cultural challenge is overcoming today's urban misconception: the belief that they need to sacrifice their starry night sky for urban light, as illustrated in Figure 1.

The presenters pointed out that stretches of highway without significant cross-traffic do not require external illumination for safe operation of vehicles. Headlamps are sufficient, and in the case of multiple lane roads with the two directions of traffic in close proximity, light baffling barricades or fencing can reduce the glare from oncoming headlights. Technological development holds real promise for motion-activated illumination at critical areas with cross-traffic or pedestrian activity.

In the end, the astronomical need for tighter control of artificial light at night may be just a corollary benefit from changes brought about by the growing realization that human health can be adversely impacted. Blue light inhibits production of melatonin; the system is presumably balanced to inhibit production during the day, while allowing production during the dark hours of night. Artificial light rich in blue light content is demonstrated to change the balance of that production rate. Since melatonin can play a role in suppressing cancerous growths, blue light at night can contribute to increased health risk. To the extent that artificial lighting disrupts the balance between insects and their predators like birds and bats, an increase in artificial lighting can increase the incidence of disease vectors, such as mosquitoes, which carry highly infectious and debilitating diseases in the developing world.

Given the mounting pressures on local and municipal budgets, economic arguments in favor of minimizing wasted light should be gaining traction. One estimate is that some 45% of outdoor lighting goes non-productively to direct uplight and overlighting. In aggregate, that amounts to 52.2 TWH per year in the US, or some 38 million tons of CO_2, a strong contribution to unwanted greenhouse gas accumulation.

Obviously there is an environmental impact associated with the wasted energy of poor quality outdoor lighting. Another perspective on the way to approach the cultural change necessary for the public to generate the demand for reduced and higher quality outdoor lighting is by reconsidering the three R's of the environmental movement. In addition to reduce, re-use, recycle, a desirable thought framework is that we have a moral duty to respect the planet and the wellbeing of humans and all biological systems. Those three R's are Respect, Responsibility, and Right. By intellectually disentangling want from need, individuals could adopt an approach of lower demand for energy use in general, and lower tolerance for wasteful excesses in outdoor lighting in particular.

2. Implementation of IAU Resolution 2009 B5 - The Right to Starlight

All of these considerations enter into the approach that astronomers and the IAU should be taking to address the goals of IAU Resolution 2009 B5, "In Defense of the Night and the Right to Starlight". It speaks to the night sky as an inspiration, to its scientific and cultural values, to the view of the night sky getting worse, to the need to educate the public, to the need to use intelligent lighting, and to astro-tourism.

Key Excerpts from the Resolution:

An unpolluted night sky that allows the enjoyment and contemplation of the firmament should be considered a fundamental socio-cultural and environmental right, and that the progressive degradation of the night sky should be regarded as a fundamental loss; IAU members [should] be encouraged to take all necessary measures to involve the parties related to skyscape protection in raising public awareness of the educational, scientific, cultural, health and recreational importance of preserving access to an unpolluted night sky for all humankind.

Protection of the astronomical quality of areas suitable for scientific observation of the Universe should be taken into account when developing and evaluating national and international scientific and environmental policies, with due regard to local cultural and natural values.

Presentation and discussion of the last session centered around the general means to accomplish these objectives and specific steps that the new IAU Commission C.B7 could take to empower astronomers to address the issues successfully.

Five general approaches were agreed to be appropriate to this worldwide task. One is engagement of more of the astronomy community in including this message in their public outreach activities. Having the IAU Commission website provide one or two effective slides could be very powerful. Second would be provision of vision and background statements, with supporting material, to those astronomers who are prepared to be more engaged in educating the public. This is where the International Dark Sky Association (IDA) can be a prime source, because of their strong approach to implementing sky protection goals.

The next general approach is working to create demand for the new products available to support astronomer goals of full cut-off and limited spectral pollution. This area is where Commission 50's highly beneficial and productive engagement with the International Commission on Illumination, CIE (http://cie.co.at/), is worth amplifying to keep the positive interaction at a high level. Since the last GA, such products are now commercially available, and as discussed above, the lighting industry will sell whatever consumers demand.

A key element is helping those astronomers who are engaged with local, national, and international authorities to protect astronomical sites. Past efforts of the previous Commission 50 and the World Heritage Working Group under Division C are beginning to pay off to get sites into the World Heritage category for protection. IAU can also provide materials and advice supporting astronomers who are engaging local and national authorities for lighting regulation to protect natural, historic, and ordinary local areas from light pollution encroachment.

The key question is how to turn those general approaches into a practical implementation plan with goals and milestones that can be tracked. Following are the specific programs that were discussed as an implementation strategy.

The imprimatur of the IAU will be particularly valuable in enunciating a clear technical standard for "astronomy friendly" lighting, including full cutoff, spectral management, and minimizing lumens required for the task.

There was unanimous agreement that the highest priority was full cutoff, with no light emitted directly above horizontal. Gaining adoption of that approach is an uphill battle, both with those who develop standards and those who implement lighting designs. The former group is dominated by engineers who deal with specular reflection and are much less familiar with the radiative transfer approach that demonstrates clearly the strong negative impact of direct upward radiation. The latter are faced with budgetary limitations that tempt them to limit the height of poles and to tilt the luminaires to cover more area.

The issue of spectral management requires sound technical advice. The availability of new LED products with low correlated color temperature (CCT), built-in full cutoff filters, and even narrow-band emission gives a number of options for choices with very low blue light content. As was shown in the presentations by the lighting engineers summarized in FM21.1, a single number like CCT is not adequate to describe color rendition or blue light suppression. Development of a clear guideline or set of acceptable technical choices is a priority goal for the new Commission with the aid of its technical members.

Incorporating the latest research on safety and night vision will be critical for developing industry standards that move in the direction of lower illumination appropriate for usage and area, as opposed to illumination levels with large margin. Curfews and

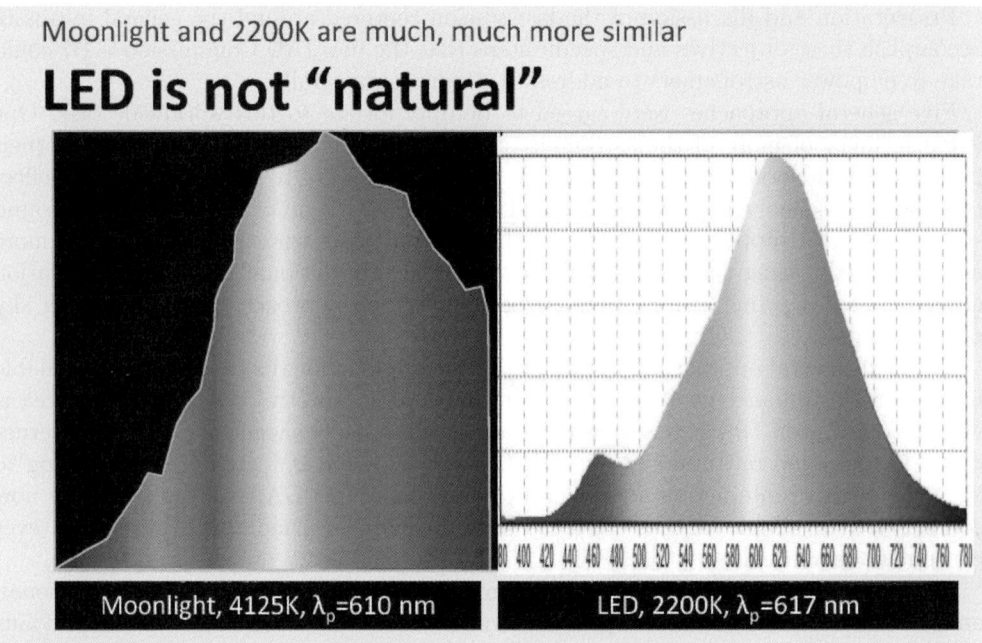

Figure 2. Correlated Color Temperature dervied from fitting a blackbody curve is not an adequate measure for characterizing the color distribution or blue light content of an artificial source like an LED. IAU through its site protection commission must develop clear recommendations for "astronomy friendly" lighting. Figure from talk by Monrad, as summarized in FM21.1

motion activation can also be part of the IAU recommendation for appropriate lighting. The challenge for this set of recommendations will be to find dark sky friendly levels that do not seem to be in contradiction with current engineering standards.

A function unique to IAU activities is to detail the standards for near zone professional observatory site protection. These would entail extremely low lumens levels per hectare and the employment of narrow-band sources, even if energy savings were not as great as with broader-band LEDs. Of course, such a reference standard would need to be modified in individual cases (e.g., existing nearby towns with commercial districts for which rural low illumination levels would be inappropriate). Nevertheless, having a recommended template as a starting point would be of value for site preservation efforts around the world.

To support the general approach of encouraging astronomers to include dark skies issues in their public lectures and talks with local policy makers, the IAU Commission should develop (i.e., write a bit and collect a lot of) information to comprise an IAU dark skies "standard" outreach package. The task is then to launch an outreach effort to professional colleagues, to get them beyond the opinion that site protection and dark sky preservation in general is a job required only of observatory directors and their designated staff.

The Cornerstone programs developed for the International Year of Light (IYL) were chosen because of their broad international impact and appeal. A very near-term goal will be to assure that IYL Cosmic Light projects and other successful IYL astronomy programs achieve sustainable status. The effort-based part will be to continue to publicize availability through the Commission website. Should some funding be needed to put a program on more solid footing, the Commission will explore with the project PIs options like crowd funding or continuing commercial sponsorship.

To take advantage of the opportunity to influence international lighting standards, we intend to re-energize active interaction with the CIE, with formal IAU liaison and practical means of participation in their technical committees defining standards. CIE has welcomed IAU involvement. With their worldwide activities and meetings, support will require a small network of technically conversant Commission members who can cover meetings on multiple continents.

A key aspect of the top-down approach to control light pollution is the drafting, implementation, and updating of outdoor lighting standards and regulations in local and regional legal codes. The Commission can provide a forum for voluntary coordination and cooperation among those working to protect key astronomical sites, as well as those concerned with preserving sky quality in natural protected regions or even urban settings. An obvious activity is exchange of information on techniques for getting local and national lighting ordinances passed.

Another way in which the reputation of the IAU can be used to good purpose is in the production of IAU letters of endorsement to relevant entities attesting to the astronomical value of a site. Observatories or regions seeking special designation, for example with the IDA, seek such letters, and they have proven to have real value in such activities. An obvious extension would be endorsement for a proposal to a locality or other government agency to retrofit lighting that would significantly improve dark sky conditions near an astronomical site.

The full scope of the IAU Resolution provides a broad charge to astronomers for dark sky preservation. In that context, the Commission should consider ways to encourage protection of natural and historic sites against light pollution encroachment. As presented in this focus meeting, the US National Park Service has been proactive in protecting dark skies as part of the unique environments of their Parks. Public viewing through small telescopes has become a major visitor attraction in some of these parks. Active encouragement of similar approaches on a worldwide basis is an obvious role for astronomers according to the Resolution.

The Site Protection Commission must continue its support of the World Heritage Working Group and UNESCO with ongoing and new projects. The immediate goal is declaration of observatory sites in Chile, and possibly in Spain and Hawaii, as World Heritage sites. The Commission can serve an advisory role in the preparation of the application materials to present the sites as uniquely well suited for astronomy in terms of their dark sky and other atmospheric qualities. Its experts (although a different set) can similarly advise UNESCO when they come to evaluate the proposal.

Finally, the Commission must define a practical means of partnering with the IDA to leverage international outreach for both organizations. Their reference and presentation material is of high value, and their international chapters can provide some local language support. Direct association with the IAU can strengthen their international reach, and help both organizations improve their support base for dark sky protection activities.

3. Afterword

This ambitious agenda was endorsed by the dedicated and passionate group participating in the Focus Meeting. The ongoing challenge is to focus the volunteer effort to best effect and grow the level of participation dramatically. Only with the conscious effort of a significant fraction of the astronomers in the IAU can we reasonably aspire to tackle the global goal of universal acknowledgement of the right to a dark night sky.

Author index

Abe, L. – 51
Abrahamyan, H. V. – 130
Adande, G. – 271
Afonso, J. – 420
Aigrain, S. – 187, 191
Almenara, J. M. – 51, 200
Alves, A. – 420
Andre, P. – 51
Andreasyan, H. R. – 130
Angus, R. – 191
Anjos, S. – 420
Antoniadou, K. I. – 38
Antonio Belmonte, J. – 102
Armstrong, D. J. – 51
Ayres, T. R. – 307
Azatyan, N. – 422
Azatyan, N. M. – 130

Ba, Y. A. – 347
Barboza, C. H. – 106
Bardin, N. – 253
Barrado, D. – 51
Barros, S. C. C. – 51
Beall, J. – 166
Beauchamp, J. L. – 327
Beiersdorfer, P. – 295
Bendjoya, P. – 51
Benitez-Herrera, S. – 399
Bennett, D. P. – 220
Benomar, O. M. – 63
Benzerara, K. – 253
Bhattacharjee, A. – 329
Biver, N. – 228, 233, 321
Bockelée-Morvan, D. – 227, 228, 233, 321
Bohlender, D. – 109
Boisse, I. – 51, 193
Boissier, J. – 233
Bonomo, A. S. – 51
Boogert, A. C. A. – 317
Bordé, P. – 198
Borovička, J. – 247
Bouchy, F. – 51
Bovaird, T. – 196
Bradley, J. – 253
Bretones, P. S. – 401
Breuer, D. – 261
Brown, D. – 51
Brown, G. V. – 295
Brun, A. S. – 14
Bruno, G. – 51

Bryson, S. – 197
Butler, K. – 297

Campante, T. L. – 71
Canas, L. – 420
Carignan, C. – 397
Char, F. – 418
Charnley, S. B. – 233, 271
Chatterjee, S. – 6, 30
Cheung, S.-l. – 403
Clementini, G. – 224
Cordiner, M. A. – 233, 271
Cortina, J. – 345
Cotte, M. – 93, 121
Coulson, I. M. – 233
Courcol, B. – 51
Coustenis, A. – 325
Crabtree, D. – 109
Crovisier, J. – 233

de Gouveia Dal Pino, E. M. – 337
de Grijs, R. – 410
Dartois, E. – 253
Delauche, L. – 253
Deleuil, M. – 51
Demangeon, O. – 51, 198
Den Hartog, E. A. – 287
DeVorkin, D. – 112
Devrikyan, V. G. – 148
Díaz, R. F. – 51, 200
Ding, H. – 209
Dluzhnevskaya, O. – 134
do Nascimento Jr. J. D. – 57
Dobrică, E. – 253
Doran, R. – 420
Dorch, B. F. – 172
Drachen, T. M. – 172
Dubernet, M. L. – 347
Duprat, J. – 253
Duriscoe, D. M. – 435

Ellegaard, O. – 172
Engrand, C. – 253
Eyer, L. – 201, 224

Fabrycky, D. C. – 40
Fares, R. – 360
Farmanyan, S. – 422
Farmanyan, S. V. – 130, 148
Federman, S. – 283
Feigelson, E. – 187
Ferraz-Mello, S. – 57

Fiksel, G. – 329
Förstel, M. – 305
Ford, E. B. – 30, 40 223
Foreman-Mackey, D. – 191
Forero-Romero, J. – 418
Fox, W. – 329
Franco, J. – 405

García, C. G. – 102
García, B. – 150
Gharib-Nezhad, E. – 307
Giampapa, M. S. – 365
Gibson, N. P. – 202
Gigoyan, K. S. – 130
Gomes, A. L. – 19
Govender, K. – 379, 382, 424
Grant, E. – 385
Green, R. – 500
Gregory, P. C. – 205
Griffin, E. – 176
Griffin, R. E. – 129
Grunblatt, S. K. – 208
Gu, L. – 291
Guinan, E. – 424
Guinan, E. F. – 390
Guy, L. – 219
Gyulzadyan, M. V. – 130

Haghighipour, N. – 3, 424
Han, B. – 209
Harakawa, H. – 63
Harmony, T. M. – 112
Haywood, R. D. – 208
Hébrard, G. – 51
He, J. – 410
Heenatigala, T. – 420
Hell, N. – 295
Hemenway, M. K. – 424
Hesser, J. E. – 109
Hilchenbach, M. – 253
Hirano, T. – 63
Hodgkin, S. – 224
Hojaev, A. S. – 392
Horvath, J. E. – 401
Hoskin, M. – 102
Howard, A. W. – 208
Hsieh, H. H. – 237
Hu, X. – 6

Ishii, H. – 253

Jafelice, L. C. – 401
Jenkins, J. M. – 210
Ji, J. – 27
Johnson, R. K. – 140
Jones, A. P. – 313

Jontof-Hutter, D. – 40
Jurua, E. – 393

Kaastra, J. S. – 291
Kaiser, R. I. – 305
Kelley, R. L. – 295
Khachatryan, K. G. – 130
Kilbourne, C. A. – 295
Kirk, J. – 51
Klahr, H. – 19
Knyazyan, A. V. – 130
Kolenberg, K. – 390
Korhonen, H. – 355
Kostandyan, G. R. – 130
Kotake, K. – 340
Kotani, T. – 213
Krantzler, S. O. – 30
Kuan, Y.-J. – 233
Kurokawa, T. – 213

Lachurié, J. C. – 51
Lam, K. W. F. – 51
Lawler, J. E. – 287
Leroux, H. – 253
Libert, Y. – 397
Lillo-Box, J. – 51
Lineweaver, C. H. – 196
Lis, D. C. – 233
Lissauer, J. J. – 40
Lyons, J. R. – 307
López, A. M. – 142
López-Coto, R. – 345

Mahelona, J. K. – 140
Mao, J. – 291
Marov, M. – 134
Martinez, P. – 51
Martins, Z. – 257
Mashonkina, L. – 283
Masuda, K. – 63
Matt, S. P. – 14
Matthes, K. – 372
McCormac, J. – 51
Mehdipour, M. – 291
Mickaelian, A. – 422
Mickaelian, A. M. – 130, 148
Mikayelyan, G. – 422
Mikayelyan, G. A. – 130
Milam, S. N. – 233, 271
Miley, G. – 380
Moda, L. F. R. – 57
Mohanty, S. – 6
Moralejo, A. – 345
Morbidelli, A. – 3
Moreau, N. – 347
Morgenthaler, S. – 201
Mori, T. I. – 319

Moutou, C. – 51
Mumma, M. J. – 233
Murakami, N. – 213
Mutondo, M. S. – 416

Nakamura, K. – 340
Naoz, S. – 65
Nava, T. S. – 427
Neumann, W. – 261
Nicoletti, J.-M. – 201
Nieva, M. F. – 297
Nikoghosyan, E. H. – 130
Nishikawa, J. – 213

Öberg, K. I. – 267, 309
Okere, B. – 395
Omiya, M. – 63
Onaka, T. – 319
Osborn, H. – 51
Oya, M. – 213

Paganini, L. – 233
Paronyan, G. M. – 130
Pereira, E. S. – 57
Pinilla-Alonso, N. – 241
Pollacco, D. – 51
Pompea, S. M. – 408
Porter, F. S. – 295
Przybilla, N. – 297

Quirico, E. – 253

Raassen, T. – 291
Rajpurohit, A. – 51
Remijan, A. J. – 233
Remusat, L. – 253
Retrê, J. – 420
Reville, V. – 14
Rey Cerda, J. – 51
Rivet, J.-P. – 51
Rogers, L. A. – 214, 223
Ros, R. M. – 150
Rowe, J. F. – 40
Ruggles, C. – 79, 89, 97, 140, 463
Russo, P. – 398

Sahlmann, J. – 217
Sakon, I. – 319
Salama, F. – 283, 327
Saletta, M. – 100
Sanhueza, P. – 114, 115
Santerne, A. – 51, 200
Sato, B. – 63
Schaffenroth, V. – 297
Sciamma-O'Brien, E. – 327
Sefako, R. – 118, 453
Shastri, P. – 406

Sheshadri, G. – 157, 179
Shimonishi, T. – 319
Shrestha, P. – 146
Sidorenko, A. – 79, 83
Silva, D. – 424
Simpemba, P. – 416
Smith, M. G. – 114, 115, 480
Smith, R. C. – 114, 115
Sneden, C. – 287
Soonthornthum, B. – 412
Spake, J. – 51
Spergel, D. – 277
Spinelli, P. F. – 399
Spohn, T. – 261
Strubbe, L. E. – 395
Strugarek, A. – 14
Suarez, O. – 51
Suzuki, D. – 220
Süveges, M. – 219

Takeda, Y. – 63
Takiwaki, T. – 340
Tamura, M. – 213
Tan, J. C. – 6
Tanaka, Y. – 213
Tekle, K. – 414
Tellez, D. – 408
Thiéblemont, R. – 372
Torres-Peimbert, S. – 405
Toublanc, D. – 51
Traub, W. A. – 221
Trimble, V. – 105
Tsantaki, M. – 51

Upton, K. T. – 327
Urdampilleta, I. – 291
Usui, F. – 319

van Dishoeck, E. F. – 299, 424
Väisänen, P. – 118
Vardanyan, A. V. – 130
Venugopal, R. – 427
Verdolini, S. – 427
Villanueva, G. – 233
Voyatzis, G. – 38

Wainscoat, R. J. – 444
Walker, C. E. – 408, 490
Walker, S. R. – 51
Wang, S. – 27
WeiPing, L. – 333
Wesley, H. – 157, 179
Williams, R. – 277
Wilms, J. – 295
Wirström, E. S. – 271
Wolfgang, A. – 223
Wolfschmidt, G. – 124

Wolter, A. – 424
Wood, M. E. – 287
Wu, R. – 319

Yan, Y. – 424
Yongheng, Z. – 473

Zhang, Y. – 209
Zhang, Z. – 410
Zhao, Y. – 209
Zhu, Z. – 6
Zucker, S. – 219, 224
Zwölf, C. M. – 347